Improving Animal Welfare

A Practical Approach

Improving Animal Welfare

A Practical Approach

Edited by

Temple Grandin

Department of Animal Sciences
Colorado State University
USA

CABI is a trading name of CAB International

CABI Head Office
Nosworthy Way
Wallingford
Oxfordshire OX10 8DE
UK

Tel: +44 (0)1491 832111
Fax: +44 (0)1491 833508
Email: cabi@cabi.org
Web site: www.cabi.org

CABI North American Office
38 Chauncey Street
Suite 1002
Boston, MA 02111
USA

Tel: +1 617 395 4056
Fax: +1 617 354 6875
Email: cabi-nao@cabi.org

A catalogue record for this book is available from the British Library, London, UK

Library of Congress Cataloging-in-Publication data

Improving animal welfare : a practical approach / edited by Temple Grandin.
p. cm.
Includes bibliographical references and index.
ISBN 978-1-84593-541-2 (alk. paper)
1. Animal welfare. 2. Animal industry—Moral and ethical aspects. I. Grandin, Temple. II. Title.

HV4708.I47 2010
179'.3--dc22

2009022046

ISBN-13: 978-1-84593-541-2

First published 2010
Reprinted 2013

Typeset by SPi, Pondicherry, India
Printed and bound in the UK by Berforts Information Press Ltd.

Contents

Contributors

Anne Marie de Passillé, Research Scientist, Pacific Agri-Food Research Centre, Agri-Food Canada, 6947 #7 Highway, PO Box 1000, Agassiz, British Columbia V0M 1A0, Canada; annemarie.depassille@agr.gc.ca

Lily N. Edwards, Assistant Professor in Animal Behavior, Department of Animal Science and Industry, Kansas State University, 248 Weber Hall, Manhattan, KS 66506-0201, USA; lilynedwards@gmail.com and lne@k-state.edu

Temple Grandin, Professor, Department of Animal Sciences, Colorado State University, Fort Collins, CO 80523-1171, USA; Cheryl.miller@colostate.edu

Camie R. Heleski, Instructor/Coordinator, Department of Animal Science, Michigan State University, East Lansing, MI 48824, USA; heleski@msu.edu

Jeff Hill, Provincial Livestock Welfare Specialist, Innovative Livestock Solutions, Blackie, Alberta T0L 0J0, Canada; Jeffery.Hill@gov.ab.ca

David C.J. Main, Senior Lecturer in Animal Welfare, Department of Clinical Veterinary Science, University of Bristol, Langford, Bristol BS40 5DU, UK; D.C.J.Main@Bristol.ac.uk

Amy K. McLean, Graduate Student, Department of Animal Science, Michigan State University, East Lansing, MI 48824, USA; mcleana5@msu.edu

David J. Mellor, Professor, Animal Welfare Science and Bioethics Centre, Institute of Food, Nutrition and Human Health, College of Sciences, Massey University, Palmerston North 4442, New Zealand; D.J.Mellor@massey.ac.nz

Bernard Rollin, Professor, Department of Philosophy, Colorado State University, Fort Collins, CO 80523-1171, USA; Bernard.rollin@colostate.edu

Jeffrey Rushen, Research Scientist, Pacific Agri-Food Research Centre, Agri-Food Canada, 6947 #7 Highway, PO Box 1000, Agassiz, British Columbia V0M 1A0, Canada; jeff.rushen@agr.gc.ca

Jan K. Shearer, Professor of Veterinary Diagnostic and Production Animal Medicine, College of Veterinary Medicine, Iowa State University, Ames, IA 50011, USA; jshearer@iastate.edu

Kevin J. Stafford, Professor Animal Welfare Science and Bioethics Centre, Institute of Veterinary, Animal and Biomedical Sciences, College of Sciences, Massey University, Palmerston North 4442, New Zealand; K.J.Stafford@massey.ac.nz

Janice C. Swanson, Professor, Department of Animal Science, Michigan State University, East Lansing, MI 48824, USA; swansoj@anr.msu.edu

Helen R. Whay, Senior Research Fellow, Department of Clinical Veterinary Science, University of Bristol, Langford, Bristol BS40 5DU, UK; bec.whay@bristol.ac.uk

Tina Widowski, Director, The Campbell Centre for the Study of Animal Welfare, Department of Animal and Poultry Science, University of Guelph, Guelph, Ontario, Canada; twidowsk@uoguelph.ca

Jennifer Woods, J. Woods Livestock Services, RR#1, Blackie, Alberta T0L 0J0, Canada; livestockhandling@mac.com

Preface

Animal welfare is an increasing concern all over the world. The World Organization for Animal Health (OIE) now has recommendations in the *Terrestrial Animal Health Code* for transport, slaughter and killing animals for disease control. In some countries, animal welfare is a new concept and this book will provide practical information which will enable veterinarians, managers and animal scientists to implement effective practical programmes to improve animal welfare. It will be especially useful for both students and training of animal welfare specialists. The emphasis is an international approach. Two of the authors worked on OIE animal welfare committees, and the most important parts of these guidelines are reviewed. The authors are from the USA, Canada, the UK and New Zealand, and they have extensive experience improving animal welfare in both the developed and the developing world. In addition to their work in North America and Europe, they have also worked in Brazil, Mali, western Africa, Uruguay, Chile, Australia, the Philippines, Mexico, China, Thailand, Argentina and New Zealand. *Improving Animal Welfare: a Practical Approach* covers how to both measure and assess welfare, plus 'how to' instructions on methods to improve practices in areas of major welfare concern such as animal handling, euthanasia, painful surgical procedures, transport, slaughter and treatment of draught animals. Another major area of emphasis will be how to use animal-based outcome measures such as scoring of body condition, lameness, lesions, behaviour and coat/feather condition. The use of numerical scoring to measure handling and stunning practices will also be covered. Measurement is essential because people manage the things that they measure. Key references to important scientific papers are included but this book is not intended to be a complete review of the literature. Additional chapters on the benefits of good stockmanship, economic factors, ethics and proven methods for motivating producers will also be helpful to bring about improvements. This book is aimed at the people who will be putting animal welfare programmes into practice to improve conditions for animals all over the world.

1 The Importance of Measurement to Improve the Welfare of Livestock, Poultry and Fish

TEMPLE GRANDIN

Colorado State University, Fort Collins, Colorado, USA

There is a need for information on how to implement auditing programmes and other strategies that will improve animal welfare. Many excellent books and articles are available that review scientific research on welfare, statistics that outline the extent of animal welfare problems, philosophical issues, animal rights and legislation. However, there is a huge need for information on how to effectively implement programmes to improve animal welfare at the practical level. This is a hands-on 'how to do it' guide that provides practical information for veterinarians, animal scientists, producers, transporters, auditors, government agencies, quality assurance managers and others who work in the field with animals. Too often legislation will be passed to ban some terrible practice but it still continues because little is done in the field to implement change.

In this book, the authors will help bridge the gap between scientific research and practical application. Recommendations on implementing animal welfare programmes are based on over 10 years of the editor's experience developing and implementing welfare auditing systems for major retailers and restaurants (Grandin, 2003, 2005). During the last 35 years the author has visited over 500 farms and slaughter plants in 25 different countries. The information in this book will help the reader use the knowledge that can be obtained from many other sources in a more effective manner to bring about real changes that will improve the treatment of livestock, poultry and fish, on farms and transport vehicles and in slaughter plants. The principles of implementing an effective animal welfare programme are the same for all species. Animal welfare is now a worldwide issue (Fraser, 2008a). The World Organization for Animal Health (OIE) now has published animal welfare guidelines for slaughter of livestock and poultry and for the transport of livestock (Petrini and Wilson, 2005; OIE 2008, 2009a, b). The welfare of farmed fish is an emerging issue and the OIE will have guidelines for humane slaughter of fish (Hastein, 2007).

Large food retailers and restaurant chains are now requiring that their suppliers comply with their animal welfare standards. The economic incentive provided by these large buyers is a major force for improving animal welfare in both the developed and the developing world. Non-governmental (NGO) animal advocacy groups are also a major factor in developing animal welfare standards and legislation. When videos of animal abuses are seen around the world on the Internet, it makes people aware of the issue and they demand improvements.

The Importance of Measurement

In this first chapter, the importance of measuring animal welfare will be discussed. It will show you how to use numerical measurements to improve both animal welfare and productivity. In order to effectively manage and improve animal welfare it needs to be measured. There are two basic types of measurements. Simple ones that are practical for farm use and more complex or expensive measurements for research or diagnostic purposes. The more complex methods should be used to validate simple practical on-farm assessment methods. Another important purpose of research is to determine the effects of common commercial practices on animal welfare. There are many fine books and journals that contain information on hundreds of research studies. Since the emphasis of this book is on the practical application of welfare programmes, the author will not attempt to do a complete review of all the research. It is also beyond the scope of this book to review all the differing animal welfare legislation around the world.

Goals of the Book

There are five main goals for this book:

1. To help people implement effective practical auditing, regulatory and assessment programmes that will improve the welfare and treatment of livestock, poultry and farmed fish.
2. To provide information that will directly improve welfare in critical areas such as slaughter, transport, handling, euthanasia and painful surgical procedures.
3. To help the reader understand the importance of animal behaviour in assessing animal welfare and its role in the design of housing and handling systems.
4. To discuss the role of ethics in animal welfare in a practical manner.
5. To understand how economic factors can be used to improve both welfare and reduce losses in farm animals. Improvements in husbandry, handling, stockmanship and transport will improve animal productivity, and reduce losses due to bruises, sickness, mortalities, lameness and other problems.

Since animal welfare is now a global issue, this book contains information that can be used by people in both developed and developing parts of the world.

Manage Things You Measure

Livestock producers routinely measure weight gain, death losses and sickness but they may not be measuring painful or distressing conditions such as lameness, bruises or electric-goad use, which severely compromises an animal's welfare. Lameness is one of the most serious welfare problems in many species of livestock and poultry. Lameness definitely causes pain because giving dairy cows the anaesthetic lidocaine reduces it (Rushen *et al.*, 2006; Flowers *et al.*, 2007). People often fail to be effective managers of conditions that they do not measure. Lameness in intensively housed dairy cattle is a good example. Over a period of many years, lameness in dairy cows housed on concrete has become steadily worse. One of the reasons why this happened was that nobody measured lameness until it became really bad. A recent British study showed that 16.2% of the dairy cows were lame (Rutherford *et al.*, 2009). Cows housed in freestall (cubicle) barns had an average of 24.6% of clinically lame cows (Espejo *et al.*, 2006). However, in the top 10% of dairies lameness was only 5.4% (Espejo *et al.*, 2006). A British survey of 53 dairies indicated that in the best 20% of the dairies, only 0–6% of the cows were lame and in the worst 20% of the dairies, 33–62% of the cows were lame (Webster, 2005a, b). This shows that good management can reduce lameness. Lameness is also a huge welfare concern in sows. In sows, 72% of the breeding animals that had to be culled were due to locomotion problems. The major causes of locomotion problems were arthritis (24%) and fractures (16%) (Kirk *et al.*, 2005).

The author observed a big increase in lame slaughter weight pigs between 1995 and 2008. A major breeder of lean rapidly growing pigs did nothing about it in the USA until in some herds 50% of the slaughter weight pigs were clinically lame. They also had very poor leg conformation. This breeder was selecting for leanness, loin-eye size and rapid growth, and over a 10-year period did not notice that there were more and more lame pigs. The increase in lameness was mainly genetic because the pigs were all housed on the same concrete slats that had been used for years. A recent US study indicated that 21% of the sows were lame (VanSickle, 2008). A study of sows in Minnesota indicated that risk of removal from the breeding herd increased when leg conformation was poor. Culling of breeding sows that was attributable to poor legs was 16.37% for the forelimbs and 12.90% for the hind limbs (Tiranti and Morrison, 2006). A study done in Spain showed that poor leg conformation was associated with higher sow culling rates (deSeville *et al.*, 2008). Selecting breeding gilts with structurally correct feet and legs will provide better welfare and productivity.

Preventing 'Bad Becoming Normal'

How could dairy cows and pigs become so severely lame before anything was done to correct it? The people who were breeding the pigs or managing the dairies were only looking at their own animals. They seldom saw other groups of pigs or cows to compare their animals to. Since the increase in lameness occurred slowly over a period of years, they failed to notice it because they were not measuring it. This is a prime example of a situation the author calls 'Bad Becoming Normal'. The reason the author noticed the increase in lame pigs when the pig breeders failed to see it was that the author had observed thousands of pigs in many different slaughter plants. Observations of slaughter weight pigs, from many different breeders, that were raised in similar buildings indicated that there were large differences between breeders in the percentage of lame pigs and leg conformation defects. To correct this problem, progressive dairy producers and pig breeders have incorporated formal lameness measurements and leg conformation evaluations into their programmes. Lameness is one of the most serious animal welfare problems. Figure 1.1 contains drawings of both correct leg conformation and different types of structurally poor legs in pigs. Charts like this can be easily made for all livestock species and poultry. They should be used when breeding stock is being selected.

Handling Practices Deteriorate Unless Measured

The author has observed that animal handling practices can slowly deteriorate and become rougher and rougher without anyone realizing it. Over the years the author has given many seminars on low-stress handling and quiet movement of pigs and cattle through corral systems on ranches, feedlots and slaughter plants. Many managers were eager to implement the new methods. Employees were taught to use behavioural principles of animal movement, stop yelling and greatly reduce the use of electric goads. A year later when the author returned to re-evaluate the handling practices, it was discouraging to observe that many employees had reverted back to their old rough ways. When the manager was informed that the animal handling methods were bad, he was surprised and upset. Because the regression back into old rough ways had happened slowly over the course of a year, the manager did not notice the slow deterioration of handling practices. Bad practices had become normal because the manager had not measured handling practices in an objective manner. In Chapters 3 and 9, an easy method to objectively measure livestock handling will be covered.

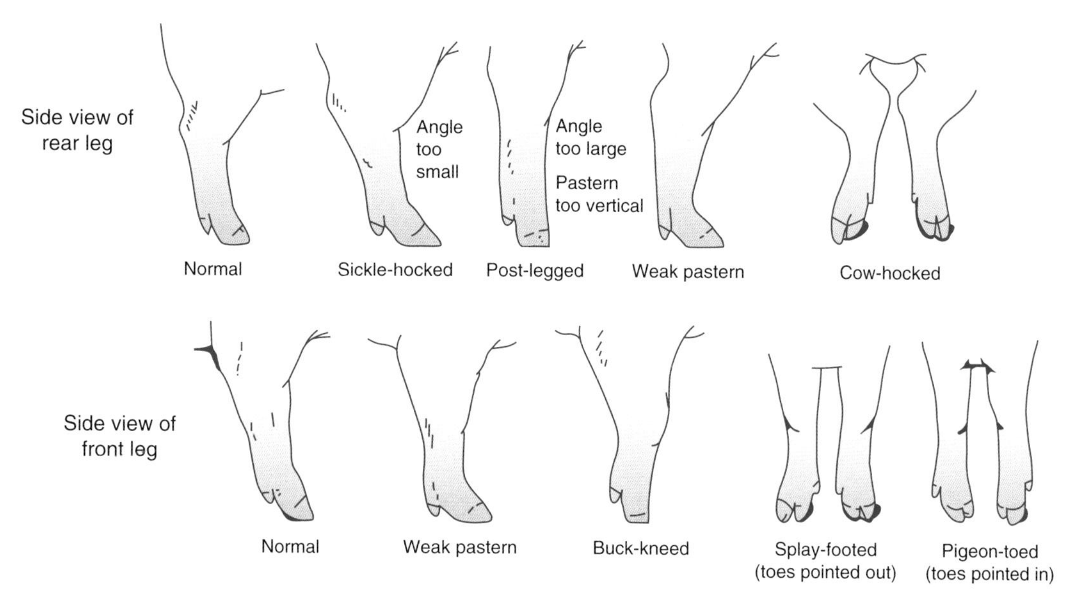

Fig. 1.1. Chart for scoring leg conformation in pigs. Plastic-laminated leg conformation charts should be used by producers when they select animals for breeding stock. Animals with poor leg conformation are more likely to become lame. The legs on the far left are normal and all the other legs are not normal. Other species such as cattle have similar leg and foot abnormalities.

The Power of Comparison Measures

When large numbers of farms are surveyed, practices and conditions can range from excellent to atrocious. Fulwider *et al.* (2007) surveyed 113 freestall (cubicle) dairies on the incidence of lesions and swellings on the dairy cow's legs. Figure 1.2 shows a cow that would be scored for a severe swelling. Table 1.1 shows the big differences between the best 20% of the farms and the worst 20%. The best 20% had 0% of the cows with swollen hocks and the worst 20% of the dairies had 7.4–12.5% of the cows had swollen hocks. In a Canadian study of 317 tie stall dairies, 26% of the dairies were well managed and cows had no open wounds on the hocks. However, 16% of the dairies

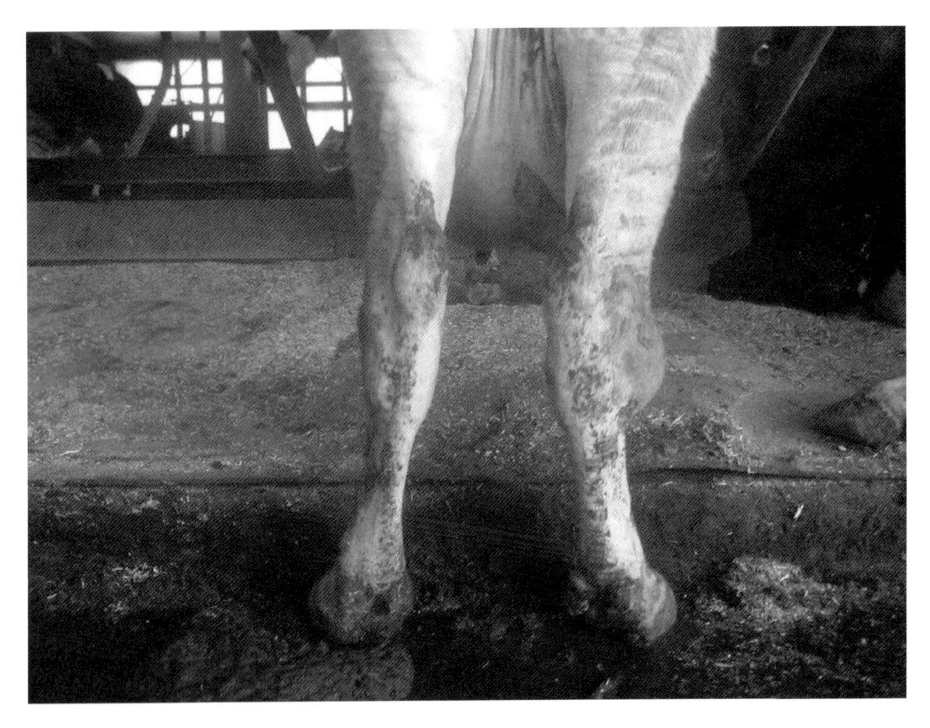

Fig. 1.2. Dairy cow with a severe leg lesion that has a diameter greater than 7.4 cm (size of a baseball). Photos like this should be put on laminated cards for scoring leg lesions on cows (photograph courtesy of Wendy Fulwider).

Table 1.1. Freestall (cubicle) dairies sorted by the best 20% to the worst 20% of farms for each welfare issue in 113 dairies (source: adapted from Fulwider *et al.*, 2007).

	Percentage of cows on				
Well-being issue	Best 20% of farms	Second best 20% of farms	Middle 20% of farms	Second worst 20% of farms	Worst 20% of farms
Hock hair loss only	0–10	10.6–20	20.8–35.8	36.2–54.4	56–96.1
Hock swelling	0	0.7–1.7	1.9–4.2	4.2–11.9	7.4–12.5
Severe swelling[a]	0	0	0	0–1.5	1.8–10.7
Dirty cows[b]	0–5	5.3–9.8	10.3–15.4	16.8–28.9	29.4–100
Thigh lesions	0	0	0	0	0–28.8

[a]Cows were rated on having a severe swelling if the worst leg had a swelling more than 7.4 cm (size of a baseball) in diameter or open or oozing injuries.
[b]Cows were rated as dirty if there was dried or wet manure on their body, belly, udder or upper portions of the leg.

were really bad and 15% or more of the cows had open hock wounds (Zurbrigg *et al.*, 2005a, b). In another study of 53 dairy farms in the UK, the best 20% had 0–13.6% lame cows and the worst 20% had 34.9–54.4% lame (Wray *et al.*, 2003). A panel of dairy veterinarians agreed that lameness and swollen or ulcerated hocks were the most serious problems that needed to be corrected (Wray *et al.*, 2003). It is likely that many of the dairy producers in the worst groups did not realize how bad they were compared to the other 80% of their colleagues. Research done in Ontario, Canada showed that 40% of the farms had 0% broken tails and the worst 20% had 5–50% of the cows with broken tails (OMAFRA, 2005).

A huge survey conducted by Knowles *et al.* (2008) in the UK showed that 27.6% of the chickens were lame with a score of 3 or higher on a six-point scale. The scores ranged from normal to down and not able to walk. A score 3 bird is mobile, but obviously lame. There was a big difference between five different chicken companies and there was also a lot of variation between the best and the worst flocks. The best flock had 0% with a score of 3 or worse lame birds, and the worst flock had 83.7% obviously lame birds. The standard deviation for data collected on 176 flocks was 24.3% for score 3 lameness. A high standard deviation indicates that there were huge differences between the best and the worst farms. Unpublished industry data in the USA from chicken farms that are being audited by major customers had only 2% obviously lame 3 kg chickens.

Numerical Scoring of Handling and Stunning

Grandin (2005, 2007) discussed the implementation of measurements of practices in slaughter plants as a method for auditing animal welfare. This measurement system is now used by McDonald's Corporation, Wendy's, Tesco, International, Burger King, Whole Foods and many other large meat-buying customers. Instead of an auditor subjectively determining whether a plant has good or bad practices, stunning and handling are evaluated with objective numerical scoring. Before the audits began, it was routine to use an electric prod multiple times on every animal and, in some plants, the stun guns were not kept repaired.

Baseline data prior to the audits indicated that only 30% of the plants could stun 95% of the cattle with a single shot from a captive bolt. After the audits began and the plants became concerned about losing a major customer, the percentage of plants that could achieve this rose to over 90% (Fig. 1.3). Both audit and survey data indicated that a lack of maintenance of the captive bolt was a major cause of poor stunning.

Similar dramatic results were obtained with the reduction of electric-goad use. Usage of electric goads dropped from multiple shocks administered to every animal to more than 75% of the cattle and pigs moving through the entire plant with no shocks (Grandin, 2005). Studies done on South American ranches have shown that training handlers and improved procedures resulted in a huge reduction in cattle trampling on top of another animal and lost vaccine. Carefully restraining each animal one at a time in the head stanchion and squeeze was compared to vaccinating a long line of animals held in a race. Restraining each animal individually reduced trampling from 10 to 0% and lost vaccine dropped from 7 to 1% (Chiquitelli Neto *et al.*, 2002; Paranhos de Costa, 2008).

Measuring Improvements within a Farm or Plant

Objective measurements can also be used by the managers of farms and plants to quantify improvements and to do continuous internal audits to prevent practices from slipping back into old sloppy or rough ways.

Measurements every week or month make it easy to determine if handling practices are improving, staying the same or becoming slowly worse. Measurements on a regular basis of welfare-related issues such as lameness or leg lesions will enable farm managers to determine if their veterinary, bedding and husbandry programmes are improving or becoming worse.

Measurements can also be used to determine if a new piece of equipment, a new procedure or a repair has made an improvement. Figure 1.4 illustrates that a simple modification such as adding a light at the entrance of a race made it possible to greatly reduce electric-prod use due to pigs baulking and refusing to enter the race. The addition of a lamp reduces baulking and fewer pigs had to be shocked. Pigs have a natural tendency to approach illuminated areas (Van Putten and Elshof, 1978; Grandin, 1982; Tanida *et al.*, 1996). Measurement also makes it possible to locate producers who

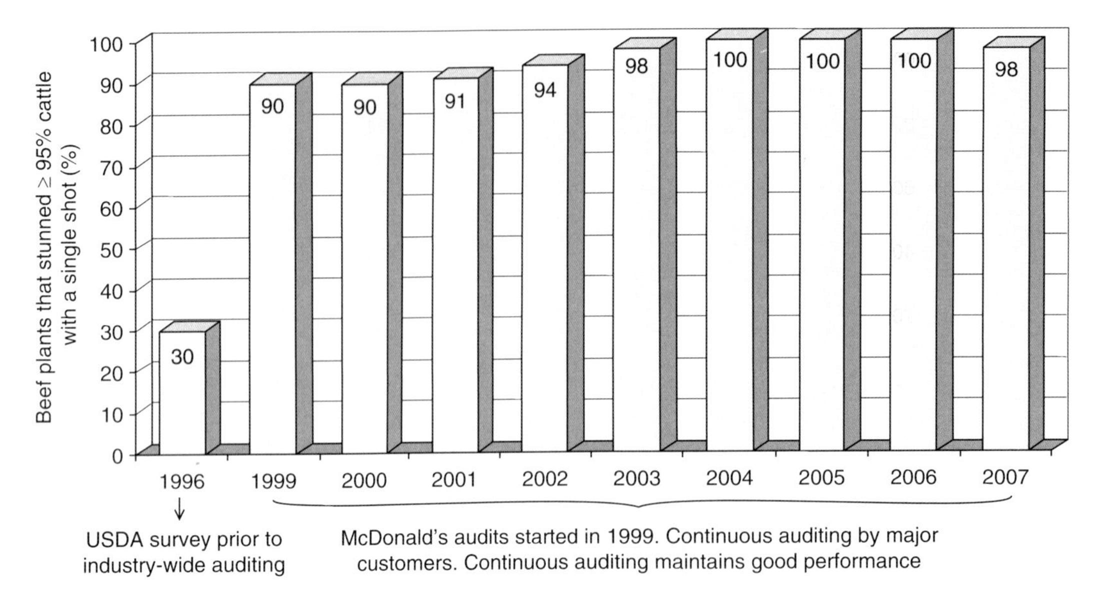

Fig. 1.3. Percentage of beef plants that stunned 95% or more cattle with the first shot. The baseline scores were measured in 1996 in ten US beef slaughter plants. After the restaurant audits started in 1999, the number of plants audited each year varied from 41 to 59. After the first 4 years of auditing, the scores further improved because the plants now had documented stunner maintenance programmes and test stands that measure the bolt velocity of the stunner.

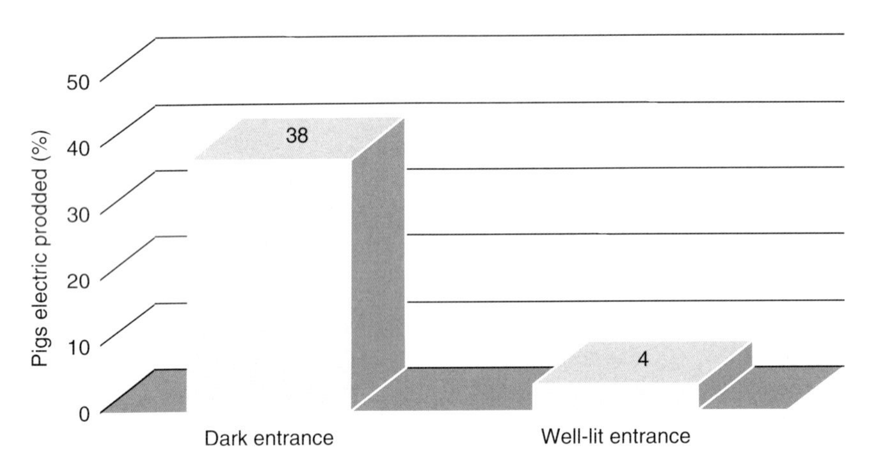

Fig. 1.4. Electric-goad use on pigs was reduced by adding lighting at the restrainer entrance. Simple changes such as placing a light at the entrance of a stunning race greatly reduced baulking and refusal to enter the race so less use of an electric goad was required. All handlers were well trained and only pigs that baulked or backed up were prodded.

have animals that are difficult to handle. Figure 1.5 shows that some groups of pigs are more difficult to drive and there is increased electric-goad use and squealing. In Chapter 3 the measuring tools that can be used to assess welfare will be covered in more detail.

Four Guiding Principles of Welfare

David Fraser (2008b) at the University of British Columbia in Canada states that from both an ethical and a scientific viewpoint, there are four guiding principles for good welfare:

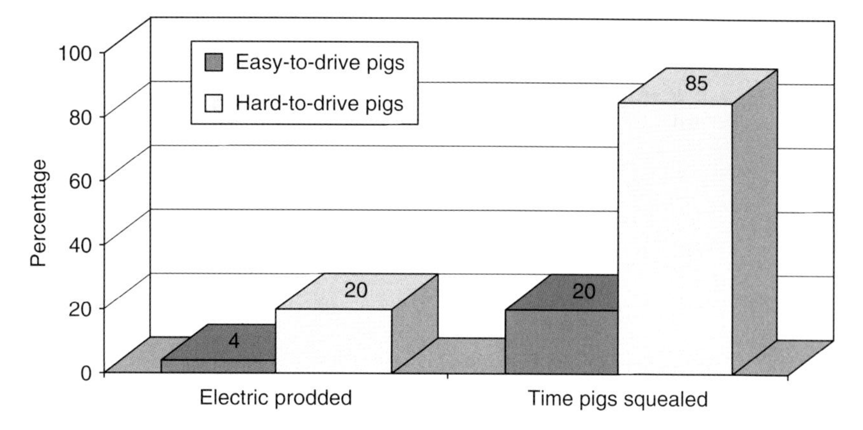

Fig. 1.5. Comparison of electric-goad use and squealing between easy-to-drive pigs and hard-to-drive pigs when moved by well-trained people who only used the electric goad on pigs that refused to move. Some pigs are more excitable and difficult to handle compared to other pigs. Measurement of each producer's pigs can be used to locate problem pigs, which baulk more and are more difficult to move through races. Squealing can be measured in small plants by counting the number of pigs that squeal. In large plants this is difficult, so the percentage of time the room was quiet was measured.

1. **Maintain basic health** – Examples: provide sufficient feed, water, vaccinations, housing and air quality to prevent disease and reduce death losses. Maintain the body condition and productivity of the animals or birds. Health is a major component of animal welfare but it is not the only factor.
2. **Reduce pain and distress** – Examples: use of anaesthetics for dehorning, preventing lameness, reducing bruises, preventing injuries and elimination of rough, stressful handling methods that cause fear or pain. Preventing hunger, thirst, heat stress and cold stress are also covered by the second principle.
3. **Accommodate natural behaviours and affective states** – Examples: provide a nest box for hens and straw for pigs to root in. The affective state is the animal's emotional state (see Chapter 8; Duncan, 1998).
4. **Natural elements in the environment** – Examples: outdoor access or natural sunlight.

A group of researchers from four European countries have proposed an alternative method for categorizing the four criteria for animal welfare (Botreau *et al.*, 2007). They are:

1. Good feeding – absence of prolonged hunger and thirst.
2. Good housing – includes thermal comfort and ease of movement.
3. Good health – includes injuries, disease and controlling pain during surgical procedures.
4. Appropriate behaviour – includes expression of social behaviour and good human animal relationship.

In the next section, the author has categorized animal welfare issues in a series of tables that will make it easier for veterinarians and managers on farms and slaughter plants to implement improvements. The first two of Fraser's guiding principles of maintaining the animal's health and preventing pain and distress cover most of the worst problems, caused by either neglect or abusive treatment. The first two guiding principles outlined by David Fraser also cover four out of the five welfare freedoms in the Brambell Report and the OIE (2008) code. The OIE code (2008) states that the five freedoms are:

1. Freedom from hunger, thirst and malnutrition.
2. Freedom from physical and thermal discomfort.
3. Freedom from pain, injury and disease.
4. Freedom to express normal behaviour.
5. Freedom from fear and distress.

The ability of an animal to express normal behaviour is important, but in some parts of the world the first priority will be to correct obvious suffering that is caused by neglect, lack of knowledge or outright abuse.

Good Health Does Not Guarantee Good Welfare

Many people mistakenly assume that good health automatically means that an animal has satisfactory welfare. Good health is a key component for an animal to have good welfare. The OIE (2009a, b) code states 'Good welfare requires disease prevention and veterinary treatment'. However, there are some situations where an animal can be healthy but its welfare may be poor. For example, a dairy cow that is healthy, disease free and providing lots of milk could have painful lesions on her legs from lying in a poorly bedded stall that did not provide enough cushioning to prevent injuries. Fulwider *et al.* (2007) found that high-producing dairy cows had more leg lesions and a shorter productive life. Some genetic lines of chickens that have been bred to grow quickly have a high rate of lameness and leg abnormalities (Knowles *et al.*, 2008). Another example of poor welfare would be healthy laying hens that keep laying eggs even though the cages are so tightly stocked that the hens cannot all lie down at the same time without being on top of each other.

Some animals that have been bred for rapid weight gain are difficult to transport and handle in a manner that ensures good welfare. Certain genetic lines of pigs that have been bred for rapid weight gain become easily fatigued and weak when grown to heavy weights of 130kg. The author has also observed healthy pigs that had been fed too much Paylean® (ractopamine) that were too weak to walk from one end of a lairage to the other. This is another example of bad conditions that some people perceived as normal. The author observed pigs 30 years ago that were strong enough to walk up long, steep ramps.

Healthy animals may also have abnormal behaviour if they are housed in an environment that does not allow them to express normal social and species-specific behaviours. Some examples of abnormal behaviour are pacing in a circle, bar biting in sows housed in gestation stalls, tail biting and pulling out feathers or hair. The behavioural needs of animals will be discussed in Chapters 8 and 15. There are also many excellent books on farm behaviour and ethology that review research on behavioural needs (Broom and Fraser, 2007; Fraser, 2008b).

Reducing Pain and Distress

Many welfare problems cause pain and distress. There are numerous scientific studies, review papers and books that document that livestock and poultry feel both pain and fear (Gentle *et al.*, 1990; Rogan and LeDoux, 1996; Grandin, 1997; Panksepp, 1998; Grandin and Johnson, 2005, 2009). Both rats and chickens will self-medicate for pain (Danbury *et al.*, 2000; Colpaert *et al.*, 2001). They will eat or drink bitter-tasting food or water containing painkillers when they are lame and have sore leg joints. To make it easier to implement a welfare programme, pain and distress items can be grouped into four categories:

1. Most severe problems with abuse or neglect that cause obvious suffering (Table 1.2) – these conditions must be corrected immediately.
2. Routine painful procedures (Box 1.1).
3. Fear stress during handling and transport – this stress can be reduced with good management (see Chapters 4, 5 and 7).
4. Overloading the animal's biological system (Box 1.2).

Painful routine procedures in livestock and poultry

The pain and distress category includes all of the painful procedures that producers routinely perform on animals (Box 1.1). Many of these procedures are done with no anaesthetics or painkillers. Numerous research studies on dehorning of livestock clearly indicate the need for pain relief (Faulkner and Weary, 2000; Stafford and Mellor 2005a). Cautery disbudding young calves is strongly recommended. Stafford and Mellor (2005b) report that removing the horns of an older animal causes greater increases in cortisol compared to disbudding calves. Other research has shown welfare benefits of providing painkillers for castration. There has been much controversy and discussion about whether or not certain procedures should be done at all. Further discussion of painful management procedures is in Chapter 6. Animal welfare legislation and animal welfare standards will vary greatly in different countries. However, most researchers and veterinary organizations agree that major surgery where the main body cavity is opened such as flank spaying a heifer requires the use of an anaesthetic.

Table 1.3 shows behaviours that are associated with pain that are easy for people to score and quantify. The behaviours in Table 1.3 occur AFTER the procedure has been done in cattle, calves, pigs and lambs. Animals should be scored

Table 1.2. The most severe animal welfare problems caused by abuse, neglect or bad management that cause obvious suffering. These conditions must be corrected immediately.

Handling and transport – prohibited practices and conditions[a]	Welfare problems caused by poor housing, environmental conditions, nutrition or health problems	Slaughter – prohibited practices[a]
• Beating, throwing or kicking animals • Poking out eyes or cutting tendons to restrain an animal • Dragging and dropping animals • Overloading trucks so tightly that a downed animal is trampled • Deliberating driving animals over the top of other animals • Poking animals in sensitive areas such as the eyes, anus or mouth • Breaking tails or legs • Overloading a draught animal and working it to exhaustion • Poking animals with pointed sticks • Conditions that cause animals to frequently fall or become injured or bruised during handling	• Starvation or allowing animals to become severely dehydrated • High ammonia levels that cause eye or lung damage • Death or severe stress from extreme heat or cold • Large swellings or other injuries caused by either a lack of bedding or poorly designed housing • Dirty animals covered with manure with no dry place to lie down • Failure to treat obvious health problems • Nutritional problems that compromise the animal's health • Conditions that cause many animals to become lame • Saddle or harness sores on a working animal	• Scalding, skinning, leg removal or other carcass dressing procedures performed on sensible, conscious animals • Immobilizing animals with an electrical current (Lambooy, 1985; Grandin *et al.*, 1986; Pascoe, 1986), not to be confused with effective electrical stunning • Puntilla method of immobilizing animals before slaughter by severing the spinal cord, which does not cause instantaneous insensibility (Limon *et al.*, 2008) • Highly stressful methods of restraining conscious animals. One example is hoisting cattle by one leg

[a]The items in the handling and transport and slaughter columns would be in violation of OIE (2009a, b) codes for slaughter and transport. The OIE standards for animal welfare are the most basic standards that everybody in both developed and developing countries should follow. To achieve a higher level of welfare will require some additional standards. Many countries have many additional standards. Disease control standards between different countries are easier to make uniform between countries than welfare standards that have more complex ethical considerations.

immediately after a procedure is done, for a minimum of 1 h, to detect signs of acute pain.

To detect signs of long-term pain, scoring can also be done over a period of days. These behaviours are associated with physiological measures of pain and stress (Molony and Kent, 1997; Eicher and Dailey, 2002; Sylvester *et al.*, 2004; Stafford and Mellor, 2005b; Vihuela-Fernandez *et al.*, 2007). The behaviours associated with different painful surgeries will vary depending on the procedure and the species. Quantifying pain-related behaviour provides an easy economical way to evaluate welfare in large numbers of animals. Animals often conceal pain-related behaviour when they see a person watching. To accurately assess the occurrence of pain-related behaviours, either the observer must be hidden from the animal's view or a remote video camera should be used.

Box 1.1. Routine painful management procedures.

- Beak trimming in poultry.
- Spaying female animals.
- Castrating male animals.
- Dehorning.
- Notching ears for identification.
- Removing tusks on boars.
- Clipping needle teeth on piglets.
- Docking dairy cow and pig tails.
- Mutilating and making big cuts in ears to identify animals.
- Mulesing sheep – cutting the skin on the rear end of a lamb to prevent flystrike.
- Wattling – cutting flaps of skin for identification.
- Tail docking.
- Hot iron branding.

Box 1.2. Welfare problems caused by overloading the animal's biological system.

- Lameness or leg abnormalities in rapidly growing pigs and poultry (Fernandez de Seville *et al.*, 2008; Knowles *et al.*, 2008).
- Increased aggression in some genetic lines of pigs or chickens (Craig and Muir, 1998).
- Future problems with animals that have been genetically modified (OIE, 2006).
- Increased excitability in some genetic lines of lean pigs bred for rapid growth.
- High rates of calving problems in cattle bred for large muscle mass (Webster, 2005a, b).
- Heat stress in cattle due to beta-agonists such as ractopamine (Grandin, 2007).
- Health problems caused by rBST (growth hormone given to dairy cows to increase milk production) in dairy cows (Willeberg 1993; Kronfield, 1994; Collier *et al.*, 2001).
- Stress gene in pigs that causes porcine stress syndrome which increases death losses (Murray and Johnson, 1998).
- Weak heavily muscled pigs due to either genetics or the use of beta-agonists such as ractopamine (Marchant-Forde *et al.*, 2003; Grandin 2007). These pigs may be reluctant to move.
- Cracks and hoof lesions in pigs fed ractopamine (Poletto *et al.*, 2009).
- High appetite drive and frustration in breeding animals selected for high weight gain when they are fed a restricted diet to prevent obesity.
- Metabolic problems that may increase death losses in poultry (Parkdel *et al.*, 2005).
- Lameness in cattle due to excessive use of beta-agonists such as ractopamine or zilpaterol.
- Increased bites from fighting in pigs caused by the beta-agonist ractopamine (Garner *et al.*, 2008).
- Lameness caused by poor leg conformation and over selection for a narrow range of production traits.
- Dairy cows that last only two lactations.

Table 1.3. Behaviours associated with pain that are easy to numerically quantify in lambs, cattle, calves, piglets and other animals. Evaluate these behaviours AFTER the painful procedure has been done (sources: Molony and Kent, 1997; Eicher and Dailey, 2002; Sylvester *et al.*, 2004; Stafford and Mellor, 2005a; Vihuela-Fernandez *et al.*, 2007).

Behaviours associated with pain	Species[a]
Time in contorted abnormal lateral or ventral recumbancy	Lambs, calves, cattle
Time in lateral recumbancy	Lambs, calves, cattle
Number of times foot stamped	Lambs, calves, cattle
Number of kicks	Lambs
Number of lip curls	Lambs
Number of ear flicks	Calves, cattle
Number of tail switches (wags)	Cattle, calves
Time standing still like a statue	Cattle, calves
Time walking (restless)	Cattle, calves
Time trembling	Calves
Time lying down in all positions	All species
Time huddling	Piglets
Time kneeling	Piglets

[a]The scientific studies were done in calves, lambs and piglets. Many of these behaviours may also occur in other species.

Vocalization scoring of painful, stressful procedures

Vocalization scoring of squeals, moos and bellows in cattle and pigs is a useful indicator that a procedure is stressful. Watts and Stookey (1998) recommend that vocalization scoring should be used to assess the stressfulness of a procedure across a group of animals and not for assessing the welfare of each individual. Watts and Stookey (1998) found that hot iron branding caused 23% of the cattle to vocalize and freeze branding caused vocalizations in only 3%. In cattle, vocalization may be especially useful for evaluating severe stress (Watts and Stookey, 1998). When low-stress weaning methods were used for weaning calves, the number of vocalizations was significantly lower compared to abrupt weaning (Price *et al.*, 2003; Haley *et al.*, 2005). The calves weaned by the low-stress method also had better weight gains.

In another study, 98% of the cattle that vocalized during handling and stunning at a slaughter plant had been subjected to an obvious aversive event such as being poked with an electric goad, ineffective stuns or excessive pressure from a restraint device (Grandin, 1998b). Vocalization scoring works well for showing how improvements in equipment and handling procedures will lower the vocalization score (Grandin, 2001). Table 1.4 shows some data where the variable of restraining the animal in the squeeze chute was separated from the variable of the stressful procedure. It also shows how a less severe electro-ejaculation method reduced the number of vocalizations in Angus beef cattle.

Vocalization scoring should be done DURING procedures such as branding, castration, weaning, restraint or handling. Vocalization during painful or stressful procedures is correlated with physiological measures of stress (Dunn, 1990; Warriss *et al.*, 1994; White *et al.*, 1995). The neuropeptide substance P is involved in pain perception. It was higher in calves with more vocalization during castration (Coitzee *et al.*, 2008).

Vocalization must not be used in sheep for scoring reactions to painful procedures or the stress of being restrained or handled. Cattle and pigs will vocalize when they are hurt or frightened, but sheep will usually remain silent.

Sheep are the ultimate defenceless prey species. They evolved to remain silent when they are hurt so they do not advertise their vulnerability to predators. However, lambs will vocalize loudly when they are separated from the mother at weaning. This is the only time that distressed sheep will vocalize. Vocalization scoring cannot be used if an animal is immobilized with electricity. Immobilization prevents vocalization. These devices are highly stressful and should not be used (see Chapter 5).

How to score vocalizations

There are two ways for scoring vocalizations that work really well in cattle and pigs. These methods can be easily used on a farm, ranch or slaughter plant. They are:

1. Score each animal as either silent or vocal to determine the percentages of animals that vocalized. This simple method works really well for assessing problems with excessive electric-prod use and other handling problems (Grandin, 1998a, 2005).
2. Score the total number of vocalizations in a group of animals. The number of vocalizations are counted and divided by the number of animals to determine an average vocalization score per animal.

Behavioural Versus Physiological Measures

Many scientists and veterinarians would prefer to use physiological assessments instead of behavioural measures. For research studies, it is recommended to use physiological and production measures combined with behavioural measures. The problem with physiological measures such as cortisol is that the lab tests are too expensive for routine farm use. The major advantage of behavioural measurements is that they can be easily implemented on the farm. Researchers should work to validate easily observable behavioural

Table 1.4. Average number of vocalizations per bull in response to restraint only or restraint plus electro-ejaculation (source: B.D. Voisinet and T. Grandin, 1997, unpublished data).

	Controls restraint in a squeeze chute with a stanchion headgate	High voltage electro-ejaculation machine	Low voltage electro-ejaculation machine
Average number of vocalizations per bull[a]	0.15 ± 0.1	8.9 ± 1.1[b]	3.9 ± 1.0[b]

[a]The total number of bellows for each treatment was tabulated and divided by the number of bulls.
[b]The difference between the high voltage and low voltage was significant at the $P \leq 0.001$ level.

measures. There is a huge need for research in species such as goats, camels, donkeys and many other animals. Research has been done that showed that the pitch of a vocalization is related to distress. Higher pitched vocalizations are associated with higher stress (Watts and Stookey, 1998). Unfortunately this method requires expensive equipment that is not practical for routine or farm use, but this research provides valuable insight into the emotional state of the animal.

Fear

Fear stress occurs when animals become agitated and excited during restraint and handling. Rough handling and multiple shocks from electric prods will increase fear stress. Stress from poor handling methods will greatly elevate stress hormones (Grandin, 1997). Pearson *et al.* (1977) found that sheep slaughtered in a small, quiet research abattoir had lower cortisol levels compared to sheep in a large, noisy commercial abattoir. Animals that are fearful of people or become agitated during handling will have lower weight gains and be less productive (Hemsworth and Coleman, 1994; Voisinet *et al.*, 1997). The fear circuits in animals' brains have been completely mapped. The literature on fear is reviewed by Rogan and LeDoux (1996) and Grandin (1997). Scientists were able to determine that the brain contains a fear centre called the amygdala. Destroying the amygdala will eliminate both learned and unlearned fear responses (Davis, 1992). Electrical stimulation of the amygdala in rats and cats raises corticosterone levels (Setckleiv *et al.*, 1961; Matheson *et al.*, 1971; Redgate and Fabringer, 1973). The amygdala is also the fear centre in humans (Rogan and LeDoux, 1996). Animals that are fearful will have poor welfare. There is a further discussion of fear stress during handling and restraint of both livestock and poultry. Chapter 4 reviews many research studies which clearly show the benefits of good stockmanship on reducing fear stress.

Tonic immobility looks calm but is really frightened

Some animals and birds will become highly agitated when they are frightened. However, others may go into tonic immobility and remain completely still and look calm. This phenomenon has been extensively researched in poultry. Tonic immobility can be induced by placing a chicken on its back in a U-shaped trough. It is held down lightly by a person for 10s (Faure and Mills, 1998). If tonic immobility is induced, it will make no attempt to get up for at least 10s (Jones, 1987). Poultry from genetic lines that are highly fearful will stay motionless for longer periods of time (Jones and Mills, 1982). The length of time a bird will remain motionless in a state of tonic immobility is used as an index of fear in chickens (Jones, 1984). Stimuli that are stressful to poultry such as electrical shocks or being kept in continuous light with no period of darkness increase the duration of tonic immobility (Hughes, 1979; Campo *et al.*, 2007). Many people mistakenly assume that chickens that go into tonic immobility are calm and relaxed. Tonic immobility testing can be used as one indicator of conditions that are stressful to poultry.

Do Fish Suffer?

Since the early 2000s, there has been an explosion of research on fish welfare. A search of the literature located many new welfare papers and lots of recent patents for equipment for stunning fish at the processing plant. The vast majority of this research has been on the teleost fish such as salmon, trout, tilapia and other finned fish that are farmed. Scientists in the UK, Canada, Norway, Brazil and other countries have recently written on the welfare of teleost fish (Chandroo *et al.*, 2004; Braithwaite and Boulcott, 2007; Lund *et al.*, 2007; Volpato *et al.*, 2007; Branson, 2008). Lund *et al.* (2007) state that fish can detect noxious stimuli and these authors conclude that 'farmed fish should be given the benefit of the doubt and we should make efforts to make sure that their welfare needs are met as well as possible'.

Research shows that fish respond to painful stimuli in a manner that is not just a simple reflex. The most convincing evidence that fish feel pain comes from the studies by Sneddon (2003), Sneddon *et al.* (2003a, b) and Reilly *et al.* (2008). Acetic acid was injected into the lips of fish to create a painful stimulus. Some of the fish engaged in weird rocking back and forth and rubbing the injected lip against the tank walls. Some individuals exhibited the behaviour and others did not. It is common in pain studies in all species to have big differences in the reaction between different individuals. There were also species differences in the occurrence of this behaviour. Zebra fish did not do it (Reilly *et al.*, 2008).

A study by Dunlap *et al.* (2005) showed that fish can be fear conditioned and that their reactions are affected in a complex manner by the presence of other fish. There is also evidence that fish react to handling stress with increases in cortisol. This would be similar to the increase in cortisol after stressful handling in mammals. The final experiment that needs to be done to verify that finned fish suffer from pain is the self-medication experiment that has clearly shown that rats and chickens will self-medicate for pain (Danbury *et al.*, 2000; Colpaert *et al.*, 2001).

From a practical standpoint, this research indicates that equipment such as a stunner should be used to render farmed fish insensible at the slaughter plant. Some of the behavioural indicators of distress that could be easily quantified on a fish farm are loss of equilibrium (fish is belly up), high respiration rate and agitated swimming (Newby and Stevens, 2008). Other researchers scored fish for fast-swimming escape responses and a tail-flip behaviour called the Mauthner-initiated startle response (Eaton *et al.*, 1977). Fish is one area where more research will be needed to develop simple on-farm assessments. There are significant species differences in a fish's reaction to stress. Specific behavioural assessments will have to be developed for each species of farmed fish.

Invertebrates

During the literature search on this section of fish welfare, the author found only one paper on possible pain perception in invertebrates such as prawns. In this experiment, acetic acid was applied to the antennae. The prawns reacted by rubbing their antennae on the sides of the tank. The research was done by Stuart Barr at the Queen's University in Belfast (Barrett, 2008). Much more research will need to be done. At the time of writing this chapter, the author recommends that welfare programmes should be initiated for farmed, teleost fish and further research is required on invertebrates.

Overloaded Biology

Overloading the animal's biological system by either genetic selection for more and more production or use of performance-enhancing substances may also cause pain and distress. This includes welfare problems caused by over-use of beta-agonists, rBST growth hormone and other performance-enhancing substances. Many of these conditions may cause serious suffering. Box 1.2 lists conditions caused by either genetic selection for a narrow range of production traits or the indiscriminant use of substances such as hormones or beta-agonists such as ractopamine. All of the conditions listed in Box 1.2 are caused by pushing the animal's biology too hard to keep producing more and more meat, eggs or milk. It is the author's opinion that overloading of the biological system is the cause of many serious welfare problems. Overloading the animal's biology is a matter of degree. Selection for higher production will usually have no detrimental effects on welfare if it is done in moderation. Careful use of low amounts of performance-enhancing substances is probably not detrimental. It is just like revving up the rpm (revolutions per minute) in a car engine. A little revving up does no harm but if the engine rpm get too high, the engine will be ruined. The problem is that breeders and producers often fail to see problems until they become very serious.

Accommodate Natural Behaviours and Affective States

Fraser (2008b) maintains that affective states are a central animal welfare principle. An affective state is the animal's emotional state. The animal's affective state provides the motivation for many natural behaviours. Scientific research clearly shows that animals are highly motivated to perform certain species-typical behaviour. Pigs are highly motivated to explore and root in soft fibrous materials such as straw, cornstalks, wood chips or other bedding materials (Van de Weerd *et al.*, 2003; Studnitz *et al.*, 2007; Day *et al.*, 2008; Van de Weerd and Day, 2009; see Chapter 8). The author has observed that fresh straw is rooted more by pigs than is old straw. After the straw is chewed up into little pieces, the pigs lose interest in it. Fraser (1975) found that providing small amounts of straw to tethered sows prevented abnormal stereotypic behaviour. Together these studies show that pigs should receive a daily ration of hay, straw or cornstalks to satisfy their rooting and chewing needs. Other behavioural needs that are strongly supported by scientific research are providing nest boxes and perches for laying hens (Duncan and Kite, 1989; Hughes *et al.*, 1993; Freire *et al.*, 1997; Olsson and Keeling, 2000;

Cordiner and Savory, 2001). A hen seeks a nest box to hide in so she has freedom from fear. Finding a secluded place to lay her eggs is an instinctual behaviour which prevented the wild ancestors of domestic hens from being eaten by predators.

Motivation can be measured in a very objective manner. Some of the methods that can be used to measure the strength of an animal's motivation to perform natural behaviours are: (i) the amount of time an animal is willing to go without feed so it can perform a behaviour; (ii) the number of times it will push a switch to get to something it wants; and (iii) weighted doors that become increasingly heavy (Widowski and Duncan, 2000; see Chapter 15).

Scientific research clearly shows that to give an animal a high level of welfare, the most highly motivated behaviours should be accommodated (O'Hara and O'Connor, 2007). Behavioural needs are important, but in places where conditions are really poor, the very serious animal welfare problems listed in Table 1.2 should be corrected first. Box 1.3 lists the most important behavioural needs.

Natural Elements and Ethical Considerations

The fourth principle about providing natural elements has little scientific basis compared to the first three principles of health, pain and distress, and natural behaviour. The first three principles are backed by many scientific studies. The fourth principle is mainly an ethical concern. Ethical concerns must not be ignored by the veterinarians, managers and other people who are in charge of implementing an animal welfare programme. Many organic programmes and large meat buyers specify that animals should be able to go outside or have daylight. A mistake that some people have made is to provide the natural elements but the animals have poor welfare due to neglect of principles 1 and 2. Health is an essential component of good welfare. The author has seen disgusting outdoor pig units full of sick pigs and excellent outdoor units with lots of healthy pigs. It is much easier to reach a consensus on welfare issues where science can provide clear answers compared to ethical concerns that have no clear answer. Ethical concerns are part of the decision-making process of legislators, animal advocacy groups and others who make policy. Lassen *et al.* (2008) provides a good summary: 'The main message of this paper for those who are professionally involved in animal production is that ethical assumptions and potential conflicts of view should be recognized and brought into the discussion of animal welfare.' Science cannot provide all the answers to ethical concerns. In some cases, ethics will overrule science. Sow gestation stalls are a good example. Research shows that sows can be highly productive in stalls, but confining a pig in a box where she cannot turn around for most of her life is not acceptable to two-thirds of the public. The author showed pictures of sow gestation stalls to passengers she sat next to on many flights. One-third of the passengers had no opinion, one-third said 'this does not seem right' and one-third hated them. One person said 'I would not keep my dogs in that'. Roughly two-thirds of the public disliked sow stalls. In 2008 in California, 63% of the voters decided to ban sow gestation stalls. Sow stalls are being phased out in both Europe and the USA. Farm animals are sentient beings and it is the author's opinion that people should provide both

Box 1.3. Basic behavioural needs that should be accommodated.

- Roughage feed for ruminant animals and equines.
- Animals should have sufficient space to be able to turn around, stand up and lie down in natural positions.
- Secluded nest boxes for poultry.
- Perches for poultry.
- Straw or other fibrous materials for pigs to root and chew on.
- Repetitive stereotyped behaviour in a barren cage or pen is an indicator of a poor environment that does not satisfy behavioural needs. Environmental enrichment should be provided to prevent abnormal repetitive behaviour (see Chapter 8).
- Opportunities for social interaction with other animals.
- An environment that helps prevent damaging abnormal behaviour such as feather pecking, wool pulling or tail biting (damage on animals can be easily quantified and measured).

food animals and working animals with a decent life that is worth living. Confining an animal for most of its life in a box in which it is not able to turn around does not provide a decent life.

Making Ethical Decisions

Scientific research shows that some behavioural needs are more important to the animal than others. For example, providing a secluded nest box is more important for a hen than providing a place to dust bathe (Widowski and Duncan, 2000). O'Hara and O'Connor (2007) state that there are priority behaviours that must be provided for in order to satisfy the minimum requirements for behavioural needs. The author recommends that the list in Box 1.3 would satisfy the minimum behavioural requirements for livestock and poultry. Higher welfare systems would provide for additional behaviours such as dust bathing in poultry or mud wallows for pigs.

Science can provide information to help people make good decisions about animal welfare. However, there are some ethical concerns that science cannot answer (see Chapter 2). To help make good decisions, many governments and large meat buyers have animal welfare advisory councils. The author has served on these councils for livestock industry associations, major retailers and restaurant chains. Most councils consist of scientific researchers in the field of welfare, animal advocacy groups and lay people. They provide advice and guidance. In Europe, advisory councils make recommendations on legislation. In the UK, the Farm Animal Welfare Council (FAWC) has been advising the government for many years. Another example is the Council on Animal Ethics at the National Veterinary Institute in Norway. It has both expert and lay members (Mejdell, 2006).

Measurements and Ethics

Numerical quantification of lameness, electric-goad use and feather pecking or other areas of welfare concern is a powerful tool for showing that practices or conditions have either improved or deteriorated. Some atrocious practices such as poking out an animal's eye or cutting leg tendons to restrain cattle should be banned. However, it is impossible to totally eliminate lameness. Chapter 3 will discuss practical ways to develop reasonable limits on the percentage of lame animals that would be permitted to pass a welfare audit. With good management, very low levels of lameness are possible.

From an ethical standpoint, interpretation of physiological measures such as cortisol levels or heart rate is more difficult. What level of cortisol should be permitted? The most practical way to help people make ethical decisions about physiological measurements is to compare the stressful or painful treatment to a control condition that most people find acceptable such as restraining the animal. It is best to evaluate physiological data by comparing them to a control condition within the same study with the same type of animals.

There are extreme levels of physiological measures that most scientific experts on an advisory council would be able to say 'This is absolutely not acceptable', for example, extremely high average levels of cortisol such as the 93 ng/ml average level reported by Dunn (1990) in cattle. The level is 30 units higher than cortisol levels due to poor handling. It is important to use the AVERAGE level of a physiological measure in a group of animals. Individual animals can have great variation in stress levels. More information on assessing stress has been reviewed in Grandin (1997). Another example of conditions that are absolutely not acceptable is the capture myopathy cases that are described in Chapter 5.

The Ethics of Animal Treatment Versus Treatment of People

The author went to Mexico and saw a man with a skinny, sickly donkey. When my host asked him about the poor condition of his donkey he pulled up his shirt to show us his skinny, bony chest with ribs that showed. He said 'I suffer too'. Obviously he cannot afford to feed his donkey because he cannot afford to feed himself. It would be unethical to ask the man to feed all his family's food to the donkey.

The most constructive way to implement improving animal welfare in this situation would be showing the man simple ways to help his donkey, to help him survive. Simple changes in the harness may prevent saddle sores, show the person how to take care of the animal's feet and work with people in the community on donkey husbandry (see Chapter 13). He cannot afford to feed it more, but some simple improvements in husbandry, such as providing plenty of water, could be taught to help the donkey live longer and be a more useful working animal. Numerical scoring of lameness, injuries and deaths in many donkeys would help show the entire community that when you take care of your donkey's welfare and you do not overload it, it will last longer. Even in the poorest country, there is never any justification for beating animals up or torturing them.

Where in the Book to Find Important Animal-based Measures for Assessing Welfare

- Body condition scoring – Chapters 3 and 13
- Lameness scoring – Chapters 1 and 7
- Condition of coat and feathers scoring – Chapter 3
- Lesion and injury scoring – Chapters 1, 3 and 13
- Handling scoring – Chapters 3, 5 and 9
- Transport losses scoring – Chapter 7
- Scoring of stunning at slaughter – Chapter 9
- Animal cleanliness scoring – Chapter 3
- Behavioural measurements – Chapters 1, 8 and 15
- Assessing pain – Chapters 1 and 6
- Vocalization scoring – Chapters 1, 5 and 9
- Panting scoring for heat stress – Chapter 7
- Ethical issues – Chapters 1, 2, 8 and 12
- Condition of pastures – Chapter 3
- Lists of practices that should be banned – Chapter 1

References

Barrett, L. (2008) Decapods feel the pinch. *Animal Behavior* 75, 743–744.

Botreau, R., Veissier, I., Butterwork, A., Bracke, M.B.M. and Keeling, L.J. (2007) Definition of criteria for overall assessment of animal welfare. *Animal Welfare* 16, 225–228.

Braithwaite, V.A. and Boulcott, P. (2007) Pain perception, aversion, and fear in fish. *Diseases in Aquatic Organisms* 75, 131–138.

Branson, E. (2008) *Fish Welfare*. Blackwell Publishing, Oxford, UK.

Broom, D.M. and Fraser, A.F. (2007) *Domestic Animal Welfare*. CAB International, Wallingford, UK.

Campo, J.L., Gil, M.G., Davila, S.G. and Munoz, I. (2007) Effect of lighting stress on fluctuating asymmetry, heterophil-lymphocyte ratio and tonic immobility duration in eleven breeds of chickens. *Journal of Poultry Science* 86, 37–45.

Chandroo, K.P., Duncan, I.J.H. and Moccia, R.D. (2004) Can fish suffer? Perspectives on pain, fear, and stress. *Applied Animal Behaviour Science* 86, 225–250.

Chiquitelli Neto, M., Paranhos de Costa, M.J.R., Piacoa, A.G. and Wolf, V. (2002) Manejo racional na vacinacio de bovinos Nelore: uma availacao preliminar da eficiencia e qualidade do trabelho. In: Josahkian, L.A. (ed.) *5th Congresso dos Racos Zebuinas*. ABCZ, Uberaba, Brazil, pp. 361–362.

Coitzee, J.F., Lubbers, B.V., Toerber, S.E., Gehring, R., Thompson, D.U., White, B.J. and Apley, M.D. (2008) Plasma concentration of substance P and cortisol in beef calves after castration and simulated castration. *American Journal of Veterinary Research* 69, 751–752.

Collier, R.J., Byatt, J.C., Denham, S.C., Eppard, P.J., Fabellar, A.C., Hintz, R.L., McGrath, M.F., McLaughlin, C.L., Shearer, J.K., Veenhuizen, J.J. and Vicini, J.L. (2001) Effects of sustained release of bovine somatotropin (sometribove) on animal health in commercial dairy herds. *Journal of Dairy Science* 84, 1098–1108.

Colpaert, F.C., Taryre, J.P., Alliaga, M. and Kock, W. (2001) Opiate self administration as a measure of chronic nocieptive pain in arthritic rats. *Pain* 91, 33–34.

Cordiner, L.S. and Savory, C.J. (2001) Use of perches and nest boxes by laying hens in relation to social status, based on examination of consistency of ranking orders and frequency of interaction. *Applied Animal Behaviour Science* 71, 305–317.

Craig, J.C. and Muir, W. (1998) Genetics and the behavior of chickens. In: Grandin, T. (ed.) *Genetics and the Behavior of Animals*. Academic Press (Elsevier), San Diego, California, pp. 265–298.

Danbury, T.C., Weeks, C.A., Chambers, J.P., Waterman-Pearson, A.E. and Kestin, S.C. (2000) Self selection of the analgesic drug carproten by lame broiler chickens. *Veterinary Research* 146, 307–311.

Davis, M. (1992) The role of the amygdala in fear and anxiety. *Annual Review of Neuroscience* 15, 353–375.

Day, J.E.L., Van deWeerd, H.A. and Edwards, A. (2008) The effect of varying lengths of straw bedding on the behavior of growing pigs. *Applied Animal Behaviour Science* 109, 249–260.

deSeville, X.F., Fagrega, E., Tibau, J. and Casellar, J. (2008) Effect of leg conformation on survivability of Duroc, Landrace, and Large White sows. *Journal of Animal Science* 86, 2392–2400.

Duncan, I.J.H. (1998) Behavior and behavioral needs. *Poultry Science* 77, 1766–1772.

Duncan, I.J.M. and Kite, V.G. (1989) Nest box selection and nest building behavior in the domestic hens. *Animal Behavior* 37, 215–231.

Dunlap, R., Millsopp, S. and Laming, P. (2005) Avoidance learning in goldfish (*Carassius auratus*) and trout (*Oncorhynchus mykiss*) and implications for pain perception. *Applied Animal Behaviour Science* 97, 2556–2571.

Dunn, C.S. (1990) Stress reaction in cattle undergoing ritual slaughter using two methods of restraint. *Veterinary Record* 126, 522–525.

Eaton, R.C., Bombardier, R.A. and Meger, D.L. (1977) The Mauthner initiated startle responses in teleost fish. *Journal of Experimental Biology* 66, 65–81.

Eicher, S.D. and Dailey, J.W. (2002) Indicators of acute pain and fly avoidance behaviors in Holstein calves following tail docking. *Journal of Dairy Science* 85, 2850–2858.

Espejo, L.A., Endres, M.I. and Salfer, J.A. (2006) Prevalence of lameness in high-producing Holstein cows housed in freestall barns in Minnesota. *Journal of Dairy Science* 89, 3052–3058.

Faulkner, P.M. and Weary, D.M. (2000) Reducing pain after dehorning in dairy calves. *Journal of Dairy Science* 83(9), 2037–2041.

Faure, J.M. and Mills, A.D. (1998) Improving the adaptability of animals by selection. In: Grandin, T. (ed.) *Genetics and the Behavior of Domestic Animals.* Academic Press (Elsevier), San Diego, California, pp. 233–265.

Fernandez de Seville, X., Febrega, E., Tibau, J. and Casellas, J. (2008) Effect of leg conformation on the survivability of Duroc, Landrace and Large White sows. *Journal of Animal Science* 86, 2392–2400.

Flowers, F.C., de Passillé, A.M., Weary, D.M., Sanderson, D.J. and Rushen, J. (2007) Softer, higher friction flooring improves gait of cows with and without sole ulcers. *Journal of Dairy Science* 90, 1235–1242.

Fraser, D. (1975) The effect of straw on the behavior of sows in tether stalls. *Animal Production* 21, 59–68.

Fraser, D. (2008a) Towards a global perspective on farm animal welfare. *Applied Animal Behaviour Science* 113, 330–339.

Fraser, D. (2008b) *Understanding Animal Welfare.* Wiley-Blackwell, Oxford, UK.

Freire, R., Applyby, M.C. and Hughes, B.O. (1997) Assessment of pre-laying motivation in the domestic hen by using social interaction. *Animal Behavior* 34, 313–319.

Fulwider, W.K., Grandin, T., Garrick, D.J., Engle, T.E., Lamm, W.D., Dalsted, N.L. and Rollin, B.E. (2007) Influence of free-stall base on tarsal joint lesions and hygiene in dairy cows. *Journal of Dairy Science* 90, 3559–3566.

Garner, J.P., Change, H.W., Richert, B.T. and Marchant-Forde, J.N. (2008) The effect of ractopamine gender and social rank on aggression and peripheral monoamine levels in finishing pigs. *Journal of Animal Science* 86 (Supplement 1), 382 (abstract).

Gentle, M.J., Waddington, D., Hunter, L.N. and Jones, R.B. (1990) Behavioral evidence for persistent pain following partial beak amputation in chickens. *Applied Animal Behaviour Science* 27, 149–157.

Grandin, T. (1982) Pig behavior studies applied to slaughter plant design. *Applied Animal Ethology* 9, 141–151.

Grandin, T. (1997) Assessment of stress during handling and transport. *Journal of Animal Science* 75, 249–257.

Grandin, T. (1998a) Objective scoring of animal handling and stunning practices in slaughter plants. *Journal American Veterinary Medical Association* 212, 36–39.

Grandin, T. (1998b) The feasibility of using vocalization scoring as an indicator of poor welfare during slaughter. *Applied Animal Behaviour Science* 56, 121–128.

Grandin, T. (2001) Cattle vocalizations are associated with handling and equipment problems at beef slaughter plants. *Applied Animal Behaviour Science* 71, 191–201.

Grandin, T. (2003) The welfare of pigs during transport and slaughter. *Pig News and Information* 24, 83N–90N.

Grandin, T. (2005) Maintenance of good animal welfare standards in beef slaughter plants by use of auditing programs. *Journal of American Veterinary Medical Association* 226, 370–373.

Grandin, T. (2007) Introduction: effect of customer requirements, international standards and marketing structure on handling and transport of livestock and poultry. In: Grandin, T. (ed.) *Livestock Handling and Transport.* CAB International, Wallingford, UK, pp. 1–18.

Grandin, T. and Johnson, C. (2005) *Animals in Translation.* Scribner (Simon and Schuster), New York.

Grandin, T. and Johnson, C. (2009) *Animals Make Us Human.* Houghton Mifflin Harcourt, Boston, Massachusetts.

Grandin, T., Curtis, S.E., Widowski, T.M. and Thurman, J.C. (1986) Electro-immobilization versus mechanical restraint in an avoid-avoid choice test. *Journal of Animal Science* 62, 1469–1480.

Haley, D.B., Bailey, D.W. and Stookey, J.M. (2005) The effects of weaning beef calves in two stages on their behavior and growth rate. *Journal of Animal Science* 83, 2205–2214.

Hastein, T. (2007) OIE involvement in aquatic animal welfare: the need for development of guideline based on welfare for farming, transport and slaughter purposes in aquatic animals. *Developmental Biology* 129, 149–161.

Hemsworth, P.H. and Coleman, G.J. (1994) *Human Livestock Interaction.* CAB International, Wallingford, UK.

Hughes, B.O., Wilson, S., Appleby, M.C. and Smith, S.F. (1993) Comparison of bone volume and strength as measures of skeletal integrity in caged laying hens with access to perches. *Research in Veterinary Medicine* 54, 202–206.

Hughes, R.A. (1979) Shock induced tonic immobility in chickens as a function of post hatch age. *Animal Behavior* 27, 782–785.

Jones, R.B. (1984) Experimental novelty and tonic immobility in chickens (*Gallas domesticus*). *Behavioral Processes* 9, 255–260.

Jones, R.B. (1987) The assessment of fear in the domestic fowl. In: Zayan, R. and Duncan, I.J.H. (eds) *Cognitive Aspects of Social Behaviour in the Domestic Fowl*. Elsevier, Amsterdam, pp. 40–81.

Jones, R.B. and Mills, A.D. (1982) Estimation of fear in two lines of the domestic chick. Correlations between various methods. *Behavioral Processes* 8, 243–253.

Kirk, R.K., Svensmark, B., Elegaard, L.P. and Jensen, H.D. (2005) Locomotion disorders associated with sow mortality in Danish pig herds. *Journal of Veterinary Medicine and Physiological and Pathology Clinical Medicine* 52, 423–428.

Knowles, T.G., Kestin, S.C., Hasslam, S.M., Brown, S.N., Green, L.E., Butterworth, A., Pope, S.J., Pfeiffer, D. and Nicol, C.J. (2008) Leg disorders in broiler chickens: prevalance, risk factors and prevention. *PLOS One* 3(2). Available at: www.pubmedcentral.nih.gov/articlerender.fegi?artid=2212134 (accessed 17 June 2009).

Kronfield, D.S. (1994) Health management of dairy herds treated with bovine somatotropin. *Journal of American Veterinary Medical Association* 204, 116–130.

Lambooy, E. (1985) Electro-anesthesia or electro-immobilization of calves, sheep and pigs by Feenix Stockstill. *Veterinary Quarterly* 7, 120–126.

Lassen, J., Sandoe, P. and Forkman, B. (2008) Happy pigs are dirty – conflicting perspectives on animal welfare. *Livestock Science* 103, 221–230.

Limon, G., Gultian, J. and Gregory, N.G. (2008) A note on slaughter of llamas in Bolivia by the puntilla method. *Meat Science* 82, 405–406.

Lund, V., Mejdell, C.M., Rockinsberg, H., Anthony R. and Holstein, T. (2007) Expanding the moral circle: farmed fish as objects of moral concerns. *Diseases in Aquatic Organisms* 75, 109–118.

Marchant-Forde, J.N., Lay, D.C., Pajor, J.A., Richert, B.T. and Schinckel, A.P. (2003) The effects of ractopamine on the behavior and physiology of finishing pigs. *Journal of Animal Science* 81, 416–422.

Matheson, B.K., Branch, B.J. and Taylor, A.N. (1971) Effects of amygdoid stimulation on pituitary adrenal activity in conscious cats. *Brain Research* 32, 151.

Mejdell, C.M. (2006) The role of councils on animal ethics in assessing acceptable animal welfare standards in agriculture. *Livestock Science* 103, 292–296.

Molony, V. and Kent, J.E. (1997) Assessment of acute pain in farm animals using behavioral and physiological measurements. *Journal of Animal Science* 75, 266–272.

Murray, A.C. and Johnson, C.P. (1998) Importance of halothane gene on muscle quantity and preslaughter death in western Canadian pigs. *Canadian Journal of Animal Science* 78, 543–548.

Newby, N.C. and Stevens, E.D. (2008) The effects of the acetic acid pain test on feeding, swimming and respiratory responses in rainbow trout (*Oncorhynchus mykiss*). *Applied Animal Behaviour Science* 114, 260–269.

O'Hara, P. and O'Connor, C. (2007) Challenge of developing regulations for production animals and produce animal welfare outcomes that we want. *Journal of Veterinary Behavior, Clinical Applications and Research* 2, 205–212.

OIE (2006) *Introduction to the Recommendations for Animal Welfare, Terrestrial Animal Health Code*. World Organization for Animal Health, Paris, France.

OIE (2008) *Resolution from the 2nd OIE Global Conference on Animal Welfare*. World Organization for Animal Health Conference, Cairo, Egypt, 20–22 October.

OIE (2009a) *Transport of Animals by Land, Terrestrial Animal Health Code*. World Organization for Animal Health, Paris, France.

OIE (2009b) *Slaughter of Animals, Terrestrial Animal Health Code*. World Organization for Animal Health, Paris, France.

Olsson, L.A. and Keeling, L.J. (2000) Nighttime roosting in laying hens and the effect of thwarting access to perches. *Applied Animal Behaviour Science* 68, 243–256.

Ontario Ministry of Agricultural Food and Rural Affairs (OMAFRA) (2005) Score Your Farm or Cow Comfort. Ontario Ministry of Agricultural Food and Rural Affairs, Ontario, Canada. Available at: www.gov.on.ca/OMAFRA/english/livestock/dairy/facts/info_tsdimen.htm (accessed 17 June 2009).

Panksepp, J. (1998) *Affective Neuroscience*. Oxford University Press, New York.

Paranhos de Costa, M. (2008) Improving the welfare of cattle, practical experience in Brazil. In: Dawkins, M.S. and Bonney, R. (eds) *The Future of Animal Farming*. Blackwell Publishing, Oxford, UK, pp. 145–152.

Parkdel, A., Van Arendonk, J.A., Vereijken, A.L. and Bovenhuis, H. (2005) Genetic parameters of ascites related traits in broilers: correlations with feed efficiency and carcass traits. *British Poultry Science* 46, 43–51.

Pascoe, P.J. (1986) Humaneness of electro-immoblization for cattle. *American Journal of Veterinary Research* 10, 2252–2256.

Pearson, A.J., Kilgour, R., deLangen, H. and Payne, E. (1977) Hormonal responses of lambs to trucking, handling and electric stunning. *New Zealand Society for Animal Production* 37, 243–249.

Petrini, A. and Wilson, D. (2005) Philosophy, policy and procedures of the World Organization for Animal

Health on the development of standards in animal welfare. *Review Science and Technology* 24, 665–671.

Poletto, R., Rostagno, M.H., Richert, B.T. and Marchant-Forde, J.N. (2009) Effects of a 'step up' ractopamine feeding program, sex and social rank on growth performance, hoof lesions, and enterobacteriaccae shedding in finishing pigs. *Journal of Animal Science* 87, 304–311.

Price, E.O., Harris, J.E., Borgwardt, R.E., Sween, M.L. and Connor, I.M. (2003) Fence line contact of beef calves with their dams at weaning reduces the negative effects of separation on behavior and growth rate. *Journal of Animal Science* 81, 116–121.

Redgate, E.S. and Fabringer, E.E. (1973) A comparison of pituitary adrenal activity elicited by electrical stimulation of preoptic amygdaloid and hypothalamic sites in the rat brain. *Neuroendocrinology* 12, 334.

Reilly, S.C., Quinn, J.P., Cossins, A.R. and Sneddon, L.V. (2008) Behavorial analysis of nocieptive event in fish: comparisons between three species demonstrate specific responses. *Applied Animal Behaviour Science* 114, 248–259.

Rogan, M.T. and LeDoux, J.E. (1996) Emotion systems. *Cells Synaptic Plasticity Cell* 85, 469–475.

Rushen, J., Pombourceq, E. and dePaisselle, A.M. (2006) Validation of two measures of lameness in dairy cows. *Applied Animal Behaviour Science* 106, 173–177.

Rutherford, K.M., Langford, F.M., Jack, M.C., Sherwood, L., Lawrence, A.B. and Haskell, M.J. (2009) Lameness prevalence and risk factors in organic and non-organic dairy herds in the United Kingdom. *Veterinary Journal* 180, 95–105.

Setckleiv, J., Skaug, O.E. and Kaada, B.R. (1961) Increase in plasma 17-hydroxycorticosteroids by cerebral cortical and amygdaloid stimulation in the cat. *Journal of Endocrinology* 22, 119.

Sneddon, L.U. (2003) The evidence for pain in fish: the use of morphine as an analgesic. *Applied Animal Behaviour Science* 83, 153–162.

Sneddon, L.U., Braithwaite, V.A. and Gentle, M.J. (2003a) Do fish have nociceptors: evidence for the evolution of vertebrate sensory system. *Proceedings of the Royal Society of London* B 270, 1115–1122.

Sneddon, L.U., Braithwaite, V.A. and Gentle, M.J. (2003b) Novel object test: examining pain and fear in the rainbow trout. *Journal of Pain* 4, 431–440.

Stafford, K.J. and Mellor, D.J. (2005a) Dehorning and disbudding distress and its alleviation in calves. *Veterinary Journal* 169(3), 337–349.

Stafford, K.J. and Mellor, D.J. (2005b) The welfare significance of the castration of cattle: a review. *New Zealand Veterinary Journal* 53, 271–278.

Studnitz, M., Jenson, M.B. and Pederson, L.J. (2007) Why do pigs root and in what will they root? A review on the exploratory behavior of pigs in relation to environmental enrichment. *Applied Animal Behaviour Science* 107, 183–197.

Sylvester, S.P., Stafford, K.J., Mellor, D.J., Bruce, R.A. and Ward, R.N. (2004) Behavioural responses of calves to amputation dehorning with and without local anesthesia. *Australian Veterinary Journal* 82, 697–700.

Tanida, H., Miura, A., Tanaka, T. and Yoshimoto, T. (1996) Behavioral responses of piglets to darkness and shadows. *Applied Animal Behaviour Science* 49, 173–183.

Tiranti, K.I. and Morrison, R.B. (2006) Association between limb conformation and reproduction of sows through the second parity. *American Journal of Veterinary Research* 67, 505–509.

Van de Weerd, H.A. and Day, J.E.L. (2009) A review of environmental enrichment for pigs housed in intensive housing systems. *Applied Animal Behaviour Science* 116, 1–20.

Van de Weerd, H.A., Docking, C.M., Day, J.E.L., Avery, P.J. and Edwards, S.A. (2003) A systematic approach towards developing environmental enrichment for pigs. *Applied Animal Behavioral Science* 84, 101–118.

Van Putten, G. and Elshof, W.J. (1978) Observations of the effects of transport on the well being and lean quality of slaughter pigs. *Animal Regulation Studies* 1, 247–271.

VanSickle, J. (2008) Sow lameness underrated. *National Hog Farmer* 15 June, p. 34.

Vihuela-Fernandez, I., Jones, E., Welsh, E.M. and Fleetwood-Walker, S.M. (2007) Pain mechanisms and their implication for the management of pain in farm and companion animals. *Veterinary Journal* 174, 227–239.

Voisinet, B.D., Grandin, T., Tatum, J.D., O'Connor, S.F. and Struthers, J.J. (1997) Feedlot cattle with calm temperaments have higher average gains than cattle with excitable temperaments. *Journal of Animal Science* 75, 892–896.

Volpato, G.I., Goncalves-de-Freitas, E. and Fernandes-deCastilho, M. (2007) Insights into the concept of fish welfare. *Applied Animal Behaviour Science* 75, 165–171.

Warriss, P.D., Brown, S.N. and Adams, S.I.M. (1994) Relationship between subjective and objective assessment stress at slaughter and meat quality in pigs. *Meat Science* 38, 329–340.

Watts, J.M. and Stookey, J.M. (1998) Effects of restraint and branding on rates and acoustic parameters of vocalization in beef cattle. *Applied Animal Behaviour Science* 62, 125–134.

Webster, J. (2005a) The assessment and implementation of animal welfare: theory into practice. *Review Science Technology Off. International Epiz.* 24, 723–734.

Webster, J. (2005b) *Animal Welfare, Limping Towards Eden.* Blackwell Publishing, Oxford, UK.

White, R.G., DeShazer, I.A., Tressler, C.J., Borcher, G.M., Davey, S., Waninge, A., Parkhurst, A.M., Milanuk, M.J. and Clems, E.T. (1995) Vocalizations and physiological response of pigs during castration with and without anesthetic. *Journal of Animal Science* 73, 381–386.

Widowski, T.M. and Duncan, I.J.H. (2000) Working for a dust bath: are hens increasing pleasure rather than relieving suffering? *Applied Animal Behaviour Science* 68, 39–53.

Willeberg, P. (1993) Bovine somatotropin and clinical mastitis: epidemiological assessment of the welfare risk. *Livestock Production Science* 36, 55–66.

Wray, H.R., Main, D.C.J., Green, L.E. and Webster, A.J.F. (2003) Assessment of welfare of dairy cattle using animal based measurements, direct observations, and investigation of farm records. *Veterinary Record* 153, 197–202.

Zurbrigg, K., Kelton, D., Anderson, N. and Millman, S. (2005a) Stall dimensions and the prevalence of lameness, injury, and cleanliness on 317 tie stall dairy farms in Ontario. *Canadian Journal of Animal Science* 46, 902–909.

Zurbrigg, K., Kelton, D., Anderson, N. and Millman, S. (2005b) Tie stall design and its relationship to lameness, injury, and cleanliness on 317 Ontario dairy farms. *Journal of Dairy Science* 88, 3201–3210.

2 Why is Agricultural Animal Welfare Important? The Social and Ethical Context

Bernard Rollin

Colorado State University, Fort Collins, Colorado, USA

There is one monumental conceptual error that is omnipresent in the agricultural industry's discussions of animal welfare – an error of such magnitude that it trivializes the industry's responses to ever-increasing societal concerns about the treatment of agricultural animals. These concerns are emerging as non-negotiable demands by consumers. Failure to respect such concerns can essentially destroy the economic base for intensive animal agriculture. When one discusses farm animal welfare with industry groups or with the American Veterinary Medical Association, one finds the same response – animal welfare is solely a matter of 'sound science'.

Those of us serving on the Pew Commission, better known as the National Commission on Industrial Farm Animal Production, encountered this response regularly during our dealings with industry representatives. This commission studied intensive animal agriculture in the USA (Pew Commission, 2008). For example, one representative of the Pork Producers, testifying before the Commission, answered that while people in her industry were quite 'nervous' about the Commission, their anxiety would be allayed were we to base all of our conclusions and recommendations on 'sound science'. Hoping to rectify the error in that comment, as well as educate the numerous industry representatives present, I responded to her as follows:

> Madame, if we on the Commission were asking the question of *how* to raise swine in confinement, science could certainly answer that question for us. But that is *not* the question the Commission, or society, is asking. What we are asking is, *ought* we to raise swine in confinement? And to this question, science is not relevant.

Judging by her 'huh', I assume I did not make my point.

Questions of animal welfare are at least partly 'ought' questions, questions of ethical obligation. The concept of animal welfare is an ethical concept to which, once understood, science brings relevant data. When we ask about an animal's welfare, or about a person's welfare, we are asking about *what* we owe the animal, and to *what extent*. A document referred to as the Council for Agricultural Science and Technology (CAST) report, first published by US agricultural scientists in the early 1980s, discussed animal welfare. It affirmed that the necessary and sufficient conditions for attributing positive welfare to an animal were represented by the animal's productivity. A productive animal enjoyed positive welfare; a non-productive animal enjoyed poor welfare (CAST, 1997).

This notion was fraught with many difficulties. First of all, productivity is an economic notion predicated of a whole operation; welfare is predicated of individual animals. An operation such as caged laying hens may be quite profitable if the cages are severely overcrowded, yet the individual hens do not enjoy good welfare. Secondly, as we shall see, equating productivity and welfare is, to some significant extent, legitimate under husbandry conditions, where the producer does well if and only if the animals do well, and square pegs, as it were, are fitted into square holes with as little friction as possible. Under industrial conditions, however, animals do not naturally fit in the niche or environment in which they are kept, and are subjected to 'technological sanders' that allow for producers to force square pegs into round holes – antibiotics, feed additives, hormones, air handling systems – so the animals do not die and produce more and more kilograms of meat or milk. Without these technologies, the animals could not be productive. We will return to the contrast between husbandry and industrial approaches to animal agriculture.

The key point to recall here is that even if the CAST report definition of animal welfare did not suffer from the difficulties we outlined, it is still an ethical concept. It essentially says 'what we owe animals and to what extent is simply what it takes to get them to create profit'. This in turn would imply that the animals are well off if they have only food, water and shelter, something the industry has sometimes asserted. Even in the early 1980s, however, there were animal advocates and others who would take a very different ethical stance on what we owe farm animals. Indeed, the famous five freedoms articulated in the UK by the Farm Animal Welfare Council (FAWC) during the 1970s (even before the CAST report) represents quite a different ethical view of what we owe animals, when it affirms that:

> The welfare of an animal includes its physical and mental state and we consider that good animal welfare implies both fitness and a sense of well-being. Any animal kept by man, must at least, be protected from unnecessary suffering.
>
> We believe that an animal's welfare, whether on farm, in transit, at market or at a place of slaughter should be considered in terms of '**five freedoms**' (see www.fawc.org.uk).
>
> 1. **Freedom from Hunger and Thirst** – by ready access to fresh water and a diet to maintain full health and vigour.
> 2. **Freedom from Discomfort** – by providing an appropriate environment including shelter and a comfortable resting area.
> 3. **Freedom from Pain, Injury or Disease** – by prevention or rapid diagnosis and treatment.
> 4. **Freedom to Express Normal Behaviour** – by providing sufficient space, proper facilities and company of the animal's own kind.
> 5. **Freedom from Fear and Distress** – by ensuring conditions and treatment which avoid mental suffering.
>
> (FAWC, 2009)

Clearly, the two definitions contain very different notions of our moral obligation to animals (and there is an indefinite number of other definitions). Which one is correct, of course, cannot be decided by gathering facts or doing experiments – indeed which ethical framework one adopts will in fact determine the shape of the science of studying animal welfare!

Your Concept of Welfare Determines What is Sound Science

To clarify: suppose you hold the view that an animal is well off when it is productive, as per the CAST report. The role of your welfare science in this case will be to study what feed, bedding, temperature, etc. are most efficient at producing the most meat, milk or eggs for the least money – much what animal and veterinary science does today. On the other hand, if you take the FAWC view of welfare, your efficiency will be constrained by the need to acknowledge the animal's natural behaviour and mental state, and to assure that there is minimal pain, fear, distress and discomfort – not factors in the CAST view of welfare unless they have a negative impact on economic productivity. Thus, in a real sense, sound science does not determine your concept of welfare; rather, your concept of welfare determines what counts as sound science!

The failure to recognize the inescapable ethical component in the concept of animal welfare leads inexorably to those holding different ethical views talking past each other. Thus, producers ignore questions of animal pain, fear, distress, confinement, truncated mobility, bad air quality, social isolation and impoverished environment unless any of these factors impacts negatively on the 'bottom line'. Animal advocates, on the other hand, give such factors primacy, and are totally unimpressed with how efficient or productive the system may be.

A major question obviously arises here. If the notion of animal welfare is inseparable from ethical components, and people's ethical stance on obligations to farm animals differs markedly across a highly diverse spectrum, whose ethic is to predominate and define, in law or regulation, what counts as 'animal welfare'? This is of great concern to the agriculture industry, worrying as they do about 'vegetarian activists hell-bent on abolishing meat'. In actual fact, of course, such concern is misplaced, for the chance of such an extremely radical thing happening is vanishingly small. By and large, however, the ethic adopted in society reflects a *societal consensus*, what most people either believe to be right and wrong or are willing to accept upon reflection.

All of us have our own personal ethics, which rule a goodly portion of our lives. Fundamental questions as to what we read, what we eat, to whom we give charity, what political and religious beliefs we hold and myriad others are answered by our personal ethics. These derive from many sources – parents, religious institutions, friends, reading books, and watching movies and television. One is certainly entitled to believe ethically, as do some People for the Ethical Treatment of Animals (PETA) members, that 'meat is murder', that one should be a vegan, it is immoral to use products derived from animal research, and so on.

Clearly, a society, particularly a free society, contains a bewildering array of such personal ethics, with the potential for significant clashes between them. If my personal ethic is based in fundamentalist religious beliefs and yours is based in celebrating the pleasures of the flesh, we are destined to clash, perhaps violently. For this reason, social life cannot function simply by relying on an individual's personal ethics, except perhaps in singularly monolithic cultures where all members share overwhelmingly the same values. One can find examples of something resembling this in small towns in rural farming areas, where there is no need to lock one's doors, remove one's keys from the car, or fear for one's personal safety. But of course such places are few, and probably decreasing in number. In larger communities of course, the extreme case being New York City or London, one finds a welter of diverse cultures and corresponding personal ethics crammed into a small geographical locus. For this reason alone, as well as to control those whose personal ethic may entail taking advantage of others, a *social consensus ethic* is required, one that transcends personal ethics. This social consensus ethic is invariably articulated in law, with manifest sanctions for its violation. As societies evolve, different issues emerge, leading to changes in the social ethic.

My claim then, is that beginning roughly in the late 1960s, the treatment of animals has moved from being a paradigmatic example of personal ethics to ever-increasingly falling within the purview of societal ethics and law. How and why has this occurred, and to what extent?

If one looks to the history of animal use in society back to the beginning of domestication some 11,000 years ago, one finds very little social ethics dictating animal treatment. The one exception to this generalization is the prohibition against deliberate, purposeless cruelty, that is needless infliction of pain and suffering or outrageous neglect, such as failing to provide food or water. This mandate is well illustrated in the Old Testament, where many injunctions illustrate its presence. For example, one is told that when collecting eggs from a bird's nest, one should leave some eggs so as not to distress the animal. Kosher and halal slaughter accomplished by a trained person using a very sharp knife were clearly intended as a viable alternative to the much more traumatic bludgeoning. (That is not, of course, to suggest that such slaughter remains welfare-friendly in high throughput industrialized slaughterhouses.) The rule of Kashrut prohibiting the eating of milk and meat – 'do not seethe a calf in its mother's milk' – seems to be aimed at avoiding loss of sensitivity to animal suffering. Raza Gharebaghi and colleagues in Iran (2007) reviewed Islamic prohibitions that forbid separating baby birds or animals from their mother. There are also many passages in the Qur'an that sustenance and water must be provided to working animals (Al Masri, 2007).

In the middle ages, St Thomas Aquinas provided a more anthropocentric reason for prohibiting cruelty, based in the prescient psychological insight that those who would abuse animals will inexorably progress to abusing humans. Aquinas does not see animals as direct objects of moral concern but, none the less, strongly prohibits their abuse.

In the late 18th century in Britain, and in subsequent years elsewhere, the prohibition against deliberate, sadistic, deviant, wilful, malicious cruelty, that is inflicting pain and suffering on animals to no reasonable purpose, or outrageous neglect such as not providing food or water, were encoded in the anti-cruelty laws of all civilized societies. While adopted in part out of a moral notion of limiting animal suffering, an equally important reason was the Thomistic one – to ferret out individuals who might graduate to harming humans; case law in the USA and elsewhere make this manifest.

In one revealing case in the 19th century, a man was charged with cruelty after throwing pigeons into the air and shooting them to demonstrate his skill. After killing the birds, he ate them. The court ruled that the pigeons were not 'needlessly or unnecessarily killed' because the killing was done 'in the indulgence of a healthful recreating during an exercise tending to promote strength, bodily agility and courage'. In discussing a similar 19th-century case of a tame pigeon shoot in Colorado, the court affirmed that

> every act that causes pain and suffering to animals is not prohibited. Where the end or object in view is reasonable and adequate, the act resulting in pain is . . . necessary and justifiable, as . . . where the act is done to protect life or property, or to minister to the necessities of man.

To the credit of the Colorado court, it did not find that such tame pigeon shoots met the test of 'worthy motive' or 'reasonable object'. Even today, however, there are jurisdictions where tame pigeon shoots and 'canned hunts' do not violate the anti-cruelty laws.

It is certainly true that cruelty to animals is closely linked to psychopathic behaviour – animal cruelty, along with fire starting, are signs of future psychopaths. The majority of children who shoot up their schools have early histories of animal abuse, as do 80% of the violent offenders in Leavenworth Prison and most serial killers. Animal abusers often abuse wives and children (Ascione *et al.*, 2007; Volant *et al.*, 2008). Most battered women's shelters must make provisions for keeping the family pet, as the abuser will hurt the animal to hurt the woman. Several studies have shown a relationship between childhood animal cruelty and violence towards people (Miller, 2001). But these laws conceptually provide little protection for animals. Animal cruelty accounts for only a tiny fraction of the suffering that animals undergo at human hands. For example, the USA produces 9 billion broiler chickens a year, and many have bruises and fractures or other skeleto-muscular injuries that occur during catching and transport. Before restaurant companies started doing animal welfare audits, careless, rough handling of chickens resulted in 5% of the birds suffering broken wings, which is a shocking 450,000,000 birds with an injury as severe as a broken arm. If even 1% of chickens are so injured (a ridiculously low proportion), then we have 90,000,000 suffering animals there alone – there is nothing like 90,000,000 incidents of cruelty, and those chickens are legally unprotected. In the USA, they are not even subject to humane slaughter law! In Europe and Canada, humane slaughter laws include poultry.

The End of Husbandry

Why was the law or societal ethic protecting animals so minimalistic historically? When I first began to work in animal ethics, I thought the answer lay in cynical exploitation of the powerless. I later realized that the answer is much more subtle, and lies in what I have called 'the end of husbandry'.

The traditional account of the growth of human civilization out of a hunter-gatherer society invariably invokes the rise of agriculture, that is the domestication of animals and the cultivation of crops. This of course allowed for as predictable a food supply as humans could create in the vagaries of the natural world – floods, droughts, hurricanes, typhoons, extremes of heat and cold, fires, etc. Indeed, the use of animals enabled the development of successful crop agriculture, with the animals providing labour and locomotion, as well as food and fibre.

This eventuated in what Dr Temple Grandin has called the 'ancient contract' with animals, a highly symbiotic relationship that endured essentially unchanged for thousands of years. Humans selected animals congenial to human management, and further shaped them in terms of temperament and production traits by breeding and artificial selection. These animals included cattle – dubbed by Calvin Schwabe (1978), a famous veterinarian, the 'mother of the human race' – sheep, goats, horses, dogs, poultry and other birds, swine, ungulates and other animals capable of domestication. The animals provided food and fibre – meat, milk, wool, leather; power to haul and plough; transportation; and served as weaponry – for example, horses and elephants. As people grew more effective at breeding and managing the animals, productivity was increased.

As humans benefited, so simultaneously did the animals. They were provided with the necessities of life in a predictable way. And thus was born the concept of husbandry, the remarkable practice and articulation of the symbiotic contract.

'Husbandry' is derived from the Old Norse words 'hus' and 'bond'; the animals were bonded to one's household. The essence of husbandry was care. Humans put animals into the most ideal environment possible for the animals to survive and thrive, the environment for which they had evolved and been selected. In addition, humans provided them with sustenance, water, shelter, protection from predation, such medical attention as was available, help in birthing, food during famine, water during drought, safe surroundings and comfortable appointments.

Eventually, what was borne of necessity and common sense became articulated in terms of a moral obligation inextricably bound up with self-interest. In the Noah story, we learn that even as God preserves humans, humans preserve animals. The ethic of husbandry is in fact taught throughout the Bible; the animals must rest on the Sabbath even as we do, one is not to seethe a calf in its mother's milk (so we do not grow insensitive to animals needs and natures); we can violate the Sabbath to save an animal. The Islamic tradition has similar instructions on animal use, 'do not ride an animal as much, as it can no longer bear your load, and be fair to animals ... If an animal is exhausted, it must take some rest' (Shahidi, 1996).

Proverbs tells us that 'the wise man cares for his animals'. The Old Testament is replete with injunctions against inflicting unnecessary pain and suffering on animals, as exemplified in the strange story of Balaam who beats his ass, and is reprimanded by the animal's speaking through the grace of God.

The true power of the husbandry ethic is best expressed in the 23rd Psalm. There, in searching for an apt metaphor for God's ideal relationship to humans, the Psalmist invokes the good shepherd:

> The Lord is my shepherd; I shall not want. He maketh me to lie down in green pastures; He leadeth me beside still waters. He restoreth my soul.

We want no more from God than what the good shepherd provides to his animals. Indeed, consider a lamb in ancient Judaea. Without a shepherd, the animal would not easily find forage or water, would not survive the multitude of predators the Bible tells us prowled the land – lions, jackals, hyenas, birds of prey and wild dogs. Under the aegis of the shepherd, the lamb lives well and safely. Shepherds and herders from many different cultures ranging from the Nues in Africa, the Incas, the Basques in Spain and Indian herders all composed poems and songs about being a shepherd (Kessler, 2009). All of these traditions promoted good husbandry and a bond with the animals and the land.

In return, the animals provide their products and sometimes their lives, but while they live, they live well. As we saw, even slaughter, the taking of the animal's life, must be as painless as possible, performed with a sharp knife by a trained person to avoid unnecessary pain. Ritual slaughter, both kosher and halal, was, in antiquity, a far kinder death than bludgeoning; most importantly, it was the most humane modality available at the time.

When Plato discusses the ideal political ruler in *The Republic*, he deploys the shepherd-sheep metaphor: the ruler is to his people as the shepherd is to his flock. Qua shepherd, the shepherd exists to protect, preserve and improve the sheep; any payment tendered to him is in his capacity as wage earner. So too it is for the ruler, again illustrating the power of the concept of husbandry on our psyches. To this day, ministers are called shepherds of their congregation, and 'pastor' derives from 'pastoral'.

The singular beauty of husbandry is that it was at once an ethical and prudential doctrine (Rollin, 2003). It was prudential in that failure to observe husbandry inexorably led to ruination of the person keeping animals. Not feeding, not watering, not protecting from predators, not respecting the animals' physical, biological, psychological needs and natures, what Aristotle called their *telos* – the 'cowness of the cow' the 'sheepness of the sheep' – meant your animals did not survive and thrive, and thus neither did you. The ultimate sanction of failing at husbandry – erosion of self-interest – obviated the need for any detailed ethical exposition of moral rules for husbandry: anyone unmoved by self-interest is unlikely to be moved by moral or legal injunctions! And thus one finds little written about animal ethics and little codification of that ethic in law before the 20th century, with the bulk of what is articulated aimed at the pathological cruelty we have discussed. Thus, lack of animal ethics and law was explained by the nature of the predominant use of animals – agriculture – and what was needed for its success – good husbandry.

In the mid-20th century, husbandry agriculture was supplanted by industrial agriculture in many countries. Ruth Harrison (1964), in her famous book *Animal Machines*, described the horrid conditions that she observed on intensive farms in the UK. Chickens crammed into tiny cages, tethered veal calves and sows that were not able to turn around. Her book was translated into seven languages and it served as an impetus for the formation of the Brambell Committee that wrote the five freedoms (McKenna, 2000; Van de Weerd and

Sandilands, 2000). Her book caused the public to become concerned about the intensification and lack of husbandry that was occurring on British poultry, veal and pig farms. Industrialization of animal agriculture in the USA occurred for a variety of understandable and even laudable reasons that are worth recounting.

1. When industrial agriculture began, roughly in the 1940s, the USA was confronted with a variety of new challenges related to food. In the first place, the economic Great Depression and Dust Bowl (severe drought) had soured many people on farming and, even more dramatically, had raised the spectre of starvation for the American public for the first time in US history. Vivid images of breadlines and soup kitchens drove the desire to assure a plentitude of cheap food. By the late 1960s and 1970s, the USA had large-scale industrialized animal agriculture with much bigger units compared to Europe.
2. Better jobs were to be found in cities, and rural people flocked to them in the hope of a better life, creating a potential shortage in agricultural labour.
3. Correlative with the growth of cities and suburbs came encroachment on agricultural land for various forms of development, raising land prices and moving acreage once available for agriculture out of that pool.
4. Many people who would otherwise have been happy with a slow, rural way of life were exposed to greater sophistication by virtue of military service in World War I and II, and thus were dissatisfied with an agrarian existence. Recall the song popular after World War I, 'How ya gonna keep 'em down on the farm (after they've seen Paree?)'. Ironically, by the late 1960s many urban people were yearning for the 'simple life' represented by the small farm. Today in developed countries, there is a huge interest in buying animal products from small, local family farms. Urban people are yearning for a return to animal husbandry.
5. Demographers predicted a precipitous and dramatic increase in population, which turned out to be accurate.
6. With the success of industrialization in new areas, notably Henry Ford's application of the concept to the automobile, it was probably inevitable that the concepts of industrialization would be applied to agriculture. (Ford himself had already characterized slaughterhouses as 'disassembly lines'.)

Thus was born an industrial approach to agriculture, with machines taking the place of labour. The traditional departments of animal husbandry in agricultural schools symbolically marked this transition by changing their names to departments of animal sciences, a field defined in textbooks as 'application of industrial methods to the production of animals'.

In this transition, the traditional bedrock values of agriculture, husbandry, sustainability, agriculture as a way of life not only a way of making a living, were transmuted into values of efficiency and productivity. With human labour replaced by machinery, in turn requiring large amounts of capital, farm units grew larger, eventuating in the mantra of the 1970s 'big or get out'. Agricultural research stressed producing cheap and plentiful food, and moved in unprecedented directions. With animals confined for efficiency and away from forage, much research was directed towards finding cheap sources of nutrition, in turn leading to feeding such deviant items to animals as poultry and cattle manure. Animals were kept under conditions alien to their natural needs for the sake of productivity. Whereas husbandry animal agriculture stressed putting square pegs into square holes, round pegs into round holes, while producing as little friction as possible, industrialized animal agriculture forced square pegs into round holes by utilizing what I have called 'technological sanders', such as antibiotics, hormones, extreme genetic selection, air handling systems, artificial cooling systems and artificial insemination to force animals into unnatural conditions while they none the less remained productive.

Consider, for example, the egg industry, one of the first areas of agriculture to experience industrialization. Traditionally, chickens ran free in barnyards, able to live off the land by foraging and express their natural behaviours of moving freely, nest building, dust bathing, escaping from more aggressive animals, defecating away from their nests, and, in general, fulfilling their natures as chickens. Industrialization of the egg industry, on the other hand, meant placing the chickens in small cages, in some systems with six birds in a tiny wire cage, so that one animal may stand on top of the others and none can perform any of their inherent behaviours, unable even to stretch their wings. In the absence of space to establish a dominance hierarchy or pecking order, they cannibalize each other, and must be 'debeaked', producing painful neuromas because the beak is innervated (Gentle *et al.*, 1990). The animal is now an inexpensive cog in a machine, part of a factory, and the cheapest part at that, and thus totally expendable. If a 19th-century farmer had attempted such a system, he would have gone broke, with the animals dead of disease in a few weeks. Some genetic lines of pigs and chickens are so highly selected for egg and meat production that they have less disease resistance. Pigs have become so susceptible to disease that some farmers have installed antibacterial filters to take germs out of the air that enters the building (Vansickle, 2008). This is a technological sander taken to the extreme.

The steady state, enduring balance of humans, animals and land is lost. Putting chickens in cages and cages in an environmentally controlled building requires large amounts of capital and energy and technological 'fixes'; for example, to run the exhaust fans to prevent lethal build-up of ammonia. The value of each chicken is negligible so one needs more chickens; chickens are cheap, cages are expensive, so one crowds as many chickens into each cage as is physically possible. The vast concentration of chickens requires huge amounts of antibiotics and other drugs to prevent wildfire spread of disease in overcrowded conditions. Breeding of animals is oriented solely towards productivity, and genetic diversity – a safety net allowing response to unforeseen changes – is lost. Bill Muir, a genetics specialist at Purdue University, found that commercial lines of poultry have lost 90% of their genetic diversity compared to non-commercial poultry (Lundeen, 2008). Dr Muir is extremely concerned about the lack of genetic diversity. Small poultry producers are lost, unable to afford the capital requirements; agriculture as a way of life as well as a way of making a living is lost; small farmers, who Thomas Jefferson argued are the backbone of society (Torgerson, 1997), are superseded by large corporate aggregates. Giant corporate entities, vertically integrated, are favoured. Manure becomes a problem for disposal, and a pollutant, instead of fertilizer for pastures. Local wisdom and know-how essential to husbandry is lost; what 'intelligence' there is hard-wired into 'the system'. Food safety suffers from the proliferation of drugs and chemicals, and widespread use of antimicrobials to control pathogens in effect serves to breed – select for – antibiotic-resistant pathogens as susceptible ones are killed off. Above all, the system is not balanced – constant inputs are needed to keep it running, and to manage the wastes it produces, and create the drugs and chemicals it consumes. And the animals live miserable lives, for productivity has been severed from well-being.

One encounters the same dismal situation for animals in all areas of industrialized animal agriculture. Consider, for example, the dairy industry, once viewed as the paradigm case of bucolic, sustainable animal agriculture, with animals grazing on pasture giving milk, and fertilizing the soil for continued pasture with their manure. Though the industry wishes consumers to believe that this situation still exists – the California dairy industry ran advertisements proclaiming that California cheese comes from 'happy cows', and showing the cows on pastures – the truth is radically different. The vast majority of California dairy cattle spend their lives on dirt and concrete, and in fact never see a blade of pasture grass, let alone consume it. So outrageous is this duplicity that the dairy association was sued for false advertising and a friend of mine, a dairy practitioner for 35 years, was very outspoken against such an 'outrageous lie'.

In actual fact, the life of dairy cattle is not a pleasant one. In a problem ubiquitous across contemporary agriculture, animals have been single-mindedly bred for productivity; in the case of dairy cattle, for milk production. Today's dairy cow produces three to four times more milk than 60 years ago. In 1950, the average dairy cow annually produced 2410 l (5314 lb) of milk per lactation. Fifty years later, it was close to 9072 l (20,000 lb) (Blaney, 2002). From 1995 to 2004 alone, milk production per cow increased 16%. The result is a milkbag on legs, and unstable legs at that. A high percentage of the US dairy herd is chronically lame and these cows suffer serious reproductive problems. Espejo *et al.* (2006) found that an average of 24.6% of dairy cows housed in cubicle (freestall) barns were lame (see Chapter 1). Whereas in traditional agriculture, a milk cow could remain productive for 10 and even 15 years, today's cow lasts slightly longer than two lactations, a result of metabolic burnout and the quest for ever-increasingly productive animals, hastened in the USA by the use of the hormone bovine somatotropin (BST) to further increase production. Such unnaturally productive animals naturally suffer from mastitis, and the industry's response to mastitis in portions of the USA has created a new welfare problem by docking of cow tails without anaesthesia in a futile effort to minimize teat contamination by manure. Still practised, this procedure has been definitively demonstrated not to be relevant to mastitis control or lowering somatic cell count (Stull *et al.*, 2002). (In my view, the stress and pain of tail amputation coupled with the concomitant inability to chase away flies may well dispose to more mastitis.) Calves are removed from mothers shortly after birth, before receiving colostrum, creating significant distress in both mothers and infants. Bull calves may be shipped to slaughter or a feedlot immediately after birth, generating stress and fear.

The intensive swine industry, which through a handful of companies is responsible for 85% of the pork produced in the USA, is also responsible for significant suffering that did not affect husbandry-reared swine. Certainly the most egregious practice in the confinement swine industry and possibly, given the intelligence of pigs, in all of animal agriculture, is the housing of pregnant sows in gestation crates or stalls – essentially small cages. The *recommended* size for such stalls, in which the sow spends her entire productive life of about 4 years, with a brief exception we will detail shortly, according to the industry is 0.9 m high by 0.64 m wide by 2.2 m long – this for an animal that may weigh 275 kg or more. (In reality many stalls are smaller.) The sow cannot turn around, walk or even scratch her rump. In the case of large sows, she cannot even lie flat, but must remain lying on her sternum. The exception alluded to is the period of farrowing – approximately 3 weeks – when she is transferred to a 'farrowing crate' to give birth and nurse her piglets. The space for her is not greater, but there is a 'creep rail' surrounding her so the piglets can nurse without being crushed by her postural adjustments.

Under extensive conditions, a sow will build a nest on a hillside so excrement runs off; forage an area covering 2 km a day; and take turns with other sows watching piglets and allowing all sows to forage. With the animal's nature thus aborted, she goes mad, exhibits bizarre and deviant behaviour such as compulsively chewing the bars of the cage; she also endures foot and leg problems, and lesions from lying on concrete in her own excrement (see Chapters 8 and 15).

These examples suffice to illustrate the absence of good welfare in confinement. Rest assured that a long litany of issues could be addressed. In general, all animals in confinement agriculture (with the sometime exception of beef cattle who live most of their lives on pasture, and are 'finished' on grain in dirt feedlots, where they can actualize much of their nature) suffer from the same generic set of affronts to their welfare absent in husbandry agriculture:

1. **Production diseases** – by definition a production disease is a disease that would not exist or would not be of serious epidemic import were it not for the method of production. Examples are liver and rumenal abscesses resulting from feeding cattle too much grain, rather than roughage. The animals that get sick are more than balanced out economically by the remaining animals' weight gain. Other examples are confinement-induced environmental mastitis in dairy cattle, weakness caused by beta-agonists to increase muscle mass in pigs and 'shipping fever' in beef cattle. There are textbooks of production diseases, and one of my veterinarian colleagues calls such disease 'the shame of veterinary medicine' because veterinary medicine should be working to eliminate such pathogenic conditions, rather than treating the symptoms.
2. **Loss of workers who are 'animal smart'** – in large industrial operations such as swine factories, the workers are minimum-wage, sometimes illegal, often migratory workers with little animal knowledge. Confinement agriculturalists will boast that 'the intelligence is in the system' and thus the historically collective wisdom of husbandry is lost, as is the concept of the historical shepherd, now transmuted into rote, cheap, labour.
3. **Lack of individual attention** – under husbandry systems, each animal is valuable. In intensive swine operations, the individuals are worth little. When this is coupled with the fact that workers are no longer caretakers, the result is obvious.
4. **The lack of attention to animal needs determined by their physiological and psychological natures** – as mentioned earlier, technological sanders allow us to keep animals under conditions that violate their natures, thus severing productivity from assured well-being.

It is absolutely essential to stress here that there was no malicious attempt to ride roughshod over animal welfare by those who created and perpetuated these systems. (Almost all animal scientists and producers in industrial agriculture came from a husbandry background.) The failure to account for animal welfare in these new systems basically arose from a conceptual error.

Let us recall that traditional agricultural success under husbandry depended upon proper treatment of animals – generally if the animals suffered the producer lost out. Under such circumstances, it was perfectly reasonable (subject to some refinements) to use animal productivity as a good (rough) criterion for welfare. The people who developed confinement agriculture continued to use this criterion, failing to mark the singular difference between industry and husbandry. In husbandry conditions, animal suffering would *usually* lead to loss of productivity. Under industry, however, the technological sanders mentioned sever the close connection between productivity and welfare, in so far as the animals may suffer harm in ways that do not impinge on economic productivity. In some cases such as the modern dairy cow, great productivity has been univocally selected into the cow's genetics, in ways that actually create new welfare problems, such as metabolic 'burn-out', foot and leg problems, and reproductive problems. In one study, the most productive cows had a higher percentage of swollen legs (Fulwider *et al.*, 2007). On many intensive dairies, a cow lasts for only two lactations. We saw this conceptual error expressed in the CAST report of 1982 (CAST, 1982) mentioned earlier.

Animal welfare is not the only problem inadvertently occasioned by the industrialization of animal agriculture. The Pew Commission on Industrial Farm Animal Production in America, on which I was privileged to serve, recently issued a report on all of the issues after over 2 years of intensive study. (The report may be found at www.ncifap.org.) We concluded that although industrial agriculture does indeed yield food that is cheap at the cash register, there are many significant costs of such agriculture that are borne by society (the consumers) that do not appear at the cash register. (Economists call this *externalization* of costs.) These costs include:

1. Environmental despoliation – in highly concentrated industrial agriculture, animal waste disposal, for example, becomes a major avenue to pollution. Air pollution is a related problem.
2. Loss of small farmers and robust rural communities to consolidation – the USA, for example, lost 85% of the swine producers it had in the early 1980s.
3. Human health danger costs – these include incubation of pathogens in high confinement; respiratory problems among those working in some high animal confinement units; low-level antibiotic non-therapeutic use in confined animals driving antibiotic resistance; and other issues of food safety.
4. Animal welfare – as discussed.

At any rate, let us return to our main topic: the notion of animal welfare implicit in emerging social ethics for animals that must be understood for animal agriculture to meet societal demands.

Increased Public Concern for Animals

It is manifest first of all that public concern for animal treatment has increased greatly over the past few years, particularly as people have become aware that much modern animal use is not a fair contract with the animals, but is patently exploitative, particularly regarding animal agriculture and animal research and testing. This has in turn led globally to a proliferation of laws attempting to address animal suffering not covered by the cruelty laws, even as the cruelty laws themselves have been strengthened and elevated to felony status in 40 US states. Over 2100 laws relevant to animal welfare were proposed in state legislatures in 2004 and laboratory animal legislation has proliferated throughout Europe and other countries. The World Organization for Animal Health (OIE) welfare codes on slaughter and transportation have now brought animal welfare issues to the attention of the developing world. Many abusive practices have been eliminated by public pressure, and consumers increasingly express their buying preferences for humanely raised foodstuffs and cosmetics derived without additional animal testing.

As media coverage of animal issues increases, the social paradigm for animals moves towards the companion animal 'member of the family', prominent people speak out for animals and books on animal ethics proliferate, people want assurance of proper animal treatment. And, as I predicted 30 years ago, with husbandry gone and most animal suffering invisible to the anti-cruelty ethic and laws, society will require a new ethic for animals and will look to extant ethics for humans, appropriately modified, to do the job. As Plato remarked, ethics proceeds from pre-existing ethics, and one develops new ethical coverage not by creating new principles (what he calls 'teaching'), but rather by *reminding* people of the logical implications of their current ethical beliefs.

The portion of our ethical beliefs being so extended to animals is patent. Every society faces a fundamental conflict of two goods – the good of the group and the good of the individual. It is, for example, in the group interest, to tax the wealthy, but not in the interest of a wealthy individual. It is in the group interests to draft young men to fight wars, but not in the interest of those individuals. Many societies, totalitarian societies, resolve this conflict by positing the primacy of the corporate entity – state, Reich, the people, church. This strategy was patent in Hitler's Germany, Stalin's Russia and Mao's China. The only other extreme alternative is to make all social decisions unanimous, a clear impossibility.

With regards to animal agriculture, the pastoral images of animals grazing on pasture and moving freely are iconic. As the 23rd Psalm indicates, people who consume animals wish to see the animals live decent lives, not lives of pain, distress and frustration. It is for this reason in part that industrial agriculture conceals the reality of its practices from a naïve public – witness Purdue's advertisements about raising 'happy chickens', or the California 'happy cow' ads. As ordinary people discover the truth, they are shocked. When I served on the Pew Commission and other commissioners had their first view of sow stalls, many were in tears and all were outraged.

Just as our use of people is constrained by respect for the basic elements of human nature, people wish to see a similar notion applied to animals. Animals, too, have natures, what I call *telos* following Aristotle – the 'pigness of the pig', the 'cowness of the cow'. Pigs are 'designed' to move about on soft loam, not to be in gestation crates. If this no longer occurs naturally, as it did in husbandry, people wish to see it legislated. This is the mainstream sense of 'animal rights'.

As property, strictly speaking, animals cannot have legal rights. But a functional equivalent to rights can be achieved by limiting property rights. When I and others drafted the US federal laws for laboratory animals, we did not deny that research animals were the property of researchers. We merely placed limits on their use of their property. I may own my car, but that does not mean I can drive it on the sidewalk or at any speed I choose. Similarly, our law states that if one hurts an animal in research, one must control pain and distress. Thus research animals can be said to have the *right* to have their pain controlled.

In the case of farm animals, people wish to see their basic needs and nature, *teloi*, respected in the systems in which they are raised. Since this no longer occurs naturally as it did in husbandry, it must be imposed by legislation or regulation. A Gallup poll conducted in 2003 showed that 62% of the public wanted legislated guarantees of farm animal welfare. This is what I call 'animal rights as a mainstream phenomenon'. Legal codification of rules of animal care respecting animal *telos* is thus the form animal welfare takes where husbandry has been abandoned.

Thus, in today's world, the ethical component of animal welfare prescribes that the way we raise and use animals must embody respect and provision for their psychological needs and natures. It is therefore essential that industrial agriculture phases out those systems which cause animal suffering by violating animals' natures and replaces them with systems respecting their natures.

As I have told many animal agriculture and veterinary groups, this does not mean the end of animal science. What it does mean is that while the fundamental values driving animal science have been efficiency and productivity, animal science must vector into the design of its systems the other values we discussed – respect for animal nature and minimizing animal pain and distress, control of environmental degradation growing out of production systems, concern for the effect of animal production on rural communities and human health, and concern for the animals' needs and natures. In short, animal science must become what the 18th century called a 'moral science'. The irreducibly moral component of the concept of animal welfare, long ignored, must be embraced and engaged by the industry.

This is why highly restrictive systems such as sow gestation stalls are being phased out. They have already been eliminated in the UK and the EU countries in Europe will ban sow gestation stalls in 2013. In the USA and Canada, pressure from consumers and animal advocacy groups has been instrumental in getting large corporations such as Smithfield and Maple Leaf to phase out sow stalls. It is being done out of response to ethical concerns of consumers.

References

Al Masri, H. (2007) *Animal Welfare in Islam.* Islamic Foundation, London.

Ascione, F.R., Weber, C.V., Thompson, T.M., Heath, J., Maruyama, M. and Hayashi, K. (2007) Battered pets and domestic violence: animal abuse reported by women experiencing intimate violence and by nonabused women. *Violence Against Women* 13, 354–373.

Blaney, D.P. (2002) The changing landscape of milk production. United States Department of Agriculture (USDA), Statistical Bulletin Number 972. USDA Agricultural Research Service (ARS), Washington, DC.

Council for Agricultural Science and Technology (CAST) (1982) *Scientific Aspects of the Welfare of Food Animals*. Report #91. CAST, Ames, Iowa.

Council for Agricultural Science and Technology (CAST) (1997) *Wellbeing of Agricultural Animals Task Force Report, Chaired by Stanley Curtis*. CAST, Ames, Iowa.

Espejo, L.A., Endres, M.I. and Salfer, J.A. (2006) Prevalence of lameness in high producing Holstein cows housed in freestall barns in Minnesota. *Journal of Dairy Science* 89, 3052–3058.

Farm Animal Welfare Council (FAWC) (2009) Available at: http://www.fawc.org.uk/freedoms.htm (accessed 19 June 2009).

Fulwider, W.K., Grandin, T., Garrick, D.J., Engle, T.E., Lamm, W.D., Dalsted, N.L. and Rollin, B.E. (2007) Influence of free stall base on tarsal point lesions and hygiene in dairy cows. *Journal of Dairy Science* 90, 3559–3566.

Gallup (2003) Available at: www.gallup.com (accessed 18 June 2009).

Gentle, M.J., Waddington, D., Hunter, L.N. and Jones, R.B. (1990) Behavioral evidence for persistent pain following partial beak amputation in chickens. *Applied Animal Behavioural Science* 27, 149–157.

Gharebaghi, R., Reza Vaez Mahdavi, M., Ghasemi, H., Dibaei, A. and Heidary, F. (2007) Animal rights in Islam. In: *Proceedings of the Sixth World Congress in Alternatives and Animal Use in the Life Sciences*, 21–25 August 2007, Tokyo, Japan, pp. 61–63.

Harrison, R. (1964) (Reprinted 1966) *Animal Machines*. Ballantine Books, New York.

Kessler, B. (2009) *Goat Song*. Scribner, New York.

Lundeen, T. (2008) Poultry missing genetic diversity. *Feedstuffs*, 1 December, p. 11.

McKenna, C. (2000) Ruth Harrison. *Guardian*, 6 July. Available at: http://www.guardian.co.uk/news/2000jul/06/guardianobituaries (accessed 20 July 2008).

Miller, C. (2001) Childhood animal cruelty and interpersonal violence. *Clinical Physiological Review* 21, 735–749.

Pew Commission (2008) Putting Meat on the Table: Industrial Farm Production in America. Pew Commission on Industrial Farm Animal Production in America. Available at: www.PCIFAP.org (accessed 18 July 2008).

Rollin, B. (2003) *Farm Animal Welfare, Social, Bioethnical, and Research Issues*. Wiley, Chichester, UK.

Schwabe, C. (1978) *Cattle, Priests, and the Progress of Medicine*. University of Minnesota Press, Minneapolis, Minnesota.

Shahidi, S.J. (1996) *Translation of Nahj-albalagheh by Iman Ali enbe Abu Talih*. Cultural and Scientific Publishing, Iran, 286 pp.

Stull, C.L., Payne, M.A., Berry, S.L. and Hullinger, P.J. (2002) Evaluation of the scientific justification of tail docking in dairy cattle. *Journal of the American Veterinary Medical Association* 220, 1298–1303.

Torgerson, R. (1997) Jeffersonian democracy and the role of cooperatives. *Rural Cooperatives* 64(3), 2.

Van de Weerd, H. and Sandilands, V. (2000) Bringing the issue of animal welfare to the public: a biography of Ruth Harrison (1920–2000). *Applied Animal Behaviour Science* 113, 404–410.

Vansickle, J. (2008) Filtration works at a cost. *National Hog Farmer*, 15 November. Available at: http://nationalhogfarmer.com/facilities-equipment/ventilation/1115-filtration-works-cost/ (accessed 13 December 2008).

Volant, A.M., Johnson, J.A., Guillone, E. and Coleman, G.J. (2008) The relationship between domestic violence and animal abuse: an Australian Study. *Journal of Interpersonal Violence* 23, 1277–1295.

3 Implementing Effective Standards and Scoring Systems for Assessing Animal Welfare on Farms and Slaughter Plants

TEMPLE GRANDIN

Colorado State University, Fort Collins, Colorado, USA

The World Organization for Animal Health (OIE) has animal welfare standards that are minimum standards every country should follow. There are three major classes of standards:

1. International standards to prevent obvious abuses such as the OIE (2009a, b) codes. Example – conscious animals should not be thrown, dragged or dropped.
2. Standards required by legislation in your country. Example – sow gestation stalls are banned in the UK on all farms.
3. Private standards for animals that are sold to restaurant chains and large supermarkets. Example – 95% or more of the cattle must remain silent and not vocalize (moo or bellow) in the stun box or while entering the stun box. If more than 5% of the cattle vocalize, the welfare audit is failed.

This includes standards for highly specialized markets such as organic, natural, grain fed or free range. Example – cattle must be kept on pasture during the growing season. For the ground to qualify as pasture, a minimum of 75% of the field or paddock the animals are in must have 75% or more of the surface covered with vegetation with a root system.

Well written private standards are compatible with OIE standards. They should be written to avoid direct conflicts with OIE. For example, numerical scoring of cattle vocalization during handling provides guidance for enforcing the OIE standard that states 'Animals should be handled in such a way to avoid harm, distress or injury' (OIE, 2009a). Standards of all types should be posted on the Internet on easily accessible web sites. This provides transparency and helps people who are writing standards to reduce conflicts between different standards.

The Need for Clear Standards

Many standards and regulations for animal welfare, disease control, food safety and many other important areas are too vague and subjective. When vague standards are used, there will be great variation in how they are interpreted. One inspector may interpret the guidelines in a very strict manner and another may allow abuse to occur and have a lax interpretation. The author has trained many auditors and inspectors to assess animal welfare in slaughter plants and on farms. These people need specific information on what conditions are acceptable and what conditions would either be a failed audit or a violation of the law. They ask very specific questions. For example, what is the size of a bruise that counts as a bruised carcass? When does nudging an animal with your foot become kicking the animal, which would be considered an act of abuse? In the poultry barn, they asked for a very specific description of acceptable and non-acceptable litter condition.

Improving Consistency of Audits, Assessments and Inspections

A good auditor training programme is essential to make an auditing programme really effective. Good training will improve consistency between different auditors. Variation in judgements between different auditors can be reduced through good training (Webster, 2005).

Inter-observer reliability of a welfare assessment can be greatly improved when the wording of the standard is clear. A measure with good inter-observer reliability will produce similar scores when it is done by different people. Data collected by the author in 66 US slaughter plants indicated that there were no significant differences between three different restaurant auditors when they had to count the percentage of cattle rendered insensible with a single shot from a captive bolt gun ($P = 0.529$) and count the percentage of cattle that vocalized (moo or bellow) in the race and stun box ($P = 0.22$). Unfortunately, there were significant differences between auditors on the percentage of cattle moved with an electric goad ($P = 0.004$). The reason for this was that the standards for stunning and vocalization were clearly written and the standards for electric-goad use were not clear. Some auditors counted all touches with the electric goad and others did not. Some auditors did not know that all touches with an electric goad should count, because it is impossible to determine accurately if the electric button was pushed. A comparison of three commonly used dairy-welfare-evaluation assessment tools indicated that they all accurately picked out the 20% of the worst dairies, but they differed greatly on other measures (Stull *et al.*, 2005).

Research done by Smulders *et al.* (2006) with pigs showed that high inter-observer reliability can be attained for evaluating welfare and behaviour on the farm. These researchers developed an easy-to-administer behaviour test and the results correlate highly with physiological measures of stress such as salivary cortisol, urinary epinephrine, norepinephrine and production traits. On the startle test, a pen of pigs was either scored as fearful or not fearful. A 21-cm-diameter yellow ball was tossed into the pen. The pen of animals was rated as fearful if more than half the animals initially ran way. Startling is influenced by both environmental enrichment and the attitude of the stockperson (Grandin *et al.*, 1987; Hemsworth *et al.*, 1989; Beattie *et al.*, 2001). Animals that have both environmental enrichment and a good stockperson caring for them will have a weaker, less fearful startle response. One of the reasons this test had such good inter-observer reliability is because it is simple.

Training auditors and types of audits

It is the author's opinion, based on in-the-field training of over 200 auditors and inspectors, that a student auditor needs to be accompanied by a highly experienced person on visits to five operations before they do audits on their own. If they are going to audit or inspect dairies they need to visit five dairies and if they are going to assess poultry slaughter, they need to visit five poultry plants. There is a need to develop simple-to-use, effective welfare auditing tools.

There is a difference between an audit that is used to screen animal welfare and measurements used for detailed scientific study or veterinary diagnosis. A screening tool has to be much simpler to make auditor training easier. The purpose of a screening tool is to determine that there is a problem. The purpose of more sophisticated measures is to diagnose and fix a problem or do scientific research.

Vague Wording Should be Eliminated to Improve Inter-observer Reliability

There are several vague words that should be eliminated from all standards and guidelines: they are 'adequate', 'proper' and 'sufficient'. One person's interpretation of proper handling may be totally different from another person's. For example, if a standard states 'minimize the use of electric goads', one person may interpret this as almost never using it and another may think that using it once on every animal is in compliance with this standard.

Some examples of vague wording in United States Department of Agriculture (USDA) standards in the USA are to avoid 'excessive electric goad use' or avoiding 'unnecessary pain and suffering'. It is impossible to train an auditor or an inspector on what is excessive use of an electric goad, and have good inter-observer reliability between different inspectors. One inspector may suspend meat inspection and shut down a plant for using an electric goad once on every animal, and another may think this is perfectly normal and does not penalize the plant. The author has repeatedly observed great inconsistency in the enforcement of vague USDA directives and standards. The use of objective numerical scoring can greatly improve agreement between different auditors or government inspectors.

In another example, a standard states that pigs should have adequate space on a truck or in a housing system. This is too vague. An example of a more effective way to write the standard is that the pigs must have enough space so they can all lie down at the same time without being on top of each other. Some other examples of clearly written standards are: ammonia levels in an animal house must not exceed 10 ppm; and tail docking of dairy cows is prohibited. Some welfare standards require that animals must be housed on pasture. There should be a good definition of the minimum requirements for pasture. At what point does barren, overgrazed ground switch from being a pasture to a dirt lot? The author suggests that to qualify as pasture, 75% or more of the area occupied by the animals must have vegetation with a root system (Fig. 3.1).

Traffic safety standards are a good example of clearly written standards that work effectively in many countries. People who write welfare standards should use traffic laws as a model. When a car is speeding, the police officer pulls it over after he has measured its speed. He does not stop cars that he thinks are speeding; he measures the car's speed. Both the drivers and the police officers know what the speed limits and rules are. In most countries, enforcement of traffic laws is effective and fairly uniform.

Unfortunately, there are some politicians and policy makers who make standards vague on purpose. A political appointee who was in charge of food inspection in a major developed country resisted the author's suggestions to make standards less vague. He admitted that his agency wanted enforcement flexibility. The reason he wanted this was so enforcement could be strengthened or weakened depending on the political conditions. His vague standards resulted in one inspector being super strict and others very lax. There was no consistency between inspectors.

Different Types of Welfare Standards

There are five basic types of standards for assessing animal welfare:

1. Animal-based measures also called performance standards or outcome criteria (Most Emphasis).
2. Practices that are prohibited (Most Emphasis).
3. Input-based engineering or design standards (Less Emphasis).
4. Subjective evaluations (Less Emphasis).
5. Record keeping, stockperson training documents and paperwork requirements. Documentation of management procedures and standard operating procedures (SOPs) (Less Emphasis).

Fig. 3.1. Pigs housed outdoors on pasture. For the ground to qualify as being pasture, 75% or more of the paddock where the animals are housed must have vegetation with a root system. The pasture in this photo has some bare spots but it would definitely be in compliance.

Animal-based standards

Animal-based measures of welfare problems that can be directly observed by an auditor when he/she visits a farm can be very effective for improving welfare. They are outcomes of poor management practices. The OIE is moving towards the use of more animal-based outcome standards. The use of outcome-based standards is recommended by many animal welfare researchers (Hewson, 2003; Wray *et al.*, 2003, 2007; Webster, 2005). The large European animal welfare assessment project is also emphasizing outcome-based measures (Linda Keeling, 2008, personal communication). She states that measures should be: (i) science based; (ii) reliable and repeatable; and (iii) feasible and practical to be implemented in the field. Other work on animal-based assessment has been done by Wray *et al.* (2003, 2007) and LayWel (2009). Directly observable conditions are easy for auditors to score and quantify. Some important examples are body condition scoring (BCS), lameness scoring, falling during handling and animal cleanliness scoring. There are many published scoring systems that have pictures and diagrams which make it easy to train auditors. Edmonson *et al.* (1989) is one example. Studies of BCS show that different observers assign similar scores if the evaluators are well trained. The correlation between different observers varied from 0.763 to 0.858 (Ferguson *et al.*, 1994). However, Kristensen *et al.* (2006) found that 51 practising veterinarians were very variable in their consistency of scoring body condition. Body score assessors should be trained and their ability to assign accurate scores validated. For a welfare audit, scoring will be easier because the assessor only has to identify animals that are too skinny (see Chapter 1 for further information on measurements).

For assessing animal handling, scoring systems where handling faults are numerically scored are easy to implement and very effective (Maria *et al.*, 2004; Grandin, 2005, 2007a; Edge and Barnett, 2008). Some of the items that are measured are the percentage of animals that fall, percentage poked with an electric goad, percentage that vocalize (moo, bellow, squeal) or the percentage that move faster than a trot or walk. Lameness scoring of dairy cows has high inter-observer repeatability of locomotion scores of individual cows and the variation between observers is low (Winckler and Willen, 2001). This indicates that a five-point lameness scale will provide reliable data. In another study, observers who visited seven dairy farms on three occasions had high inter-observer repeatability on the variables of lameness scoring, kicking and stepping during milking, cow cleanliness and avoidance (flight) distance (DeRosa *et al.*, 2003). They had similar results in water buffaloes for kicking and stepping and avoidance distance (DeRosa *et al.*, 2003). Lameness was non-existent in buffaloes and cleanliness scores are meaningless because buffaloes wallow.

The OIE codes (2009a) also support the use of numerical scoring – Chapter 7.3 states:

> Performance standards should be established in which numerical scoring is used to evaluate the use of such instruments, and to measure the percentage of animals moved with an electric instrument and the percentage of animals slipping or falling as a result of their usage.

At the slaughter plant, directly observable animal-based conditions that would be detrimental to welfare can be easily assessed. A few examples are bruises, death losses during transport, broken wings on poultry, hock burn on poultry, disease conditions, poor body condition, lameness, injuries and animals covered with manure (Table 3.1). On the farm, abnormal behaviour such as stereotypical pacing, bar biting in sows, cannibalism in poultry, excessive startle response and tail biting are also easy to numerically quantify and observe.

Not directly observable animal-based welfare criteria

Animal health measures that can be obtained from producers' records are important indicators of welfare problems. Some common examples are death losses, culling rates, animal treatment records and health records. These measures are useful, but directly observable measures should be weighted more heavily because records can be falsified. Paperwork and records are more important for tracing disease outbreaks than for assessing welfare.

Animal-based measures are continuous

It is impossible to never have a sick animal or never have a lame animal. When handling animals, doing it perfectly is not attainable. All of the animal-based measures are *continuous variables*. When a

Table 3.1. Directly observable minimum core standards (critical control points; CCPs) for assessing animal welfare on the farm for all species of animals and poultry that can be observed without looking at paperwork or records.

Directly observable animal-based standard (continuous variables)[a]	Prohibited practices (discrete variables)[b]	Input-based standard (discrete variables)[b]
• Body condition scoring (BCS) • Lameness scoring • Scoring of faecal matter on the animal's legs and body • Condition of coat or feathers • Sores and injuries • Obvious neglected health problems • Handling scoring • Abnormal behaviour scoring • Panting and other signs of heat stress • Frost bite or other signs of cold stress such as piling in piglets	• Kicking, beating, throwing or dragging animals • Prohibited surgical and management procedures • Slaughter or euthanasia methods that are prohibited • Dragging downed non-ambulatory animals. Some programmes may prohibit transport of non-ambulatory animals • Types of housing that are prohibited, such as sow gestation stalls	• Ammonia levels less than 25 ppm with a goal of 10 ppm (Kristensen and Wathes, 2000; Kristensen *et al.*, 2000; Jones *et al.*, 2005) • Minimum space requirements for housing and in vehicles • Specifications for the type of housing such as cage free or caged layers • Life support back-up requirements such as generators for totally enclosed mechanically ventilated buildings. Not required for buildings with natural ventilation • Weaning practices

[a] Continuous variables – scored as the percentage of animals in compliance with each variable.
[b] Discrete variables – scored yes/no or pass/fail for the entire farm, truck or slaughter plant.

standard is being written, a decision has to be made on the acceptable level of faults. These decisions should be based on a combination of scientific data, ethical concerns, and field data on levels that can actually be attained.

The percentage of faults that is considered acceptable may vary between different countries or customer specifications. Many people who are concerned about animal welfare have difficulty writing a standard that allows for some faults to occur. Unless a numerical limit is placed on an animal-based measure, it is impossible to enforce it in an objective manner. An example would be 5% as the maximum acceptable percentage of lame animals. Data presented from the studies reviewed in Chapter 1 show that well-managed dairies can easily achieve this. Vague terms such as minimizing lameness should not be used because one auditor may think 50% is acceptable and another may consider 5% lame animals as a failed audit.

For example, the scoring system used in slaughter plants (Grandin, 1998a, 2007b) allows 1% of the animals to fall during handling. Some people thought that allowing this level of falling was abusive. Data from audits done in many beef and pork plants indicated that for a plant to reliability pass an audit of 100 animals at the 1% level their actual falling percentages dropped to less than one in 1000. The reason why the standard has remained at 1% is that during an audit of 100 animals, a farm or plant should not be penalized for a single animal that may have jumped and fallen or was spooked by extra people getting too close to it. Putting a hard number on the allowable percentage of animals that fall has brought about remarkable improvements. The use of hard numbers also prevents gradual deterioration of practices. The OIE code (2009b) states:

> Animals should not be forced to move at a speed greater than their normal walking pace, in order to minimize injury through falling or slipping. Performance standards should be established with numerical scoring of the percentage of animals slipping or falling is used to evaluate whether animal moving practices and/or facilities should be improved. In properly designed and constructed facilities with competent animal handlers, it should be possible to move 99% of animals without their falling.

Many people are reluctant to assign hard numbers or they make the allowable numbers of bruised, injured or lame animals or birds so high that the worst operations can pass. For example, the National Chicken Council in the USA set the limit for broken wings during catching and transport at 5% of the chickens. When the more progressive managers improved their catching practices, broken wings dropped to 1% or less. The standard should be set at 1% not 5%.

All scores per animal

To simplify auditing, each animal-based welfare criterion should be scored on a yes/no, pass/fail basis per animal. For example, each animal is scored as either lame or not lame or fell or did not fall. The number of animals that have a fault is used to determine the percentage of lame animals or the percentage of animals falling. Lameness and BCS methods used for auditing must be simpler than scoring systems used in research or veterinary diagnosis. The author has learned from training many auditors that measurements must be simple to use. For welfare audits, the measure for body condition score or lameness score should be divided into two categories of pass and fail. Each animal is scored and the percentages of animals that comply are tabulated for each variable.

Prohibited practices

These standards are much easier to write because they are discrete variables. Certain practices such as sow gestation stalls or docking tails on dairy cows are prohibited. There is no room for interpretation. Certain practices are banned. The farm either has sow gestation stalls or it has group housing. Another example is specific guidelines on surgical procedures. For example, after a certain age anaesthetics would be required for castration. Severely abusive practices such as throwing animals, puntilla, poking out eyes, breaking tails and dumping or dropping animals off large trucks must be banned. The OIE codes (2009b) for slaughter have lists of practices that should not be used (see Chapter 9). These types of standards are easy to write and they usually do not have the interpretation problems that occur with vague wording. Below are OIE (2009a, b) codes for handling animals that specify practices that should not be used:

- The use of such devices (electric goads) should be limited to battery-powdered goads on the hindquarters of pigs and large ruminants, and never on sensitive areas such as the eyes, mouth, ears, anogenital region or belly. Such instruments should not be used on horses, sheep and goats of any age, or on calves or piglets.
- Conscious animals should not be thrown, dragged, or dropped.
- Animals for slaughter should not be forced to walk over the top of other animals.
- Animals should be handled in such a way as to avoid harm, distress or injury. Under no circumstances should animal handlers resort to violent acts to move animals such as crushing or breaking tails of animals, grasping their eyes or pulling them by the ears. Animal handlers should never apply an injurious object or irritant substance to animals and especially not to sensitive areas such as eyes, mouth, ears, anogenital region or belly. The throwing or dropping of animals, or their lifting or dragging by body parts such as their tail, head, horns, ears, limbs, wool, hair or feathers should not be permitted. The manual lifting of small animals is permissible.

Input-based engineering or design criteria standards

These standards tell producers how to build housing or specify space requirements. Input-based standards are easy to write clearly. They may work well with one breed of animal and poorly for another. For example, a small hybrid hen needs less space than a large hybrid. A single space guideline would not work for both large and small chickens. In many cases, input-based standards should be replaced with animal-based ones. However, there are other situations where input-based standards are recommended.

Input-based standards work well for specifying baseline minimum conditions for acceptable levels of welfare. Some examples are minimum space requirements on transport vehicles and in housing, maximum allowable ammonia levels in buildings and minimum amperage levels for effective electrical stunning. Minimum amperage levels for stunning are specified in OIE codes (OIE, 2009b; see Chapter 9). The above items can all be quantified with numbers. For these standards, charts with animal species and weights can be easily made. Animal weight is often a better variable to use because breed characteristics change. For example, Holstein and Angus cattle were much smaller in the 1960s and 1970s compared to the year 2000. Input standards work poorly when they specify exactly how specific details of animal accommodation should be designed. For example, detailed specifications on how to build dividers and neck rails on dairy cow stalls (cubicles) is not recommended because an outdated specification may block innovation and new designs. To detect a problem with poorly designed or poorly maintained stalls, animal-based measures such as injuries, swellings, sleeping posture or cow cleanliness should be used. If the stalls are either badly designed or poorly bedded, the cows will have higher percentages of swollen hocks and lameness.

To have a bare-minimum acceptable level of welfare animals must NEVER be jammed into a crate or pen so tightly that they have to sleep on top of each other. The author has observed some caged layer farms where hens had to walk on top of each other to reach the feeder. An input standard should be implemented to ban this type of abuse. However, on broiler chicken farms, both Dawkins *et al.* (2004) and Meluzzi *et al.* (2008) found that stocking density is not a straightforward indicator of welfare. Severe welfare problems such as foot pad lesions or mortality were more related to poor condition of the litter (Meluzzi *et al.*, 2008). The problems were worse in the winter when airflow through the building was reduced. There have been serious problems with conducting audits when an input standard is vague. One example was a standard for pasture; it stated that animals must have access to pasture. There was no stipulation on how much time the cows had to be on pasture.

Subjective evaluations

Unfortunately, it is impossible to eliminate all subjectivity from an animal welfare audit. Some variables that will require subjectivity to evaluate are overall maintenance of a facility and the attitude of the staff. One of the best ways for an auditor or inspector to become skilled in subjective measures is to visit many different places. This will provide a range of farms or plants ranging from excellent to bad.

Photographic Training Aids

Photographs and videos of abusive practices or poorly maintained equipment should be used for training auditors and inspectors. Photos of both poorly maintained and well-maintained equipment can be used. Common faults in equipment such as broken gates, dirty ventilation fans, dirty water troughs and worn out milking machine equipment are shown alongside properly maintained equipment. People working with animals should never kick or beat animals. A question an inspector may ask is 'When does tapping become beating?' A video could be made that shows proper handling where an animal is being moved correctly by tapping it. That could be shown alongside a video of animals being beaten. An animal must never be beaten to make the training video. Download video of animals being beaten from the Internet or videotape cardboard boxes being beaten.

Clear Comments

Auditors and inspectors must write clear comments. The author has reviewed many audits and inspection reports and there are too many vague comments or not enough comments. There should be well-written comments to describe conditions on failed audit items. Observations of both really good and bad practices should be recorded. Well-written comments will help both customers and regulatory officials to make wise decisions after an audit or inspection is failed. Some examples of vague and well-written comments are shown in Table 3.2.

Record keeping and paperwork requirements

Well-kept records are essential for an efficient farm that has high standards. They are also essential for identification and trace-back for disease control purposes. However, there is an unfortunate tendency for some regulatory agencies to turn the entire audit into a paperwork audit. This may create a situation where the paperwork is correct but

Table 3.2. Examples of vague and well-written comments.

Vague comments	Well-written comments
Rough handling of pigs	The handler kicked piglets and threw them across the room
Poor stunning	The broken stun gun misfired and failed to work about one-third of the time
Poor litter in poultry barn	The litter was clumped, chopped up newspaper. It was wet and transferred soil on to the birds

the conditions out on the farm may be poor. Records are valuable for looking at culling rates and the longevity of animals such as dairy cows and breeding sows. The longevity of an animal is one important indicator of animal welfare (Barnett *et al.*, 2001; Engblom *et al.*, 2007). On some farms, a dairy cow lasts only two lactations.

Core Criteria and Critical Control Points for Welfare

Some items on an audit are much more important than others. The situation must be avoided where the paperwork has been inspected and it passes but the farm is full of skinny, emaciated lame animals. There are certain major core standards or critical non-compliances that must be passed in order to pass an audit. Many of the most important core standards are directly observable animal-based measures that are the outcomes of many bad practices or conditions. The principle of Hazard Analysis Critical Control Points (HACCP) is used in many audit systems. The principle is to have relatively few critical control points (CCPs) or core standards that must ALL be in compliance to pass an audit. An effective CCP is an outcome measure of many poor practices. HACCP systems were originally used for food safety. When the first HACCP programmes were developed, they were often too complicated with too many CCPs. Later versions of the programmes were simplified. The numerical scoring system the author developed for slaughter plants has five core standards or CCPs that are continuous outcome-based measures that are numerically scored, one engineering or input-based standard on watering animals and one discreet core criterion that prohibits any act of abusive handling (Grandin, 1998a, 2005, 2007b). A plant has to pass on ALL seven core standards to pass a welfare audit (see Chapter 9). The HACCP principles are increasingly being applied to animal welfare audits (von Borell *et al.*, 2001; Edge and Barnett, 2008). There are a number of different ways that core standards or CCPs can be classified. The rationale behind the author's classification is for easier implementation in the field.

The core standards in Table 3.1 can be used for welfare standards for all types of systems ranging from completely free range to intensive large-scale farms. To pass a welfare audit, the farm should have an acceptable score on all these core standards. The items in Tables 3.1 and 3.3 are all things that an auditor can observe without having to look at records. For each species of animal, training materials will have to be developed for the local conditions. The reason for placing less emphasis on examining records is the author has observed that some records are faked. It is the author's opinion that auditing paperwork is an important part of a welfare audit but it should be weighted less heavily than the directly observable items on Table 3.1.

Table 3.3. Indicators of animal welfare problems that can be easily measured at the slaughter plant.

Outcomes of rough handling, transport problems and abuse	Outcomes of housing problems	Outcomes of poor management, genetic or neglected health problems
• Bruises	• Animal covered with manure	• Poor body condition
• Broken wings on poultry	• Hock burn on poultry	• Lameness
• Number dead on arrival	• Swellings on the legs of dairy cattle	• Poor coat/feather condition
• Number non-ambulatory	• Pressure sores on sows	• Difficult to handle wild animals
• Injuries such as broken horns or legs	• Serious wounds from fights	• Weak animals
• Broken tails	• Foot pad lesions on poultry	• Cancer eye
• Injuries from specific abusive practices such as poked out eyes, puncture wounds from nails or cut tendons	• Eye and lung pathology due to high ammonia levels	• External parasites
	• Breast blisters on poultry	• Internal parasites
	• Hoof problems	• Tail biting
	• Fin erosion on fish	• Twisted legs on poultry
		• Poor leg conformation
		• Prohibited practices and mutilations
		• Disease conditions
		• Dehydration
		• Horns cut on mature cattle
		• Liver abscesses – grain fed

Animal-based Measures (CCPs) that Assess Many Welfare Problems

Body condition scoring (BCS)

There are many different published scoring systems. It is often best to use charts that have been developed in your country. For American Holstein dairy cows, there are good pictures and charts in Wildman *et al.* (1982) and University of Wisconsin (2005). Both articles are free on the Internet. The BCS pictures should be on a plastic-laminated card that the assessor can always have with them. It is recommended to use a scoring system that producers in the local area are familiar with. For example cattle that starved to death in extremely cold conditions such as −18°C often did not have an emaciated body condition score of 1 (Terry Whiting, personal communication, 2008). Autopsy revealed that there was no fat around the heart and kidney. To have adequate welfare, cattle that live in very cold climates must have better (fatter) body condition than cattle that live in warmer climates. Laminated cards with photos of a rear view and side view of animals that are acceptable and not acceptable are recommended. Poor body condition can be caused by a lack of food, parasites, disease or poor management of rBST. It is an important outcome measure of poor management, nutrition and health.

Lameness scoring

Lameness is caused by many different conditions. Measuring the percentage of lame animals is an outcome of a variety of factors that cause animals to become lame. Some examples of factors that may increase the percentage of lame animals are:

- rapid growth in poultry and pigs genetically selected for growth;
- wet bedding or muddy lots;
- standing on wet concrete;
- lack of hoof trimming or foot care;
- disease conditions – example hoof rot;
- poor leg conformation;
- founder (laminitis) due to feeding high concentrates;
- lameness due to feeding high levels of beta-agonists;
- poor handling that causes animals to slip and fall; and
- poorly designed cubicles (freestalls) in dairies.

Lameness in chickens can be measured with a simple three-point scoring system (Dawkins *et al.*, 2004). The scores are: 1 – walks normally for ten paces; 2 – walks crooked and obviously lame for ten paces; and 3 – downer or not able to walk ten paces. Knowles *et al.* (2008) have developed a six-point lameness scoring system in chickens that ranges from 0 being completely normal to a 5 that is not able to stand. The Knowles *et al.* (2008) paper contains videos for training auditors that can be accessed online. Zinpro.com/lameness has excellent videos for scoring lameness in cattle with a five-point system (see Chapter 7). For welfare auditing purposes, it is most practical to classify animals as lame or not lame. Cattle, pigs and sheep that are classified as lame are those that are not able to keep up with the herd and also those that can keep up with the herd when the herd is walking, but they are obviously not walking normally. Cattle with scores of 3, 4 and 5 on a five-point scale would be classified as lame. On this scale, a 1 is completely normal and a 5 can barely stand and walk. Scoring cattle lameness with a five-point scoring system will have good correlation between different observers. The reliability correlation between different observers varied from 0.76 to 0.96 after the observers were trained and it ranged from 0.38 to 0.76 before training (Thomsen *et al.*, 2008).

Scoring faecal matter on animals

For all species, a simple 1, 2, 3, 4 scoring system can be used. This score is used for faecal matter, faecal matter mixed with earth and faecal matter mixed with bedding that has adhered to the animal or bird. This scoring system can be used for dairy cows, feedlot cattle, pigs, sheep and poultry.

1 – Completely clean legs, belly/breast and body. Birds must be completely clean; mammals may have soil on their hooves and below their knees (Fig. 3.2).
2 – Legs are soiled but belly/breast and body are clean.
3 – Legs and belly/breast are soiled.
4 – Legs, belly/breast and the sides of the body are soiled (Fig. 3.3).

Soil on animals, which is caused by an animal's attempts to cool itself, should not be confused with faecal matter on animals, which is due to a failure to provide a dry place for the animals to lie down.

Fig. 3.2. These dairy cows have been grazed on lush green pastures. Their bellies, udder and upper leg are clean. They would score a 1 for cleanliness even though their feet are soiled.

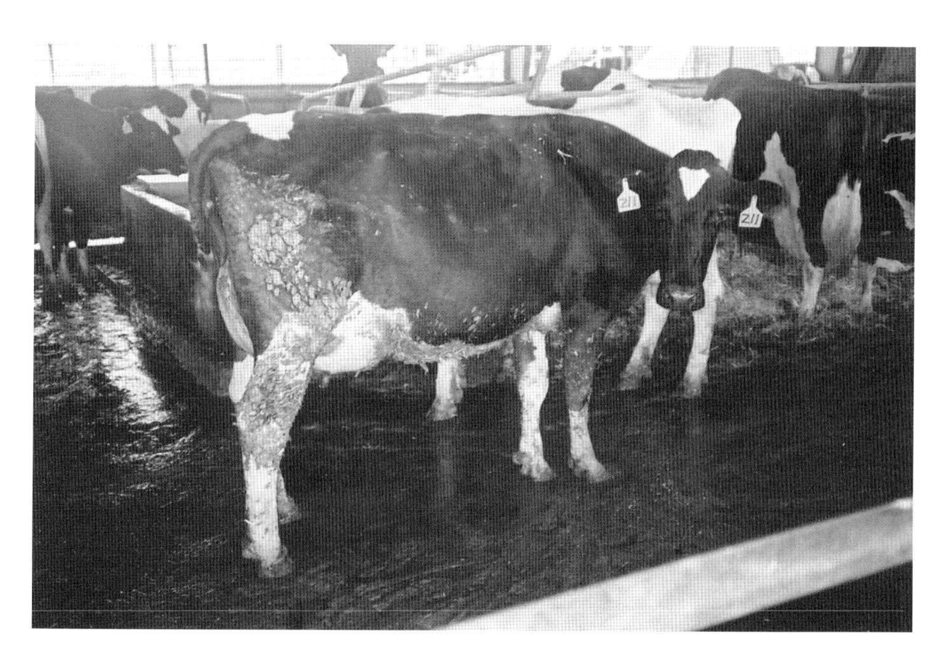

Fig. 3.3. This dirty cow has soil on her legs, udder, belly and sides of her body. She would score a 4 for being a dirty cow.

Condition of coat or feathers

When coats and feathers are being evaluated, it is important to determine the cause of the damage. The three major causes are: abrasion against a cage or feeder; damage inflicted by another bird; and animal or external parasites.

There are published scoring systems for feather condition in poultry. The best systems have separate scores for feather wear and damage caused by other birds. The LayWel web page has good pictures for scoring feather condition in poultry (Fig. 3.4) (www.laywel.eu). The type of housing system will have an effect on the pattern of feather damage. Hens on deep litter have greater damage on the head and neck and caged birds have more wing and breast damage (Bileik and Keeling, 1999; Mallenhurst *et al.*, 2005). Damage on the head is most likely due to pecking from other chickens and wing damage is more likely to be caused by abrasion against equipment.

For livestock, poor coat condition is an indicator of problems such as untreated external parasites or mineral deficiencies. These can occur in organic systems where drugs for treating parasites are not allowed. In cattle, bald spots are an indicator of untreated external parasites. This would be a severe welfare problem.

Sores and injuries

Pictures and charts for scoring will need to be provided for each species. In pigs, pressure sores on the shoulder are an indicator of poor management in gestation stalls, and bites and wounds would be scored in group housing. For dairy cows living in freestall (cubicle) housing, lesions and swellings on the legs are an important measure (see Chapter 1). Cows in cubicles that are frequently bedded have fewer lesions (Fulwider *et al.*, 2007). Freestalls that are too small also increase lesions. Leg lesions on dairy cows are an excellent outcome measure of both poor management and poor stall design. On birds, injuries caused by pecking, cannibalism or aggressive roosters can be easily quantified. In pigs, injuries such as bitten tails and ears can be quantified. In broiler chickens, lesions and burns on the feet can be scored with a chart showing different levels of severity (Fig. 3.5).

Obvious neglected health problems

Some examples would be advanced necrotic cancer eye in cattle, bald spots due to untreated parasites, large infected untreated injuries, large untreated abscesses or other obvious neglected conditions.

(a)

(b)

Fig. 3.4. LayWel feather condition scoring for laying hens. It is a four-point scale where a 1 is almost bald and a 4 is normal. This figure shows (a) score 4 and (b) score 2. The LayWel web site also has pictures for scoring breasts, backs and necks (photographs courtesy of www.laywel.eu).

Handling scoring

The following acts of abuse that can occur during handling should be grounds for an automatic audit failure. These obvious abuses are universally banned in all well-managed livestock and poultry operations.

1. Beating, kicking or throwing animals or birds.
2. Poking out eyes or cutting tendons.
3. Dragging live animals.
4. Deliberately driving animals over the top of other animals.
5. Breaking tails.
6. Poking animals in sensitive areas such as the rectum, eyes, nose, ears or mouth to move them.

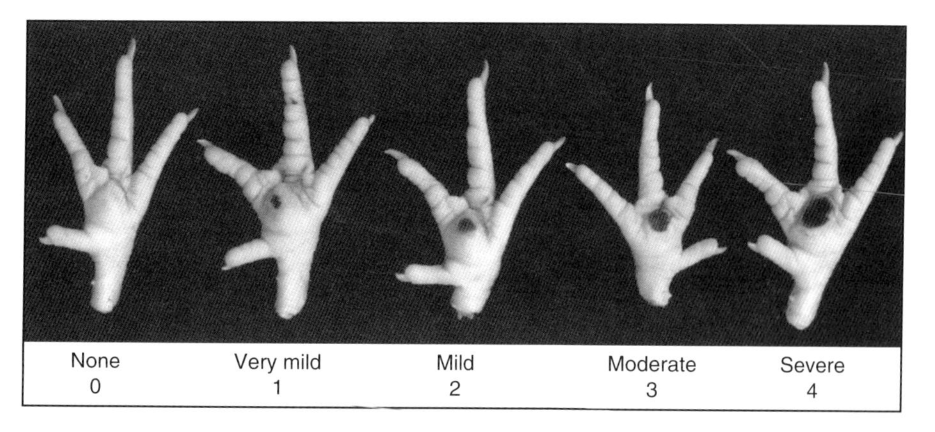

Fig. 3.5. Broiler chicken foot burn chart. Foot-pad lesions are caused by poor litter conditions. Lesions that are brown or otherwise discoloured usually occur during a later stage of growth. Lesions that are white and not discoloured at slaughter are more likely to have occurred early in the growing period.

7. Deliberately slamming gates on animals.
8. Jumping or throwing animals off a truck that has no drop-down tailgate or unloading ramp.

If possible, 100 animals should be scored as they are handled for veterinary procedures or moving to slaughter at an abattoir. Each animal is scored on a per animal yes/no basis. For vocalization and electric-prod use, each animal is scored as either silent or vocalizing or moved with an electric prod or not moved with one. The main continuous variables that are scored are the percentage of animals:

- falling;
- moved with an electric goad;
- vocalizing (moo, bellow, squeal) (do not use vocalization scoring for sheep);
- moved faster than a walk or trot; and
- that run and strike gates or fences.

(See Chapter 5 for more detail.)

All five of these handling measures are outcomes of either poorly trained people or deficiencies in the facilities. For example, falling can be caused by either rough handling methods, slick flooring (if the floor is too slick the animals will be more likely to slip and fall), or the animal not being fit for handling and transport. Vocalization scoring is a sensitive indicator of both equipment design problems and poorly trained people. Vocalizations during handling and restraint of cattle and pigs are associated with the following bad practices (Grandin, 1998b, 2001):

- excessive use of electric goads;
- slipping on the floor;
- sharp edge in a restrainer;
- malfunctioning stunning equipment;
- slamming gates on animals;
- excessive pressure applied by a restraint device; and
- being left alone in a race or stun box.

Vocalization scoring should not be used for sheep because severely abused sheep often remain silent. Many large slaughter plants have been able to easily achieve vocalization percentages of 3% or less of the cattle (Grandin, 2005). Industry data also indicate that in many plants, only 5% or less of the pigs squeal in the restrainer.

When the slaughter audits were first started in 1999, the worst plant had 35% of the cattle vocalizing (moo or bellow) during handling. Plants that use abusive methods of restraint, such as shackling and hoisting conscious animals by one rear leg, may have 40% or more of the cattle bellowing. Simple improvements that reduced electric-prod use reduced the percentage of cattle that vocalized from 8% to 0% in one plant, and reducing the pressure applied by a neck restrainer reduced vocalization from 23% to 0% (Grandin, 2001). This is just one example of how improvements can be tracked by using numerical scoring. In cattle and pigs vocalization is related to physiological measures of stress (Dunn, 1990; Warriss *et al.*, 1994; White *et al.*, 1995).

Abnormal behaviour

Feather and coat condition scoring can be used to detect damage that is caused by other animals.

Cannibalism in hens and wool pulling in sheep are two forms of abnormal behaviour that can be detected by examination for injuries. In pigs, abnormal behaviour such as biting tails and ears can be quantified by counting the percentage of pigs with injuries. The author has observed that some genetic lines of rapidly growing lean pigs have a higher percentage of tail biting compared to others. In bovines, tongue rolling and urine sucking are abnormal behaviour. Animals housed in barren intensive environments may be more likely to engage in stereotypies such as bar biting or pacing. There will be a further discussion of abnormal behaviour in Chapters 8 and 15. Most animal welfare specialists agree that lameness, disease or injury is bad for welfare but there is likely to be more disagreement on how heavily behavioural measures should be weighted. Bracka *et al.* (2007) have developed a technique where a number of different animal welfare scientists are polled about behavioural needs. Statistics were used to determine the behavioural needs that the specialists ranked as most important. The highest ranking was given to materials for pigs to root and explore and prevention of tail biting. This group consensus approach will be important for writing guidelines in areas where the science may be less clear and the decisions are based partially on ethical concepts. In 2001, the EU passed a directive that pigs must have access to fibrous materials such as straw to root and chew (see Chapter 8 on behaviour).

Heat and cold stress

Many animals have died or have had their welfare severely compromised by either heat or cold stress. The thermal neutral zone of an animal is very variable and it depends upon genetics, length of hair/coat, access to shade and many other factors. If animals or birds are panting they are heat stressed and shivering or huddling animals are cold stressed. Auditors and inspectors can easily count the number of panting animals (see Chapter 7). It is beyond the scope of this book to discuss all aspects of thermal comfort, but conditions must be corrected that have caused death from thermal stress. For heat stress, some of the corrective actions would be fans, sprinklers, shades, extra water or changing animal genetics (Fig. 3.6). For cold stress, some possible corrective actions would be shelter,

Fig. 3.6. Well-designed shades and sprinklers keep cattle cool in this feedlot located in the arid, dry south-western USA. The cattle stay clean and dry in the arid climate. Feedlot shades must be installed north and south so that the shadow will move. This prevents mud build-up under the shade. Shades should be 3.5 m (12 ft) high. Places with rainfall under 50 cm/year (20 in/year) are the best locations for feedlots. Pens should have 2–3% slope away from the feed trough to provide drainage and prevent mud build-up.

heat, additional bedding or changing animal genetics. Some of the worst problems with thermal stress occur when animals are moved between different climatic regions. It takes several weeks for an animal's body to acclimatize to either hot or cold conditions. When animals have to move from a cold area to a hotter area, it is best to transport them to the hotter region during its cool season. Transporting cold acclimatized cattle to a hot desert region may result in higher death losses.

Core Standards of Prohibited Practices and Input-based Criteria

These are easy to score because they are discrete variables. The farm is either in or out of compliance on each variable. It is simple yes/no scoring. Problems with vague guidelines are much less of a problem for these two types of standards. Some examples are acceptable and prohibited types of housing, minimum space requirements and air quality standards.

Animal-based Welfare Problems that can be Assessed at the Slaughter Plant

Many of the animal-based continuous standards in Table 3.1 that are measures of poor welfare on the farm can also be measured at the slaughter plant. Large numbers of animals and farms can be easily assessed at a single slaughter plant. Table 3.3 shows the welfare indicators that can be easily measured at the slaughter plant. There are some standards such as specifications for housing and behavioural evaluations that can only be assessed during a farm visit.

Setting Limits for Animal-based Standards

To avoid vague guidelines, numerical limits have to be put on the animal-based standards to determine passing and failing scores. The limits have to be set high enough to make the industry improve but not so high that people will either say it is impossible or fight conforming to the standard. When the author worked on auditing slaughter plants, three out of four plants failed their first audit (Fig. 3.7). They were then given time to improve their practices with no penalties. When a numerical scoring system is first introduced, baseline data will need to be taken and the limit for an acceptable rating should be set so the best 25–30% of the plants or farms will pass. The others have to be given a period of time to attain the standard. For example,

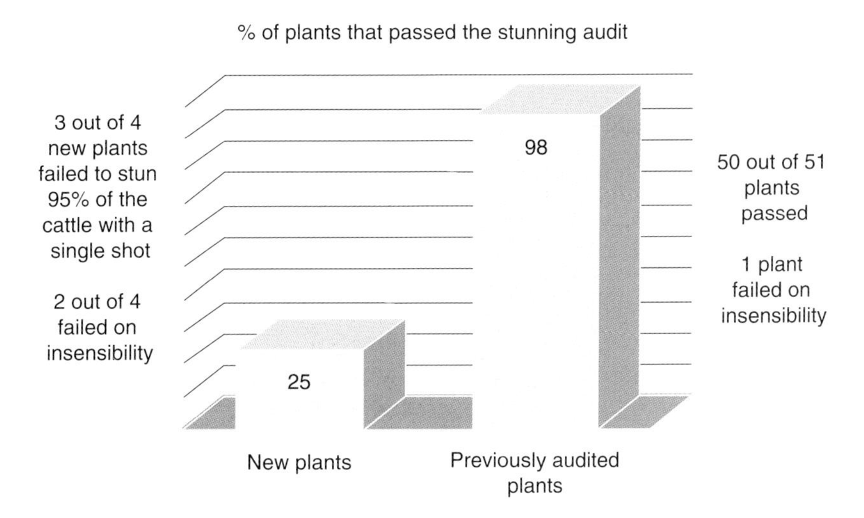

Fig. 3.7. Comparison of beef plants audited for 4 years versus new plants audited for the first time. Only 25% of the slaughter plants passed their initial animal welfare and humane slaughter audit. Plant managers did not know what to expect, and they often engaged in bad practices because they were their normal practices. An initial assessment should be considered as a training session to educate both the management and the employees.

slaughter plants should be required to correct minor problems within 30–60 days and major problems within 6 months. On farms, producers should be given up to 2 years to correct problems such as a high percentage of lame animals. This will make it easier to implement the system because the places with excellent scores can be used as examples to show the bad operators that they need to improve. In parts of the world where practices are really poor, the standards may need to be raised again after the system is implemented and the farms have improved.

Audits and Inspections are a Screening Test

An audit is a screening test to locate animal welfare problems. It is not a complete diagnosis of how to fix problems. Correcting and fixing the problems that are found during an audit will require the services of veterinarians, animal scientists, behaviour specialists and other consultants. An audit determines that there is problem. More complex tests may be needed to diagnose and fix a problem.

Guidelines Versus Standards and Legislative Codes

A guideline is a document that contains information that will be useful for training stockpeople and transporters. It contains detailed information on how to do different procedures such as handling, transport or euthanasia. Guidelines are often much longer documents because they contain lots of 'how to' information. A standard or legislative code document may contain just the standards. When standards or legislation are published without guidelines, it is often very difficult to implement them.

An example of a law or a standard would be: 'Emaciated animals should be euthanized on the farm and not transported'. A guideline that contained pictures or diagrams for BCS would be required to accurately assess body condition. A guideline would be essential for consistent implementation. This standard or law may specify the type of livestock-driving aids that are permitted, but very detailed guidelines are required to train stockpeople on how to quietly move livestock. To effectively train stockpeople may also require videos and other additional training materials in addition to standards and guidelines.

Announced Versus Unannounced Audits and Inspections

When the author first started doing audits of slaughter plants, many people did really bad things in front of her because they did not know any better. In the early years, there was no difference between announced and unannounced audits. After people learned that certain practices were wrong, some managers became very skilled at acting 'good' during an audit and then bringing back the electric goads when the audit was over. Audits and assessments of handling and transport practices are more likely to have different results between those obtained on an announced visit and those gained on an unannounced audit, whereas measures of lameness, body condition, dirty animals or maintenance of facilities are more likely to be consistent between the two types of audit. Some slaughter plants in the USA have installed video auditing that can be viewed unannounced over a secure Internet link. This solves the problem of acting 'good' during an audit and then reverting back to old, bad practices after the auditor leaves.

Combining Animal Welfare Audits with Other Audits and Inspections

Most government and industry auditing and inspection programmes have people that conduct several types of inspections. In many countries, government veterinary inspectors do both meat hygiene and animal welfare enforcement. For most of the audits done by McDonald's, Tesco and other large retailers, the same auditor has to audit both food safety and animal welfare. When farm assessments are being done, the same person may have to audit animal welfare, environmental compliance and animal drug use. Bundling auditor and inspection duties is often required to cut down on many duplicate trips. Because people often have many duties, this is another reason why clear easy-to-implement standards and guidelines are essential. To comply with EU standards, a person has to be designated as that animal welfare specialist.

Structure of Effective Audit and Inspection Systems

The most effective audit and inspection systems have a combination of internal and external audits. The best systems that have been implemented by private industry have three parts:

- Internal audits done on a daily or weekly basis by either employees at the plant or a farm's local veterinarian.
- Third-party audits done by an independent auditing company. They are conducted once or twice a year on EVERY plant and farm.
- Audits by people employed by the corporate office of a meat-buying customer. They should visit a certain percentage of their suppliers each year.

One advantage of using a third-party independent auditing company is to prevent a conflict of interest. For example, a meat buyer in the company corporate office might favour a certain plant and be less strict because they want their cheap prices. However, it is the author's opinion that a corporation should not delegate all its auditing responsibilities to a third-party auditor. Corporate personnel need to make regular visits to show suppliers that they are serious about animal welfare.

On a farm, the farm's regular veterinarian or other professional can do the 'internal' audits. Due to problems with conflict of interest, the entire responsibility for auditing should not be done by the farmer's local veterinarian. The veterinarian may be reluctant to fail a client farm because he/she would risk getting fired by the farmer. To avoid this potential conflict of interest problem, the auditor should not be the farmer's regular veterinarian. The person doing the audit should be on somebody else's payroll. They should be paid by either the government, a third-party auditing company, a meat-buying corporation such as McDonald's, a livestock association or work as field staff for a large, vertically integrated company. In many systems, farms and slaughter plants have to pay for audits. To avoid a conflict of interest, the farm or plant pays the auditing company and the auditing company pays the auditor's salary. A similar system is used for government inspection. The inspector's salary is paid by the government. In both of these systems, either the auditing company or the government makes inspector and auditor assignments. A farm or plant should never be allowed to choose their favourite inspector. To help prevent personality conflicts, a third-party auditing firm should rotate auditors to avoid having the same person always visiting the same farm or plant.

Each Plant or Farm is a Separate Unit

Big multinational companies own many plants and farms. Each plant or farm site should be treated as an independent unit regardless of ownership. It either passes or it fails an audit. A big company may have bad operations in another country. If your job is to implement a welfare audit programme in your country, it is best to concentrate on working with each individual plant or farm in your country. A plant or a farm is either taken off the approved supplier list or put back on the approved list. Meat-buying companies that have made the greatest progress in improving animal welfare have enough plant or farm suppliers so that a few can be taken off the approved supplier list and they will still have enough product. Even if all the meat comes from two big corporations, a buyer still has sufficient economic power to bring about change if he/she buys from several different plants owned by each company. In government programmes there are different levels of penalties ranging from fines to shut-down of the farm or plant.

Procedures for Handling Non-compliance

Most government and industry systems have a formal procedure that is followed when there is a non-compliance. For both minor and major non-compliance, a letter (e-mail) has to be written explaining the corrective action that will be taken to correct the problem. After a specified period of time, the farm or plant will have to have a new audit or inspection. Throughout this chapter, the author has emphasized the need for clear guidelines and standards which can be consistently applied by different auditors or inspectors.

The penalty for being one or two points below the minimum passing score should be much less severe than the penalty for a serious violation such as dragging a non-ambulatory animal down the unloading ramp. The author has worked with many large meat buyers on decisions to remove a supplier from the approved list. Usually one or two points below the minimum standard requires a corrective action letter and a re-audit. Buying from the plant would continue. Dragging the non-ambulatory animal is a serious act of abuse and the meat buyer would stop buying from the plant for a minimum of 30 days. If they had another serious violation they would be suspended for a longer time.

In conclusion of this section, the auditing and inspection process should be very objective, but sometimes great wisdom is required to determine the appropriate punishment. From 10 years of experience implementing auditing systems, the author has learned that suppliers with a poor, non-cooperative attitude need harsher punishment than more cooperative suppliers. When the original McDonald's audits were started in 1999 and 2000, out of 75 pork and beef plants, three plant managers had to be fired before improvements in animal welfare occurred. To get these three plants to conform to the standard, all purchases from them were suspended until new managers were hired. With new management, one of the plants went from being one of the worst plants in the system to one of the best. For both food safety and animal welfare, the best places have managers who want to do the right thing.

Conclusions

Implementation of an effective auditing and inspection programme can greatly improve animal welfare. Clear guidelines will improve the consistency of judgement between different people. This will help prevent the problem of one inspector being super strict and another failing to improve the treatment of animals.

References

Barnett, J.L., Hemsworth, P.H., Cronin, G.M., Jongman, E.C. and Hutson, G.D. (2001) A review of the welfare issues for sows and piglets in relation to housing. *Australian Journal of Agricultural Research* 52, 1–28.

Beattie, V.E., O'Connell, N.E., Kilpatrick, D.J. and Moss, B.W. (2001) Influence of environmental enrichment on welfare related behavioral and physiological parameters in growing pigs. *Acta Agriculturae Scandinavica: Section A, Animal Science* 70, 443–450.

Bileik, B. and Keeling, L.J. (1999) Changes in feather condition in relation to feather pecking and aggressive behaviour in laying hens. *British Poultry Science* 40, 444–451.

Bracka, M.B.M., Zonderland, J.J. and Bleumer, J.B. (2007) Expert consultation on weighing factors of criteria for assessing environmental enrichment materials for pigs. *Applied Animal Behaviour Science* 104, 14–23.

Dawkins, M.S., Donnelly, C.A. and Jones, T.A. (2004) Chicken welfare is influenced more by housing conditions than stocking density. *Nature* 427, 343–348.

DeRosa, G., Tripaldi, C., Napolitano, F., Saltalamacchia, F., Grasso F., Bisegna, V. and Bordi, A. (2003) Repeatability of some animal related variables in dairy cows and buffalos. *Animal Welfare* 12, 625–629.

Dunn, C.S. (1990) Stress reactions of cattle undergoing ritual slaughter using two methods of restraint. *Veterinary Record* 126, 522–525.

Edge, M.K. and Barnett, J.E. (2008) Development and integration of animal welfare standards into company quality assurance programs in the Australian livestock (meat) processing industry. *Australian Journal of Experimental Agriculture* 48(7), 1009–1013.

Edmonson, A.J., Lean, I.J., Weaver, L.D., Farver, T. and Webster, G. (1989) A body condition scoring chart for Holstein dairy cows. *Journal of Dairy Science* 72, 68.

Engblom, L., Lundeheim, N., Dalin, A.M. and Anderson, K. (2007) Sow removal in Swedish commercial herds. *Livestock Science* 106, 76–86.

Ferguson, J.O., Falligan, D.T. and Thomsen, N. (1994) Principal descriptors of body condition score in Holstein cows. *Journal of Dairy Science* 77, 2695.

Fulwider, W.K., Grandin, T., Garrick, D.J., Engle, T.E., Lamm, W.D., Dalsted, N.L. and Rollin, B.E. (2007) Influence of freestall base on tarsal joint lesions and hygiene in dairy cows. *Journal of Dairy Science* 90, 3559–3566.

Grandin, T. (1998a) Objective scoring of animal handling and stunning practices at slaughter plants. *Journal of American Veterinary Medical Association* 212, 36–39.

Grandin, T. (1998b) The feasibility of using vocalization scoring as an indicator of poor welfare during slaughter. *Applied Animal Behaviour Science* 56, 121–138.

Grandin, T. (2001) Cattle vocalizations are associated with handling and equipment problems in beef slaughter plants. *Applied Animal Behaviour Science* 2, 191–201.

Grandin, T. (2005) Maintenance of good animal welfare standards in beef slaughter plants by use of auditing programs. *Journal of the American Veterinary Medical Association* 226, 370–373.

Grandin, T. (ed.) (2007a) *Livestock Handling and Transport*, 3rd edn. CAB International, Wallingford, UK.

Grandin, T. (2007b) *Recommended Animal Handling Guidelines and Audit Guide*, 2007 edn. American Meat Institute, Washington, DC. Available at: www.animalhandling.org (accessed 3 January 2009).

Grandin, T., Curtis, S.E. and Taylor, I.A. (1987) Toys, mingling and driving reduce excitability in pigs. *Journal of Animal Science* 65 (Supplement 1), p. 230 (abstract).

Hemsworth, P.H., Barnett, J.L., Coleman, G.J. and Hansen, C. (1989) A study of the relationships between the attitudinal and behavioral profiles of stock persons and the level of fear of humans and reproductive performance of commercial pigs. *Applied Animal Behaviour Science* 23, 301–314.

Hewson, C.J. (2003) Can we access welfare? *Canadian Veterinary Journal* 44, 749–753.

Jones, E.K.M., Wathes, C.M. and Webster, A.J.F. (2005) Avoidance of atmospheric ammonia by domestic fowl and the effect of early experience. *Applied Animal Behaviour Science* 90, 293–308.

Knowles, J.G., Kestin, S.C., Haslam, S.M., Brown, S.N., Green, L.E., Butterworth, A., Pope, S.J., Pfeiffer, D. and Nicol, C.J. (2008) Leg disorders in broiler chickens, prevalence, risk factors and prevention. *PLOS One* 3(2). Available at: www.pubmedcentral.nih.gov/article.render.regi?artid=2212134 (accessed 3 January 2009).

Kristensen, E., Dueholm, L., Vlnk, D., Anderson, J.E., Jakobsen, E.B., Illum-Nielsen, S., Petersen, F.A. and Enevoldsen, C. (2006) Within- and across-person uniformity of body condition scoring in Danish Holstein cattle. *Journal of Dairy Science* 89, 3721.

Kristensen, H.H. and Wathes, C.M. (2000) Ammonia and poultry: a review. *World Poultry Science Journal* 56, 235–243.

Kristensen, H.H., Burgess, L.R., Demmers, T.G.H. and Wathes, C.M. (2000) The preferences of laying hens for different concentrations of ammonia. *Applied Animal Behaviour Science* 68, 307–318.

LayWel (2009) Available at: http://ec.europa.eu/food/animal/welfare/farm/laywel_final_report_en.pdf (accessed 19 June 2009).

Mallenhurst, H., Rodenburg, T.B., Bakker, E.A.M., Koene, P. and deBoer, I.J.M. (2005) On-farm assessment of laying hen welfare – a comparison of one environment based and two animal based methods. *Applied Animal Behaviour Science* 90, 277–291.

Maria, G.A., Villarrael, M. and Gebresenbet, G. (2004) Scoring system for evaluating stress to cattle during commercial loading and unloading. *Veterinary Record* 154, 818–821.

Meluzzi, A., Fabbri, C., Folegatti, E. and Sirri, F. (2008) Survey of chicken rearing conditions in Italy: effects of litter quality and stocking density on productivity, foot dermatitis and carcass injuries. *British Poultry Science* 49, 257–264.

OIE (2009a) Chapter 7.3. *Transport of Animals by Land, Terrestrial Animal Health Code*. World Organization for Animal Health, Paris, France.

OIE (2009b) Chapter 7.5. *Slaughter of Animals, Terrestrial Animal Health Code*. World Organization for Animal Health, Paris, France.

Smulders, D., Verboke, G., Marmede, P. and Geers, R. (2006) Validation of a behavioral observation tool to assess pig welfare. *Physiology and Behavior* 89, 438–447.

Stull, C.L., Reed, B.A. and Berry, S.L. (2005) A comparison of three animal welfare assessment programs in California dairies. *Journal of Dairy Science* 88, 1595–1600.

Thomsen, P.T., Munksgaard, L. and Toyersen, F.A. (2008) Evaluation of lameness scoring of dairy cows. *Journal of Dairy Science* 91, 119–126.

University of Wisconsin (2005) Body Condition Score – What is Body Condition Score? What Does it Help Us Manage? Available at: http://dairynutrient.wisc.edu/302/page.php?id=36 (accessed 27 December 2008).

von Borell, E., Bockisch, F.J., Büscher, W., Hoy, S., Krieter, J., Müller, C., Parvizi, N., Richter, T., Rudovsky, A., Sundrum, A. and Van de Weghe, H. (2001) Critical control points for on-farm assessment of pig housing. *Livestock Production Science* 72, 177–184.

Warriss, P.D., Brown, S.N. and Adams, S.J.M. (1994) The relationship between subjective and objective assessment of stress at slaughter and meat quality in pigs. *Meat Science* 38, 329–340.

Webster, J. (2005) The assessment and implementation of animal welfare: theory into practice. *Review Science and Technology Off. International Epiz.* 24, 723–734.

White, R.G., deShazer, J.A. and Tressier, C.J. (1995) Vocalization and physiological responses of pigs during castration with and without a local anesthetic. *Journal of Animal Science* 73, 381–386.

Wildman, E.E., Jones, G.M., Wagner, P.E., Boman, R.L., Troutt, H.F. and Lesch, T.N. (1982) A dairy cow body condition scoring system and its relationship to selected production characteristics. *Journal of Dairy Science* 65, 495–501.

Winkler, C. and Willen, S. (2001) The reliability and repeatability of a lameness scoring system for use as an indicator of welfare in dairy cattle. *Acta Agriculturae Scandinavica: Section A, Animal Science* 30 (Supplement 1), 103–107.

Wray, H.R., Main, D.C.J., Green, L.E. and Webster, A.J.E. (2003) Assessment of dairy cow welfare using animal based measures. *Veterinary Record* 153, 197–202.

Wray, H.R., Leeb, C., Main, D.C.J., Green, L.E. and Webster, A.J.F. (2007) Preliminary assessment of finishing pig welfare using animal based measurements. *Animal Welfare* 16, 209–211.

4 The Importance of Good Stockmanship and its Benefits for the Animals

Jeffrey Rushen and Anne Marie de Passillé

Pacific Agri-Food Research Centre, Agri-Food Canada, British Columbia, Canada

Introduction

With all of the attention that is given to intensive housing of farm animals, such as the use of battery cages for laying hens, stalls for sows, etc., we might assume that the welfare of farm animals is affected most by the way that they are housed. However, a very important component of farming that affects both animal welfare and animal productivity is the people who care for the animals. The decisions about how animals are housed, fed and managed are made by people, and it is people who actually perform procedures like breeding, farrowing, milking, debeaking, dehorning, etc. There are many ways in which the stockpeople or caretakers can affect the welfare of the animals in their care. The knowledge or technical competence of the stockperson can play a major role if it leads to improper choice of housing or poor feeding methods. The quality and diligence with which routine tasks, such as cleaning the barn, are done can also be important. In addition, research has now shown that the way that animals are handled by people can have a major effect on their welfare and their productivity. In this chapter, we review some of the research that shows the importance of good stockmanship for animal welfare, animal productivity and worker satisfaction.

Effect of Overall Stockmanship on Animal Welfare and Productivity

The stockperson can most obviously affect the welfare of animals through the way that routine animal care tasks, such as feeding, cleaning etc., are done. The care with which routine animal care tasks are performed is responsible for some of the differences between farms and farmers in the level of animal welfare. For example, dairy farms in which calves are cared for by females tend to have lower calf mortality than farms where men are responsible for care of the calves (Losinger and Heinrichs, 1997).

Lensink *et al.* (2001a) examined the role of stockmanship in affecting the health and productivity of veal calves on 50 veal farms. The farms were operated by a single company, were located in the same region, and used similar management techniques, animal feed, etc. Farmers were interviewed to assess their attitudes to the animals (e.g. whether or not they believed that calves were sensitive to human contact) and their attitudes to the work (e.g. how important cleaning procedures were). The farms were also scored for cleanliness, and the performance of various management routines was noted. High-producing farms (i.e. those with high daily weight gains, good feed conversion efficiencies and low mortality) were cleaner, had pens disinfected by an outside company, had Sunday-evening feedings of the calves, and were run by farmers whose own parents had managed a veal farm. This latter point was important because it resulted in the farmer having a greater experience in raising calves. The size of the effects on the productivity of the farms was large. The cleanliness of the barns accounted for 19% of the differences between farms in daily weight gain and 22% of the differences between farms in feed efficiency. The health of the calves was correlated with the attitudes of the farmers, for example the more the farmer believed that calves were sensitive to human contact, and the more the farmer felt that cleaning was important, the better the health of the calves. The results show the importance of general stockmanship for the welfare and productivity of the calves.

Handling and Animals' Fearfulness of People

Research has now shown quite clearly that when farm animals are frightened of the people that care for or handle them, this can have a marked effect on their welfare. This research has been reviewed a number of times previously (Hemsworth and Coleman, 1998; Rushen *et al.*, 1999b; Waiblinger *et al.*, 2004, 2006). Fear of people can be a major source of stress and a cause of lost production in most species of farm animals. There are marked differences between animals and between farms in the degree to which the animals are frightened of people, and the degree of fear of the animals on a farm is strongly associated with the productivity of the animals on the farm.

Researchers have now found large differences between farms in the level of productivity achieved by farm animals and the extent that the animals are frightened of people. In one of the first studies, Hemsworth *et al.* (1981) compared one-man pig farms and found that the degree of fear that the pigs showed to people, as measured by their readiness to approach people (Fig. 4.1), accounted for a substantial amount of the difference between farms in farrowing rate and number of piglets born (see Chapter 1 on fear). In a later study, Hemsworth *et al.* (1999) found correlations between the behavioural response of lactating sows to people and the percentage of stillborn piglets. Farms on which sows were quicker to withdraw from an approaching person (a sign of fearfulness) had a higher stillbirth rate than those who allowed the person to approach closely. The difference between gilts in

Fig. 4.1. A common way of measuring how fearful animals are of people is to measure how closely they approach people, particularly when they are feeding. The degree of fear is thought to be apparent in the extent that they withdraw from the feed when people are near. However, animals may avoid people for many reasons. For example, animals that are fed by hand are more likely to approach people than animals that are not fed by hand. However, this does not necessarily mean that their welfare is better or that they have been better handled. Caution is required in assessing the fearfulness of animals from their reactions to people when different farms are being compared (de Passillé and Rushen, 2005).

their response to the person accounted for about 18% of the difference between farms in percentage of stillborn piglets. The results indicate that high levels of fear of people by sows may strongly affect the survival of their piglets. Similarly, in poultry, feed conversion efficiency (that is the amount of feed needed to produce a certain number of eggs) has been found to be lower on poultry farms where the birds kept a larger distance from people (Hemsworth *et al.*, 1994a).

Sometimes, the effect of stockmanship may not be due to the use of rough handling but through more subtle effects. For example, Cransberg *et al.* (2000) found a higher mortality among broiler chicks on farms where the stockpeople moved quickly (presumably frightening the birds), while mortality was lower the more time the stockperson stayed still in the shed. Again, the size of the effect was large: the speed of movement of the stockperson accounted for 15% of the differences between farms in broiler mortality.

Perhaps because dairy cows are handled often, much research on how handling methods affect the fearfulness of the animals has involved dairy cattle (Fig. 4.2). Seabrook (1984) showed that the way that dairy cows were handled by the stockperson and the degree of fear the animals show towards the people can be a major factor underlying the differing productivity of different stockpeople. He observed the behaviour of cows that were being handled by 'high-producing' stockpeople (that is stockpeople whose cows produced a large amount of milk) and compared this to cows handled by 'low-producing' stockpeople. He found that the cows with the high-producing stockperson were spoken to and touched more often, appeared less frightened and were more easily moved, and were more likely to approach the stockperson.

More recently, Breuer *et al.* (2000) found substantial relationships between the levels of milk production on dairy farms, the way that the animals were handled, and the degree of fear that the

Fig. 4.2. Of all farm animals, dairy animals have the closest contact with people because of regular milking. Good handling practices, then, are particularly important for dairy animals and much research has shown that the way that dairy cows are handled can have a large impact on their productivity. There are large differences between farms and between milkers in how much milk can be obtained at milking and low milk yields have been associated with rough handling and cows that are frightened of people.

cows showed towards people. Thirty-one commercial dairy farms in Australia were visited and the stockpeople were observed while they moved and handled the animals during normal milkings. The degree to which the milkers used rough or aversive handling during milking varied widely between farms. Furthermore, the use of highly aversive handling techniques (e.g. forceful slaps, hits and tail twists) could account for almost 16% of the differences between farms in annual milk yield.

The cows' fearfulness towards people was measured by observing the time that the cows spent close to a person. Again, large differences between farms were noted in the cows' degree of fearfulness: on the farm with the least fearful cows, the cows spent six times as long close to the person compared to the farm with the most fearful cows. Milk yield was significantly lower as fearfulness of the cow increased, especially during milking. A multiple regression analysis suggested that differences between farms in the degree of fearfulness of the cows could account for 30% of the differences between farms in annual milk production, a sizeable figure! A subsequent report from the same research group (Hemsworth *et al.*, 2000) found that farms with more fearful cows had a significantly smaller proportion of cows conceiving at the first insemination. Differences between farms in the level of fearfulness accounted for 14% of the variance between farms in conception rates. Many of these results were replicated by Waiblinger *et al.* (2002) who observed the behaviour of cows and stockpeople during milking on 30 dairy farms in Austria, showing that the effects are not specific to the extensive dairy production system of Australia. Dairy cows' fear of people may also have effects on their health: Fulwider *et al.* (2008) found that dairies with cows that were most likely to approach or touch people had lower somatic cell counts.

Undesirable effects of animals' fear of people can be particularly important during preslaughter handling. A history of aversive interactions between stockpeople and veal calves has been found to affect ease of moving and transporting the animals and meat quality after slaughter. Lensink *et al.* (2001b) compared veal calves from farms where the farmer acted predominantly gently towards the animals with calves from farms where the farmer acted more roughly. Calves from farms with the rough farmers needed more effort to be loaded on to the trucks, had higher heart rates during loading and unloading (signs of stress), showed more fearful behaviour and had more traumatic accidents (e.g. falling down or hitting structures) at lairage, and had poorer meat quality. A similar relationship between farmers' behaviour and meat quality of older beef bulls (Mounier *et al.*, 2008) and of pigs (Hemsworth *et al.*, 2002a) has also been reported.

Together, these studies present convincing evidence for the major farm animal species, of a strong relationship between the methods used to handle the animals, the degree of fear shown by the animals towards people, and the level of productivity of the farm.

Effects of Aversive Handling on Production

The studies discussed above show that low levels of productivity (e.g. poor egg production, low growth rates) tend to occur on farms where the animals are handled roughly and frightened of people. However, they do not show that the rough handling was actually the cause of the fear or the reduced productivity. To establish this, researchers have experimentally altered the way that farm animals are handled and have looked at the changes in productivity that have occurred. In one of the first such studies, Hemsworth *et al.* (1981) subjected young female pigs (gilts) to either a pleasant (gentle stroking) or an unpleasant handling treatment (briefly shocked with an electric prod) for a relatively short period of time, namely three times per week for 2 min in duration from 11 to 22 weeks of age. Gilts in the unpleasant handling treatment spent less time close to the handler, showing that they were more fearful. In addition, they had a lower growth rate. A second study (Hemsworth *et al.*, 1986) showed that gilts in the unpleasant treatment had a lower pregnancy rate at the second oestrus when mated than gilts in the pleasant treatment (33.3 and 87.5%, respectively) and that boars handled in the same unpleasant way had smaller testicles at 23 weeks of age and attained a coordinated mating response at a later age than boars in the pleasant treatment (192 and 161 days, respectively). The effect of aversive handling on growth rates of pigs was confirmed by Gonyou *et al.* (1986) who noted that aversively handled pigs (i.e. those given electric shocks if they did not avoid the handler) grew during the first 3 weeks of life at only 85% of the rate of pigs that had little contact with people.

These studies show convincingly that aversive or rough handling of pigs can have effects on growth rates, reproduction and sexual development, and that the magnitude of the effects can be substantial. They support the suggestion that the relationship between animals' fear of people and low productivity, which was noted in the previous section, may be due to the way that the animals are handled by the stockpeople (see Chapter 5 for good handling methods).

Why Does Poor Stockmanship Increase Fear in Animals and Reduce Productivity?

How could poor stockmanship lead to such large reductions in productivity of farm animals? To explain these effects, Hemsworth and Coleman (1998) proposed a simple model of the relationship between the stockperson's beliefs and attitudes, their behaviour when handling animals and the impact this has on the animals, which is illustrated in Fig. 4.3. The stockperson's specific beliefs about animals have a direct influence on how the animals are handled. For example, a belief that pigs are not sensitive and are difficult to move can lead the stockperson to use rough or aversive handling. The importance of the stockperson's beliefs about animals will be discussed later. As a result of rough handling, the animals learn to associate the aversive handling with the stockperson and become frightened of him or her. This, in turn, results in the physiological changes that typically occur when an animal is stressed, which have deleterious effects on the animal's welfare and productivity. This model is a useful way of conceptualizing the sequence of events leading from the stockperson's beliefs and attitudes.

Direct evidence that aversive handling causes animals to become more fearful of people, and that this is responsible for the effects on production, comes from experimental studies on dairy cows (Munksgaard *et al.*, 1997; Rushen *et al.*, 1999a). These experiments examined whether the same cows could be made fearful of one person but not of another. In these studies, dairy cows were handled repeatedly by two people, one of whom always handled the cows gently (talking softly, patting and stroking the cow and occasionally giving feed rewards) while the other handled the cows aversively (hitting, shouting and occasional use of a cattle prod). The cows' fear of each person was tested by each person standing in front of the cow's stall and measuring how closely the cow approached. The results showed that after repeated handling the cows stood further from the aversive handler than from the gentle handler (Munksgaard *et al.*, 1997). Rushen *et al.* (1999a) also tested whether the degree of fearfulness elicited by the handling was sufficient to reduce milk yield, by milking cows in the presence of the handlers. The gentle handler stood close to the cows for one milking and the aversive handler stood close to the cows for the other milking. It is important to note that the handler did not touch or interact with the cow during the milking. Just the presence of the aversive handler during milking was sufficient to increase residual milk by 70% (which is a sign of a stress-

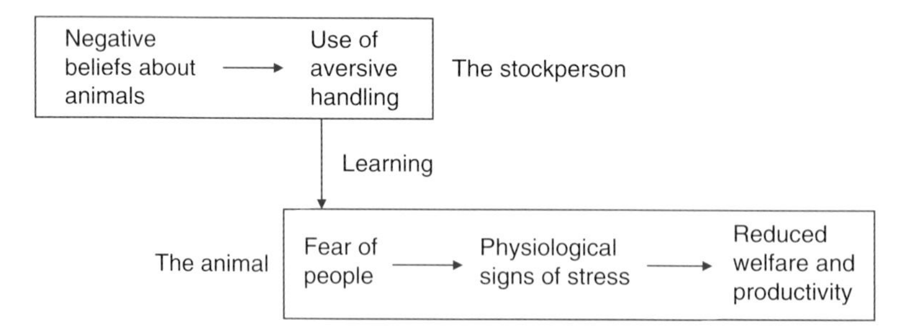

Fig. 4.3. A simple model describing the causal link between a stockperson's beliefs about animals and the effect on their welfare and productivity. According to this model, a stockperson's beliefs about animals (e.g. how easy they are to move) will influence how he/she handles the animal. Stockpeople who believe that animals are stupid, for example, are probably more likely to use rougher handling. The animal will learn to associate this rough handling with the person and thus become frightened of the person (or people in general). This fear of people will produce physiological changes associated with stress, which will reduce the animal's productivity and welfare (adapted from Hemsworth and Coleman, 1998).

induced block of milk ejection) and tended to reduce milk yield. There were also some physiological signs of fear: the presence of the aversive handler increased heart rate during milking (Table 4.1). These small-scale studies provide support for the model presented in Fig. 4.3 by showing that aversive handling will make dairy cows frightened of people, and that this fear may at times be sufficient to reduce milk yield.

An obvious question is whether an animal that is handled roughly by one person will become fearful of all people or whether they will become fearful only of that particular individual. It seems that both results are possible depending on how well the animals can recognize individual people. A number of experiments showed that farm animals respond the same way to different people. For example, Hemsworth *et al.* (1994b) found that pigs do not differentiate between different people, and that they do not behave differently with familiar and unfamiliar people. In such cases, rough handling by one person may make animals frightened of all people.

However, in some situations, farm animals can learn to recognize individual people and become frightened of particular people (reviewed in Rushen *et al.*, 2001). The studies mentioned above with dairy cows provide an example. Other examples of animals recognizing individual people come from pigs (Tanida and Nagano, 1998; Koba and Tanida, 2001) and sheep (Boivin *et al.*, 1997). What cues might the animals be using to recognize individual people? Visual cues, especially those associated with the clothing worn, seem important in recognizing people in pigs, sheep and cattle (Rushen *et al.*, 2001). Rybarczyk *et al.* (2003) showed that even very young dairy calves can distinguish between two people wearing different colours of clothing, learning to choose to approach the person who gave them milk. However, Taylor and Davis (1998) showed that cattle in fact could learn to distinguish people who wore the same colour clothes. More recently, Rybarczyk *et al.* (2001) presented evidence that at least some adult cattle can recognize people by their faces (Fig. 4.4).

Clearly farm animals show a good ability to recognize individual people using quite subtle recognition cues. If different people treat the animals in different ways then the animals will respond differently to each person.

Causes of Poor Stockmanship and Ways to Improve It

Identifying which types of handling are aversive or positive

Because it is clear that the type of handling used by the stockperson has major effects upon the fearfulness of the animals, a necessary first step in improving the relationship between animals and the stockperson is to identify the particular behaviours that the animals find aversive or rewarding.

Recently, researchers have tried to discover exactly which handling practices cattle find aversive and which are most likely to lead them to become frightened of people. Pajor *et al.* (2000) compared treatments that are often used when moving cows, such as hitting with the hand, shouting, tail twisting (but not strongly enough to break the tail!) and use of an electric prod. Cattle were placed in a runway and at the end of this they were restrained and handled. The experimenters measured the speed that the animals moved down the runway as well as the effort required by the handler to move them. Based upon these measures, all treatments appeared to be aversive to some

Table 4.1. Effects of the presence of different types of handler on milk yield and behaviour during milking (source: Rushen *et al.*, 1999a).

	Aversive handler	Gentle handler
Milk yield (kg)	18.48	19.2
Residual milk (kg)[a]	3.6	2.1
Kicks by cow/min during udder preparation	0	0.93
Heart rate change during milking (bpm)[b]	5.94	3.42

[a] Residual milk is the amount of milk a cow holds back.
[b] An increase in heart rate is a sign of stress; bpm, beats per minute.

Fig. 4.4. Research has now shown that most species of farm animals can recognize individual people, using predominantly visual cues. Some farm animals can use quite subtle cues to recognize people. Adult cows can quickly learn to approach one person, rather than another, if they receive a food reward only when they make the correct choice. Once the cows have learned to recognize the people, the cues that they use can be examined. For example, Rybarczyk *et al.* (2001) showed that when the people's faces were covered up and they were of the same height, the cows could no longer recognize them. This shows that some cows can recognize people by their faces.

extent. However, hitting with the hand and gently twisting the tail did not differ significantly from no handling, suggesting that the cattle perceived these treatments as relatively mild, although this conclusion probably depends upon the force used. Unsurprisingly, the use of the cattle prod was aversive, but interestingly shouting appeared to be as aversive as the cattle prod. A subsequent study (Pajor *et al.*, 2003), which allowed cattle to choose between pairs of treatments, confirmed these results: cattle showed no preference for shouting over the electric prod, and no preference for no treatment over tail twisting. Such methods will help us get a better understanding of which handling techniques are the most aversive to animals and will make them most afraid of people.

More recently there has been interest in identifying positive behaviours that might improve human–animal relationships. For example, Schmied *et al.* (2008) found that stroking the neck of cows reduced the degree that the cows avoided people and resulted in the cows being more likely to approach people.

Stockpeople's attitudes and beliefs

Grandin (2003) has noted that changing animal handling practices on farm can be surprisingly difficult. She has noted that people are often more willing to purchase new, expensive equipment than they are to make low-cost changes in their handling methods. This may reflect the fact that the way stockpeople handle animals is a result of long-held, deep-seated beliefs about how animals need to be handled and personality attributes in general (Hemsworth, 2003). Simply making recommendations about better ways to handle animals may not be sufficient to overcome the influence of these beliefs. A considerable amount of research has now shown that the way that stockpeople handle animals is a reflection of specific beliefs of the stockperson (Fig. 4.3), and that altering these beliefs may be an effective means of improving the way animals are handled (Hemsworth and Coleman, 1998; Hemsworth, 2003).

Coleman *et al.* (2003) examined the attitudes about pigs of handlers at an abattoir and observed how these handlers moved the pigs. Handlers that believed that pigs were greedy, aggressive and gluttons were more likely to use an electric prod when moving the animals. The use of the prod was also associated with the belief that the way that the pigs were handled had no effect upon their behaviour.

Can such beliefs be changed? Hemsworth *et al.* (2002b) examined the effect of a 'cognitive-behavioural intervention' on dairy farmers' attitudes towards cows. The intervention involved multimedia presentations emphasizing the research results, which showed the negative effects of poor handling on the fearfulness and productivity of cattle, along with some clear examples of good and poor handling techniques. The intervention clearly improved the attitude of the farmers towards dairy cattle, specifically reducing the belief that considerable force was needed to move dairy cows. Visits to the farms showed that these changes in beliefs resulted in a reduced use of aversive handling techniques, reduced fearfulness and tended to improve milk yield (Table 4.2). This study clearly shows the potential for such interventions to improve at least one component of stockmanship, and to improve both the welfare and the productivity of the cattle.

A similar study on pig farms found similar improvements in the way pigs were handled when stockmen followed a training course that challenged their attitudes and behaviour towards pigs (Coleman *et al.*, 2000). Interestingly, an unsuspected additional positive effect was found: after the completion of the study, the retention rate for employees who had participated in the training programme was higher (61%) compared to those who had not participated (47%). The results suggest that stockpeople who have been trained to better handle animals may be more likely to remain in the job, perhaps because of increased job satisfaction. The results of Maller *et al.* (2005), who found that dairy stockpeople who had positive beliefs about the ease of moving cows also tended to be more positive about working in a dairy, reinforce the link between positive attitudes towards animals and the quality of life of stockpeople. Improving stockmanship in this way can thus improve the welfare of both animals and farmers.

As well as depending on specific attitudes and beliefs about the importance of routine care, the quality of stockmanship appears to be related to general personality attributes. Seabrook (1984) used a questionnaire to examine correlations between aspects of the personality of stockpeople on a dairy farm and the level of milk production on the farms where they were working. Stockpeople who worked on dairy farms with a high level of milk production were found to be: 'not easy going, considerate, not meek, patient, unsociable, not modest, independently minded, persevering, not talkative, confident, uncooperative, and suspicious of change'! These results suggest that the personality of the stockperson can have an effect on milk production. More recently Seabrook (2005) has investigated the important personality characteristics of good pig stockpeople as rated by other stockpeople. Most considered that a good stockperson should be 'fussy', 'confident' and 'inward-looking' and not 'harsh', 'carefree', 'unsympathetic' or 'aggressive'. Coleman *et al.* (2003) found that

Table 4.2. Effect of handling on cow behaviour and milk yield (source: adapted from Table 2 and Table 3 in Hemsworth *et al.*, 2002b).[a]

	Control	Intervention to change stockpeople's attitudes and beliefs about cows
Stockperson's behaviour		
Frequency of mildly aversive handling techniques/cow/milking	0.43	0.24
Frequency of strongly aversive handling techniques/cow/milking	0.02	0.005
Frequency of gentle handling techniques/cow/milking	0.045	0.11
Cows' responses		
Flight distance (m)	4.49	4.16
Flinch, step, kick responses during milking	0.1	0.13
Milk yield (l/cow/month)	509	529

[a]The tests were done twice on 99 different farms.

handlers at an abattoir who were rated as 'tough-minded' were more likely to use an electric prod when moving pigs. Such research may have practical consequences in improving employee selection.

However, we should not focus exclusively on the stockperson's beliefs about the animals. As mentioned in the section 'Effect of Overall Stockmanship on Animal Welfare and Productivity', stockpeople's attitudes to routine tasks, such as cleanliness, can also have a major impact on animal welfare independently of how the animals are handled. Figure 4.5 shows a more complete model of the relationship between attitudes, beliefs and stockmanship.

Identifying why stockpeople mishandle animals

To improve the ways animals are handled, it helps to know what circumstances lead to animals being handled roughly. Seabrook (2000) reports on a study he undertook to try to identify why dairy farmers were aggressive towards animals. Stockpeople were asked to reflect on the times they had acted aversively towards the animals and to try to explain why they had acted that way. A number of reasons repeatedly came up. These included time pressures and the complexity of the work, frustration with the animals being 'stubborn' (e.g. kicking off the milk clusters) or with equipment not working, and family and home problems. The study further highlights the importance of general job satisfaction for good stockmanship.

Difficulties in moving animals can be a cause of frustration for stockpeople and this can be responsible for much of the rough handling that occurs. The physical design of the work environment plays an important role. Maller *et al.* (2005) noted that several aspects of the design of milking sheds (e.g. number of turns the animals had to make to exit, stall gates, etc.) were reported to be associated with difficulties in cow movement, which in turn may have led to a greater need for the dairymen to intervene (often roughly) to help move cows. Similarly, Rushen and de Passillé (2006) reported that slippery flooring, as well as slowing cows' walking speed and increasing the risk of slipping, resulted in more need for handlers to try to encourage the cows to move forward, which increased the risk of rough handling. Difficulties with moving animals can be a particular problem in slaughterhouses, and this can be a major cause of improper stunning. Grandin (2006) has outlined some of the improvements in

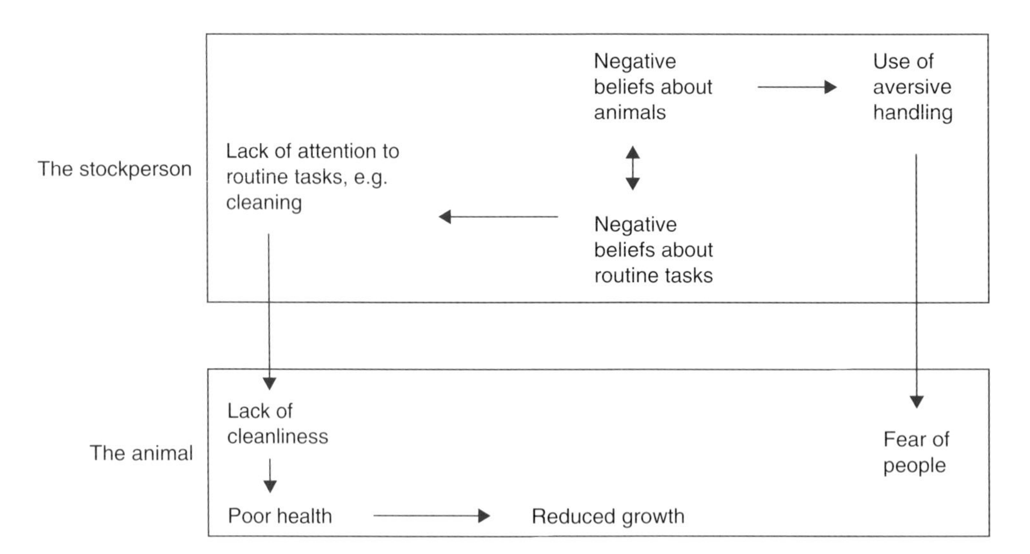

Fig. 4.5. A more complete model of the relationship between the stockperson's attitudes and beliefs and the impact on farm animals. This model is based on studies by Lensink *et al.* (2002) on veal calves. In this case, the largest effect upon the calf's growth occurs through a lack of attention to cleaning routines rather than through the animal becoming frightened of people. The calves' fear of people is a correlated response that could arise because stockpeople who have negative beliefs about the importance of routine care may also have negative beliefs about the animals, and also use aversive handling techniques (adapted from Rushen *et al.*, 2008).

installations in slaughterhouses than can improve animal movement. These include installing non-slip flooring and eliminating distractions to animals that may cause them to baulk. Good handling facilities are also particularly important for handling extensively managed animals such as beef cattle, which are more likely to be fearful of people. Grandin (1997) has produced useful guidelines for such facilities (see Chapters 5, 7 and 14). Clearly, it is essential to have well-designed areas for handling and moving animals in order to reduce both the frustration of the workers and the tendency for such workers to take out their frustrations on the animals. Finding better ways to move animals is likely to lead to significant reductions in the use of aversive handling. This could be achieved either through improved raceway design or by finding other means of improving animal movement; Ceballos and Weary (2002) found that providing small food rewards when dairy cows entered the milking parlour reduced the time the cows took to enter the parlour, and reduced the need for the stockperson to push the cows or use other aversive handling techniques.

From other professions, it is recognized that the diligence with which a job is done depends very much on the level of job satisfaction; low job satisfaction often results in careless work. Recognizing the importance of this for stockmanship, Seabrook and Wilkinson (2000) interviewed dairy stockpeople in the UK to determine what factors affected their level of job satisfaction, focusing particularly upon the routine tasks stockpeople enjoyed and those they found most unpleasant. Importantly, the stockpeople clearly valued and enjoyed their interactions with their animals, and the nature of their interactions with animals was largely responsible for the difference between a 'good day' and a 'bad day'. Milking was widely rated the most important routine task in dairying, and was rated the most enjoyable. This may, however, not be true for all countries: Maller *et al.* (2005) reported that only about a quarter of Australian dairy farmers preferred milking to other farm jobs. Possibly, this difference may relate to the differences in the design of milking sheds in the two countries.

Unfortunately, Seabrook and Wilkinson (2000) found that maintaining herd health and welfare was considered of relatively low importance, especially by younger stockpeople. Stockpeople from higher-producing herds, however, rated herd health and welfare more important than those from lower-producing herds. Interestingly, the rated importance of herd health and welfare increased when stockpeople were re-interviewed after quality assurance schemes had been introduced. The authors felt that these may have led the stockpeople to alter their priorities. Despite their importance for maintaining good animal health and welfare, cleaning buildings/parlours and foot trimming were rated the most unpleasant routine tasks. Foot trimming (Fig. 4.6) was disliked mainly because of inadequate equipment and holding facilities, and because it was generally considered a dangerous task. The authors concluded that the obvious dislike of foot trimming and cleaning routines shows the need to find improved means of cleaning cows and barns, for better designed parlours that facilitate cleaning, and for improved equipment and facilities for foot trimming. This type of research is clearly useful in showing what types of improvements are needed in order to improve job satisfaction and hence job quality.

Assessing Stockmanship on Farm Audits

Hemsworth (2007) has recently noted the need to establish appropriate standards for stockmanship, given the increasing importance of animal welfare for the public image of agriculture. Given the importance of stockmanship for the welfare of farm animals, there has been a lot of interest in including some measure of stockmanship in on-farm animal welfare assessment or audits. Usually this involves taking some measure of how fearful the animals appear in response to people. For example, Rousing and Waiblinger (2004) visited commercial dairy farms and conducted two tests: a voluntary animal-approach test based on cows' latencies to approach and touch a stationary person; and an avoidance test, based on cows' avoidance of an approaching person. The inter-observer repeatability of both tests was found to be high and the tests were highly correlated. The avoidance test was less affected by the familiarity of the people and the authors conclude that the avoidance test is valid and useful in on-farm assessment of the human–animal relationship. Other studies have shown that the distance cattle keep from people can be measured reliably on farm (Windschnurer *et al.*, 2008) and similar tests are being developed for poultry (e.g. Graml *et al.*, 2007). However, we have been critical of the use of such tests to assess stockmanship in on-farm audits (de Passillé and

Fig. 4.6. A poor facility for hoof trimming of dairy cows. Performance of routine tasks, such as routine hoof trimming of dairy cows, is an essential aspect of good stockmanship. The degree of diligence and care with which such tasks are performed can have a major effect on the welfare of the animal. For example, inadequate hoof trimming of dairy cows is known to increase the chance of cows becoming lame. However, many dairy farmers report that hoof trimming is one of the least enjoyable tasks on a dairy farm (Seabrook and Wilkinson, 2000). As a consequence, there is a danger that it will not be done as often as necessary. This situation may be improved by ensuring that farmers have the right equipment and training to do such tasks. There are many well-designed units for holding the cow that are commercially available. Replacing the dangerous set-up with good equipment will make hoof trimming easier and safer.

Rushen, 2005), not because the tests are unreliable, but because they have not been adequately validated as measures of stockmanship on farms, which is a more serious problem. First, the extent to which animals approach or avoid people can be affected by many factors other than the use of aversive handling (Boissy and Bouissou, 1988). For example, feeding animals by hand is one certain way of getting them to approach people. Animals that are not fed by hand may not as readily approach people but that does not mean that their welfare is necessarily poorer. Secondly, even if there are differences between farms in the extent that animals approach people, it is difficult to establish a clear cut-off point where we can say that the handling of the animals has been unacceptably rough. This is partly because many differences between farms will affect the precise distance that animals approach people, so it will be difficult to make precise comparisons between different farms. Thirdly, and most important, as we have discussed in this chapter, there is a lot more to good stockmanship than the way that the animals are handled (Fig. 4.5). The degree to which stockpeople perform important routine tasks, or their attitudes to animals and to such tasks may be more important to measure in on-farm audits. In other situations besides on the farm, other aspects of animal handling may be more important (and more easily measured) in an animal welfare audit than the degree of fear that the animals show towards people. For example, noting the use of electric prods to move animals and incorrect stunning procedures should be an essential part of animal welfare audits of slaughterhouses (Grandin, 1998). Thus while assessment of the quality of stockmanship is likely to be an important component of on-farm animal welfare audits, measures of the reaction of animals to people may not be the most appropriate measure as a welfare measure *between* different farms, but it would be appropriate to use it to assess improvements in stockmanshsip *within* the same farm.

Conclusions

From the research reviewed in this chapter it is clear that the people who care for the animals can have a major, and possibly a decisive, role to play in affecting their welfare and the productivity. Differences between stockpeople may be responsible for many of the differences between otherwise similar farms in the level of animal welfare and productivity. Research is beginning to provide concrete examples of how poor stockmanship can lead to poor animal welfare, and to show some of the ways that stockmanship can be improved. Notably these improvements, as well as improving animal welfare, often lead to substantial benefits to the farmers themselves, either through increased health and productivity of the animals, through increased efficiency and safety of operations that involve handling the animals, or through increased job satisfaction and probably self-esteem. Research to date has focused upon the more obvious aspects of stockmanship: how the animals are handled and how they become fearful of people. However, stockmanship involves much more, and the relationships that can develop between people and animals can be quite subtle. Research needs to consider a broader range of qualities associated with stockmanship. Seabrook (2000: 30) concludes:

> The welfare of farm animals is crucially dependent upon the actions of the stockpersons who day in day out handle, observe and monitor the animals in their charge. The animals' well-being and performance depend as much o[n] the stockperson as the building and the space allocation. Aversive action can arise through ignorance, but usually staff know what to do to handle animals correctly. They know what to do for good welfare but sometimes they do not act correctly and in their words '*let ourselves and our animals down*'.

Acknowledgements

We thank all of our colleagues and collaborators for useful discussions on this topic, in particular: Paul Hemsworth, Lene Munksgaard and Hajime Tanida. Our research in this area was supported by grants from the Natural Sciences and Engineering Research Council of Canada, Dairy Farmers of Canada, Novalait Inc. and Agriculture and Agri-Food Canada.

References

Boissy, A. and Bouissou, M.F. (1988) Effects of early handling on heifers' subsequent reactivity to humans and to unfamiliar situations. *Applied Animal Behaviour Science* 20, 259–273.

Boivin, X., Nowak, R., Desprès, G., Tournadre, H. and Le Neindre, P. (1997) Discrimination between shepherds by lambs reared under artificial conditions. *Journal of Animal Science* 75, 2892–2898.

Breuer, K., Hemsworth, P.H., Barnett, J.L., Matthews, L.R. and Coleman, G.J. (2000) Behavioural response to humans and the productivity of commercial dairy cows. *Applied Animal Behaviour Science* 66, 273–288.

Ceballos, A. and Weary, D.M. (2002) Feeding small quantities of grain in the parlour facilitates pre-milking handling of dairy cows: a note. *Applied Animal Behaviour Science* 77, 249–254.

Coleman, G.J., Hemsworth, P.H., Hay, M. and Cox, M. (2000) Modifying stockperson attitudes and behaviour towards pigs at a large commercial farm. *Applied Animal Behaviour Science* 66, 11–20.

Coleman, G.J., McGregor, M., Hemsworth, P.H., Boyce, J. and Dowling, S. (2003) The relationship between beliefs, attitudes and observed behaviours of abattoir personnel in the pig industry. *Applied Animal Behaviour Science* 82, 189–200.

Cransberg, P.H., Hemsworth, P.H. and Coleman, G.J. (2000) Human factors affecting the behaviour and productivity of commercial broiler chickens. *British Poultry Science* 41, 272–279.

de Passillé, A.M. and Rushen, J. (2005) Can we measure human-animal interactions in on-farm animal welfare assessment? Some unresolved issues. *Applied Animal Behaviour Science* 92, 193–209.

Fulwider, W.K., Grandin, T., Rollin, B.E., Engle, T.E., Dalsted, N.L. and Lamm, W.D. (2008) Survey of dairy management practices on one hundred thirteen north central and northeastern United States dairies. *Journal of Dairy Science* 91, 1686–1692.

Gonyou, H.W., Hemsworth, P.H. and Barnett, J.L. (1986) Effects of frequent interactions with humans on growing pigs. *Applied Animal Behaviour Science* 16, 269–278.

Graml, C., Niebuhr, K. and Waiblinger, S. (2007) Reaction of laying hens to humans in the home or novel environment. *Applied Animal Behaviour Science* 113, 98–109.
Grandin, T. (1997) The design and construction of facilities for handling cattle. *Livestock Production Science* 49, 103–119.
Grandin, T. (1998) Objective scoring of animal handling and stunning practices at slaughter plants. *Journal of the American Veterinary Medical Association* 212, 36–39.
Grandin, T. (2003) Transferring results of behavioral research to industry to improve animal welfare on the farm, ranch and the slaughter plant. *Applied Animal Behaviour Science* 81, 215–228.
Grandin, T. (2006) Progress and challenges in animal handling and slaughter in the US. *Applied Animal Behaviour Science* 100, 129–139.
Hemsworth, P.H. (2003) Human-animal interactions in livestock production. *Applied Animal Behaviour Science* 81, 185–198.
Hemsworth, P.H. (2007) Ethical stockmanship. *Australian Veterinary Journal* 85, 194–200.
Hemsworth, P.H. and Barnett, J.L. (1992) The effects of early contact with humans on the subsequent level of fear of humans in pigs. *Applied Animal Behaviour Science* 35, 83–90.
Hemsworth, P.H. and Coleman, G.J. (1998) *Human–Livestock Interactions.* CAB International, Wallingford, UK.
Hemsworth, P.H., Brand, A. and Willems, P. (1981) The behavioural response of sows to the presence of human beings and its relation to productivity. *Livestock Production Science* 8, 67–74.
Hemsworth, P.H., Barnett, J.L. and Hansen, C. (1986) The influence of handling by humans on the behaviour, reproduction and corticosteroids of male and female pigs. *Applied Animal Behaviour Science* 15, 303–314.
Hemsworth, P.H., Coleman, G.J., Barnett, J.L. and Jones, R.B. (1994a) Behavioural responses to humans and the productivity of commercial broiler chickens. *Applied Animal Behaviour Science* 41, 101–114.
Hemsworth, P.H., Coleman, G.J., Cox, M. and Barnett, J.L. (1994b) Stimulus generalization: the inability of pigs to discriminate between humans on the basis of their previous handling experience. *Applied Animal Behaviour Science* 40, 129–142.
Hemsworth, P.H., Pedersen, V., Cox, M., Cronin, G.M. and Coleman, G.J. (1999) A note on the relationship between the behavioural response of lactating sows to humans and the survival of their piglets. *Applied Animal Behaviour Science* 65, 43–52.
Hemsworth, P.H., Coleman, G.J., Barnett, J.L. and Borg, S. (2000) Relationships between human–animal interactions and productivity of commercial dairy cows. *Journal of Animal Science* 78, 2821–2831.
Hemsworth, P.H., Barnett, J.L., Hofmeyr, C., Coleman, G.J., Dowling, S. and Boyce, J. (2002a) The effects of fear of humans and pre-slaughter handling on the meat quality of pigs. *Australian Journal of Agricultural Research* 53, 493–501.
Hemsworth, P.H., Coleman, G.J., Barnett, J.L., Borg, S. and Dowling, S. (2002b) The effects of cognitive behavioral intervention on the attitude and behavior of stockpersons and the behavior and productivity of commercial dairy cows. *Journal of Animal Science* 80, 68–78.
Koba, Y. and Tanida, H. (2001) How do miniature pigs discriminate between people? Discrimination between people wearing coveralls of the same colour. *Applied Animal Behaviour Science* 73, 45–58.
Lensink, B.J., Veissier, I. and Florand, L. (2001a) The farmers' influence on calves' behaviour, health and production of a veal unit. *Animal Science* 72, 105–116.
Lensink, B.J., Fernandez, X., Cozzi, G., Florand, L. and Veissier, I. (2001b) The influence of farmers' behavior on calves' reactions to transport and quality of veal meat. *Journal of Animal Science* 79, 642–652.
Lensink, B.J., Boivin, X., Pradel, P., LeNeindre, P. and Vessier, L. (2002) Reducing veal calves reactivity to people by providing additional human contact. *Journal of Animal Science* 78, 1213–1218.
Losinger, W.C. and Heinrichs, A.J. (1997) Management practices associated with high mortality among preweaned dairy heifers. *Journal of Dairy Research* 64, 1–11.
Maller, C.J., Hemsworth, P.H., Ng, K.T., Jongman, E.J., Coleman, G.J. and Arnold, N.A. (2005) The relationships between characteristics of milking sheds and the attitudes to dairy cows, working conditions, and quality of life of dairy farmers. *Australian Journal of Agricultural Research* 56, 363–372.
Mounier, L., Colson, S., Roux, M., Dubroeucq, H., Boissy, A. and Veissier, I. (2008) Positive attitudes of farmers and pen-group conservation reduce adverse reactions of bulls during transfer for slaughter. *Animal* 2, 894–901.
Munksgaard, L., de Passillé, A.M.B., Rushen, J., Thodberg, K. and Jensen, M.B. (1997) Discrimination of people by dairy cows based on handling. *Journal of Dairy Science* 80, 1106–1112.
Pajor, E.A., Rushen, J. and de Passillé, A.M.B. (2000) Aversion learning techniques to evaluate dairy cattle handling practices. *Applied Animal Behaviour Science* 69, 89–102.
Pajor, E.A., Rushen, J. and de Passillé, A.M.B. (2003) Dairy cattle's choice of handling treatments in a Y-maze. *Applied Animal Behaviour Science* 80, 93–107.
Rousing, T. and Waiblinger, S. (2004) Evaluation of on-farm methods for testing the human–animal relationship in dairy herds with cubicle loose housing

systems – test-retest and inter-observer reliability and consistency to familiarity of test person. *Applied Animal Behaviour Science* 85, 215–231.

Rushen, J. and de Passillé, A.M. (2006) Effects of roughness and compressibility of flooring on cow locomotion. *Journal of Dairy Science* 89, 2965–2972.

Rushen, J., de Passillé, A.M.B. and Munksgaard, L. (1999a) Fear of people by cows and effects on milk yield, behavior and heart rate at milking. *Journal of Dairy Science* 82, 720–727.

Rushen, J., Taylor, A.A. and de Passillé, A.M.B. (1999b) Domestic animals' fear of humans and its effect on their welfare. *Applied Animal Behaviour Science* 65, 285–303.

Rushen, J., de Passillé, A.M.B., Munksgaard, L. and Tanida, H. (2001) People as social actors in the world of farm animals. In: Gonyou, H. and Keeling, L. (eds) *Social Behaviour of Farm Animals*. CAB International, Wallingford, UK.

Rushen, J., de Passillé, A.M., von Keyserlingk, M. and Weary, D.M. (2008) *The Welfare of Cattle*. Springer, Dordrecht, The Netherlands, 303 pp.

Rybarczyk, P., Koba, Y., Rushen, J., Tanida, H. and de Passillé, A.M. (2001) Can cows discriminate people by their faces? *Applied Animal Behaviour Science* 74, 175–189.

Rybarczyk, P., Rushen, J. and de Passillé, A.M. (2003) Recognition of people by dairy calves using colour of clothing. *Applied Animal Behaviour Science* 81, 307–319.

Schmied, C., Boivin, X. and Waiblinger, S. (2008) Stroking different body regions of dairy cows: effects on avoidance and approach behavior toward humans. *Journal of Dairy Science* 91, 596–605.

Seabrook, M.F. (1984) The psychological interaction between the stockman and his animals and its influence on performance of pigs and dairy cows. *Veterinary Record* 115, 84–87.

Seabrook, M.F. (2000) The effect of the operational environment and operating protocols on the attitudes and behaviour of employed stockpersons. In: Hovi, M. and Bouilhol, M. (eds) *Human–Animal Relationship: Stockmanship and Housing in Organic Livestock Systems*. Proceedings of the Third Network for Animal Health and Welfare in Organic Agriculture (NAHWOA) Workshop, Clermont-Ferrand, France, 21–24 October 2000. University of Reading, Reading, UK, pp. 21–30. Available at: http://www.veeru.reading.ac.uk/organic/ProceedingsFINAL.pdf (accessed 23 June 2009).

Seabrook, M.F. (2005) Stockpersonship in the 21st century. *Journal of the Royal Agricultural Society of England* 166, 1–12.

Seabrook, M.F. and Wilkinson, J.M. (2000) Stockpersons' attitudes to the husbandry of dairy cows. *Veterinary Record* 147, 157–160.

Tanida, H. and Nagano, Y. (1998) The ability of miniature pigs to discriminate between a stranger and their familiar handler. *Applied Animal Behaviour Science* 56, 149–159.

Taylor, A. and Davis, H. (1998) Individual humans as discriminative stimuli for cattle (*Bos taurus*). *Applied Animal Behaviour Science* 58, 13–21.

Waiblinger, S., Menke, C. and Coleman, G. (2002) The relationship between attitudes, personal characteristics and behaviour of stockpeople and subsequent behaviour and production of dairy cows. *Applied Animal Behaviour Science* 79, 195–219.

Waiblinger, S., Menke, C., Korff, J. and Bucher, A. (2004) Previous handling and gentle interactions affect behaviour and heart rate of dairy cows during a veterinary procedure. *Applied Animal Behaviour Science* 85, 31–42.

Waiblinger, S., Boivin, X., Pedersen, V., Tosi, M.V., Janczak, A.M., Visser, E.K. and Jones, R.B. (2006) Assessing the human–animal relationship in farmed species: a critical review. *Applied Animal Behaviour Science* 101, 185–242.

Windschnurer, I., Schmied, C., Boivin, X. and Waiblinger, S. (2008) Reliability and inter-test relationship of tests for on-farm assessment of dairy cows' relationship to humans. *Applied Animal Behaviour Science* 114, 37–53.

5 How to Improve Livestock Handling and Reduce Stress

TEMPLE GRANDIN

Colorado State University, Fort Collins, Colorado, USA

Handling animals quietly during milking, vaccinating and truck loading has many benefits. Calm animals are easier to handle, sort and restrain than frightened, excited animals. If cattle become severely agitated, it takes 20–30 min for them to calm back down and have their heart rate return to normal (Stermer *et al.*, 1981). When they are brought into the corrals from pasture, it is often best to let them calm down and settle before working them in the race for veterinary procedures. If a horse becomes severely agitated during a veterinary procedure, it may be easier to handle if it is given 30 min to calm down before the procedure is attempted again. Research has shown that yelling and whistling at cattle is really stressful (Waynert *et al.*, 1999; Pajor *et al.*, 2003). People handling animals should be mostly silent except for a little *shh shh* sound or soft, soothing speech.

Benefits of Good Handling

Livestock that remain calm while they are restrained and walk calmly during handling will have better weight gain and better meat quality compared to animals that become agitated during handling. Cattle that struggle when they are restrained or run wildly out of a race after handling had lower weight gains, higher cortisol, more dark cutters and tougher meat (Voisinet *et al.*, 1997a, b; Petherick *et al.*, 2002; Curley *et al.*, 2006; King *et al.*, 2006). Dark cutting beef is a serious quality defect that makes meat darker and drier than normal. Cattle that were prodded six times with a battery-operated electric goad within 15 min of slaughter produced beef that was rated as tougher by a taste panel (Warner *et al.*, 2007; Ferguson and Warner, 2008). Work by Australian researcher Paul Hemsworth with pigs and dairy cows showed that negative handling reduced milk production in dairy cows and reduced weight gain in pigs (Barnett *et al.*, 1992; Hemsworth *et al.*, 2000) (see Chapter 4). Handling animals quietly will also help to reduce sickness and improve reproductive performance. Stress lowers immune function (Mertshing and Kelly, 1983). Stressful handling practices shortly after breeding will also lower conception rates in sheep, cattle and pigs (Doney *et al.*, 1976; Hixon *et al.*, 1981; Fulkerson and Jamieson, 1982). Sows that were afraid of people and backed away from a person had fewer piglets (Hemsworth, 1981). Stress from poor handling practices can also make animals more susceptible to disease. Immune function, which helps animals fight disease, was reduced. Restraint stress reduced immune responses in pigs (Mertshing and Kelly, 1983). Multiple shocks with an electric goad is very detrimental to pigs. Pigs that were shocked multiple times were more likely to become non-ambulatory and not able to walk (Benjamin *et al.*, 2001). McGlone (personal communication, 2005) reported that the occurrence of non-ambulatory pigs was four times higher when an electric goad was used on over 50% of the pigs compared to 10%. The observations were made in a large commercial slaughter plant that processed large, 115-kg pigs. Electric goads also cause large increases in lactic acid and glucose in the pig's blood (Benjamin *et al.*, 2001; Ritter *et al.*, 2009). Edwards *et al.* found that the use of electric goads in the stunning race raised lactate levels and lowered the pH of the meat (L.N. Edwards, T. Grandin, T.E. Engle, M.J. Ritter, A. Sosnicke and

D. Anderson, 2009, unpublished results). Low pH meat is more likely to be pale soft exudative (PSE) meat. This is a severe quality defect.

Downed animals should never be dragged. Pigs can become non-ambulatory due to either the porcine stress syndrome (PSS) (a genetic condition) or because they have been fed too high a dose of beta-agonists such as ractopamine (see Chapters 1 and 7). Careful, quiet handling will reduce the number of non-ambulatory animals.

Another advantage of low-stress, quiet cattle handling is improved safety for both people and animals. Both animals and people will have fewer injuries. A 10-year analysis of workman compensation claims indicated that claims for injuries associated with cattle handling represented the highest percentage of costly severe injuries (Douphrate *et al.*, 2009). Cattle that are handled quietly during loading into a truck will also have half as many bruises (Grandin, 1981). Improvements in cattle handling in Brazil reduced bruising from 20% of the cattle to 1.3% (Paranhas de Costa, personal communication, 2006). Good handling will also help prevent pig death losses. A Spanish survey indicated that death losses increased when the time taken to load the pigs decreased (Averos *et al.*, 2008). When people hurry, handling usually gets rougher.

Behavioural Principles for Handling Extensively Raised Animals that Are Not Completely Tame

Handlers need to understand the behavioural principles of the flight zone and point of balance for moving cattle, pigs, sheep and other animals. To keep animals calm and move them easily, the handler should work on the edge of the flight zone (Grandin, 1980a, b, 2007a; Grandin and Deesing, 2008) (Fig. 5.1). The flight zone is the animal's personal space and the size of the flight can vary from 0 m from a person to over 50 m. Hedigar (1968) defined the process of taming an animal as removal of the flight zone to the point that the animal allows people to touch it. When a person penetrates the flight zone, the animals will turn and move away. When the animals turn and face the handler, the person is outside their flight zone. The size of the flight zone will vary depending on how wild or tame the animal is. Completely tame animals that have been trained to lead have no flight

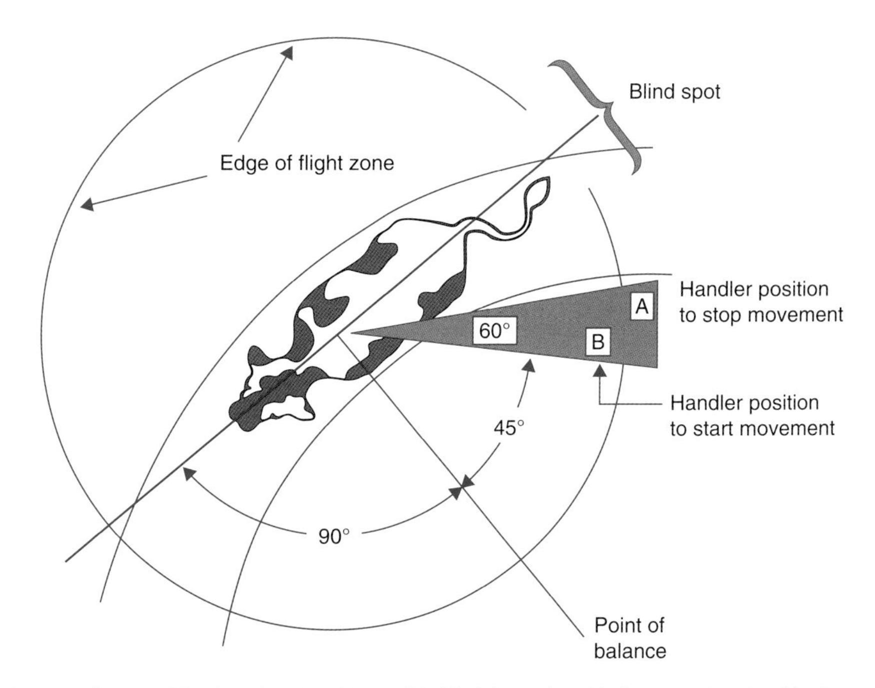

Fig. 5.1. Flight zone diagram. The handler should stand behind the point of balance at the shoulder to make an animal go forward (positions A and B) and in front of the point of balance at the shoulder to make an animal back up. Handlers should work on the edge of the flight zone.

zone and they should be led instead of being driven. Extensively raised animals that seldom see people will have a larger flight zone than animals that see people every day. When cattle are handled quietly, their flight zone will become smaller.

The size of the flight zone is determined by three factors. They are: (i) the amount of contact with people; (ii) the quality of the contact with people, calm and quiet versus shouting and hitting; and (iii) animal genetics. Animals that see people every day when they are fed will usually have a smaller flight distance than animals on extensive pasture that see people only a few times a year. Genetic differences in the animal's temperament will also affect the flight distance. There are individual animals in every herd that will have a more flighty, excitable temperament and a larger flight zone. When raised under extensive conditions, British breeds such as Shorthorns were less likely to become agitated during handling compared to Brahman crosses (Fordyce *et al.*, 1988). Within a breed of livestock, temperament and the tendency to startle is definitely heritable (Hearnshaw and Morris, 1984). There are also genetic differences in the tendency of cattle, pigs or sheep to bunch and flock together. Merino and Rambouillet sheep will flock more tightly together than Suffolks, Hampshires and Cheviots.

To move extensively raised animals in a calm, controlled manner, the handler should work on the edge of the flight zone. The handler penetrates the flight zone to make the animals move and he backs up if he wants them to stop moving. The handler should avoid the blind spot behind the animal's rear. Deep penetration of the flight zone should be avoided because continuous deep penetration of the flight zone will cause the animals to run. The handler should also practise the principle of pressure and release. When the animals are moving in the desired direction, the handler should back out of the flight zone. When the animals slow down or stop moving, the flight zone should be re-entered.

Animals may become agitated when a person is inside their personal space and the animal is unable to move away due to being in a race or small pen. If animals turn back and run past the handler while they are being driven down a drive alley in the corrals, overly deep penetration of the flight zone is a likely cause. The handler should back up and retreat from inside the flight zone when the animals give the slightest indication that they will turn back.

Flight zone diagram

The flight zone diagram (Fig. 5.1) shows the correct positions for the handler to move livestock. To make an animal go forward, he should work on the edge of the flight zone in positions A and B. The handler should stand behind the point of balance at the shoulder to make an animal go forward and in front of the point of balance at the shoulder to make an animal back up (Kilgour and Dalton, 1984; Smith, 1998; Grandin, 2007a). The most common mistake that people make is to stand in front of the point of balance and poke the animal on the rear to make it go forward. This gives the animal conflicting signals and confuses it. The handler must be behind the shoulder when he wants to urge the animal forward.

Point of balance

The second diagram (Fig. 5.2) illustrates how to use the point of balance at the animal's shoulder to move extensively raised livestock forward in a race. When a person quickly walks back past the point of balance in the OPPOSITE direction of desired movement, the animals will move forward. The principle is to walk quickly in the opposite direction of desired movement inside the flight zone and outside the flight zone in the same direction. When animals are out on pasture, a basic principle is that walking in the opposite direction of cattle movement speeds up herd movement, and walking alongside the herd in the same direction tends to slow the herd down. This principle applies to the movement of a person who is alongside a herd and not behind it.

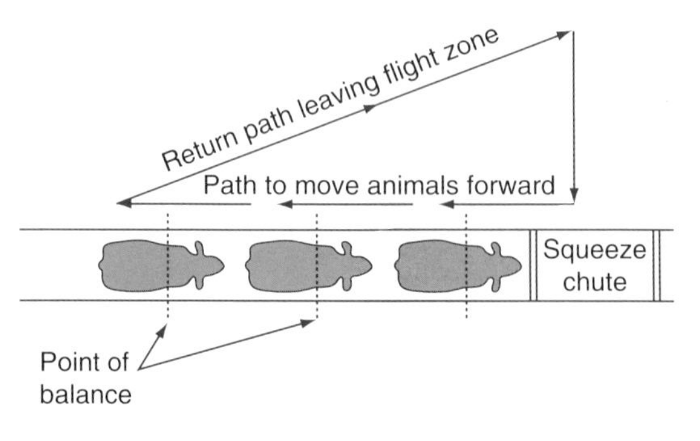

Fig. 5.2. Point of balance diagram. Walking quickly past the point of balance in the opposite direction of desired movement will make the animals move forward.

Fill the crowd pen half full

One of the most common mistakes when handling cattle and pigs is overloading the crowd pen that leads to a single-file race or loading ramp. The crowd pen should be filled HALF full so that animals will have room to turn. Figure 5.3 shows a crowd pen leading to a truck loading ramp that has been overloaded with too many cattle. All livestock will follow the leader and handlers can take advantage of this natural behaviour to move animals easily. Animals will enter a single-file race more easily if it is allowed to become partially empty before attempting to fill it. A partially empty race takes advantage of following behaviour because there is enough space for between four and ten animals to enter while following the leader. Small bunches of cattle and pigs should be moved. If your crowd pen holds 20 animals when it is stuffed full, you should bring in only ten animals so it will be half full. Sheep have such intense following behaviour that they can be moved in one continuous large group. The principle of moving sheep is to never have a break in the flow. It is analogous to siphoning water. Once the flow is going, it should be kept going because restarting the flow may be difficult. Be careful with a single animal that is left by itself. Animals often panic when they are alone. A single animal left alone and separated from its herd mates will often become agitated. It may run into a fence or run over a person. If a single animal becomes agitated, some other animals should be put in with it. Grazing animals are herd animals and being alone is frightening and stressful. The author has observed that an isolated, lone bovine that is highly agitated and attempting to return to its herd mates is the cause of many injuries to people during handling of extensively raised cattle.

Follow the leader

Extensively raised sheep with a large flight zone will follow a trained lead animal. Both goats and sheep work well as lead animals. Figure 5.4 shows a goat that is leading sheep through a stockyard. Trained sheep work very well for leading sheep on and off trucks and through the lairage at slaughter plants. Cattle can also be easily trained to follow a

Fig. 5.3. Too many cattle are jammed in this crowd pen and they have turned around. Cattle will enter the truck more easily if smaller groups are moved. This is a well-designed loading ramp but it is not being used correctly. The cattle will move into the truck more easily if they can be moved through the crowd pen without stopping. Cattle that wait in the crowd pen often turn around.

Fig. 5.4. Trained goats or sheep will lead sheep through stockyards or lairages. Lead animals also work really well for moving sheep on and off trucks. A specialist lead sheep or goat should be used for different handling procedures. Training the lead animals will be easier if one leader is trained to unload trucks, another is trained to load trucks and a third animal is trained to lead sheep out of pens and through the stockyards. In confined spaces such as trucks and small lairage pens at the slaughter plant, lead animals are recommended instead of dogs. Lead sheep have been working very successfully for many years at the largest sheep slaughter plant in the USA.

lead person or vehicle when they are being rotated to another pasture. When animals become very tame, the principles of using the flight zone and the point of balance will no longer work. Tame animals MUST be led.

Hearing in Farm Animals

Farm animals have sensitive hearing and are sensitive to high-pitched sounds. They can hear high frequency sounds that people cannot hear. The human ear is most sensitive to sounds in the 1000–3000 hz range and cattle and horses are most sensitive to frequencies at 8000 hz and above (Heffner and Heffner, 1983). Sheep can hear up to 10,000 hz (Wollack, 1963). High-pitched intermittent sounds are also more likely to cause cattle or pigs to react (Talling *et al.*, 1998; Lanier *et al.*, 2000). Handlers should watch where animals point their ears. Horses, cattle, sheep and many other animals will point their ears towards things that attract their attention.

Visual Perception

Grazing animals are very sensitive to rapid movement. Sudden rapid movement may frighten cattle (Lanier *et al.*, 2000). Cattle, sheep and horses have wide-angle vision and they can see all around themselves without turning their heads (Prince, 1970; Hutson, 1980; Kilgour and Dalton, 1984). Grazing animals have depth perception (Lemmon and Patterson, 1964). It is likely that their depth perception is poor because animals will stop and put their heads down when they see a shadow on the floor.

Grazing animals are dichromats. The retina of cattle, sheep and goats is most sensitive to yellowish-green (552–555 nm) and bluish-purple light (444–455 nm) (Jacobs *et al.*, 1998). The dichromatic vision of the horse is most sensitive at 428 nm and 539 nm (Murphy *et al.*, 2001). Dichromatic vision and the absence of a retina receptor for red may explain why livestock are so sensitive to sharp contrasts of light and dark such as shadows or shiny reflections on handling equipment.

Poultry appear to have excellent vision. Chickens and turkeys possess four cone cell types in the retina giving them tetrachromatic colour vision, as compared to the human trichromatic vision based on three cone cell types (Lewis and Morris, 2000). The spectral sensitivity of chickens is greater than that of humans from 320–480 nm to 580–700 nm. Their maximum sensitivity is in a similar range (545–575 nm) to humans (Prescott and Wathes, 1999). The broader spectral sensitivity of poultry may make them perceive many light sources as being brighter than what a human would see. Poultry may be more docile during handling in blue light spectra (Lewis and Morris, 2000). Lighting conditions have a large effect on chicken behaviour when they are shackled for slaughter (Jones *et al.*, 1998). During handling of poultry, the occurrence of flapping should be minimized. Changes in lighting may be used as one tool to keep birds calmer during handling.

Remove distractions from handling facilities

Extensively raised livestock and animals that are trained to lead may often stop and refuse to move past little distractions that people do not notice. Animals are very sensitive to both high contrasts of light and dark and rapid movement. They notice visual detail in their environment that people often fail to notice. A calm animal will look right at a distraction such as a shirt hung on a corral fence. A frightened animal that is being forced to move towards the shirt will often turn back and try to run back past the handler. People need to be aware of the distractions that frighten animals. Handlers should walk through their races and corrals and look for distractions that can stop cattle movement. Distractions cause the most problems with animals that are not familiar with the handling facilities. Below is a list of common distractions that can cause animals to stop and refuse to move through a race or other handling facility.

- Shadows or bright sunbeams of light are often a problem on sunny days. Livestock may refuse to walk over them (Fig. 5.5). In Fig. 5.6 the pigs are avoiding walking on the shadows and they are baulking at the metal strip. Allow the leader time to put its head down to investigate a shadow. After it has determined that the shadow is safe, it will lead the others over the shadow. If the handlers rush the leader's investigation of a

Fig. 5.5. Calf baulking at a bright spot of sunlight as it exits the squeeze chute (crush). Note how the animal is pointing both its eyes and ears towards the sunbeam. A calm animal will look directly at the distraction. This facility has an excellent non-slip floor made from woven tyre treads.

Fig. 5.6. The pig is baulking and refusing to walk over a metal strip. Animals will often avoid walking on shadows or beams of light.

shadow, the animals may turn back on the handlers. When an animal is being led, allow it time to put its head down and investigate the shadow before urging it to walk over it.

- Objects on the fence or in the race such as loose chains or shirts should be removed. In Fig. 5.7 the animal is refusing to move past a dangling rope. Cattle may refuse to move past a shirt that is hung on a fence. A little, loose chain that swings may also stop cattle movement. Bright yellow clothes and objects are especially bad. Objects with high contrasts of light and dark cause the most problems.
- Vehicles parked outside the corrals may also cause problems. A shiny reflection on the bumper of a car may stop animal movement. The vehicles should be moved.
- Livestock will often refuse to approach visible people who are standing up ahead. The people should either move or a shield can be installed so that the cattle do not see the people. Installing solid fences to block the animal's vision of distractions outside the facility often improves animal movement (Grandin, 1996). The materials must be stiff and not flap. Flapping materials will scare the animals.
- Driving livestock directly into either the rising or setting sun is difficult because the animals are blinded. The best way to fix this problem is to change the time of day that the livestock are handled. When new facilities are built, avoid pointing races or truck loading ramps towards the sun.
- Animals may refuse to enter a dark building where a race is located (Fig. 5.8). They will enter more easily if they can see daylight through the other side of the building (Fig. 5.9). Removing a side wall may help. In new facilities, installing white translucent skylights to admit lots of shadow-free daylight will usually improve animal movement. Cattle and pigs have a natural tendency to move from a dark place to a brighter place unless they are approaching a blinding run (van Putten and Elshof, 1978; Grandin, 1982, 2007a; Tanida *et al.*, 1996). At night, a lamp can be used to attract animals into buildings or trucks. The light MUST NOT shine directly

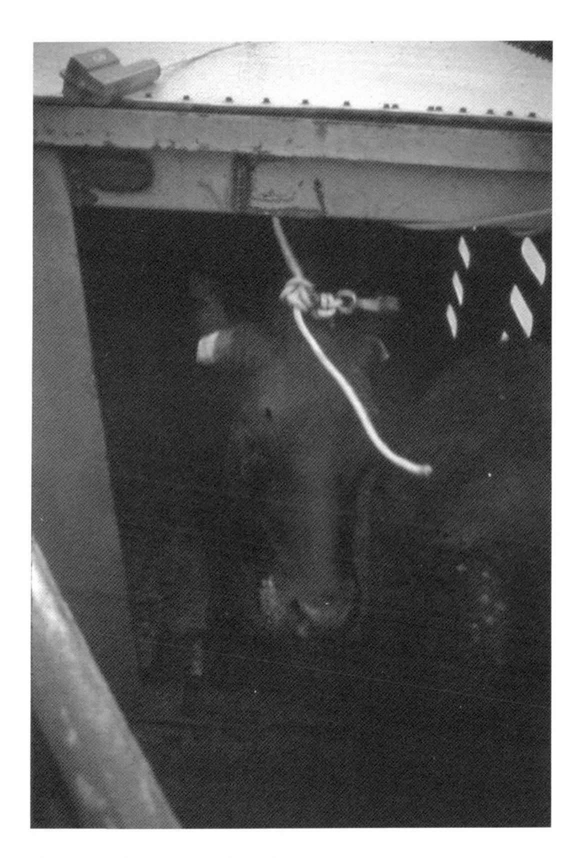

Fig. 5.7. This animal is refusing to walk past the dangling rope. Remove ropes and chains that hang down in front of animals. Note that the truck driver left his electric goad on the top of the truck. If the dangling rope had been removed, the electric goad would not be needed. Cattle will unload easily from a truck and electric goads can be eliminated if all the distractions are removed.

in their eyes. It should provide even, bright, indirect light.

- Remove paper cups, old feed bags, plastic bags or other rubbish that is lying on the ground because animals may refuse to walk over them.
- Reflections in a puddle of water in a corral or race can stop animal movements. Either fill in the puddles with dirt or change the time of day that the animals are handled. In indoor facilities, changing the position of lamps will often eliminate a reflection. One person should get down in the race to see an animal's eye view while the other changes the position of lamps.
- Abrupt changes in flooring will also stop animal movement. Examples are moving from a dirt floor to a concrete floor or walking up a wooden loading ramp into a vehicle with a metal floor. Spreading some dirt, hay or other material on the junction between the two flooring types will remove the contrast. Often a very small amount will work. In indoor facilities, a drain grating or metal drain cover may stop animal movement. When facilities are being designed, drains should be outside the areas where animals will walk. Even animals that are completely tame and trained to lead can panic and become agitated if they are forced to approach a strange thing. This is especially a problem if it is something they have never seen before. Horses may fear llamas, and pigs are scared of bison.

Non-slip Flooring

A non-slip surface for animals to stand on is essential (Grandin, 1983, 2007b). Animals panic and can become agitated when they slip. Some of the most important places to have non-slip flooring are in the single-file race, truck floors, veterinary facilities, stun boxes and the crush. Small repeated slips where one hoof moves back and forth rapidly are really scary for animals. A floor can be made non-slip by covering it with gravel, sand or dirt. In facilities with metal or concrete floors, a grating made from steel bars can be made (Fig. 5.10). This works really well in high traffic areas such as stun boxes, loading ramps and scales. It is best to use heavy rods with a minimum diameter of 2 cm for cattle and other large animals. The rods can be welded to make a grid in a 30 cm × 30 cm square pattern. The mat of rods MUST LIE FLAT. Do NOT criss-cross the rods on top of each other. The gaps can injure hooves. Non-slip flooring is essential for good animal welfare. Both handling facilities and trucks should have non-slip floors. Grates made from steel rods work well for providing non-slip flooring in vehicles. A rough broom finish on concrete will quickly wear smooth and animals will slip and fall. For pigs and sheep, a good finish for concrete is stamping the wet concrete with expanded metal mesh. This provides a rougher surface that is still reasonably easy to clean. Deeper grooves will be required for larger animals.

Driving Aids

An electric goad should never be used as a person's primary driving tool and the World Organization

Fig. 5.8. Animals may refuse to enter this dark building, Opening up doors or windows to admit light into the building will attract the animals in. This is especially a problem on bright, sunny days. At night, lamps providing indirect lighting can be used to attract the animals into the building.

Fig. 5.9. Animals will enter this veterinary building more easily because they can see through it and see daylight on the other side.

for Animal Health (OIE, 2009a, b) code recommends that only battery-operated goads should be used. The OIE (2009a, b) code also states that electric goads should not be used on sheep, horses and very young animals such as piglets. The goad should only be picked up when other non-electrified methods fail to move an animal. Research has shown that the attitude of the handlers towards the

Fig. 5.10. A grating constructed from steel rods will provide a non-slip surface on unloading docks, scales, stun boxes and other high animal-traffic areas. The rods must not be criss-crossed on top of each other. The rod mat should lay completely flat. For large animals, the rods are laid out in a 30 cm × 30 cm square pattern.

animal was better when shocks were not being used (Coleman *et al.*, 2003). A brief shock on the hind-quarters is preferable to hard tail twisting or beating an animal. After the stubborn animal is moved with the electric goad, the goad should be put away. A plastic bag taped to the end of a stick or a flag is an excellent tool for moving cattle (Fig. 5.11). Other commonly used tools are plastic paddles. For pigs, panels work really well (McGlone *et al.*, 2004). Some other innovative tools for moving pigs are long strips of opaque plastic and the 'witches' cape'. The pigs respect the solid barriers and they

Fig. 5.11. Handlers in Uruguay use flags to move cattle. To promote the concept of animal welfare, the flags were provided by a local meat company and had the words 'animal welfare' printed on them in Spanish.

Fig. 5.12. Pigs can be easily moved with a plastic cape. The top of the cape is stiffened with a rod. This alternative driving tool makes it possible to greatly reduce electric-prod use.

will not run through them (Fig. 5.12). Driving aids must never be applied to sensitive parts of the animal such as the eyes, ears, nose, rectum, genitals or udder. The OIE (2009a, b) code states:

> Painful procedures (including whipping, tail twisting, use of nose twitches, pressure on eyes, ears or external genitalia) or the use of goads or other aids which cause pain and suffering (including large sticks, sticks with sharp ends, lengths of metal piping, fencing wire or heavy leather belts) should not be used.

Measure Animal Handling

People manage the things that they measure. Throughout a long career, the author has worked with many ranchers and feedlots on cattle handling. The handling practices become really good while the author is there, but when the author returned a year later, the yelling and screaming and electric-prod use have often increased. In many cases, people did not realize this had happened because reverting back to old, rough practices often happens slowly. To prevent return of old, rough practices, handling should be measured with numerical scoring. This makes it possible to determine if practices are improving or getting worse. To measure handling, keep score on the number of animals that have the following faults and determine the percentages:

- Percentage of animals that fall during handling – score a fall if the body touches the ground. The OIE (2009a, b) codes state that numerical scoring should be used to measure falling. Corrective action should be taken to improve either the handling practices or the facility if falling exceeds 1% of the animals (Grandin, 2007b).
- Percentage of animals that are moved faster than a trot – animals should NOT be running and jumping. They should not be forced to move faster than a normal walking speed.
- Percentage of animals that run into a gate or fence.
- Percentage of animals moved with an electric goad – the OIE (2009a, b) codes state that numerical scoring should be used to measure electric-goad use. On a farm or ranch, an electric goad should be used on 5% or less of the animals.
- Percentage of animals that vocalize (moo, bellow or squeal) when they are caught by the head or body restraint – when the quality of handling is assessed, do not score after ear tagging or other procedures start.

Measurement of Improvements in Handling

If the handlers are doing a good job of low-stress handling, the percentage of animals that have any one of these faults will be very low. By keeping score every time you handle livestock, you can determine if handling is getting better or becoming worse.

Falling should be extremely infrequent. If more than 1% of the animals fall, there is a serious problem with either a slick floor or rough handling. Falling is a very good measure because it is the outcome of either a slippery floor or poor handling methods.

Research the author conducted on vocalization showed that 99% of the time that cattle vocalized during active handling at a slaughter plant was due to a frightening or painful event (Grandin, 1998). Some of the events that caused vocalization were slipping on the floor, electric goads, sharp edges on a restraint device or being squeezed too hard by a restrainer or head stanchion. When a head restraint that applied excessive pressures was loosened, the percentage of cattle that vocalized was reduced from 23% to 0% (Grandin, 2001). The goal is to have 5% or less of the animals vocalize due to handling and restraint problems. On farms, vocalization to assess the quality of handling should only be scored when animals are being actively moved in the single-file race and during restraint. If the bovine vocalizes the moment it is caught by either a head or body restraint, it is being hurt by the device. Do not score vocalization from weaning animals or procedures such as branding when you are assessing handling. However, when you are measuring stress from a procedure such as weaning or branding, animal vocalizations associated with these procedures should be scored. Vocalizations caused by handling and veterinary procedures should be scored separately from vocalizations caused by weaning stress. Vocalization is an outcome measure of an event that either frightens the animal or causes pain. Cattle and sheep left alone will both vocalize because they are scared. If a head restraint has a sharp edge, that could also cause vocalization. Vocalization scoring is a useful measure of animal distress. Vocalization in both cattle and pigs is associated with physiological measures of stress (Dunn, 1990; Warriss *et al.*, 1994; White *et al.*, 1995). Procedures that cause either pain or fear will usually increase vocalization in cattle and pigs (Lay *et al.*, 1992a, b; White *et al.*, 1995; Watts and Stookey, 1999). Vocalization scoring cannot be used for assessing stressful or painful procedures in sheep, because sheep will not vocalize when they are hurt. Sheep are defenceless prey animals. In the wild they do not want to advertise to predators that they are hurt. Sheep vocalize when they become isolated because they are seeking the safety of the flock, but they seldom vocalize when they are being handled or restrained.

Bull Behaviour

Dairy bulls have a bad reputation for attacking humans, possibly due to the differences in the way beef bulls and dairy bulls are raised in some countries. Bulls are responsible for about half of the fatal accidents with livestock (Drudi, 2000). In India, bulls injure many people. Wasadikar *et al.* (1997) reported that a rural medical college treated 50 cases of bull-horn injuries over a 5-year period. Dairy bull calves are often removed from the cow shortly after birth and raised in individual pens, whereas beef bull calves are reared by the cow.

Price and Wallach (1990) found that 75% of Hereford bulls reared in individual pens from 1 to 3 days of age threatened or attacked the handlers, whereas only 11% threatened handlers when they were hand reared in groups. These authors also report that they have handled over 1000 dam-reared bulls and have experienced only one attack. Bull calves that are hand reared in individual pens may fail to develop normal social relations with other animals and they possibly view humans as a sexual rival (Reinken, 1988). It is important for young bulls to learn that they are cattle. Problems with bull attacks are a social-identity issue, not a tameness issue. If a young bull is going to start being aggressive towards people, the first attack is likely to occur at 18–24 months of age.

Both dairy and beef breed bulls will be safer if bull calves are raised on a cow and kept in groups with other cattle. This provides socialization with their own species and they will be less likely to direct attacks towards people. Similar aggression problems have also been reported in hand-reared male llamas (Tillman, 1981) and orphan buck deer. In one sad case, an orphan buck attacked and killed the person who raised him. Fortunately, hand rearing usually does not cause aggression problems in females or castrated animals. It will make these animals easier to handle. More information on bulls can be found in Smith (1998).

Before a bull attacks, he will turn sideways and do a broadside threat to show how big he is. Bulls that display a broadside threat towards humans are really dangerous and should either be slaughtered or sent to a secure bull stud. If a bull threatens a person, the person should slowly back away. Handlers must be instructed to NEVER turn their back on a bull. He is more likely to attack when the person is not looking at him. To help prevent dangerous attacks by intact males, orphan bull calves should either be castrated at an early age or raised on a nurse cow.

Preventing Fear Memories

It is very important that an animal's first experience with something new should be a good experience. When a new corral or other animal handling facility is built, the animal's first experience with it should be a positive one such as feeding them in it. If the first experience is really bad, it may be difficult to get the animal to enter the corrals in the future. Sheep remembered a single stressful handling experience a year later (Hutson, 1985). Painful procedures should NEVER be associated with an animal's first experience with a new person or a new piece of equipment. Miller (1960) reported that if a rat receives a severe shock the first time it enters a new arm of a maze, it will never go in that arm again. However, if its first experience in the new arm resulted in delicious food treats, the rat would tolerate gradually increasing shocks and keep re-entering the same arm to get the food reward. This same procedure may also apply to livestock. Hutson (2007) recommends that the first few times the animals are handled in the facility, low-stress procedures such as sorting or weighing should be done first. Doing this may help make the animals more willing to re-enter the facility when a painful procedure has to be done. Frightening or painful first experiences with new things can create a permanent fear memory.

A question that often gets asked is: will doing a painful procedure on a 2-day-old calf or young lamb have an effect on handling in the future? Practical experience has shown that very young infant animals probably do not remember the experience and make an association with the person. When the calves and lambs get older, they have excellent memories for associating painful or frightening experiences to specific people, places, sounds or sights.

Fear memories of experiences that were really frightening or painful can never be erased (LeDoux, 1996; Rogan and LeDoux, 1996). Even if an animal is trained to tolerate something it had previously been afraid of, the fear memory can sometimes suddenly reappear. This is most likely to occur in animals that startle easily such as Arabian horses or Saler cattle. This is the reason why rough, harsh training methods are especially harmful to breeds such as Arabians. If an Arabian horse is abused, it may no longer be a safe riding horse because it cannot stop being frightened. Calmer breeds of livestock such as draught horses or Hereford cattle have a greater ability to suppress past memories of scary or painful events. The fear circuits in an animal's brain have been fully mapped. The brain has a fear centre called the amygdala. Scientists proved that a fear centre exists in the brain because destroying the amygdala completely stops fear (Kemble *et al.*, 1984; Davis, 1992).

Sensory-based Memory

Animals store memories as pictures, sounds or other sensory impressions (Grandin and Johnson, 2005). Since they think in sensory-based memories, their memories are very specific. Research with horses showed that a horse sometimes failed to recognize a large children's toy when the toy was rotated to a different position (Hanggi, 2005). When the toy was rotated, it created a different picture that the horse failed to recognize. Visual picture memory is a basic function of the brain. Even ants have the ability to remember visual images (Judd and Collett, 1998). For example, a cow or horse may fear men with beards or people with black hats (Grandin and Johnson, 2005). This occurs because the animal was looking at the hat or beard when a really painful or frightening event occurred. Animals tend to associate a bad experience with an obvious feature on a person such as a white coat, beard or blond hair. They also can learn to fear certain sounds with a painful or scary past experience. Animals can learn the voices of a trusted stockperson and the voice of a person who has hurt them (see Chapter 4). To help prevent fear of the regular stockpeople, painful procedures should be done: (i) by different people; or (ii) by people wearing distinctive clothes that are reserved for painful procedures; or (iii) in a specific place that is reserved for painful procedures.

Stress of Handling and Procedures

Often the fear stress caused by restraining an extensively raised animal may be almost as stressful as hot iron branding, ear tagging or injections. Extensively raised cattle that were restrained in a squeeze chute had similar cortisol levels as cattle branded with a hot iron (Lay *et al.*, 1992a). When the same experiment was repeated with tame, dairy animals, the hot-iron-branded animal had significantly higher cortisol (Lay *et al.*, 1992b). When more severe procedures such as dehorning are done, the procedure causes an elevation of cortisol for a longer period compared to handling (see Chapter 6). This occurs in both extensively and intensively raised animals. Therefore, stress from handling may often exceed the stress caused by minor procedures such as ear tagging and injections. For more severe procedures such as dehorning or castration, pain-relieving medication will reduce stress (Grandin, 1997a) (see Chapter 6). Isolating sheep and leaving them tied up for long periods is very stressful and will cause very high cortisol levels (Apple *et al.*, 1993).

Train Animals to Tolerate New Experiences

A common complaint that the author often hears is 'my cattle/my horse was gentle and calm at home, but my animals went crazy and wild at a show-ground or auction'. This problem occurs because the animals are suddenly confronted with many new, scary things that they never saw on the home farm. To prevent this problem, the animals must be acclimatized to the sights and sounds of a fairground or auction BEFORE they go there. Flags, bikes and balloons can be really frightening even though an animal is completely tame and trained to lead. The author has observed tame show animals that panicked and ran through the fairgrounds, knocking people over. Animals should be carefully habituated to items such as flags and bikes on the home farm.

The best way to teach animals to tolerate moving, scary things is to allow the animals to voluntarily approach flags or balloons that are tied to a pasture fence. An important principle is that new novel things are scary when an animal is forced to approach them but they are attractive if the animal can explore them voluntarily. Animals with a high-strung, nervous temperament are more likely to get scared when novelty is suddenly introduced. Objects that move rapidly can cause severe problems. Most animals have probably learned that large moving vehicles are safe but a flag is something new. New moving objects that the animals have never seen before can be really frightening.

A large panel such as a sheet of plywood that is moved rapidly can also cause panic. The author observed a tame heifer on a lead rope that attempted to break away from her handler when a large 1.2 m × 3 m white panel was suddenly moved. When the panel was still, she ignored it. Training livestock to tolerate a large object that suddenly moves is highly recommended. Grazing animals have an instinctual fear of things that move rapidly. In the wild, predatory animals such as lions move rapidly. Animals can also get really scared and react violently when a novel object rapidly moves near their face while they are eating. Extensively raised cattle that had a calm temperament reacted violently when a ball was dropped near their head while they were licking salt in a single-animal salt feeder (Sebastian, 2007). In this type of feeder, the animal puts its head under a plastic hood that keeps rain off the salt. When the ball fell from the ceiling of the feeder, some of the animals may have instinctually reacted to get a 'predator' off their head.

In developing countries, cattle and water buffaloes often graze along the highways. They see lots of new things from a very early age and they will be less likely to be frightened by an object such as a large piece of plywood. These animals have seen every kind of object passing by them on the road. The mother animal teaches her calf not to be afraid of rapid movement. The calf sees his mother calmly grazing by the highway so he will continue to graze beside her.

Ried and Mills (1962) were two of the first researchers to introduce the idea that sheep could be acclimatized to changes in routine. It is important to gently teach cattle that new people and new vehicles are safe. It is recommended to have different people and vehicles involved in feeding cattle. If the cattle are fed by only one person, they are more likely to panic if they have to be handled or fed by a new person. Animals also perceive a person on a horse and a person on the ground as two different things. It is important for cattle to learn how to move in and out of pens in the corral by BOTH a person on foot and a person on a horse. The author has seen cattle that were gentle and easy to handle by a person on a horse but they became agitated and dangerous the first time a person on foot tried to move them out of a pen. Learning in animals is very specific. A person on a horse and a person walking on the ground are two different visual images.

Signs of Fear and Agitation in Grazing Animals

People working with animals need to be able to recognize the behavioural signs that an animal is becoming more and more agitated and fearful. Learning to recognize these signs will help prevent both people and animals from getting injured.

Tail switching

Horses and cattle will switch their tails when no flies are present. The speed of tail switching will increase as the animal gets more and more agitated. Failure to heed this warning signal may result in people getting kicked. Bison hold their tails straight up.

Defecation

Scared animals with high anxiety will defecate more. Loose diarrhoea in a healthy animal may be a sign of emotional stress unless it has been grazing on lush green fields. When animals are handled quietly, there is often less manure left in the facility to clean up.

Whites of the eyes show

Bulging eyes where the whites show is a sign of distress. Research shows that visible eye white is a sign of emotional distress because the response can be blocked in cattle with the anti-anxiety drug diazepam (Valium®) (Sandem *et al.*, 2006). The percentage of eye white showing is highly correlated with scores on standard cattle temperament tests (Core *et al.*, 2009).

Sweating with minimal physical exertion

This occurs mainly in horses.

Quivering skin

Animals that quiver all over when they are touched and have no flies on them may be becoming fearful.

Nostrils flaring

This is more visible in horses compared to cattle.

Both ears laid back

When an animal lays both its ears back it is a sign of either fear or aggression. Calm, alert animals will point their ears forwards towards people or other animals.

Head held up high

Both cattle and horses will raise their heads up high when they are fearful. This is an instinctual behaviour that is used for looking for predators.

Tonic immobility

This occurs mainly in poultry. The bird stays still and does not move. The length of time that a bird will stay still after it has been held by a person is used as a measure of fearfulness in poultry. Different genetic lines can be assessed for fearfulness by measuring the length of time they will remain in tonic immobility (Faure and Mills, 1998). The author has observed that some *Bos indicus* cattle and American bison may also lie down and become immobile if they are shocked multiple times while they are held in a race from which they cannot escape.

Principles of Training Animals to Cooperate

Animals can be trained to completely cooperate with veterinary procedures such as blood sampling. When an animal cooperates, the animal's cortisol levels will remain very low. Sheep, pigs, cattle and wild ungulates such as antelopes can be easily trained to voluntarily enter a restraint device (Panepinto, 1983; Grandin, 1989a, b; Grandin *et al.*, 1995). Almost baseline levels of cortisol (stress hormone) and glucose were obtained from trained antelopes (Phillips *et al.*, 1998). Cortisol levels from wild antelopes captured in the bush were three to four times higher. The fear stress of being trapped and restrained can kill a wild animal by causing capture myopathy. Sometimes the animal dies right away and other times it may take up to 2 weeks to die. Chalmers and Barrett (1977) and Lewis *et al.* (1977) contain excellent photos and descriptions for diagnosing capture myopathy in elk and pronghorn antelopes. A basic principle is: when an animal voluntarily cooperates, fear stress will be very low. In all these cases, the animals were eagerly lining up to get the delicious treats.

Animals with a flighty, excitable temperament such as antelopes will take longer to train and must be trained more slowly compared to a calmer species such as cattle or sheep. Great care had to be taken to carefully habituate the antelopes to the sights and sounds of the new equipment. If the animal is frightened early in the habituation phases, it may become so afraid of the equipment that training will be impossible. With flighty antelopes, it took 10 days to habituate them to the sound of a sliding door opening. In the early stages of habituation, the animal was never pushed past the orienting stage. The orienting stage occurs when the animal points its eyes and ears towards a sight or sound. When it oriented towards the door, the trainer stopped moving it. Each day the door was moved a little more. After the 10-day habituation period, the door could be opened quickly. After careful habituation, food rewards were used to train the animal to stand still.

Obviously it is not practical to train large numbers of cattle and pigs to the point where there is total voluntary cooperation. However, livestock will be easier to handle if people carefully work with them. Sheep will move through a race more quickly if they get a feed reward (Hutson, 1985). Binstead (1977), Fordyce (1987) and Becker and Lobato (1997) all found that quietly moving extensively raised young calves through the handling races, or walking among them, produced calmer adult animals. These handling sessions were done every day for 10 days. Pigs that have been walked in the aisles of their barn or had people quietly walking through their pens were easier to drive and move in the future (Abbott *et al.*, 1997; Geverink *et al.*, 1998; Lewis *et al.*, 2008). Poultry that are socialized and habituated to people at a young age were easier to handle, gained more weight and had better immune function (Gross and Siegel, 1982). In all species, numerous positive experiences with handling at a young age will produce calmer, easier-to-handle adult animals.

Principles of Restraint

When restraining animals, either with a mechanical apparatus such as a squeeze chute (crush) or holding a small animal in a person's hand, the same behavioural principles of restraint apply. Below is a list of restraint principles to keep animals calmer and minimize struggling or vocalization. Animals should be restrained in a comfortable, upright position.

Non-slip flooring

This is essential because the fear of falling is a primal fear. Repeated small slips where the animal repeatedly slips and then pulls its foot back into position to maintain its balance will cause an animal to get frightened. Repeated slipping is likely to occur if an animal will not settle down while it is standing in a race or in a head stanchion.

Avoid sudden jerky motion

Avoid sudden jerky motion of either people or equipment. Smooth movements by people, ropes or equipment will keep animals calmer (Grandin, 1992).

Support the body

When an animal is restrained with its feet off the ground, it will remain calmer if its body is fully supported. This applies to a small animal held in a person's hand or a larger animal held in a mechanical restraint device. When the body is fully supported, the animal will be comfortable and the fear of falling will be eliminated. Restraint devices such as the Panepinto sling (Panepinto, 1983) and the double rail restrainer (Westervelt *et al.*, 1976; Grandin, 1988) employ the principle of full body support in a balanced position (see Chapter 14).

Even pressure

Animals are calmed by even pressure over a wide area of the body (Ewbank, 1968). It is important that there are no concentrated pressure points or pinch points. The author fixed a restrainer for pigs that was causing them to squeal by replacing a narrow bar that hurt their backs with a wide board that spread the pressure evenly along the pig's back. This stopped the pigs from squealing.

Optimum pressure

Pressure should not be too loose or too tight. An animal needs to be held tight enough to give it the feeling of being held, but not so tight that it feels pain. Excessive pressure will cause struggling (see Chapter 9 for further information).

Blocking vision

This is not required for completely tame animals with no flight zone. A blindfold made from completely opaque material will keep both extensively raised cattle and bison calmer (Mitchell *et al.*, 2004). It is essential that the material is completely opaque. If the animal can see moving shadows through it, it will not work. Solid sides installed on a race or a fully enclosed dark box will have a calming effect on cattle with a large flight zone (Grandin, 1980a, b, 1992; Hale *et al.*, 1987;

Pollard and Littlejohn, 1994; Muller *et al.*, 2008). Species such as bison, antelopes and wild horses will often rear up in a race. A solid top on the race will stop rearing.

Avoid painful methods

Animals remember painful experiences. Nose tongs hurt and it is strongly recommended to use a halter (head collar) when the head has to be restrained for veterinary procedures (Sheldon *et al.*, 2006). When more comfortable restraint methods are used, the animals will be more willing to enter the veterinary race in the future.

Electrical Immobilization is Highly Aversive

Devices that restrain animals by paralysing and freezing the muscles are highly detrimental to animal welfare. These devices should not be confused with electrical stunning that produces instant insensibility. OIE (2009b) codes state that these devices should not be used. Numerous scientific studies have shown that restraining an animal with these devices is very stressful. Jephcott *et al.* (1986) found that immobilization significantly increased beta-endorphin levels, which is an indicator of pain. Another study showed that animals were more averse to electrical immobilization than to restraint in a tilting restraint table. When sheep were given a choice between the two methods of restraint, the tilting restraint table was preferred (Grandin *et al.*, 1986).

A series of studies done by Rushen (1986a, b) and Rushen and Congdon (1986a, b) all showed that electro-immobilization was highly aversive and should not be used. Other researchers also came to the same conclusion that electrical immobilization is a method that should not be used (Lambooy, 1985; Pascoe, 1986). The author immobilized her own arm with three different brands of electrical immobilizers. The author reported that 'it felt like sticking my arm into an electrical socket'. Four different research groups independently came to the same conclusion that electrical immobilization is very stressful.

The Use of Dogs

In confined areas such as trucks, lairages at slaughter plants, and stockyards, the use of dogs is not recommended. The use of trained lead sheep or goats is strongly recommended for moving sheep in confined areas. Kilgour and de Langen (1970) found that dogs biting sheep was highly stressful. The author has observed that cattle that have been bitten by dogs in confined places, such as races where they are not able to move away, are more likely to kick at people. The author recommends that dogs should be limited to pastures, large pens and other open areas where the animals have room to move away.

Are Facilities Appropriate for the Type of Animal?

Simple facilities that will work for tame animals that are trained to lead are not appropriate for handling untamed, extensively raised animals with a large flight zone. One example is slaughter facilities that are used in the Middle East for extensively raised Australian sheep. Because the facilities were originally constructed for use with local sheep that are accustomed to close contact with people, they were not appropriate for extensively raised animals. When wild Australian sheep were brought to these facilities, animal welfare was poor because no races were available for handling the wild sheep. This resulted in pile ups, sheep flipping over and people roughly grabbing them. The welfare of extensively raised wild cattle in a facility designed for tame animals may be even worse. Since there is no single-file race or stun box, handlers may use cruel methods to restrain animals such as poking out eyes or cutting tendons. Good animal welfare in these facilities will be impossible until equipment that is suitable for handling wild, extensively raised animals is installed. Information on the design of suitable equipment for extensively raised animals can be found in Grandin (1997b, 2007b) and Grandin and Deesing (2008)(see Chapters 9 and 14). If the animals are completely tame and trained to lead, no restraining or handling equipment may be required. The one essential component for good welfare with all types of animals is to have a non-slip surface for the animal to stand on.

Handling Facilities for Extensively Raised Livestock

Race and corral layouts that include curves often work more efficiently than layouts with straight races (Grandin, 1997b, 2007a, b) (Figs 5.13, 5.14 and 5.15). A curved single-file race is efficient for two reasons:

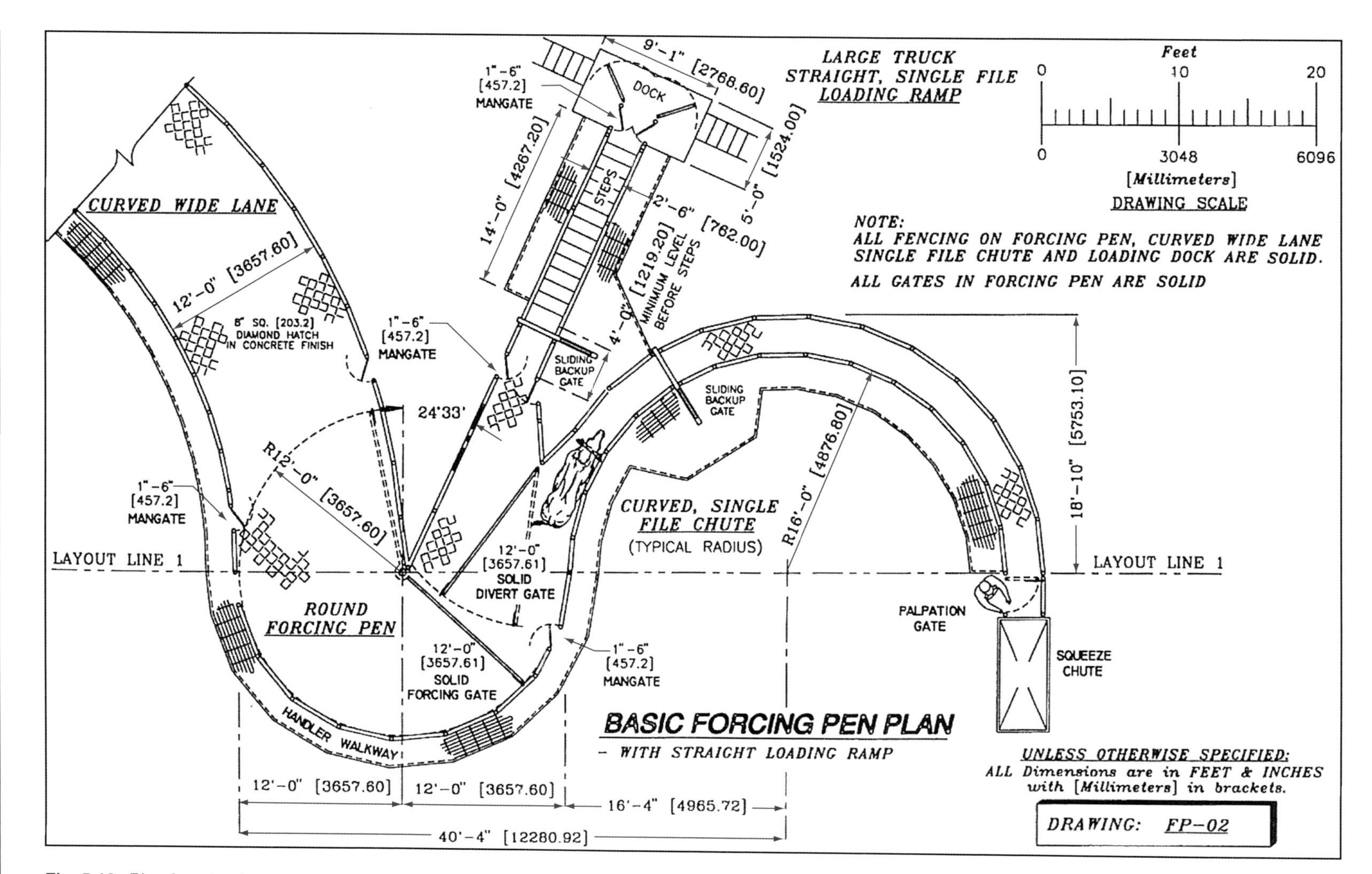

Fig. 5.13. Plan for a basic round crowd pen and curved race system for handling cattle and loading trucks. To facilitate cattle movement through this system, the junction between the round forcing pen, the loading ramp and the single-file chute must be laid out exactly as shown. There is a drawing of a cow at the entrance of the single-file race. This layout makes it possible for her to see two body lengths up the race. The cow must be able to see a clear pathway up ahead before the race bends.

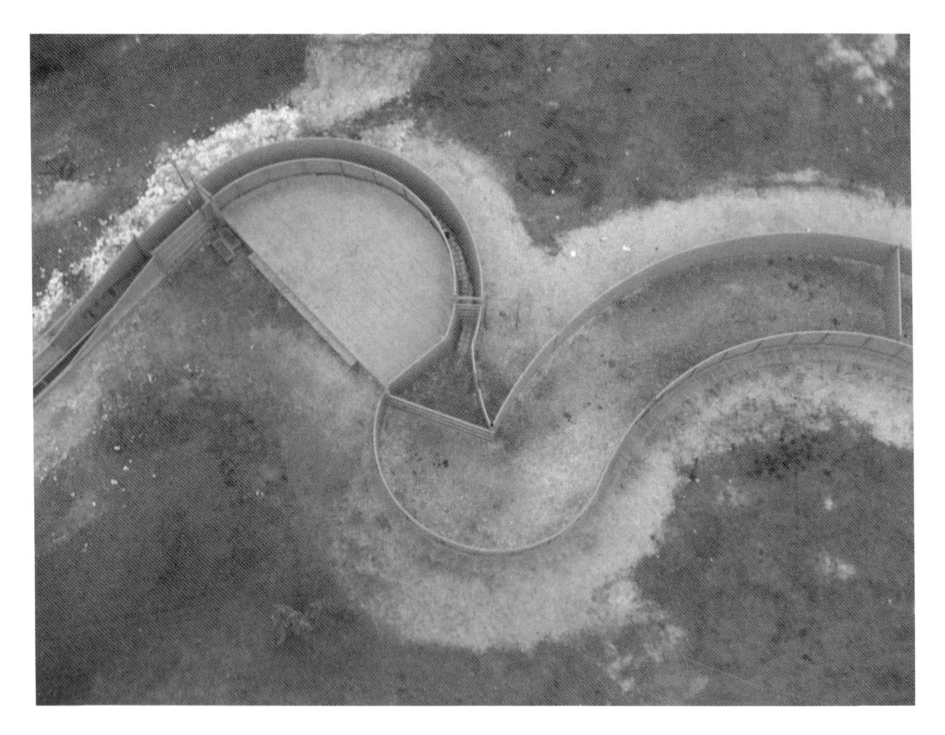

Fig. 5.14. Curved cattle handling facility for handling large numbers of extensively raised animals. This facility is recommended for handling large numbers of cattle for dipping, vaccinations, truck loading or movement into a slaughter plant.

1. Animals entering the race from the crowd pen do not see the people standing at the stun box or head stanchion.
2. Animals have a behavioural tendency to go back to where they came from. A curved race takes advantage of this tendency.

For these systems to work efficiently, they must be laid out correctly. The most common mistake is bending the single-file race too sharply where it joins on to the crowd pen. Figure 5.13 illustrates how to lay out the system and Fig. 5.14 is an aerial photo of a curved system for large numbers of extensively raised cattle. Figure 5.15 shows a curved system that will work well for extensively raised sheep. Well-designed facilities will make handling safer for both people and animals.

Conclusions

Learning the behavioural principles of handling livestock will have many advantages. These will be improvements in welfare, productivity, fewer accidents and improved meat quality. Handlers need to understand basic concepts such as the point of balance and flight zone. Stress can also be reduced by carefully habituating animals to novel stimuli.

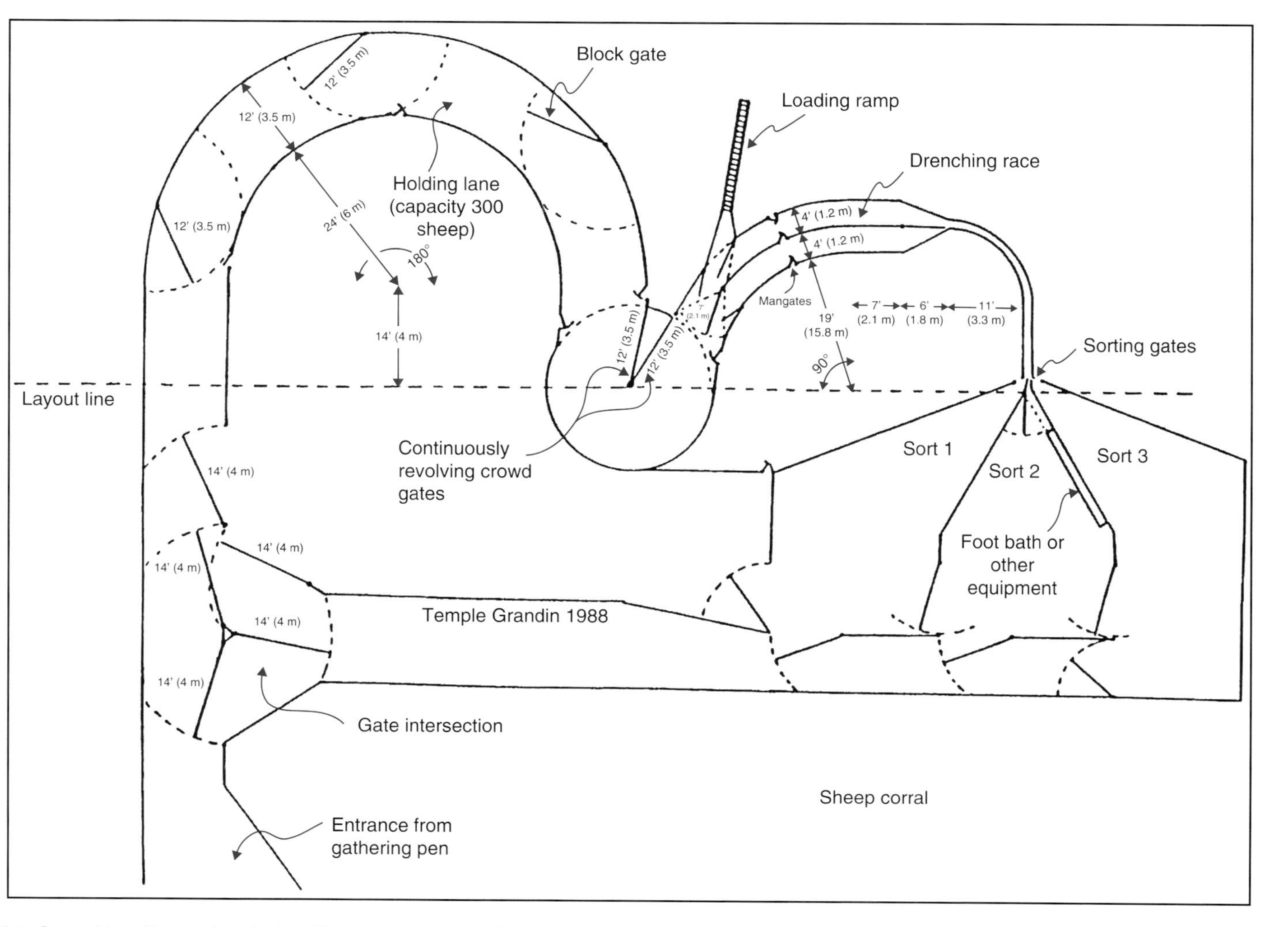

Fig. 5.15. Curved handling system for handling large numbers of sheep.

References

Abbott, T.A., Hunter, E.J., Guise, J.H. and Penny, R.H.C. (1997) The effect of experience of handling on pig's willingness to move. *Applied Animal Behaviour Science* 54, 371–375.

Apple, J.K., Minton, J.E., Parsons, K.M. and Unruh, J.A. (1993) Influence of repeated restraint and isolation stress an electrolyte administration on pituitary secretions, electrolytes, and blood constituents of sheep. *Journal of Animal Science* 71, 71–77.

Averos, X., Knowles, T.G., Brown, S.N., Warriss, P.D. and Gusalvez, L.F. (2008) Factors affecting mortality in pigs being transported to slaughter. *Veterinary Research* 163, 386–390.

Barnett, J.L., Hemsworth, P.H. and Newman, E.A. (1992) Fear of humans and its relationships with productivity in laying hens at commercial farms. *British Poultry Science* 33, 699–710.

Becker, B.G. and Lobato, J.F.P. (1997) Effect of gentler handling on the reactivity of Zebu crossed calves to humans. *Applied Animal Behaviour Science* 53, 219–224.

Benjamin, M.E., Gonyou, H.W., Ivers, D.L., Richardson, L.F., Jones, D.J., Wagner, J.R., Seneriz, R. and Anderson, D.F. (2001) Effect of handling method on the incidence of stress response in market swine in a model system. *Journal of Animal Science* 79 (Supplement 1), 279 (abstract).

Binstead, M. (1977) Handling cattle. *Queensland Agriculture Journal* 103, 293–295.

Chalmers, G.A. and Barrett, M.W. (1977) Capture myopathy in pronghorns in Alberta, Canada. *Journal of the American Veterinary Medical Association* 171, 918–926.

Coleman, G.J., McGregory, M., Hemsworth, P.H., Boyce, J. and Dowling, S. (2003) The relationship between beliefs, attitudes, and observed behaviors in abattoir personnel in the pig industry. *Applied Animal Behaviour Science* 82, 189–200.

Core, S., Miller, T., Widowski, T. and Mason, G. (2009) Eye white as a predictor of temperament in beef cattle. *Journal of Animal Science* 87, 2174–2178.

Curley, K.O., Pasqual, J.C., Welsh, T.H. and Randel, R.D. (2006) Technical note: exit velocity as a measure of cattle temperament is repeatable and associated with serum concentration of cortisol in Brahman bulls. *Journal of Animal Science* 84, 3100–3103.

Davis, M. (1992) The role of the amygdala in fear and anxiety. *Annual Review of Neuroscience* 15, 353–375.

Doney, J.M., Smith, W.F. and Gunn, R.G. (1976) Effect of post mating environmental stress and administration of ACTH on early embryonic loss in sheep. *Journal of Agricultural Science* 87, 133.

Douphrate, D.I., Rosecrance, J.C., Stallones, L., Reynolds, S.J. and Gilkey, D.P. (2009) Livestock handling injuries in agriculture: an analysis of Colorado workers' compensation data. *American Journal of Industrial Medicine*, 5 February (E. Pub.). Available at: http://www.ncbi.nlm.nih.gov/pubmed/19197949/ordinalpos=4&citool=EntrezSystem2.PEntr (accessed 3 April 2009).

Drudi, D. (2000) *Are Animals Occupational Hazards? Compensation and Working Conditions.* US Department of Labor, Washington, DC, pp. 15–22.

Dunn, C.S. (1990) Stress reactions of cattle undergoing ritual slaughter using two methods of restraint. *Veterinary Record* 126, 522–525.

Ewbank, R. (1968) The behavior of animals in restraint. In: Fox, M.W. (ed.) *Abnormal Behavior in Animals.* W.B. Saunders, Philadelphia, Pennsylvania, pp. 159–178.

Faure, J.M. and Mills, A.D. (1998) Improving the adaptability of animals by selection. In: Grandin, T. (ed.) *Genetics and the Behavior of Domestic Animals.* Academic Press (Elsevier), San Diego, California, pp. 233–265.

Ferguson, D.M. and Warner, R.D. (2008) Have we underestimated the impact of pre-slaughter stress on meat quality in ruminants? *Meat Science* 80, 12–19.

Fordyce, G. (1987) Weaner training. *Queensland Agricultural Journal* 113, 323–324.

Fordyce, G., Dot, R.M. and Wythes, J.R. (1988) Cattle temperaments in extensive herds in northern Queensland. *Australian Journal of Experimental Agriculture* 28, 683–688.

Fulkerson, W.J. and Jamieson, P.A. (1982) Pattern of cortisol release in sheep following administration of synthetic ACTH or imposition of various stress agents. *Australian Journal of Biological Science* 35, 215.

Geverink, N.A., Kappers, A., van de Burgwal, E., Lambooij, E., Blokhuis, J.H. and Wiegant, V.M. (1998) Effects of regular moving and handling on the behavioral and physiological responses of pigs to preslaughter treatment and consequences for meat quality. *Journal of Animal Science* 76, 2080–2085.

Grandin, T. (1980a) Livestock behavior as related to handling facilities design. *International Journal for the Study of Animal Problems* 1, 33–52.

Grandin, T. (1980b) Observations of cattle behavior applied to the design of cattle handling facilities. *Applied Animal Ethology* 6, 9–31.

Grandin, T. (1981) Bruises on southwestern feedlot cattle. *Journal of Animal Science* 53 (Supplement 1), 213.

Grandin, T. (1982) Pig behaviour studies applied to slaughter-plant design. *Applied Animal Ethology* 9, 141–151.

Grandin, T. (1983) Welfare requirements of handling facilities. In: Baxter, S.H., Baxter, M.R. and MacCormack, J.A.D. (eds) *Farm Animal Housing and Welfare.* Martinus Nihoff, Boston, Massachusetts, pp. 137–149.

Grandin, T. (1988) Double rail restrainer for livestock handling. *Journal of Agricultural Engineering* 41, 327–338.

Grandin, T. (1989a) Behavioral principles of livestock handling. *Professional Animal Scientist* 5(4), 1–11.

Grandin, T. (1989b) Voluntary acceptance of restraint by sheep. *Applied Animal Behaviour Science* 23, 257.

Grandin, T. (1992) Observation of cattle restraint devices for stunning and slaughtering. *Animal Welfare* 1, 85–91.

Grandin, T. (1996) Factors that impede animal movement at slaughter plants. *Journal of the American Veterinary Medical Association* 209, 757–759.

Grandin, T. (1997a) Assessment of stress during handling and transport. *Journal of Animal Science* 75, 249–257.

Grandin, T. (1997b) The design and construction of handling facilities for cattle. *Livestock Production Science* 49, 103–119.

Grandin, T. (1998) The feasibility of using vocalization scoring as an indicator of poor welfare during slaughter. *Applied Animal Behaviour Science* 56, 121–128.

Grandin, T. (2001) Cattle vocalizations are associated with handling at equipment problems in beef slaughter plants. *Applied Animal Behaviour Science* 71, 191–201.

Grandin, T. (ed.) (2007a) Behavioural principles of handling cattle and other grazing animals under extensive conditions. In: Grandin, T. (ed.) *Livestock Handling and Transport.* CAB International, Wallingford, UK, pp. 44–64.

Grandin, T. (2007b) Handling and welfare of livestock in slaughter plants. In: Grandin, T. (ed.) *Livestock Handling and Transport.* CAB International, Wallingford, UK, pp. 329–353.

Grandin, T. and Deesing, M. (2008) *Humane Livestock Handling.* Storey Publishing, North Adams, Massachusetts.

Grandin, T. and Johnson, C. (2005) *Animals in Translation.* Scribner (Simon and Schuster), New York.

Grandin, T., Curtis, S.E., Widowski, T. and Thurmon, J.C. (1986) Electro-immobilization versus mechanical restraint in an avoid-avoid choice test. *Journal of Animal Science* 62, 1469–1480.

Grandin, T., Rooney, M.B., Phillips, M., Irlbeck, N.A. and Graham, W. (1995) Conditioning of nyala (*Tragelaphus angasi*) to blood sampling in a crate using positive reinforcement. *Zoo Biology* 14, 261–273.

Gross, W.B. and Siegel, P.B. (1982) Influence or sequences or environmental factors on the response of chickens to fasting and to *Staphylococcus aureus* infection. *American Journal of Veterinary Research* 43, 137–139.

Hale, R.H., Friend, T.H. and Macaulay, A.S. (1987) Effect of method of restraint of cattle on heart rate, cortisol, and thyroid hormones. *Journal of Animal Sciences* 65 (Supplement 1), 217 (abstract).

Hanggi, E.B. (2005) The thinking horse: cognition and perception reviewed. In: Broken, T.D. (ed.) *Proceedings of the 51st Annual Convention of the American Association of Equine Practitioners*, Seattle, Washington, 3–7 December 2005. American Association of Equine Practitioners, Lexington, Kentucky, pp. 246–255.

Hearnshaw, H. and Morris, C.A. (1984) Genetic and environmental effect on temperament score in beef cattle. *Australian Journal of Agricultural Research* 35, 723.

Hedigar, H. (1968) *The Psychology and Behavior of Animals in Zoos and Circuses.* Dover Publications, New York.

Heffner, R.S. and Heffner, H.E. (1983) Hearing in large mammals: horse (*Equis caballus*) and cattle (*Bos taurus*). *Behavioral Neuroscience* 97(2), 299–309.

Hemsworth, P.H. (1981) The influence of handling by humans on the behavior, growth and corticosteroids in the juvenile female pig. *Hormones and Behaviour* 15, 396–403.

Hemsworth, P.H., Coleman, G.J., Barnett, J.L. and Borg, S. (2000) Relationships between human–animal interactions and productivity of commercial dairy cows. *Journal of Animal Science* 78, 2821–2831.

Hixon, D.L., Kesler, D.J., Troxel, T.R., Vincent, D.L. and Wiseman, B.S. (1981) Reproductive hormone secretion and first service conception rate subsequent to ovulation control with Synchromate B. *Theriogenology* 16, 219–229.

Hutson, G.D. (1980) Visual field restricted vision and sheep movement in laneways. *Applied Animal Ethnology* 6, 175–187.

Hutson, G.D. (1985) The influence of barley feed rewards on sheep movement through a handling system. *Applied Animal Behaviour Science* 14, 263–273.

Hutson, G.D. (2007) Behavioural principles of sheep handling. In: Grandin, T. (ed.) *Livestock Handling and Transport.* CAB International, Wallingford, UK, pp. 155–174.

Jacobs, G.H., Deegan, J.F. and Neitz, J. (1998) Photo pigment basis for dichromatic colour vision in cows, goats, and sheep. *Visual Neuroscience* 15, 581–584.

Jephcott, E.H., McMillen, J.C., Rushen, J., Hargreaves, A. and Thorburn, G.C. (1986) Effect of electro-immobilization on ovine plasma concentrations of β-endorphin, β-lipotrophin, cortisol and prolactin. *Research in Veterinary Science* 41, 371–377,

Jones, R.B., Satterlee, D.G. and Cadd, G.G. (1998) Struggling responses of broiler chickens shackled in groups on a moving line: effects of light intensity hoods and curtains. *Applied Animal Behaviour Science* 38, 341–352.

Judd, S.P.D. and Collett, T.S. (1998) Multiple stored views and landmark guidance in ants. *Nature* 392, 710–714.

Kemble, E.D., Blanchard, D.C., Blanchard, R.J. and Takushi, R. (1984) Taming in wild rats following medical amygdaloid lesions. *Physiology and Behaviour* 32, 131.

Kilgour, R. and Dalton, C. (1984) *Livestock Behavior: A Practical Guide.* Westview Press, Boulder, Colorado.

Kilgour, R. and de Langen, H. (1970) Stress in sheep from management practices. *Proceedings, New Zealand Society of Animal Production* 30, 65–76.

King, D.A., Schuchle-Pletter, C.E., Randel, R., Welsh, T.H., Oliphant, R.A. and Baird, B.E. (2006) Influence of animal temperament and stress responsiveness on carcass quality and beef tenderness of feedlot cattle. *Meat Science* 74, 546–556.

Lambooy, E. (1985) Electro-anesthesia or electro-immobilization of calves, sheep, and pigs by Feenix Stockstill. *Veterinary Quarterly* 7, 120–126.

Lanier, J.L., Grandin, T., Green, R., Avery, D. and McGee, K. (2000) The relationship between reaction to sudden intermittent movements and sounds to temperament. *Journal of Animal Science* 78, 1467–1474.

Lay, D.C., Friend, T.H., Bowers, C.C., Grissom, K.K. and Jenkins, O.C. (1992a) A comparative physiological and behavioral study of freeze and hot iron branding using dairy cows. *Journal of Animal Science* 70, 1121–1125.

Lay, D.C., Friend, T.H., Randel, R.D., Bowers, C.C., Grissom, K.K. and Jenkins, O.C. (1992b) Behavioral and physiological effects of freeze branding and hot iron branding on crossbred cattle. *Journal of Animal Science* 70, 330–336.

LeDoux, J.E. (1996) *The Emotional Brain.* Simon and Schuster, New York.

Lemmon, W.B. and Patterson, G.H. (1964) Depth perception in sheep: effects of interrupting the mother-neonate bond. *Science* 145, 835–836.

Lewis, C.R.G., Hulbert, C.E. and McGlone, J.J. (2008) Novelty causes elevated heart rate and immune changes in pigs exposed to handling alleys and ramps. *Livestock Science* 116, 338–341.

Lewis, P.D. and Morris, T.R. (2000) Poultry and coloured light. *World Poultry Science Journal* 56, 189–207.

Lewis, R.J., Chalmers, G.A. and Barrett, M.W. (1977) Capture myopathy in elk in Alberta, Canada. *Journal American Veterinary Medical Association* 171, 927–932.

McGlone, J.J., McPherson, R. and Anderson, D.L. (2004) Case study: moving devices for market-sized pigs: efficacy of electric prod, board, paddle or flag. *Professional Animal Scientist* 20, 518–523.

Mertshing, H.J. and Kelly, A.W. (1983) Restraint reduces size of the thymus bland and PHA swelling in pigs. *Journal of Animal Science* 57 (Supplement 1), 175.

Miller, N.E. (1960) Learning resistance to pain and fear effects over learning, exposure, and rewarded exposure in context. *Journal of Experimental Psychology* 60, 137.

Mitchell, K., Stookey, J.M., Laturnar, D.K., Watts, J.M., Haley, D.B. and Huyde, T. (2004)The effects of blindfolding on behaviour and heart rate in beef cattle during restraint. *Applied Animal Behaviour Science* 85, 233.

Muller, R., Schwartzkopf-Genswein, K.S., Shah, M.A. and von Keyserlinkg, M.A.G. (2008) Effect of neck injection and handler visibility on behavioral reactivity of beef steers. *Journal of Animal Science* 86, 1215–1222.

Murphy, C.J., Neitz, C.J., Hoever, N.M. and Neitz, J. (2001) Photopigment basis for chromatic color vision in the horse. *Journal of Vision* 1, 80–87.

OIE (2009a) Chapter 7.3. *Transport of Animals by Land, Terrestrial Animal Health Code.* World Organization for Animal Health, Paris, France.

OIE (2009b) Chapter 7.5. *Slaughter of Animals, Terrestrial Animal Health Code.* World Organization for Animal Health, Paris, France.

Pajor, E.A., Rushen, J. and dePaisille, A.M.B. (2003) Dairy cattle choice of handling treatments in a Y maze. *Applied Animal Behaviour Science* 80, 93–107.

Panepinto, L.M. (1983) A comfortable minimum stress method of restraint for Yucatan miniature swine. *Lab Animal Science* 33, 95–97.

Pascoe, P.J. (1986) Humaneness of an electrical immobilization unit for cattle. *American Journal of Veterinary Research* 10, 2252–2256.

Petherick, J.C., Holyroyd, R.G., Doogan, V.J. and Venus, B.K. (2002) Productivity, carcass, meat quality of feedlot *Bos indicus* cross steers grouped according to temperament. *Australian Journal of Experimental Agriculture* 42, 389–398.

Phillips, M., Grandin, T., Graffam, W., Irlbeck, N.A. and Cambre, R.C. (1998) Crate conditioning of bongo (*Tragelephus eurycerus*) for veterinary and husbandry procedures at Denver Zoological Gardens. *Zoo Biology* 17, 25–32.

Pollard, J.C. and Littlejohn, R.P. (1994) Behavioral effects of light conditions on red deer in holding pens. *Applied Animal Behaviour Science* 41, 127–134.

Prescott, N.B. and Wathes, C.M. (1999) Spectral sensitivity of domestic fowl (*Galus g. domesticus*). *British Poultry Journal* 40, 332–339.

Price, E.O. and Wallach, S.J.R. (1990) Physical isolation of hand reared Hereford bulls increases their aggressiveness towards humans. *Applied Animal Behaviour Science* 27, 263–267.

Prince, J.H. (1970) The eye and vision. In: Swenson, M.J. (ed.) *Dukes Physiological of Domestic Animals.* Cornell University Press, New York, pp. 696–712.

Reinken, G. (1988) General and economic aspects of deer farming. In: Reid, H.W. (ed.) *The Management and Health of Farmed Deer.* Springer, London, pp. 53–59.

Ried, R.L. and Mills, S.C. (1962) Studies in carbohydrate metabolism in sheep, XVI. The adrenal response of sheep to physiological stress. *Australian Journal of Agricultural Research* 13, 282–294.

Ritter, M.J., Ellis, M., Anderson, D.B., Curtis, S.E., Keffaber, K.K., Killefer, J., McKeith, F.K., Murphy, C.M. and Peterson, B.A. (2009) Effects of multiple concurrent stressors on rectal temperature, blood acid base status and longissimus muscle glycolytic potential in market weight pigs. *Journal of Animal Science* 87, 351–362.

Rogan, M.T. and LeDoux, J.E. (1996) Emotion, systems cells, and synaptic plasticity. *Cell* 85, 469–475.

Rushen, J. (1986a) Aversion of sheep to electro-immobilization and physical restraint. *Applied Animal Behaviour Science* 15, 315–324.

Rushen, J. (1986b) Aversion of sheep for handling treatments: paired-choice studies. *Applied Animal Behaviour Science* 16, 363–370.

Rushen, J. and Congdon, P. (1986a) Relative aversion of sheep to simulated shearing with and without electro-immobilization. *Australian Journal of Experimental Agriculture* 26, 535–537.

Rushen, J. and Congdon, P. (1986b) Sheep may be more averse to electro-immobilization than to shearing. *Australian Veterinary Journal* 63, 373–374.

Sandem, A.I., Janczak, A.M., Salle, R. and Braastad, B.O. (2006) The use of diazepam as a pharmacological validation of eye white as an indicator of emotional state in dairy cows. *Applied Animal Behaviour Science* 96, 177–183.

Sebastian, T. (2007) Temperament in beef cattle: methods of measurement, consistency and relationship to production. University of Sasketchewan Library. Available at: http://library2.usask.ca/theses/available/etd-12112007–213618/ (accessed 30 November 2008).

Sheldon, C.C., Sonesthagan, T. and Topel, J.A. (2006) *Animal Restraint for Veterinary Professionals*. Mosby/Elsevier, St Louis, Missouri.

Smith, B. (1998) *Moving Em: a Guide to Low Stress Animal Handling*. Graziers Hui, Kamuela, Hawaii.

Stermer, R., Camp, T.H. and Stevens, D.C. (1981) Feeder cattle stress during transportation. American Society of Agricultural Engineers Paper No. 81–6001. American Society of Agricultural Engineers, St Joseph, Michigan.

Talling, J.C., Waran, N.K., Wathes, C.M. and Lines, J.A. (1998) Sound avoidance in domestic pigs depends on characteristics of the signal. *Applied Animal Behaviour Science* 58, 255–266.

Tanida, H., Miura, A., Tanaka, T. and Yoshimoto, T. (1996) Behavioral responses of piglets to darkness and shadows. *Applied Animal Behaviour Science* 49, 173–183.

Tillman, A. (1981) *Speechless Brothers: the History and Care of Llamas*. Early Winters Press, Seattle, Washington.

van Putten, G. and Elshof, W.J. (1978) Observations of the effects of transport on the well being and lean quality of slaughter pigs. *Animal Regulation Studies* 1, 247–271.

Voisinet, B.D., Grandin, T., Tatum, J.D., O'Connor, S.F. and Struthers, J.J. (1997a) Feedlot cattle with calm temperaments have higher average daily gains than cattle with excitable temperaments. *Journal of Animal Science* 75, 892–896.

Voisinet, B.D., Grandin, T., Tatum, J., O'Connor, S.F. and Struthers, J.J. (1997b) *Bos inicus*-cross feedlot cattle with excitable temperaments have tougher meat and a higher incidence of borderline dark cutters. *Meat Science* 46, 367–377.

Warner, R.D., Ferguson, D.M., Cottrell, J.J. and Knee, B.W. (2007) Acute stress induced by preslaughter use of electric prodders causes tougher meat. *Australian Journal of Experimental Agriculture* 47, 782–788.

Warriss, P.D., Brown, S.N. and Adams, S.J.M. (1994) Relationship between subjective and objective assessment of stress at slaughter and meat quality in pigs. *Meat Science* 38, 329–340.

Wasadikar, P.P., Paunikar, R.G. and Deshmukh, S.B. (1997) Bull horn injuries in rural India. *Journal of the Indian Medical Association* 95(1), 3–4, 16.

Watts, J.M. and Stookey, J.M. (1999) Effects of restraint and branding on rates and acoustic parameters of vocalization in beef cattle. *Applied Animal Behaviour Science* 62, 125–135.

Waynert, D.E., Stookey, J.M., Schwartzkopf-Gerwein, J.M., Watts, C.S. and Waltz, C.S. (1999) Response of beef cattle to noise during handling. *Applied Animal Behaviour Science* 62, 27–42.

Westervelt, R.G., Kinsman, D., Prince, R.P. and Giger, W. (1976) Physiological stress measurement during slaughter of calves and lambs. *Journal of Animal Science* 42, 833–834.

White, R.G., DeShazer, J.A., Tressler, C.L., Borcher, G.M., Davey, S., Warninge, A., Parkhurst, A.M., Milanuk, M.J. and Clens, E.T. (1995) Vocalization and physiological response of pigs during castration with and without anesthetic. *Journal of Animal Science* 73, 381–386.

Wollack, C.H. (1963) The auditory acuity of the sheep (*Ovis aries*). *Journal of Auditory Research* 3, 121–132.

6 Painful Husbandry Procedures in Livestock and Poultry

KEVIN J. STAFFORD AND DAVID J. MELLOR

Massey University, Palmerston North, New Zealand

Introduction

The management of farm livestock, poultry and other domesticated animals is based on the ability of people to control them and many aspects of their environments. Animal husbandry involves the provision of food and shelter, the control of reproduction, and the use of handling and restraint methods that are safe for both the stock handlers and the animals. It also includes measures designed to promote animal health by the prevention and treatment of disease. Some of these husbandry objectives are met by practices that are painful for the animals concerned. For each species several such practices may be used (Table 6.1) for a range of reasons that include the following:

- to minimize the risk of injury to animals and people (e.g. dehorning);
- to reduce aggressive behaviour and make male animals easier to handle (e.g. castrating oxen);
- to prevent carcass damage such as bruising (e.g. dehorning);
- to enhance carcass quality (e.g. castration);
- to reduce the risk of flystrike (e.g. mulesing in sheep);
- to allow other husbandry practices (e.g. shearing) to be undertaken more quickly and efficiently (e.g. tail docking);
- to prevent damage to the environment (e.g. nose ringing in pigs);
- to aid in identification (e.g. ear marking or notching, branding); and
- to enable the harvesting of products (e.g. velvet antler removal in deer).

Likewise, poultry are subjected to painful management practices such as beak trimming to minimize feather pecking and cannibalism, and trimming their wattles and comb (dubbing, desnooding) to minimize pecking injury by other birds.

The type of farming system determines when these procedures are carried out. For instance, animals managed indoors throughout life can be handled and treated at any time, whereas animals held on extensive pastoral properties may be several months of age before they are brought into yards for the first time and may thereafter be handled infrequently throughout life. The value of individual animals is also a factor as it influences the time that will be allotted to each one undergoing such procedures and it also affects assessments of the cost-effectiveness of using anaesthetic and analgesic drugs, veterinarians or veterinary technicians. In addition the availability of analgesics and veterinary expertise may be important. For instance, in some developed countries analgesic drugs may not be directly available to farmers due to legal restrictions. A veterinarian would be required to administer them. In other countries, a farmer could buy the drugs if he had a prescription from a veterinarian. The situation is more difficult in some developing countries where pain medicines may not even be available in rural communities for the people.

Most of the husbandry procedures used today were developed many years ago. They were usually selected to be carried out quickly, easily, cheaply and with inexpensive equipment, and to be generally safe for the animals and people involved (Stafford and Mellor, 1993). At the time of their development, little significance was attributed to any associated animal pain, apart from completing the procedure quickly, and scientific understanding

Table 6.1. Routine husbandry practices carried out on farm livestock.

Procedure	Cattle	Sheep	Goats	Horses	Pigs
Castration	+++[a]	+++	++	+++	++
Docking	+	+++		+	+++
Disbudding	+++		+		
Dehorning	+++	+			
Spaying	+				
Ear notching	+++	+++	++		+
Ear tagging	+++	+++	++		+
Branding – hot	++			+	
Branding – freeze	++			++	
Mulesing		+			
Teeth grinding		+			
Teeth clipping					+
Tusk grinding					+
Nose ringing	+				+
Tail nicking				+	

[a]+, In some management systems; ++, in many management systems; +++, in virtually all management systems.

of pain and its management in animals was poor. This situation has now changed.

General Aspects of Animal Pain

In general terms, there are three main categories of pain associated with husbandry procedures that cause tissue damage – they are acute, chronic and pathological pain (Flecknell and Waterman-Pearson, 2000; Gregory, 2004). Acute pain is generated by the initial act of damaging the tissues and by subsequent chemical changes within those tissues. Acute pain is linked to characteristic behavioural and physiological responses that are used to indicate its presence and demonstrate the efficacy of alleviation methods (Lester *et al.*, 1996; Mellor *et al.*, 2000). Chronic pain, which is usually less intense than acute pain, persists for the days or weeks it takes for the tissue damage to resolve and healing to be complete (Molony *et al.*, 1995; Sutherland *et al.*, 2000). The assessment of low-level chronic pain and differentiating it from irritation are problematical, and knowledge about the alleviation of chronic pain is poor. Pathological pain occurs when barrages of nerve impulses in pain pathways during the acute pain phase modify the operation of those pathways by effects at the injury site itself or in the spinal cord or brain (Mellor *et al.*, 2008). Animals affected by pathological pain are more sensitive to both painful and otherwise non-painful stimulation in the area of the injury or elsewhere. Pathological pain may last for weeks, months or years. It is not clear whether tissue-damaging husbandry procedures may cause pathological pain in farm livestock, but there is some evidence that castration of lambs soon after birth might have such an effect.

Behavioural and physiological indicators of pain

Pain-inducing stimuli elicit a range of physiological and behavioural responses. They allow the presence or absence of significant pain to be assessed and strategies for alleviating the pain to be developed (Mellor *et al.*, 2000). However, as pain is a subjective experience, which by definition is not quantifiable in absolute terms, all such variables are indirect indices of pain. Cautious interpretation is therefore required. Nevertheless, both physiological and behavioural indices have been used successfully in experimental settings to study animal pain responses. Physiological indices include changes in heart rate, blood pressure, rectal and skin temperatures, plasma levels of stress hormones and related metabolites, and brain electrical activity. Useful behaviours, which may be species-specific and/or injury-specific, enable acute, chronic and pathological pain to be distinguished, and they indicate when pain is present and, importantly, when it is at very low levels or absent. Some behaviour is overt and obvious, such as the initial marked restlessness of lambs after application of rubber castration rings (Lester *et al.*, 1996). Vocalization (bellows, moos or squeals) in cattle and pigs are associated with pain and distress during a procedure such as castration or branding. Cattle that were hot iron branded had more vocalizations compared to controls that were restrained and not branded (Lay *et al.*, 1992a, b; Watts and Stookey, 1999). Vocalization (squeals) in piglets was reduced during castration when anaesthetics were used (White *et al.*, 1995). Less overt behaviour such as the repetitive tail swishing in calves persists for about 8 h after dehorning (Sylvester *et al.*, 2004). Ear flicking after a procedure is another indicator of pain. Calves that had their tails docked had more ear twitching (Eicher and Dailey, 2002). In clinical settings, behaviour, demeanour and appearance are used more commonly because they are more immediately observable than are some key physiological indices (Mellor *et al.*, 2008). Changes in stress hormone concentrations or pain-related features of brain electrical activity require subsequent biochemical or mathematical analysis.

Grazing animals may hide pain

Older animals that are not completely tame, such as 4–6-month-old extensively raised calves and sheep, may not exhibit obvious symptoms of pain after castration or dehorning. A common comment by ranchers is: 'They must have little pain because my cattle ate and drank immediately after the operation'. Grazing animals are prey species that often hide obvious pain behaviours to avoid being eaten by predators. This is most likely to occur in animals with a large flight zone that have some fear of people (T. Grandin, personal communication, 2008). Observation of 8-month-old bulls castrated with heavy rubber bands illustrates this problem (Grandin and Johnson, 2005). Some bulls acted normally, whereas others repeatedly stamped their feet and a few of them lay on the ground in strange contorted postures. These pain-related behaviours disappeared when a person walked up to their pen, and animals that were rolling on the ground jumped up and ran over to their herd mates when they saw a person. In order to see the pain-related behaviours, the observer had to be hidden so the animals did not know they were being observed. For all species, the most accurate way to study pain-related behaviour after a procedure is to hide the observer or use a video camera. Sheep are the most defenceless prey species animal. They may be more likely to cover up and hide pain-related behaviour especially when they see people watching.

Alleviation of Animal Pain

The behavioural and physiological responses, and by inference the pain, caused by particular husbandry procedures may be reduced by choosing methods that cause less pain, by carrying out the procedures in young rather than older animals and by using pain-relieving drugs. A fuller analysis of matters that should be considered when deciding whether or not, and how, to undertake particular painful husbandry procedures is presented in Box 6.1 and discussed in the final section of this chapter.

Box 6.2 lists words associated with pain relief.

Box 6.1. Primary and secondary questions relevant to deciding whether or not, and how, to undertake a particular painful husbandry procedure (source Mellor *et al.*, 2008).

- *First*, is it necessary to perform the procedure?
  - What are the anticipated benefits of the anatomical modification?
  - Does the modification achieve the desired benefits?
  - Does the modification benefit a significant proportion of the treated animals?
  - How significant are the benefits (i.e. how pressing is the need to make the modification)?
  - Can the same benefits be achieved in other, less invasive ways?
- *Second*, what harms are caused by the procedure?
  - Does the procedure cause transient, short-term harm, such as acute pain and distress, when it is carried out?
  - Does the procedure cause longer-lasting harm, such as chronic pain and distress, during recovery from it?
  - Does the procedure itself cause harms that persist beyond recovery from the procedure, such as persistent adverse behavioural or functional changes?
  - What are the magnitudes of such transient, longer-lasting and/or persistent harms – how bad are they?
  - In what proportion of animals do such harms occur to a significant extent?
  - Are there effective ways of reducing any such significant harms?
- *Third*, do the benefits of the procedure outweigh the harms?
  - Do the act and the anatomical modification cause greater harms to the animals (individuals and groups) than they prevent?
  - Alternatively, are there sufficient direct benefits to the animal?
  - Are there sufficient indirect wider benefits, for instance, commercial, educational, recreational, scientific or social benefits, to offset the harms caused by the procedure itself and the associated anatomical modification to individuals or groups?

Box 6.2. Glossary of words associated with pain relief.

Alpha-2 adrenoreceptor agonists – a class of drug with sedative and analgesic effects (e.g. xylazine)
Analgesia – a state in which pain is absent or at very low levels
Analgesic – a drug that provides pain relief, usually given by injection or orally
Epidural anaesthesia – local anaesthetic is injected into the spinal canal
Local anaesthetic – drug used to prevent nerve transmission from a painful location
NSAID (non-steroidal anti-inflammatory drug) – a drug that may have anti-inflammatory effects at the site of injury only or that may, in addition, have analgesic effects on the spinal cord or within the brain (e.g. ketoprofen)
Systemic analgesia – analgesic drugs administered by various routes carried around the body in the bloodstream

Local anaesthetic nerve blockade

Local anaesthetics produce their effects by disrupting the function of nerve membranes and thereby preventing electrical impulses from passing along nerves (Flecknell and Waterman-Pearson, 2000; Mellor *et al.*, 2008). They are usually injected close to nerves in order to block sensations in particular parts of the body. The size of the area affected determines the name given to the type of nerve blockade, such that a 'local' blocks a relatively small area (e.g. the scrotum and testicles), a 'regional' blocks a larger area (e.g. parts of the flank and abdominal contents) and a 'spinal' or 'epidural' blocks a much larger region of the body (e.g. hind legs, perineal region, flank, uterus and some abdominal contents). The effectiveness of local, regional or epidural anaesthetics can easily be demonstrated by the absence of escape behaviour during tissue-damaging procedures (i.e. surgery).

Lignocaine is the most common local anaesthetic drug in worldwide veterinary use. It is used to alleviate the acute pain experienced by animals during and for 1–2 h after a number of painful procedures, including some livestock husbandry practices (Mellor and Stafford, 2000; Stafford and Mellor, 2005a, b). It is a short-acting local anaesthetic that is usually cleared from the site of injection quickly enough for its effects to last for about 60–120 min (Table 6.2). Lignocaine blockade of the cornual nerve prior to dehorning of calves virtually eliminates the plasma cortisol response indicative of pain for about 2 h (Stafford and Mellor, 2005a). There are concerns about a carcinogenic metabolite of lignocaine, called 2,6 xylidine, because in a 2-year study of rats about 50% of those receiving a high dose of the metabolite developed papillomata or carcinomata of the nasal cavity. In the UK, lignocaine has effectively been withdrawn from use in food animals, as the maximum, safe, residual levels for it have not yet been established. There are other local anaesthetics (Table 6.2) such as bupivacaine and mepivacaine, with different characteristics, that may be useful in livestock. For instance, bupivacaine blockade of the cornual nerve of calves virtually eliminates the cortisol response to dehorning for about 4 h (Stafford and Mellor, 2005a). However, most such local anaesthetics are not licensed for livestock.

Epidural nerve blockade is produced by injecting local anaesthetic into the epidural space of the spinal cord (Flecknell and Waterman-Pearson, 2000; Mellor *et al.*, 2008). Effective epidural analgesia can also be achieved by injecting alpha-2 agonists such as xylazine epidurally and this is used when castrating adult cattle because it has the added advantage of being accompanied by sedation. A lignocaine-xylazine mixture produces effective epidural analgesia in cattle castrated using a castration clamp (Burdizzo®) and extends the duration of analgesia. Although epidural injections were once considered to have practical difficulties and side effects that made them unattractive when castrating

Table 6.2. Local anaesthetics.

	Time to onset of action (min)	Duration of activity (h)
Lignocaine[a]	1–2	1–2
Bupivacaine	5–10	4–12
Mepivacaine	1–5	1–2
Procaine	5	1

[a] Lignocaine is generally licensed for use in animals, is cheap and short acting.

large numbers of bulls, this technique is now used in cattle veterinary practice.

Systemic analgesia

Systemic analgesics are carried in the bloodstream throughout the body and thus have the capacity to act generally or on specific target tissues depending on the biological actions of the particular drug (Flecknell and Waterman-Pearson, 2000; Stafford *et al.*, 2006; Mellor *et al.*, 2008). They may be administered orally, topically, per rectum or vagina, or by intravenous, intramuscular or intraperitoneal injection. They are appropriate for the treatment of pain caused by disease, trauma or surgery. Systemic analgesics are used infrequently in farm livestock at present. Their use is limited by concern about residues and cost.

In the past, when pain alleviation was applied it was focused on the surgical act itself, making surgery easier and safer by using local or regional anaesthesia together with sedation. Post-operative pain was given little or no attention because of a dearth of effective analgesics. Three major groups of systemic analgesics are available. They are the opioids, alpha-2 adrenoreceptor agonists, and non-steroidal anti-inflammatory drugs (NSAIDs). Most of the research on the effect of these analgesics on ruminants has involved sheep rather than cattle.

The opioids have long been used in animals but are not very effective in ruminants. Clinically, alpha-2 agonists are effective analgesics in sheep. Xylazine given intramuscularly produces significant analgesic effects in adult sheep but it appears to be less effective in cattle. Xylazine given alone to calves reduces the plasma cortisol response to dehorning but not to the degree achieved with a lignocaine blockade of the cornual nerve.

The development of NSAIDs with analgesic effects has been a major breakthrough in pain relief in cattle and sheep as these were the first really useful analgesics for these species. The analgesic effects of these drugs were considered to be due to their peripheral anti-inflammatory actions but some of them have central nervous system effects as well. The effectiveness of any particular NSAID as an analgesic will depend on its rate of clearance from the body, on the cause of pain and on poorly understood characteristics of the drug itself. Some NSAIDs, notably phenylbutazone, are more anti-inflammatory than analgesic; others such as carprofen are more analgesic than anti-inflammatory. The longer period of activity of some NSAIDs may make them more useful than opioids for treating non-ruminants. Some of the NSAIDs are very effective analgesics in cattle. The NSAID anti-inflammatory drug meloxicam given in conjunction with a local anaesthetic provides additional reduction in the physiological stress response after hot iron disbudding of calves (Heinrich *et al.*, 2009; Stewart *et al.*, 2009).

Most NSAIDs are strongly protein bound and tend to have long meat withholding times. However, ketoprofen has a very short biological half-life and there are no significant residues in the meat and milk of cows given ketoprofen intravenously. Used with local anaesthesia it virtually eliminated the cortisol response, and by inference markedly reduces the pain, for at least 12 h following dehorning and surgical castration of calves. Also, when given alone, ketoprofen significantly reduces the plasma cortisol response, and by inference the pain, in the hours following dehorning, surgical castration and Burdizzo® (or clamp) castration of cattle.

Some NSAIDs are effective analgesics in horses and have been used in the treatment of colic for years. Their use for post-operative pain has become popular recently. Opioids are also used in horses, and pethidine provides analgesia for a few hours if given intramuscularly. Alpha-2 adrenoreceptor agonists are also useful in horses but they are mainly used as sedatives in this species. The use of analgesics in pigs is poorly studied but some NSAIDs are effective for post-surgical pain control, while opioids are generally short acting in pigs and not very useful clinically. Our knowledge of analgesia in poultry is limited, but NSAIDs apparently reduce articular pain in chickens.

The future

Since the mid-1990s, many advances have been made in the field of pain research but much remains to be done. There are two major problem areas. One is scientific and involves determining what pain animals experience especially in the longer term. The other is practical and relates to what pain-relieving drugs livestock farmers can afford and will be permitted to use within existing regulatory controls. The now well-established physiological and behavioural responses of farm livestock to procedures such as dehorning and castration will enable these procedures to be used as models for the routine testing of the efficacy of currently available and new analgesics. These models could also be used to screen the efficacy of analgesics for treating the pain caused by different types of injuries.

The painful husbandry procedures discussed below are covered on a species basis with the more important procedures receiving the greatest attention. The *alleviation* of pain may be easy to bring about but the *elimination* of pain is a much more difficult proposition.

Cattle

Cattle are subjected to five common painful husbandry procedures (disbudding, dehorning, castration, branding and ear notching or tagging: Table 6.1). In addition, some heifers may be spayed, and dairy cattle may have their tails docked. Cattle are also subjected to procedures that may be stressful, but are not painful, so that these procedures will not be discussed here. They include pregnancy diagnosis, vaccination, different processes around artificial breeding, and treatment with anthelmintics (orally, by injection or transcutaneous) and ixodicides (by shower or dipping).

Disbudding and dehorning

Although the terms disbudding and dehorning are considered to be synonyms by some people, here disbudding is used to mean the prevention of horn growth before it has become advanced, and dehorning means the amputation of horns at any stage after their growth has progressed beyond the early budding stage. Disbudding of calves at a very young age is recommended. Disbudding calves by cauterizing the horn bud caused a significantly lower cortisol response compared to horn amputation (Stafford and Mellor, 2005a).

A summary of the types of procedures used for disbudding and dehorning is provided in Box 6.3.

General considerations

In general, cows with horns are more dangerous and difficult to manage, so many cattle routinely have their horns removed (Fig. 6.1). On farms where predation is a problem, cows may be left with horns to protect their calves and themselves. Likewise on organic dairy farms, horns may not be removed, and this can result in injuries during aggressive interactions between cows. Horned cattle are also more dangerous to people. In some countries it is forbidden to transport cattle with horns because they may injure other animals during the journey. Alternatively, before transport some cattle may have the tips cut off their horns, a process called 'tipping'. Cattle with large horns do not fit easily into head bails, and this makes restraint for oral drenching, ear-tag placement and other procedures difficult. Cattle without horns require less trough space. In some cows horn growth can be circular, which necessitates tipping every few years to prevent horns from growing back into the animal's head. Tipping should not be so extensive as to affect the pain-sensitive core of the horn (Fig. 6.2).

Although horn removal is a common practice, many breeds, particularly beef breeds, are now bred not to produce horns, that is they are polled.

Box 6.3. Disbudding and amputation dehorning.

Cautery disbudding – this is carried out on calves from birth to until about 6 weeks of age. A bar, often concave in shape, is heated and placed against the horn bud and rotated for sufficient time to destroy the horn bud germinal tissue.
Caustic chemical disbudding – the use of caustic chemicals such as potassium hydroxide in a stick form to destroy the horn buds. The stick is rubbed on to the bud on very young calves.
Surgical disbudding – a knife is used to cut off the horn bud or a scoop dehorner is placed around the base of the horn bud and closed, removing the bud and a 5 mm strip of skin around the horn bud.
Amputation dehorning – the horn and a 1 cm strip of tissue around the horn are removed by saw, embryotomy wire, guillotine shears or scoop.
Haemostasis – may be by cautery, a tourniquet, haemostatic powder, or the major blood vessels may be identified, picked up by forceps and twisted. A figure-of-eight tourniquet may be tied around the horns to prevent bleeding.
Local anaesthesia – may be by cornual nerve block using 5–10 ml of lignocaine injected midway along the lateral edge of the crest of the frontal bone. A ring block may also be used.

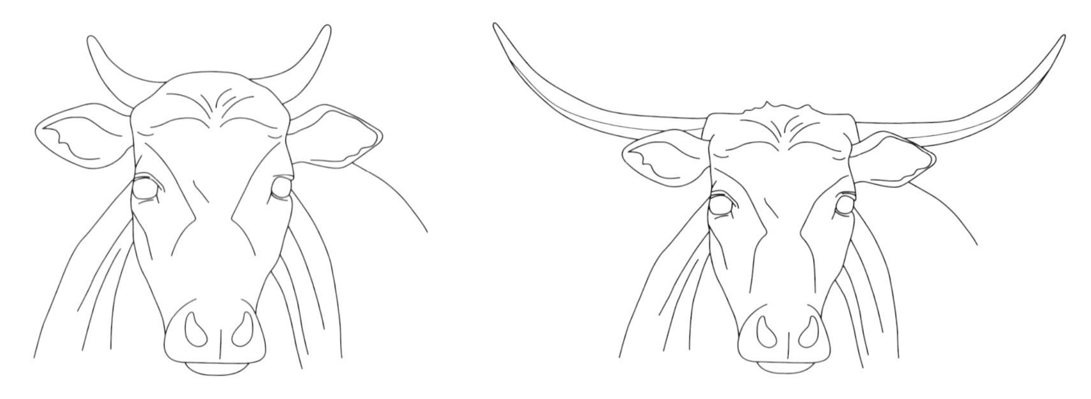

Fig. 6.1. Horns of cattle.

The genetics for polled characteristics in *Bos taurus* cattle are well understood but those for *Bos indicus* cattle are more complex. Crossbred cattle bred from polled and horned cattle or mixtures of them may or may not produce horns. The best way to reduce the pain and distress caused by horn injuries or horn removal would be to select for polled cattle. At present the common dairy breeds, Friesian and Jersey, are horned. It may be possible for such breeds to be bred polled if an effort were made to do so and this would completely remove the necessity for disbudding or dehorning.

Disbudding

In developed dairy industries the horn buds of calves are routinely destroyed by application of a caustic paste or stick (sodium hydroxide, potassium hydroxide, or colloidion) or by cautery (Stafford and Mellor, 2005a). Horn buds and the skin immediately surrounding them may also be removed surgically using a scoop or knife (Fig. 6.3). The ideal age for destroying the horn buds is during the first few weeks after birth. When caustic chemicals (paste or stick) are used, caution is required because housed calves may lick the caustic material from the horn buds of other calves and outdoors rain may wash the caustic material into the eyes of treated calves. When a cautery iron (heated electrically or by gas) is used it must be hot enough and applied for long enough to destroy the generative tissue in the skin around the base of the horn bud. If all of this tissue is not destroyed then horns may grow in a lopsided or distorted manner. Successful surgical disbudding is achieved by ensuring that each horn bud is removed together with the skin around its margin, which contains the generative tissue. Disbudding is carried out on dairy calves worldwide.

Fig. 6.2. Tipping horns.

The use of potassium hydroxide caustic stick to destroy the horn buds in 4-week-old calves causes intense pain for about 4h. As judged by cortisol stress responses and calf behaviour, both cautery disbudding and surgical (scoop) disbudding cause pain immediately and for some hours afterwards, but the amount of acute pain caused is apparently less with cautery than with surgical disbudding or destruction by caustic chemicals (Morisse *et al.*, 1995; Petrie *et al.*, 1996a). Thus cautery disbudding is preferable, but extreme escape behaviour of the calf shows that application of the hot iron is painful. This behaviour, and the often less obvious response that occurs with surgical disbudding, can

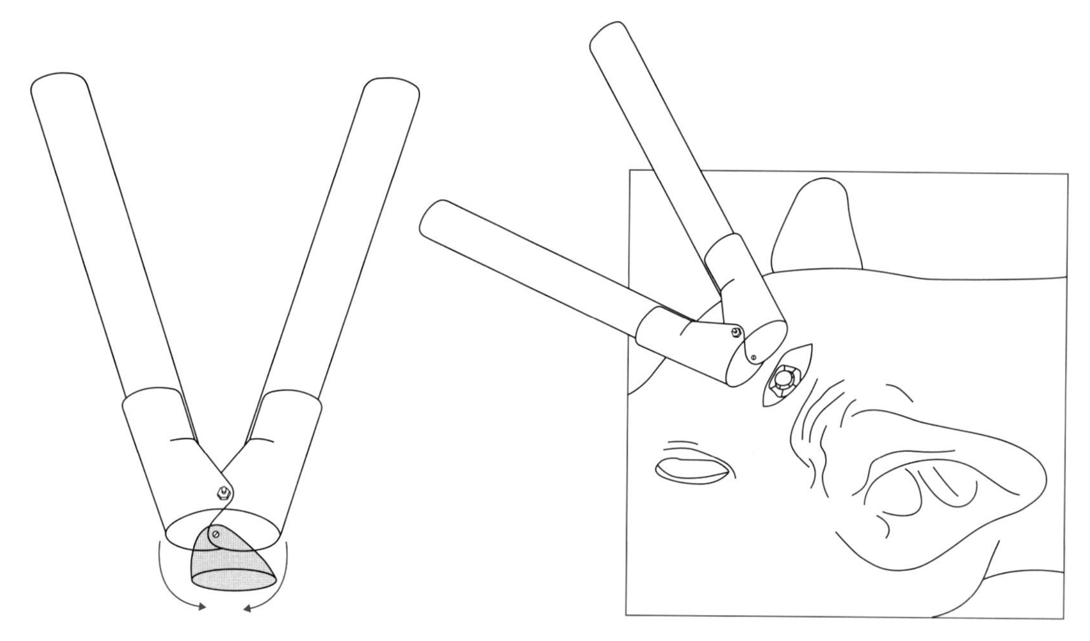

Fig. 6.3. Scoop dehorner.

both be eliminated using local anaesthetic blockade of the corneal nerve or a ring block around the base of each horn bud.

Dehorning

Removal of horns from mature cattle without anaesthetics and post-operative analgesia is very painful. Cutting horns on mature animals may be more painful than most other procedures. Crude methods of dehorning mature cattle, such as chopping off with a machete, should never be used.

Horns are removed from calves, weaners or adult cattle by amputation using a variety of tools (Stafford and Mellor, 2005a), including saws (manual and electric), embryotomy wire, guillotine shears and scoops (Fig. 6.4). During such dehorning, the horn and a small area of skin surrounding the horn base are removed. In larger cattle the wound may open up the frontal sinus. It is a painful procedure and cattle attempt to escape during it. Effective restraint is required if the amputation is done without local anaesthesia, and dehorning is often carried out with the animal held in a head bail or with its legs tied.

Haemorrhage will occur after dehorning. This may be minimized by using an embryotomy wire for dehorning rather than a saw, guillotine shears or scoop. Dehorning by embryotomy wire is harder work and takes longer than with a saw, but allows better control of the wound size. Guillotine shears are easy to use but if the head shifts during the procedure the wound may be larger than required and the frontal bone may fracture. The size of scoops makes them useful on smaller cattle but not

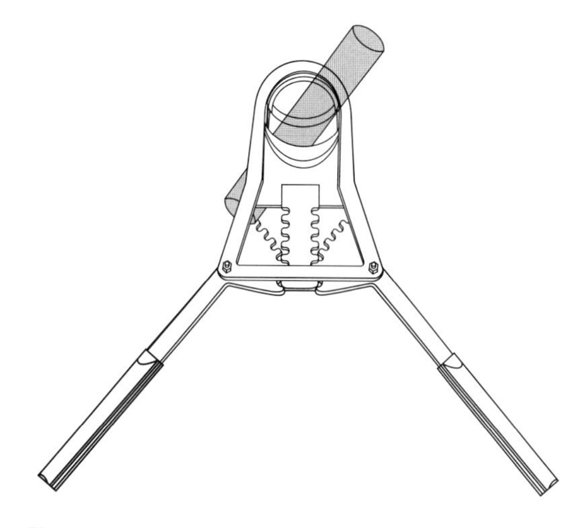

Fig. 6.4. Dehorning shears.

large cattle. Haemorrhage may be controlled by identifying the bleeding arteries and twisting or cauterizing them, or by ligating them. Haemostatic pads may be used, as may haemostatic powder. A tourniquet may be applied before amputation using a figure-of-eight pattern around the horns.

Wounds may become infected after dehorning and in summer they may become flystruck or be attacked by screwworm. Wounds need to be treated to prevent these problems from developing. In larger cattle the wound may open into the frontal sinus. If the cattle are housed, dirt may get into the sinuses and cause sinusitis. This may result in a purulent discharge from the wound and the sinus may have to be drained by trephination. With a large wound or when cattle are held indoors it is wise to cover the wounds with pads containing antibiotic powder. Wound healing after amputation dehorning takes several weeks.

The behavioural response to the pain caused by the act of amputating the horn can be eliminated using local anaesthetic blockade of the cornual nerve or by a ring block around the horn base. Cornual nerve blockade is not always successful and it is wise to test the skin around the base of the horn by needle prick before cutting begins to check whether more local anaesthetic may be required. The plasma cortisol response follows a defined pattern after amputation dehorning (Fig. 6.5). The concentration initially rises rapidly to a peak, then declines to a plateau and finally returns to pre-treatment levels after about 8 h (McMeekan *et al.*, 1998a). In one trial the size of the wound did not affect the magnitude of the cortisol response, which suggests that within the small range studied the wound size does not affect the acute pain caused by the procedure (McMeekan *et al.*, 1997).

Effective cornual nerve blockade using local anaesthetic eliminates the initial peak of the plasma cortisol response and cortisol concentrations remain at pre-treatment levels for about 2 h. Thereafter the concentrations increase for about 6 h before returning to pre-treatment levels (McMeekan *et al.*, 1998a). This is interpreted to indicate that effective corneal nerve blockade eliminates pain for about 2 h, after which some pain is experienced. This delayed pain is probably dull and not as acutely sore as the initial pain of horn amputation. To eliminate the acute cortisol response for at least 12 h the dehorned animal needs to be given a systemic analgesic along with the local anaesthetic; a combination of an NSAID with a local anaesthetic is effective in this regard (McMeekan *et al.*, 1998b; Stafford *et al.*, 2003; Milligan *et al.*, 2004; Stewart *et al.*, 2009). Long-lasting systemic analgesics have the potential to alleviate the pain for up to 3 days. Alternatively, when local anaesthetic is given before dehorning, cautery of the wound to control haemorrhage also reduces the plasma cortisol response, and by inference the pain experienced by cattle, for at least 24 h (Sutherland *et al.*, 2002). Dehorning alters feeding and rumination behaviour for 2 or 3 days, which suggests

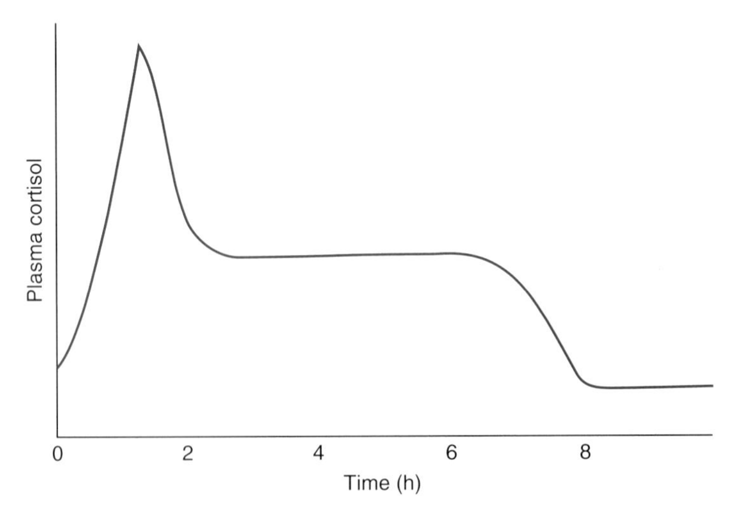

Fig. 6.5. Plasma cortisol response to amputation dehorning.

that the animals experience pain during this period. There is limited evidence that dehorning may not cause protracted pathological pain.

Castration

General considerations

Castration is a standard husbandry practice in most cattle production systems. It makes male animals safer and easier to handle, and reduces the likelihood of unwanted breeding by poorer quality bulls. Cattle are also castrated to become oxen or bullocks for ploughing and traction. Castration modifies the fat content of the beef carcass. Castrated cattle are sometimes given growth hormones to increase growth rates and accelerate finishing. In some countries, however, bulls are reared for beef. In New Zealand, for example, Friesian bull calves are reared intact for beef production and are usually slaughtered before they reach 2 years of age.

Methods

Castration is carried out using a variety of techniques (Stafford and Mellor, 2005b), including application of tight rubber rings or larger bands, surgery or a bloodless castration clamp (Burdizzo®, Fig. 6.6) (Box 6.4). Small rubber rings are used on calves and larger rubber bands on large cattle (Fig. 6.7). Surgical castration entails removing the distal scrotal wall and then either pulling the testicles and spermatic cords until the latter break or cutting through the spermatic cords. Ligatures or clamps may be placed on the spermatic cords before they are severed, or an emasculator may be used to sever the cords (Fig. 6.8). Bloodless castration clamps (Burdizzo®), which are applied across the spermatic cords, are available in different sizes.

When analgesia is not available, it is probably best for animal welfare to castrate young calves with a rubber ring. This method is easy, cheap and safe for the calves and the stock handlers using it. However, wounds formed by the rings are present for several weeks before the scrotum becomes detached. There are few side effects of using rings on small calves, but tetanus may occur if the scrotum is dirty when rings are put in place. Band castration of older post-pubertal bulls is carried out because it is believed that these animals have growth advantages over animals castrated early in life. With grass-fed beef cattle, however, there is little evidence to support this idea, and it is not recommended (Knight *et al.*, 2000). Bands may cause large wounds in the abdominal wall if placed too close to the abdomen. Banding 6-month-old bulls may cause pain and reduce growth at the time the testicles fall off (L.A. Gonzales, personal communication, 2009).

Clamp castration results in more failures than either rubber-ring or surgical castration (Kent *et al.*, 1996; Stafford *et al.*, 2002). This is important when steers are run with heifers, as the unintended presence of fertile males may result in poorly developed heifers being bred accidentally with all the potential welfare problems that entails. Failure occurs when one or both spermatic cords are not crushed sufficiently. Misapplication of the clamp may also cause crushing of the penis and damage to the urethra leading to urine leakage.

Surgical castration may be followed by haemorrhage, but this is seldom serious. The surgical wound may become dirty and infected. However, if the hole in the scrotum is large enough, drainage

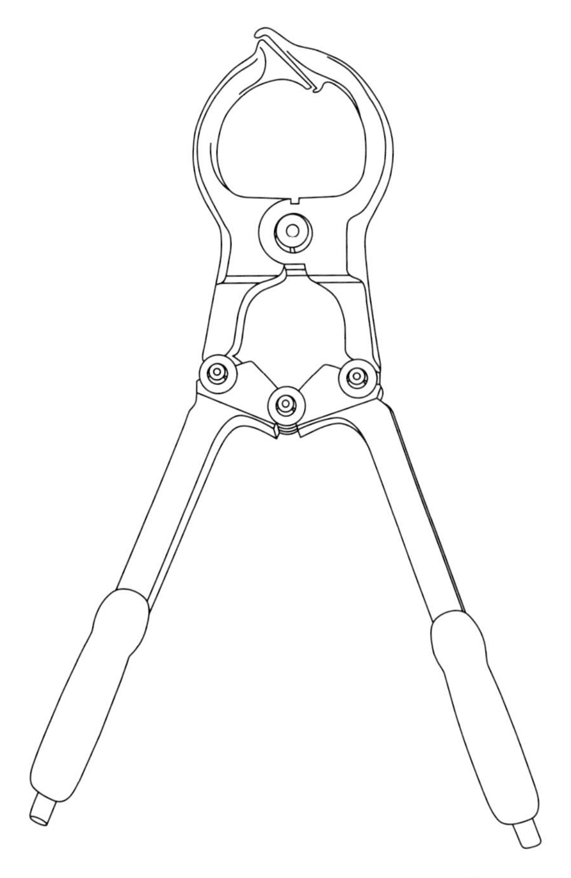

Fig. 6.6. Bloodless castration clamp – Burdizzo®.

Box. 6.4. Methods of castrating cattle.

Rubber ring – a rubber ring is placed around the scrotum dorsal to the two testicles using an applicator that expands the rubber ring and permits its placement on smaller calves.

Band – a large band is placed in a similar location to the rubber ring, but is looped around the neck of the scrotum and then tightened and tied with a specific applicator. Since this method is usually used on larger cattle, it is essential to apply the band very tightly to prevent swelling. The animals must be observed 12–24 h after application for swelling. The bands must be re-applied on animals that have swollen testicles. Failure to inspect animals for swelling can cause serious welfare problems. Some veterinarians also recommend vaccinating for tetanus when this method is used.

Clamp (Burdizzo®) – the clamp jaws are applied to the scrotum and one spermatic cord. The clamp is closed and left hanging from the scrotum for up to 1 min. This may be repeated 1 cm distal to the initial placement. Then it is repeated on the other side. An area of scrotal skin must be left intact between the crushed tissues on both sides.

Surgery – the distal scrotum is removed and each testicle and spermatic cord is dissected free. The spermatic cord is then either drawn out until it breaks or it is severed using an emasculator or by scraping a scalpel blade across it.

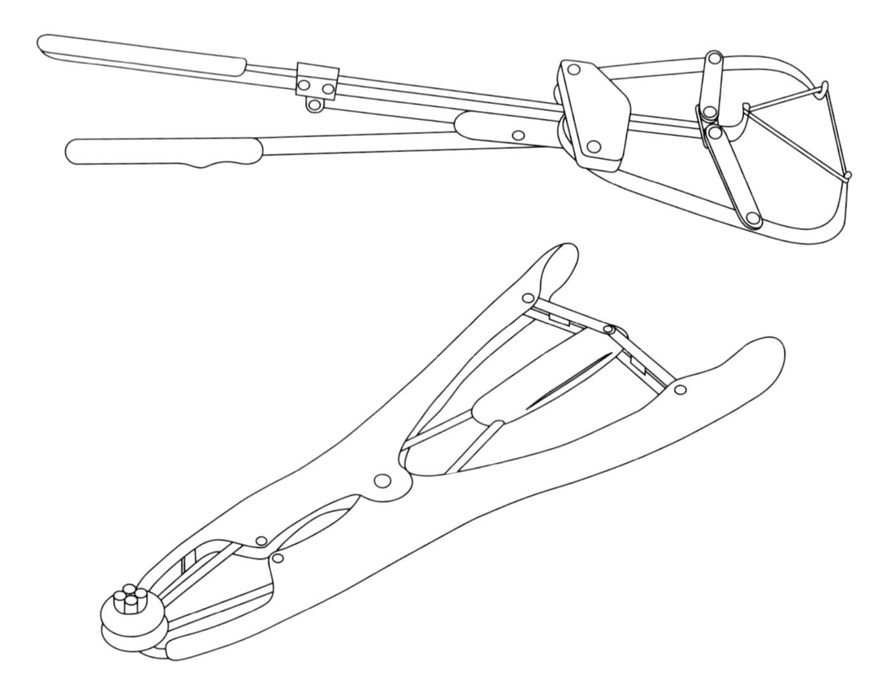

Fig. 6.7. Rubber rings used for castration.

will be sufficient to keep the wound clean. Tetanus may also occur if drainage is inadequate.

Pain alleviation

Castration is painful (Stafford and Mellor, 2005b; Stafford *et al.*, 2006). When no form of analgesia is used, cattle respond behaviourally in a manner that suggests pain. In addition, plasma cortisol concentrations increase over the first hour or so after castration and then return to pre-treatment levels after about 3–4 h (Fig. 6.9). The cortisol response to clamp castration is less than the response to rubber-ring or surgical castration, which suggests that the clamp is less painful (Stafford *et al.*, 2002). Local anaesthesia placed in the distal scrotum and

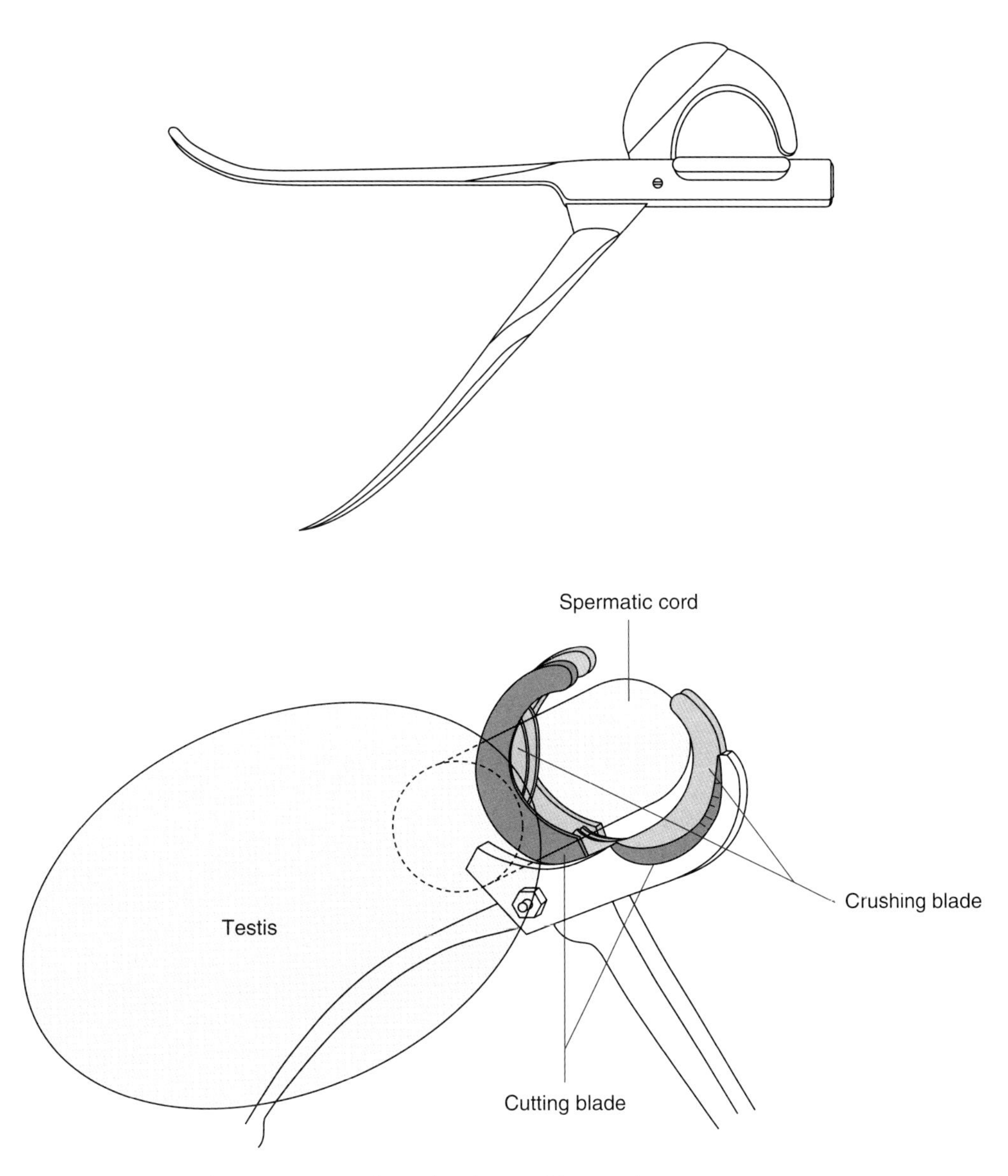

Fig. 6.8. Emasculator used to sever spermatic cord and prevent haemorrhage.

testicles eliminates or reduces the pain-related behaviours seen at castration by these methods. It eliminates the plasma cortisol response to castration by rubber ring or band. This is because the tight ring or band stops blood and lymph perfusion of the testicles and scrotum so that the local anaesthetic remains in those tissues beyond the time required for anoxic death of the pain receptors and associated nerves. In contrast, the local anaesthetic reduces the cortisol response to the clamp castration and has little overall effect on the cortisol response to surgical castration. The latter method, particularly when the spermatic cords are stretched or torn, results in damage to tissues in the inguinal canal and abdomen, areas that are not affected by local anaesthetic placed in the testicles and scrotum. To eliminate the plasma cortisol response and related pain due to clamp and surgical castration it is

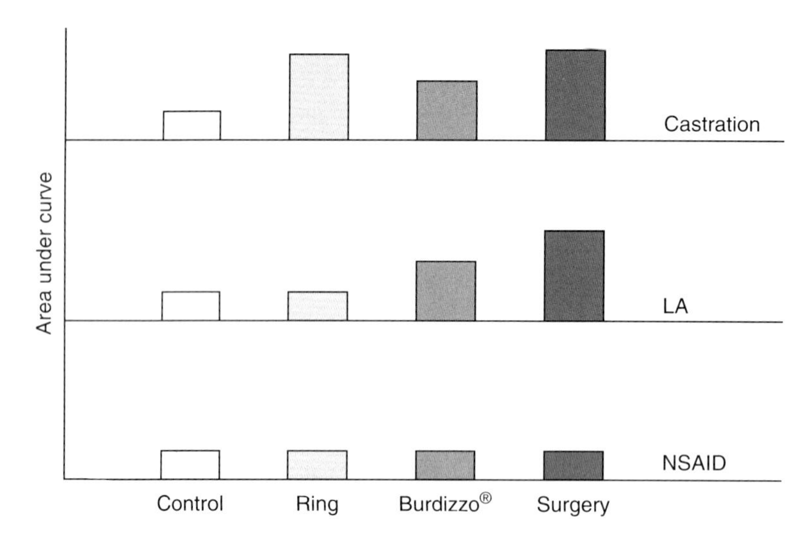

Fig. 6.9. Plasma cortisol response to different methods of castration. The area under the curve for 8h after castration in cattle castrated by ring, Burdizzo® or surgically is shown under three treatments: without local anaesthetic (Castration); with local anaesthetic (LA); and with local anaesthetic and a non-steroidal anti-inflammatory drug (NSAID). The control animals were handled as for castration but were not castrated.

necessary to combine administration of local anaesthetic with a systemic analgesic such as an NSAID (Stafford *et al.*, 2002, 2006; Ting *et al.*, 2004).

There is no obvious alternative to castration in the production of draught oxen, but for meat production, bulls can be reared and finished intact as seen in management systems in New Zealand and Europe. Another possibility is the use of immunocastration. The vaccines used for this are now available, and it is a matter of seeing whether this practice is acceptable to producers and consumers.

Spaying cattle

Heifers are spayed to prevent mis-mating and to prevent mating of cull cows in extensive farming systems. The ovaries are removed either through the vagina in larger cows and heifers following epidural anaesthesia or in smaller heifers through an incision in the flank (see Ohme and Prier, 1974). Special surgical instruments have been developed for cutting through the flank and for removing the ovaries. A flank incision is certainly painful and warrants local anaesthesia and probably post-surgical analgesia. Rendering heifers infertile is now possible immunologically and this will become popular if it is shown to be practicable and acceptable to consumers and producers.

Tail docking

Tail docking is carried out on dairy heifers either as calves or after calving to make access to the cow's udder easier during milking in a herringbone or rotary dairy shed. Docking has no beneficial effect on milk quality, udder cleanliness or on the incidence of mastitis or leptospirosis (see Stull *et al.*, 2002). Animals that had docked tails had more flies on them, and more fly avoidance behaviours such as foot stamping (Eicher *et al.*, 2001; Eicher and Dailey, 2002). Docking young calves by rubber ring or cautery is not especially painful, and local anaesthetic either given as a tail ring block or an epidural can be used to alleviate the pain (Petrie *et al.*, 1996b). Rubber-ring docking is preferable as haemorrhage may be significant after cautery docking. It is a serious and more painful procedure in heifers and is not recommended for them. Tails may be docked short at the level of the vulva or the tip of the tail together with the hair switch may be removed. The latter approach retains most of the tail but reduces the likelihood of stock handlers being hit by the tail hairs, with or without dung contamination, while accessing the udder. A non-surgical option is to trim the tail hairs to reduce the likelihood of the milker being hit by a tail covered with faeces. Tail docking of dairy cows is prohibited in some countries and in many natural and organic programmes. It is a procedure that should be phased out, because there is no scientific justification for it.

Ear tagging or ear notching

Ear tagging is a common way of identifying individual animals and a range of tags have been developed. Farmers often use large plastic ear tags to aid recognition at a distance. Individual identification facilitates management by allowing the easy recognition of superior animals for use in breeding programmes. The emphasis on trace-back of meat products and on disease control has led many countries to make identification of individual cattle mandatory. Such identification requires an easily recognized ear tag, made of plastic or metal, to be placed in one or both ears. Ear tags or ear notches are used also to indicate the ownership of animals grazed on common ground or those likely to stray on to other properties, and also help to safeguard against theft. Ear notches of different shapes and/or placed in different positions are also used to mark the year of birth, vaccine status and selection status for breeding.

The ear is a sensitive organ and ear tagging and notching are likely to be painful. Ear tagging side effects include infection, tears and loss of tags that then have to be replaced. Tag placement involves either punching a hole in the ear with a specific tool and then placing the tag, or the use of a tagging device that both punches the hole and places the tag. The tag contains a stud that pierces the ear. This type of tag usually makes a smaller hole and may be less painful because each tag contains a new sharp ear-piercing stud. Failure to sharpen the tool used to install the other type of tag may increase pain. Little work has been done on reducing the pain caused by these procedures, and as it is not easy to anaesthetize the ears using local anaesthetic, systemic analgesia would be preferable.

Ear tagging is definitely preferable from a welfare standpoint compared to wattling or ear splitting. In some countries, cattle ears are slit and long strips of skin are left dangling. When wattling is done, strings of hide and skin are cut and left dangling from the neck. Both of these procedures would be banned in many countries and they violate the voluntary industry welfare guidelines of many countries. These extreme cuts should not be confused with ear notching where small notches are made in the ear.

Branding

Branding by hot iron or freeze branding are common means of identifying cattle both for ownership and for recognizing individual animals (Lay *et al.*, 1992a, b). It is a common practice in countries with extensively raised cattle that are pastured on huge areas of land. It is used in Australia, the western USA, Central America, South America and many other countries. Hot branding causes definite escape behaviour in cattle and its cauterizing action is obviously painful. Freeze branding using irons first immersed in liquid nitrogen causes the destruction of pigment-producing melanocytes in the skin. The brands are placed on the skin and held there for an appropriate time (seconds rather than minutes). The skin freezes and when the brand is removed the skin warms up and some time later a white mark is left. The animals probably experience pain during thawing of the skin and reperfusion of it with blood, but hot branding is probably more painful than freeze branding. Little work has been done on reducing the pain caused by these procedures. As multiple injections of local anaesthetic in the area to be branded are impractical, systemic analgesia would probably reduce but not eliminate the pain experienced following branding. Animals should never be branded on sensitive parts of the body such as the face.

Microchips to identify cattle are becoming a feasible option to replace branding for identification and their placement will be less painful than branding. However, they are not really suited to everyday identification and it will take some time before hand-held identity chip readers become a widely used tool on cattle farms. Another alternative for animal identification is DNA testing. The cost of DNA testing is steadily falling and it may become economically feasible in the future.

Nose rings or ropes

Nose rings or nose ropes are placed in the nasal septum of cattle and water buffalo to help control fractious animals, especially bulls. They are usually placed while the animals are securely restrained or sedated, with or without local anaesthetic. Placement without local anaesthetic is likely to cause pain and related behavioural responses.

Sheep

The three most common painful husbandry procedures used in sheep are castration, tail docking and ear tagging or notching. Merino sheep in Australia may be subject to mulesing if flystrike is a problem. These procedures are generally carried out on lambs. Shearing of the whole body or to remove faecal pads (dagging) or wool from the perianal region (crutching), as well as pregnancy diagnosis, bathing or showering (backlining) with insecticides, oral drenching with anthelmintics and vaccination are stressful procedures but are not painful unless animals are injured by the procedure (e.g. cuts during shearing).

Castration

General considerations

Most male lambs in many countries are castrated. One reason is to prevent unwanted breeding of ewe lambs that are to be reared and finished for slaughter after puberty, the other is to prevent meat taint. Ram lambs that will be slaughtered before 5 or 6 months of age may not need to be castrated for these reasons, but may still be castrated to improve hygiene at slaughter, because in some breeds the intact scrotum may be large, woolly and contaminated with faecal matter.

Where ewes are managed indoors during lambing, castration and tailing can be carried out at any age from the second day after birth and certainly before lambs and ewes are let out on pasture. Avoiding the first day after birth promotes successful ewe–lamb bonding and a good colostrum intake. Maintaining clean deep litter to reduce floor contamination is important for indoor systems to minimize neonatal and post-castration infections. When lambing is managed outdoors over a 3-week period, it may be possible to muster the ewes and lambs into yards and castrate the lambs at 4 or 5 weeks of age without problems of mismothering or abandonment of lambs. In contrast, when ewes lamb over 6–8 weeks in flocks that have not been divided according to lambing dates, they should not be yarded for castration and related procedures until the danger of mismothering of the youngest lambs has passed. In addition, it is important that when castration is carried out outdoors it should be in dry yards or in temporary grass yards on fine days, without rain, sleet or snow, as muddy conditions may predispose lambs to tetanus or other infections.

In New Zealand and Australia the scrotum is removed from many lambs using a rubber ring applied below the testicles, leaving them intact. The lambs treated thus are called either cryptorchid lambs or short scrotum lambs (Fig. 6.10). In New Zealand 40% of male lambs are short scrotum lambs, 20% are castrated and 40% are left intact. The intact animals are slaughtered as prime lamb early in life, but short scrotum lambs may be kept as stores and slaughtered after puberty. A small proportion of short scrotum lambs may be fertile and behave sexually like a ram, but they are not usually considered a nuisance. In addition, short scrotum lambs grow better than wethers.

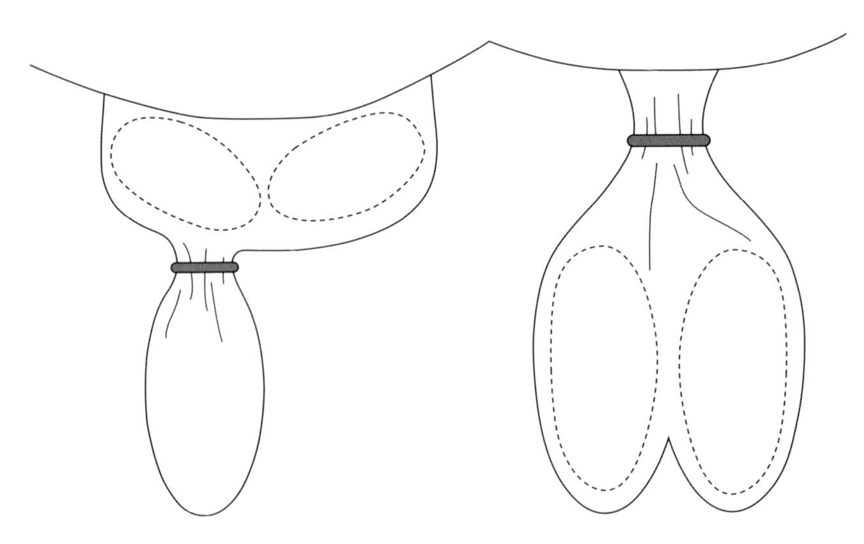

Fig. 6.10. Short scrotum and castration in lambs using a rubber ring.

Methods

Castration is carried out using a variety of techniques, including application of tight rubber rings, surgery or a bloodless castration clamp (Burdizzo®) (Mellor and Stafford, 2000). These techniques are essentially the same as those described above for cattle. For full castration the ring is placed around the neck of the scrotum above both testicles, although it is sometimes difficult to find both testicles in really young lambs as one may still be in the inguinal canal. With the short scrotum technique the ring is applied to the scrotum below the testicles to push them against the entrance to the inguinal canals. This makes the majority of the lambs infertile by elevating intratesticular temperature, but testosterone secretion continues, thereby explaining why such lambs retain the growth and behavioural characteristics of uncastrated males. Rubber-ring placement in young lambs is simple and straightforward. The rings used on young lambs may not be appropriate for older, larger lambs.

Surgical castration involves removal or incision of the distal scrotal sac so that the testicles and spermatic cords may be drawn out of the scrotum until the cords break. A number of knives have been designed to carry out this surgery. In older lambs the cords may be severed by knife or an emasculator rather than by traction, and in mature rams the cords may be ligated to prevent haemorrhage.

With clamp castration, the clamp is applied either individually across each spermatic cord from outside the scrotum, or across the full width of the scrotal neck including both spermatic cords at once. The latter approach is not recommended unless a ring is applied to the scrotal neck proximal to the clamp cut. In any case, if both rings and a clamp are available, the method of choice, overall, is rings. Clamp castration is easier on larger lambs with longer spermatic cords but, as described for calves, it is technically more difficult than either rubber-ring or surgical castration.

Pain alleviation

All physical forms of castration cause pain (Mellor and Stafford, 2000). Ring castration causes a smaller acute cortisol response than surgical castration but lamb behaviour is move active following ring castration than after surgical castration. This has led to disagreement as to which of these techniques causes more acute pain, but as the behavioural responses are specific to each technique it not possible to use behaviour in a comparative way. However, the acute cortisol responses, and by inference the pain, are markedly lower and shorter lived with ring than with surgical castration, so that ring castration is preferable in these terms (see Mellor and Stafford, 2000). Yet wound healing takes longer following ring castration than surgical castration. Lambs castrated by rubber ring still turn to look at their scrotal area about 42 days after castration, unlike surgically castrated lambs (Molony *et al.*, 1995). This suggests that they may still be experiencing some irritation or low-level chronic pain associated with ring-induced skin lesions before the scrotum finally becomes detached. There is evidence that ring castration of lambs within 3 days of birth causes less acute pain than subsequent castration, and, paradoxically, that it may increase the sensitivity of these lambs to other painful procedures conducted about 1 month later. These findings need to be elucidated.

Local anaesthetic injected into the scrotum and testicles eliminates the plasma cortisol response and by inference the acute pain caused by rubber-ring castration (Dinniss *et al.*, 1997). Kent *et al.* (1998) found that lignocaine injected into the neck of the scrotrum was more effective than injection into the testicles, but Dinniss *et al.* (1997) showed that intratesticular injections were just as effective but practically much easier to execute than scrotal neck injections. Local anaesthetics have less impact on responses to surgical or clamp castration, and with these techniques a systemic analgesic such as an NSAID is required together with local anaesthesia to eliminate the pain caused by castration. Squirting local anaesthetic into the scrotal neck and on to the spermatic cords before they are severed towards the inguinal area may reduce the pain experienced by lambs following surgical castration where the spermatic cords are drawn out and broken by traction. Following surgical castration the wound area must be treated to prevent screwworm infestation in areas where this parasite occurs.

Whenever possible, ram lambs should be left intact, but if they must be castrated then it is important to determine whether the short-scrotum procedure, which causes a lower physiological stress response than ring or surgical castration (Lester *et al.*, 1991), is acceptable to farmers, processor and consumers. If acceptable, this should be carried out in preference to full castration. Immunocastration vaccines for lambs are likely to become available in the near future and they could be used to eliminate the pain caused by castration if their use is acceptable throughout the production and consumption chain and provided that they are cheap enough.

Tail docking

Tail docking is carried out on woolly sheep and not on hair sheep. It is one method of reducing the incidence of flystrike (external myiasis) by decreasing the likelihood of faecal pads (dags) adhering to wool in the perineal region. It is also easier to shear and to crutch (removal of wool from the perineal area) sheep that have been docked. Tails may be cut short on lambs that are reared for meat. A longer tail, perhaps to the tip of the vulva, should be left on breeding animals as this reduces the likelihood of dags developing by maintaining a channel along which urine and faeces flow. Extremely short tail docking, flush with the body, must be avoided because it increases the incidence of rectal prolapse (Thomas *et al.*, 2003).

Tail docking may be carried out using a rubber ring, a cautery iron or bar, or a sharp knife (Mellor and Stafford, 2000). Rubber-ring and cautery tail docking elicit similar acute plasma cortisol responses, suggesting that they cause similar amounts of acute pain. In contrast, the use of a cold, sharp knife produces a substantially greater acute plasma cortisol response and probably significantly greater pain than the other two methods. Local anaesthesia may be injected as a ring block around the tail (subcutaneously), but this is difficult and time consuming. The use of local anaesthetics greatly reduced behavioural indicators of pain (Kent *et al.*, 1998). Unshorn lambs living in environments where flystrike or screwworm occurs must be treated with anti-flystrike and anti-screwworm medication following docking.

It is possible to breed sheep with reduced wool in the perineal area and for shorter and less woolly tails (Scobie *et al.*, 2007). This would reduce the necessity of docking to reduce flystrike, but shearing would still be impeded by the presence of the tail.

Ear tagging or ear notching

These procedures are similar to those carried out in cattle. Lambs' ears are smaller however, and relatively more damage is done during tagging or marking (Fig. 6.11). Identity microchips may reduce the need for ear marking and tagging in the future, but they are economically and practically difficult to use on large commercial sheep farms at present.

Mulesing

Mulesing is a surgical procedure used to remove wrinkled skin from the breech of some Merino sheep by cutting away skin from the perianal region. Care is taken to avoid the underlying muscle tissue. In addition, the tail will be docked. The procedure is typically performed by trained personnel with wool trimming shears, or modified versions of them, while the animal is restrained in a cradle. Carried out in Australia, it is generally recommended as essential to prevent flystrike (myiasis) in Merino sheep. It is usually done at marking time, that is when lambs are castrated, tail docked and ear tagged at 1–2 months of age. The procedure in its radical form helps to prevent flystrike

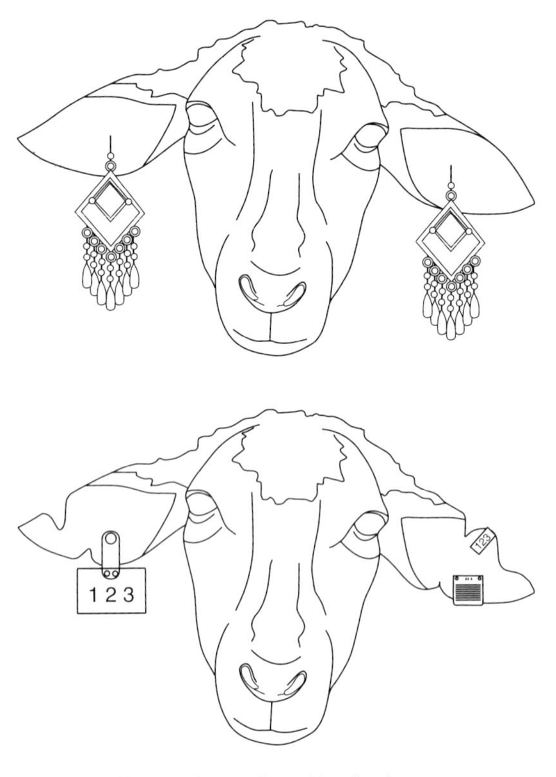

Fig. 6.11. Ear tagging and marking in sheep.

because wool-free skin around the anus (and vulva in females) is pulled tight and the cuts heal by forming scar tissue that does not get fouled with faeces and urine.

It is difficult to eliminate the pain caused by mulesing because an injectable local anaesthetic would have to be given epidurally and this is not feasible in young lambs under farm conditions. Applying local anaesthetic to the wounds topically after the operation does reduce the pain (Lomas *et al.*, 2008). The use of both a topical local anaesthetic and a long-acting NSAID would improve welfare. Carprofen plus a topical anaesthetic abolished the cortisol response (Paull *et al.*, 2008). Carprofen alone was not effective for reducing the cortisol response. Other less painful ways of removing the breech folds or wool are currently being researched in Australia. These include an injectable protein product that, when injected subcutaneously in the perineal region, may have these desired effects, as may the placement of clips specially designed to remove excess skin. Breeding Merino sheep with fewer breech folds or less wool is also being attempted. The length of tail left after mulesing/docking is important: a short tail is more likely to result in faecal pads (dags) and flystrike, whereas medium-length docking appears to reduce the likelihood of dags forming possibly by leaving a channel for urine and faeces to travel through.

The Australian wool industry aims to phase out mulesing by 2010. This will be facilitated by breeding Merino sheep with fewer wrinkles in the perineal area and the use of anti-fly vaccination or long-lasting insecticides (Elkington and Mahony, 2007; Rothwell *et al.*, 2007; Scobie *et al.*, 2007).

Teeth grinding

It was thought that shortening the incisor teeth of older ewes could extend their grazing efficiency and prolong their productive life. The teeth were ground down towards the gum using an electrical grinder. Although the procedure was shown not to be particularly painful, its use was discontinued in Australia where it was shown not to confer any significant production benefits.

Marking and processing

'Marking and processing' is the term used in some countries to describe the combination of procedures carried out on lambs when they are first brought into yards; the procedures may include castration, docking, ear tagging/notching, vaccination and, in Australia at present, mulesing. Except for vaccination, these procedures cause pain and it is difficult to provide analgesia sufficient to eliminate it. It may be worthwhile considering eliminating the pain caused by some procedures, for example, by using local anaesthetic for ring castration, and reducing the pain caused by other procedures, for example, by using a systemic analgesic for ear tagging/notching.

Local anaesthetic and systemic analgesic might be given together some time before the painful procedures are carried out. This would be possible with small flocks or when small numbers of lambs are marked at any one time such as with flocks lambed indoors. However, when hundreds or thousands of lambs are being marked outdoors it is not so feasible to handle them twice; that is to catch lambs to administer a local anaesthetic and systemic analgesic and then release them, and then catch again to complete the painful procedures. One option in such flocks may be to give systemic analgesia when the procedures are being carried out. This will alleviate some of the pain but not eliminate it. More research is required to assess the beneficial effects of such approaches. Where flystrike occurs many farmers spray lambs after docking with an anti-fly ectoparasiticide. If this spray were mixed with local anaesthetic, it may reduce, but would probably not eliminate, the pain caused by castration and tail docking.

Goats

Disbudding

Disbudding is carried out on young goats before the horn bud becomes too large. The horn-generating tissue is destroyed using either a heated iron that is placed over the horn bud or by a chemical agent that corrodes the tissue. The germinal tissue may also be removed surgically using a purpose-designed scoop or knife, similar to those used for cattle, or a knife or scissors. Disbudding goats by cautery may cause injury to the brain by heat transference through the skull, so the heated iron should not be applied for more than 5 s, allowing the same interval between successive applications if they are necessary. Local anaesthetic is often used when disbudding goats.

Castration and ear tagging or ear notching

These painful procedures and the use of analgesia are conducted in goats in a manner similar to those used in cattle and sheep.

Pigs

The management of pigs ranges from intensive indoor to extensive outdoor approaches. Piglets are subject to a number of painful procedures including castration, teeth clipping, tail docking and, if they are going to live outdoors, rings may be inserted into their noses. Generally, it is advised that these procedures be conducted as close to birth as practical, usually within 3–8 days, but not during the first 6 h.

Castration

Traditionally, pigs have been castrated surgically to prevent boar taint. In some production systems, pigs are slaughtered at lighter weights before boar taint becomes a problem and so castration is not required. The physiological and behavioural (Taylor and Weary, 2000) responses to castration indicate that it is certainly painful and pain relief is recommended. Local anaesthetics prevented pain-related behaviours in 2-week-old piglets, but not 7-week-old piglets (McGlone and Heilman, 1988). In Europe, general anaesthetic is now being recommended for pigs being castrated and some retailers intend accepting for sale only meat from pigs that have not been castrated. Gaseous anaesthesia technology suited to pig farms is being developed and it may soon become normal to give piglets a general anaesthetic and then to castrate them, clip teeth and dock tails at one time. A simple inhaler device can be built for US$100 and the cost of isoflurane was US$0.02–0.03 per pig (Hodgson, 2006). In older pigs, general anaesthetic before castration can be produced in a number of ways. A common cheap practice is to inject intra-testicular pentobarbitone sodium. The pig becomes unconscious and as the source of the drug is removed when the testicles are excised, they regain consciousness soon afterwards. However, intra-testicular injections of large volume of barbiturate may be painful. Immunocastration is a possible alternative to surgical castration (Dunshea *et al.*, 2001; Thun *et al.*, 2006) and may become more a widely accepted practice. In Brazil, immunocastration with a vaccination has become a common practice, but in Europe it is not very popular.

Teeth clipping

A common procedure in intensive pig systems is to clip or grind off the canine teeth of piglets to reduce facial damage to litter mates and damage to the sow's udder. The significance of the damage done by piglets to litter mates and the sow's udder varies between indoor and outdoor management systems. Some large pig farms have eliminated the practice. If clipped close to the gum the tooth may split and infection may enter the gum and tooth root area. Teeth clipping may not be needed in small litters. It is a painful procedure and there is no easy way to eliminate the pain unless it is carried out under general anaesthesia when castration is done. Grinding resulted in greater increases of cortisol compared to clipping (Marchant-Forde *et al.*, 2009).

Tail docking

Docking is carried out to prevent tail biting. It is often carried out before 8 days of age without pain relief, and involves removing one-third to one-half of the tail using clippers, a searing iron, knife or some other instrument that severs the tail immediately. Neuromas may develop and these may be painful (Simonsen *et al.*, 1991). The effect of tail docking on tail biting is questioned. Different breeds may be more prone to tail biting than others and modifying the environment, for example by providing straw bedding and more space, may reduce the problem. Tail docking is painful and local anaesthetic could be used to alleviate the pain in piglets that are not being castrated under general anaesthesia. When docking was done without anaesthesia, hot clipping resulted in higher frequency squealing and reduced growth compared to cold clipping (Marchant-Forde *et al.*, 2009), although hot clipping (cautery) caused a lower cortisol response, suggesting that pain was less 60 min after the procedure (Sutherland *et al.*, 2008).

Ear notching

Ear notching is carried out to identify piglets. Microchip identification technology will probably eliminate the need for ear notching in large pig farms.

Castration, teeth clipping, tail docking and ear notching are carried out at one time on young male piglets. The development of general anaesthesia for castration will reduce the pain experienced by male piglets during these procedures, but female piglets will probably have their teeth clipped, tails docked and ears notched without any anaesthesia. The three procedures probably cause moderate but transient pain (Noonan *et al.*, 1994; Prunier *et al.*, 2005). The distress caused by general anaesthesia may be more unpleasant than the pain caused by these procedures, but this would be exceptionally difficult, if not impossible to quantify. If general anaesthesia is not used then alternative protocols to alleviate pain should be developed and used.

Tusk grinding

Tusk grinding is undertaken in mature animals to reduce damage caused during fighting between aggressive boars. The canine teeth (tusks) are shortened using rotary grindstones, saws or embryotomy wire. It may need to be repeated every few years. Protruding tusks on boars can also cause very serious injuries to people.

Nose ringing

Pigs farmed outdoors, especially sows, often have rings, clips or pieces of wire placed in their noses to prevent digging and rooting (Fig. 6.12). Sows housed on pasture will destroy pasture land unless rings are used to reduce rooting. Pastures with ringed sows had significantly more grass cover compared to pastures with control sows (Erickson *et al.*, 2006). The rings may be purpose-designed for pigs and placed in the nasal septum (as with a bull ring), or clips or pieces of wire placed through the cartilage at the top of the snout. Rings are placed when the animal is restrained securely, often using a nose loop. Local anaesthesia and sedation could be used to reduce the pain caused by ring placement or general anaesthesia could be used. As rooting is probably independent of hunger, and is an activity pigs favour, the pain induced by movement of the rings during foraging is probably quite significant. Some welfare specialists prefer septum rings. A popular high welfare programme in the USA permits septum rings and forbids the use of rings in the outer rim of the nose. A ring in the septum would permit gentle rooting. The use of a single septum ring may be a reasonable trade-off between the welfare of the sow and protecting the pasture when sows are housed outdoors.

Deer

Velvet antler removal

The growing antler of deer is called 'velvet' because it is covered by soft hair that looks like the fabric of that name. Velvet is living tissue supplied with

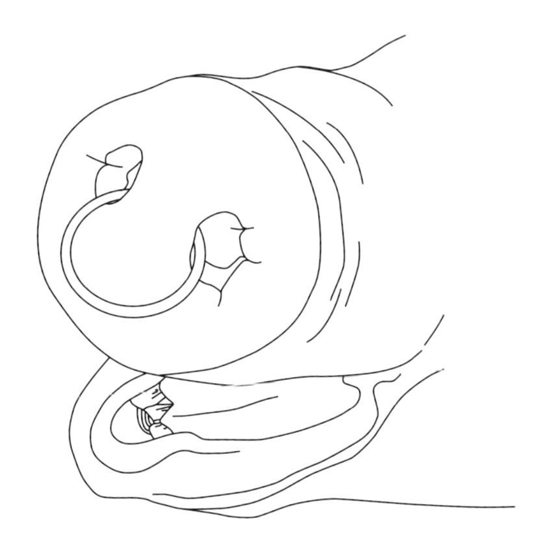

Fig. 6.12. Pig with a single nose ring in the septum (upper illustration); pig with multiple rings on the outer rim of the nose (lower illustration).

pain nerves and its removal is forbidden in some European countries. However, velvet is a widely used medicine in Korea and China. It is harvested from farmed and wild deer throughout Asia and from farmed deer in New Zealand, usually in early summer before the antler hardens and its nerve supply becomes non-functional. Antlers are quite dangerous weapons and some people recommend that they be removed from breeding stags for the safety of stock handlers and other animals. They may also get caught in fences and cause injury to the stag itself. However, many stags are reared and farmed only for velvet production, a farming practice that has been questioned for welfare reasons. In New Zealand, velvet harvesting is controlled by a national authority. This authority issues permits to veterinarians who supervise the farmers undertaking the procedure, and audits the veterinarians and the farmers. During velvet removal the deer are restrained either in a crush or by sedation, usually with xylazine. Prior to velvet removal local anaesthesia must be applied either to specific nerves, or better still as a ring block, to prevent or alleviate the acute pain caused by the procedure (Wilson and Stafford, 2002). The use of a compression device to reduce pain is not recommended – lidocaine anaesthesia is superior (Woodbury *et al.*, 2002; Johnson *et al.*, 2005). However, the presence of chronic pain has not been examined and it is not mandatory to use a systemic analgesic to alleviate this, if it exists, after the effects of the local anaesthetic wear off.

Horses

Horses are subject to only a few painful procedures, including castration. In some countries docking is allowed, as is tail nicking.

Castration

Many horses are castrated to make them easier and safer to handle and to prevent unwanted mating. Castration is carried out in many countries under general anaesthesia, or standing under sedation and with local anaesthesia (see Ohme and Prier, 1974), but in some countries neither anaesthesia nor analgesia is used, and horses are tied down during the operation. In some countries only registered veterinarians can castrate horses. The age at castration will vary depending on the breed and the use to which the animal will be put, but most horses are castrated at about 12 months of age. The side effects include haemorrhage, eventration (prolapse of a part of the small intestine into the inguinal canal), infection and tetanus. Post-operative analgesia is being used increasingly, and if an NSAID were used this may also reduce post-castration swelling.

Tail docking

Draught horses were often docked in the Middle Ages, as were hunters. This made it easier to keep draught horses clean and prevent tangling with harnesses and reins. Docking was also carried out to improve the appearance of the hindquarters and it was fashionable to dock hackney horses. Traditionally, docking shears were used and these were sometimes heated in a forge to sear the tissues.

The decline in the use of working horses resulted in a decline in this practice and in some countries it has been banned. Indeed, in the late 1930s the practice was condemned as cruel unless carried out because of irreparable tail damage or a malignant tumour. Docking horses is still practised in the USA for cosmetic reasons, and draught horses are shown there with docked tails. The tail is usually docked at the level of the ventral commissure of the vulva or just below the ischial arch in stallions and geldings.

There are two basic techniques used to dock horses. In very young foals breeders place a rubber ring on the tails. The hair is clipped to increase the speed of constriction and the loss of the tail. The alternative is to remove the tail surgically and this is described in many modern American texts on veterinary surgery and older British texts. The horse is sedated and an epidural anaesthetic given. The proposed amputation site is clipped and a tourniquet is placed proximal to it. Two large flaps of skin and coccygeal muscle are made on the ventral and dorsal side of the tail. The tail is amputated at an intracoccygeal joint and the flaps sutured together. The horses are immunized against tetanus.

Tail nicking

Tails are modified in hackney horses and American Saddlebreds to induce high tail carriage by a technique called nicking or tail setting. The procedure, apparently introduced into England in the early 18th century from the Low Countries of Europe, involves transecting the sacro-coccygeal muscles under the tail that depress it, so that the horse thereafter has to hold his tail aloft. The procedure is not described in modern veterinary surgery texts, but as long ago as the mid-1920s it was called a brutal operation. It is still carried out as a cosmetic exercise in the USA, but is banned in many other countries. Horses that have had this operation have difficulty swishing their tails to remove flies.

Poultry

Poultry are subjected to a few procedures that are painful. Laying hens may have their beaks trimmed as young chicks. Breeding birds may have their wattles removed (dubbing) and their claws removed. Turkeys, especially breeding turkeys, may have their snoods removed.

Beak trimming

Beak trimming is the most important amputation carried out on chickens and turkeys (Fig. 6.13). Its purpose is to prevent cannibalism, vent pecking and feather pecking in laying birds and broiler breeder flocks. Young broilers grown for meat are not beak trimmed. It is undertaken on birds managed under all conditions. For instance, in Switzerland in 2000, where cages for layer chickens had not been used since 1992, 59% of flocks (61% of hens) were beak trimmed (Hane *et al.*, 2000). These birds were kept in deep litter, grid-floor or aviary systems, many with access to outdoor runs. Thus, even in non-cage-layer systems beak trimming is common.

Beak trimming is carried out using a cauterizing blade or infrared beam, and chickens are trimmed either during the first 10 days after hatching or later at 16–18 weeks of age when moved from the rearing house into the laying house. The operator should not remove more than one-half of the upper beak and one-third of the lower beak, being about 3 mm and 2.5 mm in 1-day-old chicks, respectively. In 10-day-old birds, not more than 4.5 mm and 4 mm should be removed from the upper and lower beaks, respectively. In large commercial hatcheries, beaks are trimmed on egg-layer chicks at 1 day of age. The infrared-beam treatment is a highly automated system that uses an infrared beam and is used in many countries (Goran and Johnson,

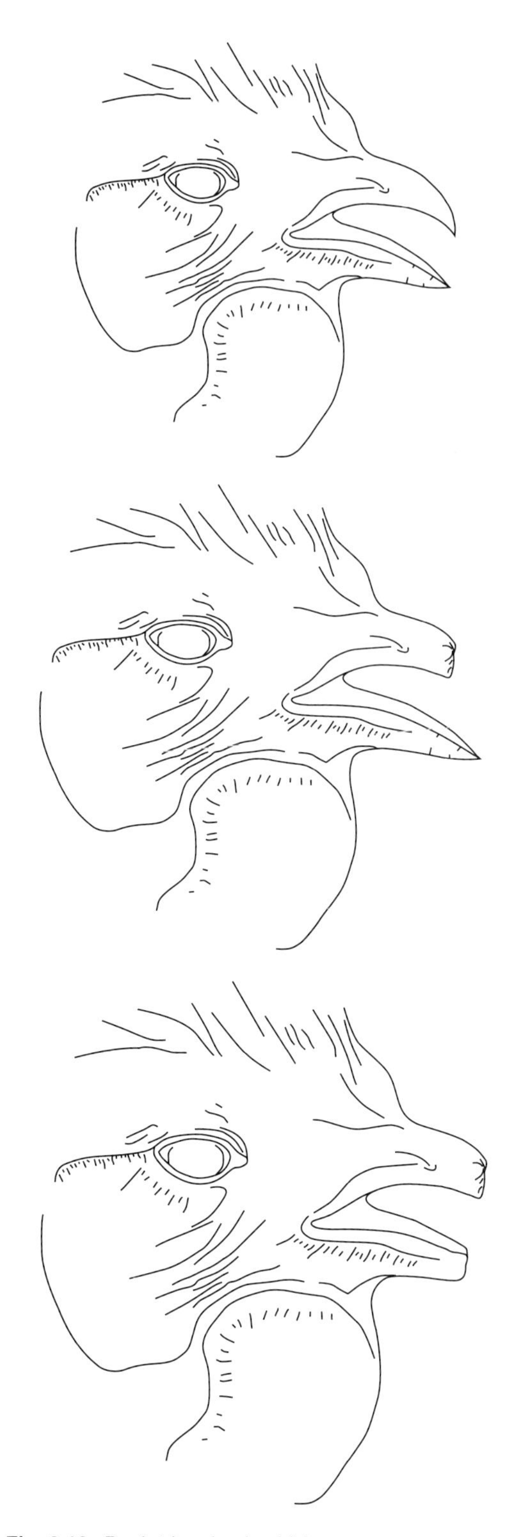

Fig. 6.13. Beak trimming in chicks.

2005). Proper maintenance and adjustment of the machine are essential. Unpublished industry data initially indicated that chickens that received the infrared-beam treatment performed more poorly than chicks trimmed with a hot blade until the machines were properly adjusted. When the infrared-beam treatment machine is working correctly, it is superior to the hot blade. One advantage of the machine is that its operational effectiveness remains consistent throughout its use. Birds trimmed with a hot blade had more deviations of the beak compared to birds subjected to the infrared beam (Marchant-Forde *et al.*, 2008). Both beak-trimming methods resulted in weight loss, with the hot-blade birds having more weight loss (Gentle and McKeegan, 2007). Both beak-trimming methods resulted in acute pain (Marchant-Forde *et al.*, 2008). Heart rates were higher immediately after the hot-blade compared to the infrared-beam treatment (T. Grandin, 2008, personal communication). Beak trimming may have to be repeated in older birds to prevent cannibalism. There is a great need for research on the use of pain-relief medications for beak trimming.

The beak is a sensitive organ with many sensory receptors. Beak trimming is acutely painful and food intake levels decrease following it (Glatz, 1987). Behavioural and physiological evidence suggests that pain and beak sensitivity persists for weeks or months after trimming (see Craig and Swanson, 1994). Neuromas can form leading to pain (Gentle, 1986; Gentle *et al.*, 1990). To prevent the formation of neuromas less than 50% of the beak should be removed (Kuenzel, 2007). However, it may have long-term beneficial effects in that the weight of the adrenal glands and the plasma corticosterone levels of trimmed birds suggest that they are less stressed than birds with intact beaks. A local analgesic (bupivacaine and diethyl sulfoxide) administered to the cut beak of 6-week-old chicks after trimming prevents at least some of the acute pain and the reduction in feed intake usually seen in the first 24 h after trimming.

It is suggested that the pain of trimming is outweighed by the likelihood of beak-related injuries and the suffering caused by cannibalism, vent pecking and feather plucking. It is strongly recommend that beak trimming be carried out only on chicks less than 10 days of age. It is possible to breed against the tendency to feather peck (Craig and Muir, 1993). Some birds that are highly selected for egg production are more prone to peck each other. Specific environmental features, such as perches, reduce the likelihood of cannibalism (Gunnarsson *et al.*, 1999) and giving hens the opportunity to forage may reduce feather pecking (Blokhuis, 1986). Thus, a combination of using specific genetic lines of laying hens that have a low tendency to feather peck, and the use of particular environmental features, might reduce cannibalism to the degree that beak trimming is not necessary.

Dubbing

Dubbing is the surgical removal of the comb and wattles using scissors. Male breeding birds are usually dubbed to prevent serious injuries to the comb and wattles, which, when left intact, are targeted during fighting between birds. It should be carried out during the first few days after hatching. Local anaesthetic can be used to reduce the acute pain caused by the procedure in older birds and in these haemorrhage may be a problem.

Desnooding of turkeys

Amputating the snood may help to prevent head injuries from pecking and fighting. It is also thought to reduce the risk of frostbite and the spread of erisiphelas. The snood is removed from 1-day-old birds using the thumbnail and finger and in older birds it can be cut off using scissors. Desnooding is apparently a less common practice today than it once was.

Claw removal

The last joint of the inside toes of male breeding birds is removed to prevent injuries to the hens during mating. This should be done at an early age, probably within 72 h of hatching. For older birds, only the nail should be trimmed. However, spurs may be removed or trimmed from roosters to reduce injuries to handlers and to other birds during fighting. Toes may be clipped off each foot of 1-day-old turkeys to prevent scratched and torn backs which are important causes of carcass downgrading. These amputations are carried out without local anaesthetic using sharp scissors or shears. Haemorrhage is not usually much of a problem. Genetic selection for lower aggression could be one method that could be used to eliminate the need for this procedure. There are genetic effects on aggression in roosters (Millman and Duncan, 2000).

Concluding Remarks

The decision to use a painful husbandry procedure and whether or not to use pain relief may be facilitated by reference to a series of questions (Box 6.1). In essence these questions are asked: Is the procedure justified in terms of direct benefits to the animals and/or by other benefits such as to the farming enterprise? What harms are caused, how bad are they and can they be avoided or reduced? One harm is pain, so that providing relief from it helps to address our responsibility to take all practical steps to minimize that harm. In this context, additional issues need to be considered such as the effectiveness, availability and cost of pain-relieving drugs, their ease of administration, possible side effects and residue concerns. In some countries, many drugs have to be administered by veterinarians and this raises the matter of the availability of veterinarians and the cost. A further matter relevant to cost is whether it will be recovered in farm-gate prices, or whether in the world of today, the cost of providing pain relief is becoming a farm overhead that must be absorbed in order to do business with trade partners in welfare-sensitive markets.

References

Blokhuis, H.J. (1986) Feather-pecking in poultry: its relation with ground pecking. *Applied Animal Behaviour Science* 16, 63–67.

Craig, J.V. and Muir, W.M. (1993) Selection for the reduction of beak inflicted injuries among caged hens. *Poultry Science* 72, 411–420.

Craig, J.V. and Swanson, J.C. (1994) Welfare perspectives on hens kept for egg-production. *Poultry Science* 73, 921–938.

Dinniss, A.S., Mellor, D.J., Stafford, K.J., Bruce, R.A. and Ward, R.N. (1997) Acute cortisol responses to lambs to castration using rubber ring and/or a castration clamp with or without local anaesthetic. *New Zealand Veterinary Journal* 45, 114–121.

Dunshea, F.R., Corantoni, C., Howard, K., McCauley, I., Jackson, P., Long, K.A., Lopaticki, S., Nugent, E.A., Simms, J.A., Walker, J. and Hennessy, D.P. (2001) Vaccination of boars with GnRH vaccine (Improvac) eliminates boar taint and increases growth performance. *Journal of Animal Science* 79, 2524–2535.

Eicher, S.D. and Dailey, J.W. (2002) Indicators of acute pain and fly avoidance behaviors in Holstein calves following tail docking. *Journal of Dairy Science* 85, 2850–2858.

Eicher, S.D., Morrow-Tesch, J.L., Albright, J.L. and Williams, R.E. (2001) Tail docking alters numbers of fly avoidance behaviors and cleanliness, but not physiological measures. *Journal of Dairy Science* 84, 1822–1828.

Elkington, R.A. and Mahony, T.J. (2007) A blowfly strike vaccine requires an understanding of host–pathogen interactions. *Vaccine* 25(28), 5133–5145.

Erickson, J., Studnitz, M., Strudsholm, K., Kongsted, A.G. and Hermansen, J.J.E. (2006) Effect of nose ringing and stocking rate of pregnant and lactating sows on exploratory behavior, grass cover and nutrient loss potential. *Livestock Science* 104, 91–102.

Flecknell, P. and Waterman-Pearson, A. (2000) *Pain Management in Animals.* W.B. Saunders, London.

Gentle, M.J. (1986) Neuroma formation following partial beak amputation (beak trimming) in the chicken. *Research in Veterinary Science* 41, 383–385.

Gentle, M.J. and McKeegan, E. (2007) Evaluation of the effects of infrared beak trimming in broiler chicks. *Veterinary Record* 160, 145–148.

Gentle, M.J., Waddington, J.D., Hunter, L.N. and Jones, R.B. (1990) Behavioral evidence for persistent pain following partial beak amputation in chickens. *Applied Animal Behaviour Science* 27, 149–157.

Glatz, P. (1987) Effects of beak trimming and restraint on heart rate, food intake, body weight and egg production in hens. *British Poultry Science* 28(4), 601–611.

Goran, M.S. and Johnson, S.C. (2005) Beak treatment with tongue protection. US Patent 7,363,881B2, Patent Office, Washington, DC.

Grandin, T. and Johnson, C. (2005) *Animals in Translation.* Scribner (Simon and Schuster), New York.

Gregory, N.G. (2004) *Physiology and Behaviour of Animal Suffering.* Blackwell Publishing, Oxford, UK.

Gunnarsson, S., Keling, L.J. and Svedberg, J. (1999) Effects of rearing factors on the prevalence of floor eggs, cloacal cannibalism and feather pecking in commercial flocks of loose housed laying hens. *British Poultry Science* 40, 12–18.

Hane, M., Huber-Eicher, B. and Frohlich, E. (2000) Survey of laying hen husbandry in Switzerland. *World Poultry Science Journal* 56, 21–31.

Heinrich, A., Duffield, T.F., Lissemore, K.D., Squires, E.J. and Millman, S.T. (2009) The impact of meloxicam on postsurgical stress associated with cautery dehorning. *Journal of Dairy Science* 92, 540–547.

Hodgson, D.S. (2006) An inhaler device using liquid injection of isoflurane for short-term anesthesia of piglets. *Veterinary Anaesthesia and Analgesia* 33, 207–213.

Johnson, C.B., Wilson, P.F., Woodbury, M.R. and Calkett, N.A. (2005) Comparison of analgesic techniques for antler removal in halothane anesthetics red deer (*Cervus elaphus*) electroencephalo-graphic responses. *Veterinary Anaesthesia and Analgesia* 32, 61–71.

Kent, J.E., Thrusfield, I.S., Robertson, I.S. and Molony, V. (1996) Castration of calves: a study of the methods

used by farmers in the United Kingdom. *Veterinary Record* 138, 384–387.
Kent, J.E., Molony, V. and Graham, M.J. (1998) Comparison of methods for the reduction of acute pain produced by rubber ring castration of week-old lambs. *Veterinary Journal* 155, 39–51.
Knight, T.W., Cosgrove, G.P., Death, A.F. and Anderson, C.B. (2000) Effect of method of castrating bulls on their growth rate and liveweight. *New Zealand Journal of Agricultural Research* 43, 187–192.
Kuenzel, W.J. (2007) Neurological basis of sensory perception: welfare implications of beak trimming. *Poultry Science* 86, 1273–1282.
Lay, D.C., Friend, T.H., Randel, R., Bowers, C.C., Grissom, K.K. and Jenkins, O.C. (1992a) Behavioral and physiological effects of freeze and hot iron branding on crossbred cattle. *Journal of Animal Sciences* 70, 330–336.
Lay, D.C. Jr, Friend, I.S., Bowers, C.L., Grisson, K.K. and Jenkins, O.C. (1992b) A comparative physiological and behavioural study of freeze and hot-iron branding using dairy cows. *Journal of Animal Science* 70, 1121–1125.
Lester, S.J., Mellor, D.J., Ward, R.N. and Holmes, R.J. (1991) Cortisol responses of young lambs to castration and tailing using different methods. *New Zealand Veterinary Journal* 39, 134–138.
Lester, S.J., Mellor, D.J., Holmes, R.J., Ward, R.N. and Stafford, K.J. (1996) Behavioural and cortisol responses of lambs to castration and tailing using different methods. *New Zealand Veterinary Journal* 44, 45–54.
Lomas, S., Sheil, M. and Windsor, P.N. (2008) Impact of topical anesthesia on pain alleviation and wound healing in lambs after mulesing. *Australian Veterinary Journal* 86, 159–168.
Marchant-Forde, R.M., Fahey, A.G. and Cheng, H.W. (2008) Comparative effects of infrared and one-third hot blade trimming on beak topography behaviour and growth. *Poultry Science* 87, 1474–1483.
Marchant-Forde, J.N., Lay, D.C., McMunn, K.A., Cheng, H.W., Pajor, E.A. and Marchant-Forde, R.M. (2009) Post natal piglet husbandry practices and wellbeing: the effect of alternative techniques delivered separately. *Journal of Animal Science* 87(4), 1479–1492.
McGlone, J.J. and Heilman, J.M. (1988) Local and general anesthetic effects on the behaviour and performance of two and seven week old castrated and uncastrated piglets. *Journal of Animal Science* 66, 3049–3058.
McMeekan, C.M., Mellor, D.J., Stafford, K.J., Bruce, R.A., Ward, R.N. and Gregory, N.G. (1997) Effect of shallow and deep scoop dehorning on plasma cortisol concentrations in calves. *New Zealand Veterinary Journal* 45, 72–74.
McMeekan, C.M., Mellor, D.J., Stafford, K.J., Bruce, R.A., Ward, R.N. and Gregory, N.G. (1998a) Effects of local anaesthesia of 4 or 8 hours duration on the acute cortisol response to scoop dehorning in calves. *Australian Veterinary Journal* 76, 281–285.
McMeekan, C.M., Stafford, K.J., Mellor, D.J., Bruce, R.A., Ward, R.N. and Gregory, N.G. (1998b) Effect of regional analgesia and/or a non-steroidal anti-inflammatory analgesic on the acute cortisol response to dehorning in calves. *Research in Veterinary Science* 64, 147–150.
Mellor, D.J. and Stafford, K.J. (2000) Acute castration and/or tailing distress and its alleviation in lambs. *New Zealand Veterinary Journal* 48, 33–43.
Mellor, D.J., Cook, C.J. and Stafford, K.J. (2000) Quantifying some responses to pain as a stressor. In: Moberg, G.P. and Mench, J.A. (eds) *The Biology of Animal Stress: Basic Principles and Implications for Welfare.* CAB International, Wallingford, UK, pp. 171–198.
Mellor, D.J., Thornber, P.M., Bayvel, A.C.D. and Kahn, S. (eds) (2008) Scientific assessment and management of animal pain. *OIE* [World Organization for Animal Health] *Technical Series* 10, 1–218.
Milligan, B.N., Duffield, T. and Lissemore, K. (2004) The utility of ketoprofen for alleviating pain following dehorning of young dairy calves. *Canadian Journal of Veterinary Research* 45, 140–143.
Millman, S.T. and Duncan, I.J.H. (2000) Strain differences in aggressiveness of male domestic fowl in response to a male model. *Applied Animal Behaviour Science* 66, 217–233.
Molony, V., Kent, J.E. and Robertson, I.S. (1995) Assessment of acute and chronic pain after different methods of castration of calves. *Applied Animal Behaviour Science* 46, 33–48.
Morisse, J.P., Cotte, J.P. and Huonnic, D. (1995) Effect of dehorning on behaviour and plasma cortisol responses in young calves. *Applied Animal Behaviour Science* 43, 239–247.
Noonan, G.J., Rand, J.S., Priest, J., Ainscow, J. and Blackshaw, J.K. (1994) Behavioural observations of piglets undergoing tail docking, teeth clipping, and ear notching. *Applied Animal Behaviour Science* 39, 203–213.
Ohme, F.W. and Prier, J.E. (1974) *Large Animal Surgery.* Williams and Wilkins, Baltimore, Maryland.
Paull, D.R., Lee, C., Atkinson, S.J. and Fisher, A. (2008) Effects of meloxicam or tolfenamic acid administration on the pain and stress responses of Merino lambs to mulesing. *Australian Veterinary Journal* 86, 303–311.
Petrie, N., Mellor, D.J., Stafford, K.J., Bruce, R.A. and Ward, R.N. (1996a) Cortisol responses of calves to two methods of disbudding used with or without local anaesthetic. *New Zealand Veterinary Journal* 44, 9–14.
Petrie, N., Mellor, D.J., Stafford, K.J., Bruce, R.A. and Ward, R.N. (1996b) Cortisol responses of calves to two methods of tail docking used with or without local anaesthetic. *New Zealand Veterinary Journal* 44, 4–8.

Prunier, A., Mounier, A.M. and Hay, M. (2005) Effects of castration, tooth resection, or tail docking on plasma metabolites and stress hormones in young pigs. *Journal of Animal Science* 83(1), 216–222.

Rothwell, J., Hynd, P., Brownlee, A., Dolling, M. and Williams, S. (2007) Research into alternatives to mulesing. *Australian Veterinary Journal* 85, 94–97.

Scobie, D.R., O'Connell, D., Morris, C.A. and Hickey, S.M. (2007) A preliminary genetic analysis of breech and tail traits with the aim of improving the welfare of sheep. *Australian Veterinary Journal* 58, 161–167.

Simonsen, H.B., Klinken, L. and Bindseil, E. (1991) Histopathology of intact and docked pigtails. *British Veterinary Journal* 147, 407–412.

Stafford, K.J. and Mellor, D.J. (1993) Castration, tail docking and dehorning – what are the constraints? *Proceedings of the New Zealand Society of Animal Production* 53, 89–195.

Stafford, K.J. and Mellor, D.J. (2005a) Dehorning and disbudding distress and its alleviation in calves. *Veterinary Journal* 169, 337–349.

Stafford, K.J. and Mellor, D.J. (2005b) The welfare significance of the castration of cattle: a review. *New Zealand Veterinary Journal* 53, 271–278.

Stafford, K.J., Mellor, D.J., Todd, S.E., Bruce, R.A. and Ward, R.N. (2002) Effects of local anaesthesia or local anaesthesia plus a non-steroidal anti-inflammatory drug on the acute cortisol response of calves to five different methods of castration. *Research in Veterinary Sciences* 73, 61–70.

Stafford, K.J., Mellor, D.J., Todd, S.E., Ward, R.N. and McMeekan, C.M. (2003) Effects of different combinations of lignocaine, ketoprofen, xylazine and tolazoline on the acute cortisol response to dehorning in calves. *New Zealand Veterinary Journal* 51, 219–226.

Stafford, K.J., Chambers, J.P. and Mellor, D.J. (2006) The alleviation of pain in cattle: a review. *Perspectives in Agriculture, Veterinary Science, Nutrition and Natural Resources* 1, No. 032. Available at: http://www.cababstractsplus.org/cabreviews (accessed 1 June 2008).

Stewart, M., Stookey, J.M., Stafford, K.J., Tucker, C.B., Rogers, A.R., Dowling, S.K., Verkerk, G.A., Schaefer, A.L. and Webster, J.R. (2009) Effects of local anaesthetic and a nonsteroidal anti-inflammatory drug on pain responses of dairy calves to hot-iron dehorning. *Journal of Dairy Science* 92, 1512–1519.

Stull, C.L., Payne, M.A., Berry, S.L. and Hullinger, P.J. (2002) Evaluation of the scientific justification for tail docking in dairy cattle. *Journal of the American Veterinary Medical Association* 220, 1298–1303.

Sutherland, M.A., Stafford, K.J., Mellor, D.J., Gregory, N.G., Bruce, R.A. and Ward, R.N. (2000) Acute cortisol responses and wound healing in lambs after ring castration plus docking with or without application of a castration clamp to the scrotum. *Australian Veterinary Journal* 78, 402–405.

Sutherland, M.A., Mellor, D.J., Stafford, K.J., Gregory, N.G., Bruce, R.A. and Ward, R.N. (2002) Effects of local anaesthetic combined with wound cauterisation on the cortisol response to dehorning in calves. *Australian Veterinary Journal* 80, 165–167.

Sutherland, M.A., Bryer, P.J., Krebs, N. and McGlone, J.J. (2008) Tail docking of pigs: acute physiological and behavioural responses. *Animal* 2, 292–297.

Sylvester, S.P., Stafford, K.J., Mellor, D.J., Bruce, R.A. and Ward, R.N. (2004) Behavioural responses of calves to amputation dehorning with and without local anaesthesia. *Australian Veterinary Journal* 82, 697–700.

Taylor, A.A. and Weary, D.M. (2000) Vocal responses of piglets to castration: identifying procedural sources of pain. *Applied Animal Behavioural Science* 70, 17–26.

Thomas, D.L., Waldron, D.F., Lowe, G.D., Morrical, D.G., Meyer, H.H., High, R.A., Berger, Y.M., Clevenger, D.D., Fogle, G.E., Gottfredson, R.G., Loerche, S.C., McClure, K.E., Willingham, T.D., Zartman, D.L. and Zellinsky, R.D. (2003) Length of docked tail and the incidence of prolapse in lambs. *Journal of Animal Science* 81, 2225–2232.

Thun, R., Gajewski, Z. and Janett, F. (2006) Castration of male pigs: techniques and animal welfare issues. *Journal of Physiological Pharmacology* 57 (Supplement 18), 189–194.

Ting, S.T., Earley, S., Hughes, J.M. and Crowe, M.A. (2003) Effect of ketoprofen, lidocaine local anesthesia and combined xylazine and lidocaine caudal epidural anesthesia during castration of beef cattle on stress responses, immunity, growth, and behavior. *Journal of Animal Science* 81, 1281–1293.

Watts, J.M. and Stookey, J.M. (1999) Effects of restraint and branding on rates and acoustic parameters of vocalization in beef cattle. *Applied Animal Behaviour Science* 62, 125–135.

White, R.G., DeShazer, J.A., Tressler, C.L., Borcher, G.M., Davey, S., Warninge, A., Parkhurst, A.M., Milanuk, M.J. and Clens, E.T. (1995) Vocalization and physiological response of pigs during castration with and without anesthetic. *Journal of Animal Science* 73, 381–386.

Wilson, P.R. and Stafford, K.J. (2002) Welfare of farmed deer in New Zealand. 2: velvet antler removal. *New Zealand Veterinary Journal* 50, 221–227.

Woodbury, M.R., Caulkett, N.A. and Wilson, P.R. (2002) Comparison of lidocaine and compression for velvet antler removal in Wapiti. *Canadian Veterinary Journal* 43, 869–875.

Further Reading

Anonymous (2005) *Animal Welfare (Painful Husbandry Procedures) Code of Welfare 2005*. National Animal Welfare Advisory Committee (NAWAC), c/o Ministry of Agriculture and Forestry, Wellington, New Zealand, ISBN 0-478-29800-5, pp. 1–36.

Appleby, M.C., Hughes, B.O. and Elson, H.A. (1992) *Poultry Production Systems: Behaviour Management and Welfare.* CAB International, Wallingford, UK.

Faucitano, L. and Schaefer, A.L. (2008) *Welfare of Pigs.* Wageningen Academic Publishers, Wageningen, The Netherlands.

Hall, L.W. (1971) *Veterinary Anaesthesia and Analgesia.* Bailliere Tindall, London.

Knight, T.W., Cosgrove, G.P., Death, A.F. and Anderson, C.B. (2000) Effect of method of castrating bulls on their growth rate and liveweight. *New Zealand Journal of Agricultural Research* 43,187–192.

Mellor, D.J. and Stafford, K.J. (1999) Assessing and minimising the distress caused by painful husbandry procedures. *In Practice* 21, 436–446.

Stafford, K.J. and Mellor, D.J. (2006) The assessment of pain in cattle: a review. *Perspectives in Agriculture, Veterinary Science, Nutrition and Natural Resources* 1, No. 013. Available at: http://www.cababstractsplus.org/cabreviews (accessed 1 June 2008).

Stafford, K.J., Mellor, D.J. and McMeekan, C.M. (2000) A survey of methods used by farmers to castrate calves in New Zealand. *New Zealand Veterinary Journal* 48, 16–19.

Weaver, A.D. (1986) *Bovine Surgery and Lameness.* Blackwell Scientific Publications, London.

7 Welfare During Transport of Livestock and Poultry

TEMPLE GRANDIN

Colorado State University, Fort Collins, Colorado, USA

The single most important factor for the welfare of animals during transport is to put an animal that is fit for transport on the truck (lorry), airplane or ship. The World Organization for Animal Health (OIE) (2009a, b) states that animals that are unfit for travel include, but may not be limited to:

- those that are sick, injured, weak, disabled or fatigued;
- those that are unable to stand unaided or bear weight on each leg;
- those that are blind in both eyes;
- those that cannot be moved without causing them additional suffering;
- newborns with an unhealed navel;
- females that have given birth within the previous 48 h travelling without young;
- pregnant animals that would be in the final 10% of their gestation period at the planned time of unloading;
- those whose body condition would result in poor welfare because of expected climatic conditions (land transport only); and
- animals with unhealed wounds from recent surgical procedures such as dehorning.

Risks during transport can be reduced by selecting animals best suited to the conditions of travel and those that are acclimatized to the expected weather conditions.

Animals at particular risk of suffering poor welfare during transport and which require special conditions (such as in the design of facilities and vehicles and the length of the journey) and additional attention during transport may include:

- very large or obese individuals;
- very young or old animals;
- excitable or aggressive animals;
- animals subject to motion sickness;
- animals that have had little contact with humans; and
- females in the last third of pregnancy or in heavy lactation.

Hair or wool length should be considered in relation to the weather conditions expected during transport.

Unfit Cull Animals are a Major Problem Area

Some of the worst welfare problems during transport occur when old cull breeding animals of little economic value are shipped. Cull breeding stock and old animals should be marketed when they are still fit. The OIE (2009a, b) states that weak, injured or fatigued animals are unfit for travel. To avoid totally different interpretations by different people, the author recommends using body condition scoring (BCS), lameness scoring and injury scoring to determine fitness for travel. The advantage of these measures is greater consistency of judgement between the different people who will be making decisions on fitness to travel. BCS and lameness scoring charts will need to be developed with photos of the local types of animals in each country (see Chapters 1 and 3).

Below is a simple lameness scoring system that can be used by transporters.

1 – Normal walking
2 – Slight lameness, may have an arched back
3 – Obviously limping, but able to keep up with its herdmates when the group is walking
4 – Not able to keep up with its herdmates because lameness is more severe
5 – Can barely walk and likely to become non-ambulatory

Animals with a score of 1, 2 or 3 would definitely be fit for transport. Score 4 animals should be hauled only a short distance for slaughter or veterinary treatment. Score 5 animals should be either euthanized or slaughtered on the farm. A score 5 would fit the OIE (2009b) definition of an unfit animal. The code states unfit animals are: 'Those that are unable to stand unaided or bear weight on each leg'.

Animals that are emaciated or severely lame, with a lameness score of 4 or 5 on a five-point scale are more likely to fall down and get trampled during transport compared to stronger animals. Audits at slaughter plants that process cull dairy cows indicated that timely marketing was a major problem (Gary Smith, personal communication, 2008). Some dairies waited until their cows had an emaciated body score of 1 before shipping them. Similar problems have occurred with sows. Observations by the author in slaughter plants indicate that 90% of the emaciated, weak cull cows and sows originate from about 10% of the worst producers.

Problems in the Developing World with Old Cull Animals

In some parts of the developing world, the treatment of old cull animals is atrocious. Cull buffaloes in India are beaten and dragged by workers in the markets where they are loaded on to trucks for transport to slaughter (Chandra and Das, 2001). This is in violation of the OIE (2009a) code that states 'conscious animals should not be thrown, dragged, or dropped'. Of the bruises, 43% were on the hind limbs and 21% were on the abdomen and udder (Chandra and Das, 2001). Bruises in these locations are due to abuse because animals in countries with good practices seldom have bruises in these locations. Minka and Ayo (2006) studied the incidence of injuries in Nigeria. Animals on long journeys had more injuries. It is clear that transport practices need to be improved.

Effective Methods for Stopping Abuse

One of the most effective ways to get these abusive practices to stop is economic accountability for damage (see Chapter 11 on the effect of economics on animal welfare). It is also obvious from research carried out by Chandra and Das (2001) that the loading facilities were not adequate. Two to three people were required to load each animal. A combination of economic accountability and better loading facilities will improve animal welfare in many countries. The author has observed that livestock markets in many countries have more problems with abusive handling compared to farms. This happens because markets are not held accountable for losses.

Proper Preparation of Animals for Transport

Preparing animals for transport will help to prevent sickness or death associated with transport. Below is a list of preparation methods that will help improve welfare and reduce losses caused by lack of animal preparation.

Introduce animals to the ship's feed before loading

This is especially important for sheep travelling long distances on ships. Sheep should be fed the ship's pelleted feed at an assembly yard before transport to get them accustomed to eating it. Failure to train the sheep to eat the new feed may result in high death losses because many animals may refuse to eat on the ship. Sheep need a minimum of 1 week in an assembly yard to get them behaviourally and physiologically adjusted to the ship's pelleted feed (Warner, 1962; Arnold, 1976).

Shear sheep before transport to a hot climate

This is especially important for sheep that will be travelling to a destination with much hotter temperatures. Wool reduces the ability of sheep to dissipate heat from their bodies (Marai *et al.*, 2006).

Ship acclimatized animals

Death losses will be reduced if cold-acclimatized animals are shipped to a cold climate and hot-weather-acclimatized animals are shipped to a hot climate. Norriss *et al.* (2003) reported that death losses in cattle shipped by sea to the hot Middle East were lower when animals from the hot northern part of Australia were shipped. Heat stress or cold stress deaths at an animal's destination are more likely to occur in animals that are not acclimatized to the conditions at their destination. The author has observed this problem in many countries where animals were shipped long distances either north or south.

Feed colostrum to neonatal calves

Young calves will be less likely to get sick and die after transport if they have been fed colostrum within 6 h after birth.

Vaccinate and pre-wean weaner calves

Vaccinating and pre-weaning 6-month-old beef calves before they are shipped a long distance to a feedlot reduced sickness after arrival (Swanson and Morrow-Tesh, 2001; Loneragen *et al.*, 2003; Lalman and Smith, 2005). Unvaccinated calves shipped the same day they were weaned had more death losses and respiratory illness (Fike and Spire, 2006). Calves should be weaned and vaccinated at least 45 days before they leave the ranch (Powell, 2003).

Some animals should be fasted or fed different feed before transport

Cattle from lush green pastures should be brought in and fed dry hay prior to transport. This will prevent runny diarrhoea. A survey of 37 slaughter plants in the EU showed that failure to fast pigs before loading resulted in higher mortalities (Averos *et al.*, 2008). Ideally no more than 12 h should elapse between feed removal and stunning. For all species, water should never be restricted before or after transport.

Prevent weakness caused by performance growth promoters

Excessive use of beta-agonist feed additives to increase muscle mass may cause weakness or increased problems with heat stress, unless they are used very carefully. The author has observed healthy pigs and cattle that were lame, stiff or weak from excessive doses of ractopamine. Heavy 125-kg pigs fed high doses of ractopamine were weak and many animals were not able to walk from the lairage pen to the stunner. Marchant-Forde *et al.* (2003) also reported that pigs on ractopamine were more difficult to handle and more likely to become fatigued. The effects are dose and time dependent. Feeding 200 mg of ractopamine/day for 28 days to beef cattle did not compromise welfare (Baszczack *et al.*, 2006). These cattle were raised in a temperate climate in Colorado, USA, and they were not exposed to extremely hot conditions. They were transported for 60 min to a slaughter plant. Higher doses for longer periods of time or longer journeys may be very detrimental to welfare in hot climates. At three slaughter plants the author observed grain-fed cattle that were lame, stiff and heat stressed due to feeding ractopamine. The dose was unknown.

Choose appropriate breeds or crossbreeds for transport to different climates

Ship heat-tolerant breeds or crossbreeds to hot areas and cold-tolerant breeds to cold areas. Heat-tolerant *Bos indicus*-cross cattle had fewer death losses when shipped from Australia to the Middle East (Norriss *et al.*, 2003). *B. indicus* cattle have fewer pronounced changes in their physiology when they are exposed to prolonged heat and humidity (Beatty *et al.*, 2006). Highly improved breeds, such as Holsteins, should be crossed with *B. indicus* in hot climates. This is a common practice in tropical parts of Mexico. In extremely cold climates *B. indicus* breeds will have more difficulty. Brown-Brandt *et al.* (2001) reported that newer lines of lean pigs were more susceptible to heat stress. Their total heat production was up to 20% higher than published standards based on older genetic lines.

Habituate animals to people walking through them

Both pigs and cattle will be easier to drive if they have become habituated to people walking through them (Fig. 7.1). Animals differentiate between a person in their pen and a person in the alley. The person must get in the pen and train the animals to move away in an orderly manner as he/she walks through them. Cattle handled exclusively on horseback may be difficult and dangerous to drive by people on foot. These cattle should get some

Fig. 7.1. Pigs housed in this slatted floor pen differentiate between a person in the pen and a person in the aisle. Pigs will be easier to load and unload on to trucks if a person walks through their pens every day during the fattening period. The animals need to learn how to move away from the person in an orderly manner as he walks through them. This building in the USA has a well-designed floor where the surface is over 80% solid. It also has lots of natural light. White translucent curtains admit light. Raising pigs in dark or dimly lit buildings may also make them more difficult to handle and load.

experience being moved in and out of pens by people on foot before they are transported to a new location (see Chapter 5 for additional information).

Genetic Problems that Reduce Fitness for Transport

Pigs with the porcine stress syndrome (PSS) stress gene

Pigs that have the PSS stress gene will have much higher transport death losses. Death losses during transport were 9.2% in homozygous positive, 0.27% in carriers, 0.05% in homozygous negative pigs and 0.1% in pigs where the gene was removed (Murray and Johnson, 1998; Holtcamp, 2000). Pigs with the stress gene are often more lean and will often contain Pietrain genetics. When pigs are purchased from large breeding companies, it is recommended to obtain information on whether or not they have the PSS gene.

A study done by Ritter *et al.* (2009a) of dead-on-arrival and downed non-ambulatory fatigued pigs showed that over 95% of these compromised pigs were free of the PSS stress gene. A total of 2019 pigs were sampled at a US commercial slaughter plant. The sample of pigs came from over 100,000 animals from over 400 farms. Ritter *et al.* (2009a) concluded that the PSS stress gene (HAL-1843 mutation) was not a major cause of losses. The author has observed that problems with non-ambulatory pigs have worsened since beta-agonists were approved in the USA. The pigs in the Ritter *et al.* (2009a) study were grown to heavy weights and beta-agonists such as ractopamine may have contributed to the pigs becoming non-ambulatory.

Weak fast-growing animals

Some pigs and poultry bred for rapid weight gain are weaker compared to less highly selected animals. Problems with weaknesses can occur in pigs that are free of the PSS stress gene. Pigs from rapidly growing genetic lines grown on full feed may be weaker and more likely to go down during transport. In one study, heavy weight 130-kg market pigs had a trend for higher death losses (Rademacher and Davis, 2005). Heavy, fast-growing lines of poultry are also weaker and get fatigued more easily. Genetics and environment interact. Chicken producers have increased the strength and livability of fast-growing poultry genetic lines by changing nutrition so that the younger chicks grow more slowly. In the final days of the growing programme, the older birds are put on full feed for maximum growth. When the animals are going to be transported and grown in parts of the world where modern veterinary and nutritional services are not available, the use of hardy local breeds may result in both better animal welfare and fewer death losses.

Animals with poor leg conformation

Animals with structurally poor legs are more likely to become non-ambulatory during transport. Animals that are being selected for breeding should be evaluated for both lameness and leg conformation. Animals with strong legs will have better welfare during transport. Simple leg conformation charts that show examples of good and bad legs can be used for choosing breeding animals (National Hog Farmer, 2008; see Chapter 1).

Livestock Loading Facilities

Vehicles in Europe and some developed countries have loading ramps built into the vehicle. These systems either have a drop-down tailgate or a hydraulic lift. They work extremely well in technologically advanced countries. In many parts of the world, these systems are not practical because the local people will have difficulty maintaining them. Even a drop-down ramp is problematic because the springs that are used for raising the tailgate are made from special alloy that may not be available locally. During visits to developing parts of the world, the author has observed many rusting pieces of imported equipment that the local people could not maintain.

A lack of good loading ramps is a serious problem in many parts of the world. This results in animals being thrown or forced to jump off a vehicle that is 1.2 m off the ground. Lack of good ramps to get animals on to trucks is also a major problem. In some facilities, ramps have been built but they are either poorly designed or too steep.

There is a great need for the construction of good ramps for loading and unloading vehicles. Stationary ramps can be easily built from locally available materials such as wood, concrete or steel (Fig. 7.2). Many people in developing countries are very good carpenters, welders and masons. Loading ramps can be easily built from the same materials that people use to build houses. Concrete blocks, cement, steel reinforcing rods and wood are available in developing countries. In some situations, a ramp can be made by piling up an earthen berm for the vehicle to back up against. Another alternative is a portable ramp that can be moved from farm to farm. In the developed world, portable ramps are commercially available. In the developing world, the axles and wheels for the portable ramp could be salvaged from an old car. The author has observed that the welders and iron workers in the developing world are masters at turning old cars into donkey carts and many other things. A smart welder could easily make a portable ramp if he was given a few pictures printed from the Internet of a commercially made portable livestock ramp.

Recommended Design of Truck Loading Ramps

For cattle and pigs, a non-slip ramp built on a 20° angle or less is recommended (Grandin, 1990) (Fig. 7.2). Sheep can easily move up and down steeper ramps. The best non-slip footing on concrete ramps is provided by a stair-step design (Fig. 7.3). Steps are more effective than a grooved ramp and they provide non-slip footing even when they are old and worn. Grooves wear down quickly and become slick. Stair steps are a proven design used on many ramps around the world. Figure 7.3 shows the recommended step dimensions for a concrete ramp for cattle, horses, buffaloes, camels and other large livestock. The recommended dimensions are a maximum 10 cm rise and a 30 cm or longer tread length. Two grooves at least 2.5 cm deep should be made in each step. For pigs, a shorter 8 cm rise is recommended.

For ramps constructed from either steel or wood, cleats are normally used. Figures 7.4, 7.5 and 7.6 show the correct and wrong spacing of cleats to provide non-slip traction. Spacing the cleats too far apart is a common mistake and can cause slipping. When cleats are too close together, the hooves slide over the top and may provide poor traction. During unloading, cleats spaced too far apart caused dewclaw injuries in piglets. The feet of the piglets slide and the dewclaws catch on the cleats. For both

Fig. 7.2. This excellent truck loading ramp was built with locally available materials. In South America, most cattle ranches have their own loading ramps. This ramp has a level dock at the top which helps prevent slipping and falling during unloading. Adding completely solid sides may improve cattle movement (see Chapter 5).

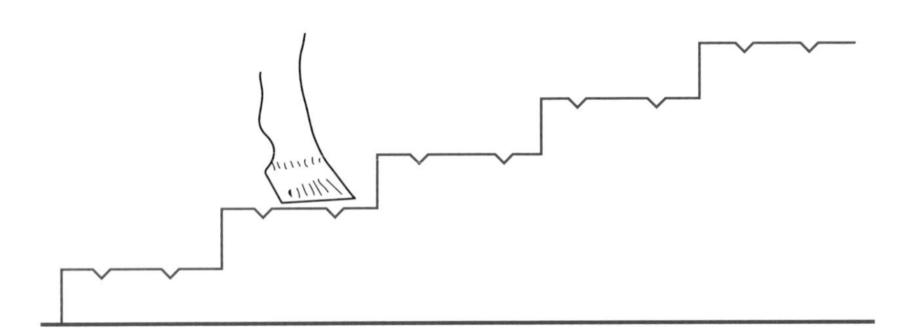

Fig. 7.3. When a loading ramp is constructed from concrete, stair steps are recommended. The steps should have a 10 cm rise and a 30–60 cm tread length. Do not exceed a 10 cm rise. The minimum tread length is 30 cm. The steps must be grooved to provide non-slip footing. These dimensions are recommended for cattle, water buffalo, horses and other large animals. For pigs, a smaller 8 cm rise is recommended (from Grandin, 2008).

small and large livestock species, the cleats should be spaced so the hoof of the animal will fit easily between them (Fig. 7.4). For cattle, 20 cm of clear space between the cleats is recommended. Smaller animals require closer spacing. The stair step and cleat spacing recommendations are the same for both animals trained to lead with a halter, and extensively raised animals with a large flight zone.

Cleats can be made from many different materials. On wooden ramps for cattle and other large livestock, 5 cm × 5 cm cleats made from hardwood work well. Use 2.5 cm × 2.5 cm cleats for smaller species. Steel concrete reinforcing rods make good cleats on steel ramps. Steel reinforcing rods are readily available in all countries where homes are built from masonry or concrete block. Pipe or

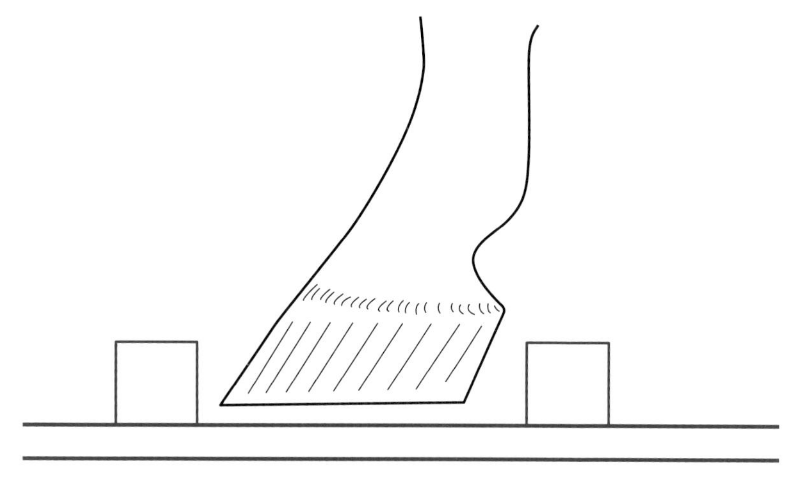

Fig. 7.4. This is the correct cleat spacing for all livestock. The cleats should be spaced so that the hooves fit easily between them (from Grandin, 2008).

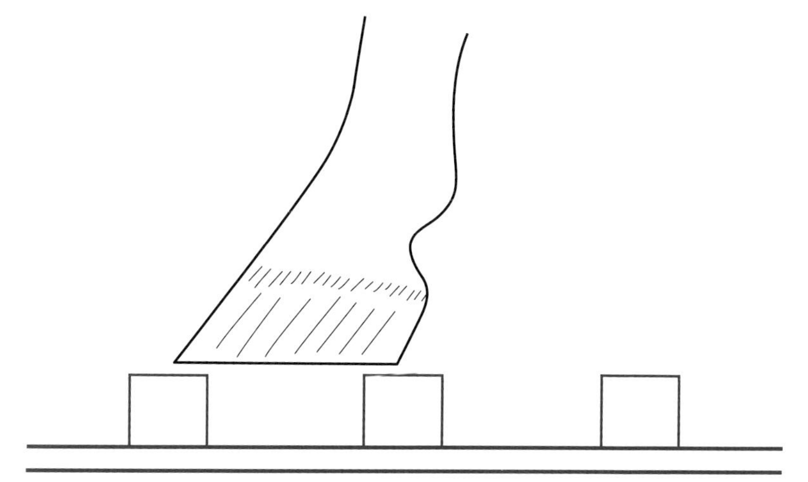

Fig. 7.5. Wrong cleat spacing for a loading ramp – the cleats are too close together and the foot may slide and slip over the top of the cleats (from Grandin, 2008).

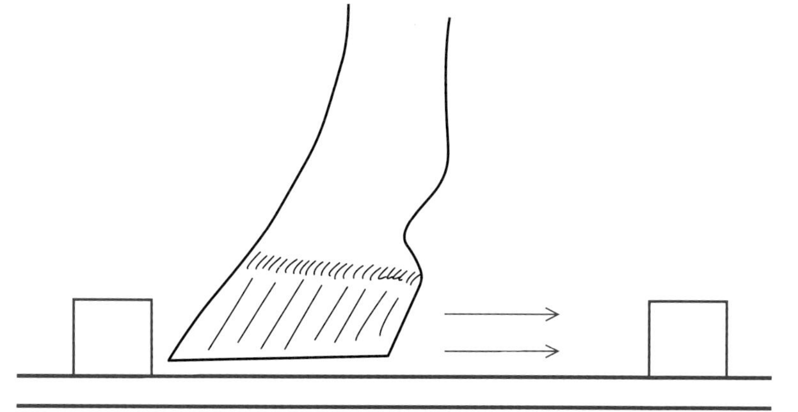

Fig. 7.6. Wrong cleat spacing for a loading ramp – the cleats are too far apart and the animal will slip. Animals are more likely to fall down if the cleats are too far apart (from Grandin, 2008).

square steel tube can also be used. Creative local people may invent other surfaces that are effective and provide good non-slip footing.

Evaluation of Loading and Unloading Facilities

Ramps and other parts of unloading and loading facilities should be evaluated with numerical scoring. If more than 1% of the animals fall down during either loading or unloading, the ramp or floor is either not adequate or the animals are being handled roughly. If more than 1% continues to fall after handling is improved then the ramp or floor surface will need to be improved to provide better non-slip footing. When facilities are being evaluated use animal-based, numerical scoring instead of input-based engineering standards.

Handling and Transport of Poultry

One study showed that machine catching of chickens reduced stress (Duncan *et al.*, 1986). One of the reasons is that in a well-designed automated catching machine, the birds are not turned upside down. A more recent study by Nijdam *et al.* (2005) at eight commercial farms found no difference in the percentage of bruises between hand-caught and machine-caught birds. Both Nijdam *et al.* (2005) and unpublished industry data from the USA showed that machine catching had a higher number of deaths. In the USA, chickens are grown where it is very hot and machine catching increased the number of deaths because it took longer to catch the birds unless two machines were used simultaneously. Longer catching times resulted in the ventilation system being disabled for a longer period of time. In Europe, catching machines are popular due to high labour costs.

In areas of the world where labour is inexpensive, the lowest level of broken wings that the author has observed was in Brazil. Each bird was carefully lifted by the body and placed in the coop (crate). Only 0.25% of these carefully hand-caught birds had broken wings (Fig. 7.7). In the USA, the best hand catching of heavy (>3 kg birds) produced

Fig. 7.7. On this farm in Brazil, a combination of economical labour and good management produced excellent chickens. Careful hand catching, where each bird is held upright, resulted in only 0.25% of the birds having broken wings. The birds are raised in simple, naturally ventilated buildings. The building has open sides with window screens to keep out wild birds. There are no mechanized feeders or fans. Litter condition is excellent and the birds are completely clean.

0.86% of birds with broken wings. Before restaurant companies started auditing their chicken suppliers, broken or dislocated wings were present on 5–6% of the birds. Both hand catching and machine catching can be done with an acceptable level of welfare. To prevent catching from becoming rough and sloppy, the percentage of deaths, bruises and broken wings should be continuously measured. Putting workers on incentive pay for low levels of broken wings, deaths and bruises will also help improve handling (see Chapter 11).

Solving Bruise Problems in Large Animals

Animals that are moved through many dealers and markets will have more bruises. Both Hoffman *et al.* (1998) and Weeks *et al.* (2002) found that cattle moved through markets had more bruises than cattle sold for slaughter directly from the farm or ranch. Rough handling that causes cattle to run and become excited will almost double the amount of bruising (Grandin, 1981). Handling animals more quietly will reduce bruising. Many people do not realize that the hide does not have to be damaged to have a bruise underneath. An animal can have a huge bruise under a hide that has completely normal hair and no sign of injury. Grazing animals have tough hair and hide on the outside but underneath they are as fragile as soft fruit.

Several surveys have shown that horned cattle have twice as many bruises compared to polled or dehorned cattle (Ramsey *et al.*, 1976; Shaw *et al.*, 1976). A few horned cattle can greatly increase bruising. Contrary to popular belief, tipping (cutting off parts of the horn) does not reduce bruising (Ramsey *et al.*, 1976). Tipping horns on heavy-weight steers is painful and reduces live weight gain (Winks *et al.*, 1977). Ranchers should remove horns from young calves or breed polled cattle.

Bruises are still a big problem in the developed countries. Data collected in 2005 in the USA indicated that 9.4% of feedlot cattle had multiple bruises (Garcia *et al.*, 2008). The author has observed that slaughter plants that have an effective bruise prevention programme have a much lower level of bruises.

Equipment problems

Edges with a small diameter such as angle irons, channels, truck doors and the exposed end of a sharp edge of a pipe are most likely to bruise cattle. Bumping into a smooth, wide surface such as a flat wall or a large, round, 15-cm-diameter post is less likely to bruise. Broken boards and protruding gate latches can also cause bruises. Gates should be equipped with tiebacks to prevent them from swinging out into an alley. It is also important to provide non-slip flooring. Slick floors that cause animals to fall cause many bruises and injuries.

Troubleshooting bruises

When bruising is observed at a slaughter plant, a determination must be made to find out where the bruises are occurring. If the bruises occur on animals from many different origins, then it is likely that the problem is at the slaughter plant. If bruises occur on animals from one origin, then the problem is likely to be occurring on the truck or on the farm. A sudden occurrence of bruises is usually caused by a change in personnel or broken equipment. If faulty equipment is causing the bruises, there will be sharp edges that have been rubbed shiny from animals hitting them. Determining the precise age of a bruise is difficult, but it is possible to differentiate old from fresh bruises. Fresh bruises are red and bruises a few days old often have yellowish secretions on them. This secretion will not be present on fresh bruises.

Loin bruises

A major cause of loin bruises is rough handling during truck loading or unloading. This causes cattle that are too excited to wedge themselves in the truck doorway. Loin bruises can be reduced by having full-width doors on trucks. Horns also greatly increase loin bruises. Loin bruises can also be caused by slamming gates on cattle. The loin becomes bruised when the animal becomes wedged between the end of the gate and an alley fence. Protruding gate latches also cause many loin bruises.

Shoulder bruises

Shoulders are often bruised on sharp edges such as protruding gate latches or broken boards. If the bruises are occurring at the slaughter plant, broken parts in the restrainer conveyor entrance may be to blame. Another cause of shoulder bruises is rolling out of the stunning box. Stunned cattle can still be bruised until they are bled (Meischke and Horder, 1976). Other causes are a sliding gate track that is not recessed and broken sideways one-way flapper gates in a single-file chute. Rough handling and horns also cause many shoulder bruises.

Back bruises

Back bruises are most likely to be caused by equipment problems. Being hit with a stick will also bruise backs (Weeks *et al.*, 2002). One common cause is tall cattle hitting the bottom of the top deck when they exit from the bottom compartment of a truck. These bruises can be reduced by unloading cattle at a slow walk. Animals that jump up are more likely to hit the upper deck. Another major cause is careless use of vertical slide gates. The bottom edge of a vertical slide gate should be constructed from 10-cm-diameter round pipe covered with rubber. Another cause of back bruises is improperly adjusted one-way gates. Cattle can get severe back bruises if they back up under a one-way gate that is set too high.

Hind legs, udder and abdomen bruises

Abusive beating, dragging or kicking is usually the cause when this occurs on fully ambulatory cattle that have not fallen in a truck.

Bruising over large areas of the body

The major cause of extensive bruising where large areas of the body are bruised is due to the animal being trampled in a truck. If a truck is overloaded, an animal may fall down and it will not be able to get back up. In sheep, extensive bruising is often caused by grabbing sheep by their wool. Handlers should never grab or lift sheep by their wool.

Transporting Tame Animals

Miriam Parker has done extensive work in the developing world on the handling and transport of cattle, water buffalo, donkeys and other tame animals that can be led with a head collar (Ewbank and Parker, 2007). If the animal has a nose ring or a lead threaded through its nose (Fig. 7.8), it must

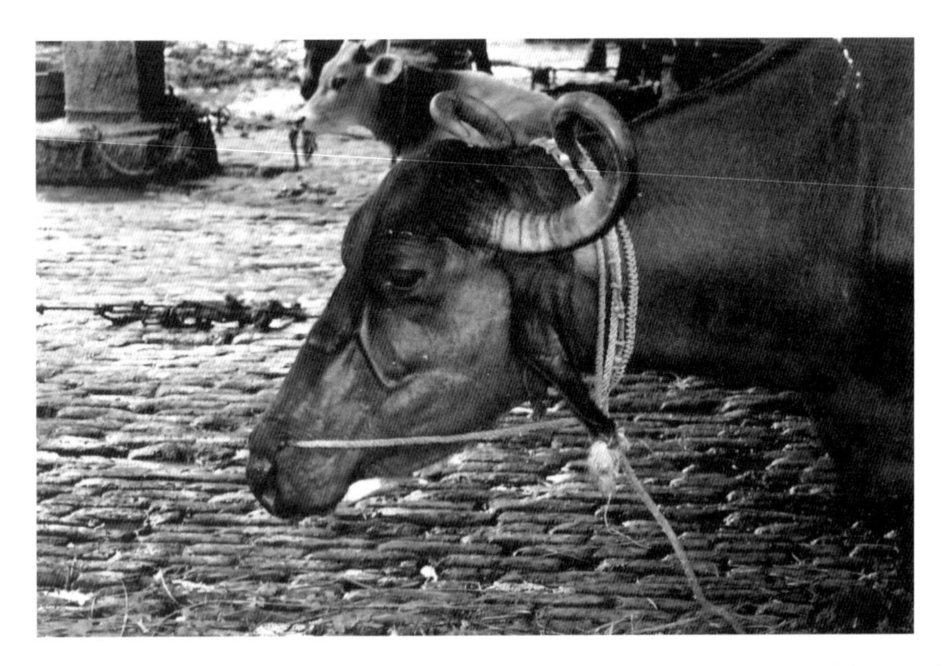

Fig. 7.8. A typical nose lead that is used in Asia and India. This design incorporates the nose lead into a neckband so that the animal can be safely tied up when grazing. It must never be tied up with just the nose lead. However, when the animal is being transported in a vehicle, the use of a halter (head collar) in addition to this rig is strongly recommended to prevent pulling on the sensitive nose during the journey (photograph courtesy of Miriam Parker).

NEVER be tied in a vehicle by the nose. If it falls or becomes frightened, it may tear its nose. If it is tied for transport, a head collar (halter) must be used. Animals of similar size should travel together. If the animals are tied, each animal should have its own lead rope and not be tied to another animal (Fig. 7.9). Figure 7.10 shows tame cattle being transported on the deck of a ship in the Philippines.

Solving Bruise and Injury Problems in Poultry

Bruised legs

The author has observed that rough shackling is a major cause of bruised drumsticks. The people doing the shackling squeeze the legs too hard when they put the birds on the shackles. An understaffed shackle line where people have to hurry is one cause of bruised legs.

Bruised chicken breasts

There are two major causes of bruised chicken breasts. One cause is jamming birds too quickly through the small opening in the top of the coop. The other cause results from machine-catching systems when birds are jammed against the coop door by a conveyor that is out of alignment.

Smashed heads

This injury is most likely to occur when drawer systems are used for transporting chickens, where the birds are put in trays that are slid into rack-like dresser drawers (Fig. 7.11). Drawer systems that have been redesigned so that there is a gap between the top of the drawer and the rack frame seldom have this problem. This gap prevents the head of a chicken from being smashed when the drawer full of chickens is slid back into the rack.

Broken wings

The number one cause of this problem is rough handling during catching. Chickens should NEVER be picked up by a single wing. Measuring the percentage of broken wings is a sensitive measure of how the people are handling the birds. Broken wings can also occur in systems where birds have to be removed from the coop through a small door. When individual coops are used, the best ones have a small door for loading the birds and the entire

Fig. 7.9. Training materials designed for training people who cannot read, showing the right and wrong way to tie up animals. The two illustrations on the left side show the correct way and the two illustrations on the right side show the wrong way (reproduced with permission of Miriam Parker).

Fig. 7.10. Tame cattle that are trained to lead are being transported on the deck of a ship in the Philippines. They remain calm because they have been exposed to many novel new things long before they got on the ship (see Chapter 5). The animals' welfare is good on this ship. Cattle that are trained to lead but have not been exposed to many new things would probably panic and jump overboard. The animals on this ship have lived most of their lives grazing along a busy road. This is an example of a situation where under one set of conditions the animals' welfare would be bad and under another set of conditions it is good.

Fig. 7.11. Drawer system for transporting chickens in Denmark and Canada. In the left-hand photo, chickens are being conveyed into the drawers from a mechanical catching machine. When the drawer is full, the catcher must carefully slide the drawer into the rack to prevent the heads of the chickens from being caught. A gap between the top edge of the drawer and the rack helps prevent heads or wings from being damaged. This is an important design feature that helps prevent injuries.

top opens to remove the birds for shackling at the plant.

Table 7.1 shows a big difference in the percentage of chickens with broken wings between the best and the worst plants. Some ethicists are hesitant to state an acceptable level of broken wings because that translates into thousands of birds with broken wings. The author has observed that when numerical standards were introduced for measuring broken wings, they were greatly reduced. In the

Table 7.1. Percentage of chickens with broken or dislocated wings in 22 US and Canadian broiler chicken processing plants.[a]

Birds with broken wings (%)[b]	Number of plants	Plants (%)
≤1	8	36
1.01–2	6	27
2.01–3	6	27
>3	2	9

[a] Average bird weight = 2.75 kg (6 lb). Birds were handled in the dump module system and all data were collected when the birds still had their feathers on before picking (defeathering equipment). Broken wings were scored after the live birds were hung on the shackles. All birds were electrically stunned. All data were collected in 2008.
[b] The average percentage of birds with broken wings in all plants was 1.67%. The best plant had only 0.20% of birds with broken wings and the worst plant had 3.80% of birds with broken wings.

USA before measurements started, 5–6% of the birds had broken wings. With the present data today, a reasonable goal would be 1% and plants with 3% are clearly not acceptable.

Broken legs

In spent hens, weak bones are a major contributor to fractures (Webster, 2004). In broiler chickens, rough handling is a major cause. There is much controversy among welfare specialists on the correct way to pick up chickens during catching. Some specialists state that they should never be picked up by a single leg. In some countries, this is the normal catching method. The author has observed one-legged catching where the number of birds injured was very low. The coops were brought into the barn close to the catcher. The catcher never walked more than 3 m to load the chickens into the coops.

Recommendations on catching methods

It is the author's opinion that instead of arguing over whether poultry are hand-caught or machine-caught, caught by one leg or two legs, or lifted by the entire body, the best approach is to measure injuries and deaths. These measurements can be made by measuring the percentage of:

- broken, dislocated wings (count with feathers on);
- deaths;
- broken legs;
- bruised legs; and
- bruised breasts.

These percentages will measure the outcome of poor handling. Broken wings should be counted when the feathers are on to prevent counting breakage from the feather-removal picking machine. The broken wing score should include both breaks and dislocations.

Hide Damage and Branding

Branding damages the hide. A rib brand put on a small calf will ruin an entire side of leather because the brand grows with the calf. Freeze branding is less painful than hot iron branding (Lay *et al.*, 1992), but unfortunately it still damages the hide. The leather is more likely to crack where the hide was freeze branded. In countries where animals are extensively raised and have to be branded to prevent theft, both hide damage and welfare can be improved by moving rib brands to the rear of the animals. There has been much concern from animal advocacy groups about branding animals on the face. This stressful practice has been stopped in some countries. The face is a sensitive part of the animal and moving brands to the less-sensitive hindquarters is strongly recommended. Branding on the most posterior part of the hindquarters will minimize hide damage.

Poking animals with sticks studded with nails will ruin the top grain surface of the hide. Unfortunately this is still a problem in many countries. When an animal absolutely refuses to move, a brief shock from a battery-operated goad is preferable to a stick with a nail on the end. The author has seen hides where the entire side was ruined by pointed sticks. Other things that are unrelated to transport and handling can severely damage hides. Lice will ruin the top grain of the leather. If a young animal gets lice, the hide may still have a damaged surface.

Space Requirements for Livestock and Poultry

For all modes of transport – land, sea and air transport – the OIE (2009a, b) code states:

> The amount of headroom necessary depends on the species of animal. Each animal should be able to assume its natural standing position for transport (including during loading and unloading) without coming into contact with the roof or upper deck of the vehicle and there should be sufficient headroom to allow adequate airflow over the animals.

For sea journeys where animals have to be able to lie down, the OIE (2009b) code states: 'When animals lie down, there must be enough space for every animal to adopt normal lying posture'. The OIE transport code applies to cattle, horses, sheep, camelids, pigs and goats. The code does not apply to poultry. Practical experience, industry guidelines and research show that all species should have sufficient space for all animals to lie down at the same time without lying on top of each other. Poultry are routinely transported in lying positions in containers where they are not able to stand and they must have sufficient space to all lie down at the same time without being on top of each other.

During truck transport, most large animals such as cattle, horses and water buffaloes will remain standing. Unfortunately there are no space allowance charts in the OIE (2009a) code for animals that remain standing on a truck. This is probably due to the fact that there is controversy among researchers on the correct amount of floor space for each animal. Most researchers will agree that jamming the maximum number of animals that can be forced on to a vehicle is wrong. Death losses, bruises and injuries will increase in severely overloaded trucks (Eldridge and Winfield, 1988; Tarrant *et al.*, 1988; Valdes, 2002) (Table 7.2). Too many birds loaded into each compartment was a major cause of deaths in chickens (Nijdam *et al.*, 2005). For large animals such as cattle, bruises and injuries increase on an overloaded truck because an animal that goes down is not able to get up. This results in it getting trampled. Cattle and other large animals usually remain standing in a truck. Pigs need space to lie down unless they are hauled a very short distance (Ritter *et al.*, 2007). A study by Ritter *et al.* (2008) showed that for large 129-kg pigs transported for under 4 h, there was a critical space requirement to prevent death losses and greater numbers of non-ambulatory pigs. Each large pig needed 0.462 m^2 per pig or greater (Ritter *et al.*, 2008). Pigs were transported at six different space allowances ranging from 0.396 to 0.520 m^2. Good information on space requirements for weaner pigs can be found in Sutherland *et al.* (2009). For people who are implementing an animal welfare programme, the author recommends that you use your local, national or industry regulations and guidelines for space allowances on trucks and carefully measure outcomes such as bruises, death losses and non-ambulatory animals.

Many countries have either legislation or guidelines for space requirements for truck floor space when animals are being transported. The OIE (2009a, b) codes are a minimum standard that all countries should follow to provide a basic level of animal welfare. The legislation and standards in many countries will be much stricter. Reviews of scientific studies on transport can be found in *Livestock Handling and Transport* (Grandin,

Table 7.2. Effect of varying transport journey times and stocking densities (kg/m^2) on bruising in steers (source: Valdes, 2002).

	Journey (3 h)				Journey (16 h)			
	400 kg/m^2		500 kg/m^2		400 kg/m^2		500 kg/m^2	
	n	%	*n*	%	*n*	%	*n*	%
Total carcasses	28	100	32	100	28	100	32	100
Bruised carcasses	10	35.7	11	34.3	12	42.8	18	56.2
Grade 1 (subcutaneous tissue compromise only)	8	28.5	10	31.3	11	39.2	14	43.8
Grade 2 (muscle tissue compromise)	2	7.1	1	3.1	1	3.5	4	12.5

2007). Some of the major journal article reviews on transportation are: Knowles (1999) (cattle), Fike and Spire (2006) (cattle), Knowles *et al.* (1998) (sheep), Hall and Bradshaw (1998) (sheep and pigs), Warriss (1998) (pigs), Ritter *et al.* (2009b) (pigs), Weeks (2000) (deer) and Warris *et al.* (1992) (chickens).

Databases for Welfare Information

Research on animal transport, animal welfare and animal behaviour can be accessed on the Internet at the following databases:

- PubMed (good for veterinary research);
- www.scirus.com (good for behaviour research);
- Google Scholar;
- www.vetmedresource.org (good for author searches); and
- CAB Abstracts.

Summaries of research can be acquired for free. The easiest way to find these databases is to type their names into a search engine such as Google. When you get to the correct web page, a new search box will appear on the page. Type keywords into the new search box and you will locate abstracts (summaries) of scientific studies. Use every possible keyword to find information. For example, many different words are used for cattle such as cows, bulls, cattle, steers, bullocks, oxen and calves. Do a separate transport search with each word on each database. Some full-text papers will be available for free and others will only give you a free abstract (summary). Often the full papers can be obtained for free if you go to a library at a major university or you can e-mail the author and request a copy.

Rest Stops During Road Transport

Another contentious issue within both the research community and the animal advocacy non-governmental organizations (NGOs) is the length of time an animal is allowed to travel by road transport before it has to have a rest stop. People who are implementing an animal welfare programme should follow the government regulations and industry guidelines in their country. The OIE (2009a) code does not address this issue, because the member countries could not agree on standards. For further information on long-distance transport issues, refer to Appleby *et al.* (2008).

In many countries, 48 h is the maximum length of time that animals can remain on a vehicle before they have to be unloaded. The EU (2005) has strict standards. Cattle are allowed to travel for 14 h and if they have a 1 h rest stop they are allowed to travel for another 14 h. If the second part of such a trip (i.e. that following the rest stop) brings them within 2 h of their destination, the second half of the trip can be extended to 16 h. For longer trips, a 24 h rest stop is required. Pigs can travel for 24 h without stopping if they have continuous access to water. For chickens, 12 h is the maximum time they should be held in the transport containers (Mohan Raj, personal communication, 2009). There are lots of rules and they are constantly being changed. Transporters who ship livestock or poultry out of their own country must make sure they know the latest regulations.

Rest stop research

Rest stops that are too frequent may be ‘stress’ stops instead of rest stops. This is more likely to be a bigger problem when extensively raised animals with large flight zones are being handled. These animals are more likely to become stressed during loading and unloading compared to tame livestock. Sheep need long rest stops to get a benefit. They often will not drink until they have eaten. They need a longer rest stop compared to cattle in order for them to eat and drink. A short 3 h rest period does not provide enough time for drinking. On a trip that took 24 h, sheep arrived in better condition with no rest stop (Cockram *et al.*, 1997). Australian research showed that sheep can easily tolerate trips up to 48 h (Ferguson *et al.*, 2008). In cattle, the maximum length of a non-stop trip should be 24–32 h for 6-month-old to yearling calves (Grandin, 1997; Schwartzkopf-Genswein, personal communication, 2009). This recommendation is for transportation when the temperature is less than 30°C. Grain-fed cattle should definitely be rested after 48 h.

Methods to Improve Welfare and Prevent Losses During Transport

Measure losses

One of the most effective ways to bring about improvements in transport is to measure losses. Bruises, injuries, death losses, sickness and other losses should be counted and tabulated. This makes it possible to identify drivers and producers who are having high losses.

Hold sellers accountable for losses

Sellers should receive bonuses or fines based upon the condition of the animals and the incidence of sickness a few weeks after arrival at their destination (see Chapter 11 for further information).

Better management of livestock markets

Lack of refrigeration is one reason why cull animals are transported long distances in the developing world. This may make it impossible to eliminate long-distance transport in some parts of the world. Both improved facilities and management expertise will be required to correct the atrocious conditions reported by Chandra and Das (2001) because sellers have no economic interest in old cull spent animals. The author has observed that markets are the most difficult segment of the industry to improve because middlemen and dealers pass on losses to the next segment in the marketing drain (see Chapter 11). NGO animal advocacy groups could work to find a corporation or wealthy donor to buy the market, upgrade the facilities and employ competent managers. The management component is essential. The author has seen many modern facilities built in developing countries fall into disrepair because they lacked competent management. Finding a wealthy donor is a real option and not as far-fetched as it sounds. In the USA, the wife of T. Boone Pickens, a billionaire oilman, offered to pasture 30,000 wild horses to prevent them from being slaughtered. After the facility was rebuilt, the cost of paying a competent manager and employees would be essential to maintain it. The ethics of doing this would be greatly debated in the developing world where people are not able to attain essential needs. Many people would consider this unethical when there are people starving.

Change insurance policies

Insurance policies on livestock should only cover catastrophic losses such as a truck rollover. They should have a high deductible fee to prevent careless or abusive treatment of animals from being rewarded. For example, a driver should not receive an insurance payment for two or three dead pigs.

Careful driving

Smooth starts and careful braking will help prevent animals from falling down during transport. Careful driving is essential. Cockram *et al.* (2004) reported that 80% of the time that an animal lost its balance followed a driving event such as cornering or braking. Unpublished industry data indicated that certain truck drivers had higher percentages of bruised cattle. McGlone (2006) reported on data collected on 38 drivers who transported over a million pigs. The best drivers had 0.3% dead or non-ambulatory pigs and the worst drivers had twice as many.

Driver fatigue – a major cause of accidents

A survey conducted by Jennifer Woods in Canada showed that a high percentage of livestock truck rollovers are due to driver fatigue. A total of 415 commercial truck wrecks were tracked in the USA and Canada. Of these, 85% of the wrecks were due to driver error (Woods and Grandin, 2008). Most people assumed that cold, icy weather would be a major cause of truck wrecks. However, the highest percentage of the wrecks occurred in October when the roads were clear of ice. Another indicator that fatigue was the cause of many of the accidents was that only 20% of the livestock truck wrecks involved another vehicle and 59% of the wrecks occurred during the early morning hours between 12:00 midnight and 9:00 in the morning. In North America, vehicles are driven on the right-hand side of the road and when a driver falls asleep, the vehicle usually drifts to the right. Of the trucks that tipped over, 84% fell on their right side.

Never pay transporters based on truck weight

This method of payment provides an economic incentive to overload trucks. Transporters should be paid a negotiated rate for each trip. The rate should specify the number of animals of different types to put on the truck.

Never pay livestock handlers based on speed of handling

Rewarding people for handling more and more animals per hour will motivate abusive handling. Provide incentive pay for reducing bruises and deaths. Unpublished industry data showed that providing financial bonuses greatly reduced the number of broken wings in poultry and deaths in pigs.

Farm of origin effects

The author has observed that about 10% of the farms are responsible for about 90% of the worst emaciated cull dairy cows (Grandin, 1994). Both a Spanish and a US study had similar conclusions. Fitzgerald *et al.* (2008) found that out of nine farms, the worst one had 0.93% more losses on each load of market pigs. In Spain, pigs from certain farms had more deaths and carcass problems (Gosalvez *et al.*, 2006). A large Canadian study also showed that farm of origin was a major factor in predicting the percentage of dead and non-ambulatory market pigs (Tina Widowski, personal communication, 2009). To reduce problems caused by poor producers, a system of economic penalties and rewards should be implemented. At one large slaughter plant, problems with weak pigs caused by feeding excessive doses of ractopamine was stopped by charging a US$25 handling fee for each non-ambulatory pig. A survey of slaughter horses indicated that conditions caused by neglect at the farm of origin create some of the worst welfare problems (Grandin *et al.*, 1999).

Documentation for Transport

The OIE (2009a) code for land transport lists the documentation which should accompany every consignment of livestock. Each country or region within a country may require additional documents to prove ownership or enable trace-back after a disease outbreak. The OIE (2009a) code states:

> Article 7.3.6.
>
> Documentation
>
> 1. Animals should not be loaded until the documentation required to that point is complete.
> 2. The documentation accompanying the consignment should include:
>   a. journey travel plan and an emergency management plan;
>   b. date, time and place of loading and unloading;
>   c. veterinary certification, when requested;
>   d. animal welfare competencies of the driver (under study); [The author recommends using training materials developed in each country.]
>   e. animal identification to allow animal traceability to the premises of departure and, where possible, to the premises of origin;
>   f. details of any animals considered at particular risk of suffering poor welfare during transport (point 3e of Article 7.3.7);
>   g. documentation of the period of rest, and access to feed and water, prior to the journey;
>   h. stocking density estimate for each load in the consignment;
>   i. the journey log – daily record of inspection and important events, including records of morbidity and mortality and actions taken, climatic conditions, rest stops, travel time and distance, feed and water offered and estimates of consumption, medication provided and mechanical defects.
> 3. When veterinary certification is required to accompany consignments of animals, it should address:
>   a. fitness of animals to travel;
>   b. animal identification (description, number, etc.);
>   c. health status including any tests, treatments and vaccinations carried out;
>   d. when required, details of disinfection carried out.

At the time of certification, the veterinarian should notify the animal handler or the driver of any factors affecting the fitness of animals to travel for a particular journey. The documentation required for sea and air transport is very similar.

Delays at Country Borders

Some of the most serious delays that are detrimental to animal welfare are delays at the borders between different countries. Government officials, livestock producer organizations and NGO advocacy groups should work to solve this problem. In some cases, more customs officials may need to be hired. In other cases, paperwork procedures may need to be changed or a special lane for livestock could be introduced. Forcing a truckload of livestock or poultry to wait hours at a border is a problem that needs to be corrected.

Training of Handlers and Drivers

Many livestock associations, federal governments and state and provincial governments have developed programmes for training handlers and drivers. Some large slaughter plants require all the drivers that enter the premises to be trained. In many countries, the training documents for animal welfare are part of industry-sponsored livestock quality assurance programmes that also cover disease control and food safety. These types of programmes are already in place in Europe, the USA, Australia, Canada and many other developed countries. Argentina, Chile, Uruguay and Brazil have also developed programmes. A good driver-training programme should cover the following areas:

- Behavioural principles for moving animals – such as explanation of the flight zone and point of balance. This can be omitted if all animals are tame and trained to lead (see Chapter 5).
- A description of the distractions that can frighten animals – drivers need to understand how shadows, reflections or the sun's position may affect animal movement (see Chapter 5).
- Biosecurity rules for disease control.
- Health regulations for transport across the borders of countries, regions, state or provincial lines.
- Methods to prevent bruises and injuries.
- How to determine fitness of an animal for travel.
- Emergency procedures after an accident.
- Proper use of electric goads and other implements for moving animals.
- How to protect animals from heat stress or cold stress – these must be written specifically for the breeds of animals and climatic conditions in each country.

Emergency Plans

Truck drivers should be equipped with an emergency plan so they will know what to do after an accident. Often police and fire departments do not know how to handle livestock or poultry that get loose on a highway. Sometimes they respond in an inappropriate manner which results in terrified cattle or sheep charging down a highway. Chasing scared frightened animals is the worst thing to do. When police arrive at an accident scene, they should concentrate on directing traffic instead of chasing animals. In some countries, police and fire departments receive specific training on how to respond to livestock truck accidents. The author recommends that programmes for training police and fire departments should be implemented. People with livestock expertise should be called to the accident scene to gather up loose animals. The driver should always carry with him the phone numbers of both police and people who know how to handle animals. Phone numbers for getting another vehicle should also be included on the list of emergency numbers. The driver needs to have contact information for different people along the route that can assist. The locations of places where animals can be unloaded in an emergency should also be carried with the driver. These locations could be farms, stockyards or auction markets. The phone numbers of people who can quickly bring a portable loading ramp and portable fence panels should also be carried by the driver. Emergency response plans work best if plans on how to respond in the event of an accident have been made before an accident happens.

Euthanizing Animals After an Accident

Chapter 10 covers euthanasia of different types of animals. Animals that are severely injured should be euthanized at the site of the accident to stop continued suffering. A plan for making euthanasia decisions should be made ahead of time. In some cases, there may be legal problems with the owner of an animal when it has survivable injuries and it has been euthanized. Large corporations who transport their own animals should have policies that state that animals that are severely injured in an accident will be immediately euthanized at the accident site. Some examples of injuries where animals should be euthanized at the accident site are a broken leg, if the internal organs are exposed or the animal is not able to stand.

Heat Stress

Different animals have greater tolerance for heat stress than others. In poultry, researchers in the Czech Republic found that death losses were greater in summer (Vecerek *et al.*, 2006). Heat stress may be responsible for up to 40% of chicken mortalities (Bayliss and Hinton, 1990). When the temperature was over 23°C, the percentage of chicken mortalities was almost seven times higher in modern broilers. Heat stress is a much greater problem for chickens than cold stress unless the birds are subjected to extreme cold. McGlone

(2006) analysed data from 2 million pigs that arrived at a large slaughter plant in the Midwestern USA. Death losses started to increase when the temperature was over 23°C (75°F) and at 32°C (90°F) death losses doubled.

Breeds of cattle, sheep and other animals that come from the hot regions of the world can tolerate more heat compared to breeds developed in more temperate climates. Heat can build up rapidly in a parked vehicle. This is especially going to be a problem if the vehicle has solid sides and a solid roof. Vehicles and ships where animals are dependent on mechanical ventilation can heat up to lethal levels in less than 1 h if the ventilation system fails. Failure of the ventilation system is a major cause of death losses in animals that are transported below the deck on ships.

When vehicles have to be parked during hot weather there should be provisions for heat relief. When the temperature is 29°C (85°F), the air inside the vehicle can quickly reach 35°C (95°F) and 95% relative humidity (McGlone, 2006). At poultry plants, trucks loaded with chickens are parked in a shed equipped with a bank of fans. Some pork slaughter plants also have banks of fans for cooling parked trucks. One of the simplest ways to keep a hot truck ventilated is to keep it moving.

Assessing heat stress

Panting in both mammals and birds is a sign that the animals are becoming heat stressed. In cattle, panting scores are a reliable indicator of heat stress (Mader *et al.*, 2005). Mader developed a simple five-point panting score for cattle:

0 – Normal respiration
1 – Elevated respiration
2 – Moderate panting or presence of drool
3 – Heavy open-mouth breathing
4 – Severe open-mouth breathing with protruding tongue. Neck may be extended forward.

The animal's welfare would be severely compromised when the score reaches 3 or 4. At score 2, provide heat relief. Under very hot conditions, purebred Brahmans had lower respiration rates than crosses with Herefords (Gaughan *et al.*, 1999). For sheep and cattle, additional information on assessing heat stress can be found in Mader *et al.* (2005) and Gaughan *et al.* (2008). For pigs, one of the easiest to use heat stress charts is shown in Fig. 7.12. A combination of high heat and high humidity can be dangerous for pigs. The chart provides guidelines on the combinations of heat and humidity where extra precautions to prevent heat stress must be

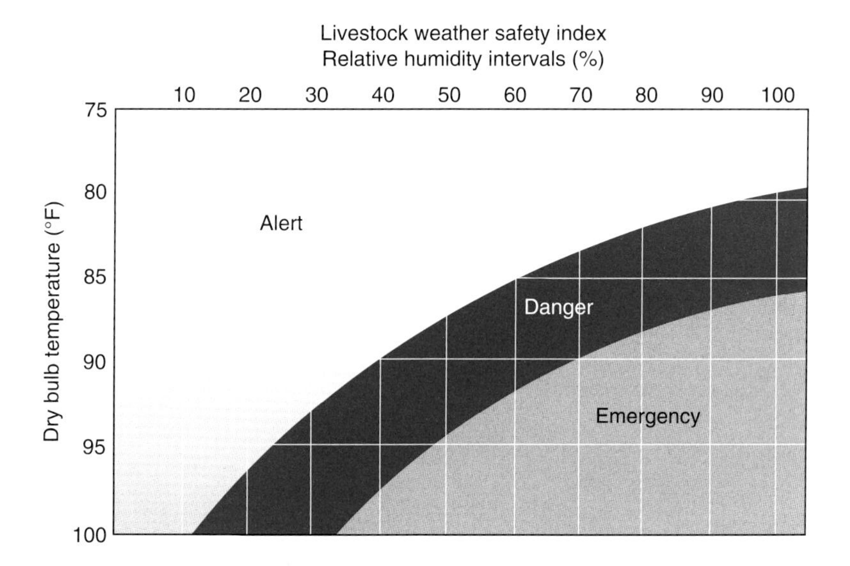

Fig. 7.12. Heat stress chart for pigs. A combination of both high heat and high humidity is dangerous for pigs. When the temperature and humidity is in the danger and emergency zone, it is recommended to transport the pigs at night or in the very early morning. Temperature conversions: 75°F (24°C); 80°F (27°C); 85°F (29°C); 90°F (32°C); 95°F (35°C); 100°F (38°C).

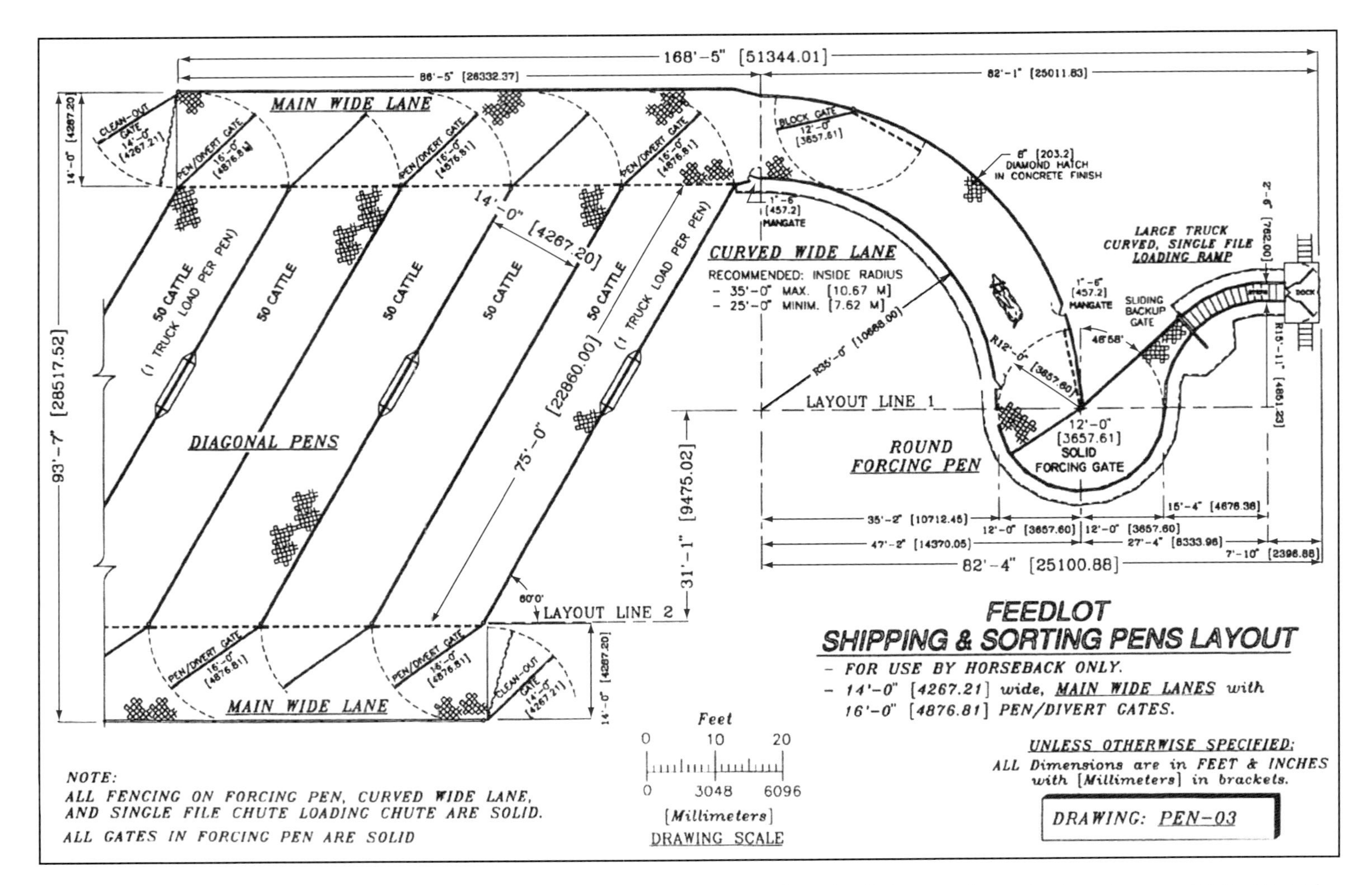

Fig. 7.13. Shipping facility for a large feedlot market. Diagonal pens provide easy one-way flow of cattle because they enter through one end of the pen and leave through the other end. Sharp 90° corners are eliminated.

taken. In pigs, respiration is a major indicator of heat stress (Brown-Brandt *et al.*, 2001). An in-depth view of stress physiology during transport can be found in Knowles and Warriss (2007).

Cold Stress

Transporters must be trained to be careful to avoid the effects of wind chill. Freezing rain at 0°C that wets an animal's coat can be deadly because the animal's coat loses the ability to insulate. Driving a truck through freezing rain is often more dangerous for livestock than colder dry weather where animals remain dry. In countries with cold weather, guidelines should be developed on when animals in vehicles need to be protected from cold by covering a vehicle with tarps or boards. In the northern regions of Europe and North America, great care has to be taken to prevent frostbite or death losses due to cold stress in pigs. For pigs, deep straw bedding will help protect the animals. The author has observed that pigs can easily get frostbite if they lie against a bare metal floor or side of a vehicle, when the temperature is −18°C. Piglets weighing 7kg should be transported in a warm vehicle but the maximum temperature should not exceed 30°C (Lewis, 2007). Now that animal welfare is a concern all over the world, published standards for acceptable ranges of temperature may be different due to both animal genetics and different conditions in tropical countries.

Vehicle Requirements

In the EU, there are no regulatory requirements for truck design for trips under 8h. For trips over 8h, fans for ventilation are required. These systems should be designed so that the animals are not totally dependent on the mechanized ventilation system. If the mechanical ventilation fails, the truck or trailer should be designed so that passive systems such as removable panels or curtains can be opened. This would provide ventilation to prevent death losses from heat stress. During cold weather, curtains or panels can be used to close up ventilation openings to prevent frostbite and deaths from wind chill. Many livestock associations have recommendations for the use of panels and curtains at different temperatures. It is important to develop specific guidelines for your local area that are based on collecting data from many truckloads. The vehicle shown in Fig. 7.2 is a typical truck used all over South America, and many other countries. In the parts of the world where the temperature is warm year-round, a vehicle like the one shown in Fig. 7.2 provides lots of natural ventilation.

Stockyard Layout

Angled pens in the shipping area at large feedlots, slaughter plants and markets allow movement of animals by eliminating sharp 90° corners. Figure 7.13 shows an angled pen layout with a curved loading ramp. This design works really well with extensively raised or feedlot animals that are not trained to lead. It is especially suited for large operations where many trucks have to be quickly loaded.

Conclusions

To have an adequate level of animal welfare during transport, fit animals must be loaded in the vehicle. Transporting unfit animals is a cause of major welfare problems during transport. To maintain transport practices at a high standard, continuous measurement of animal-based outcome measures is essential.

References

Appleby, M.C., Cussen, V., Garces, L., Lambert, L.A. and Turner, J. (2008) *Long Distance Transport and the Welfare of Farm Animals.* CAB International, Wallingford, UK.

Arnold, G.W. (1976) Some factors influencing feeding behaviour of sheep in pens. In: *Proceedings of the Sheep Assembly and Transport Workshop.* Western Australian Department of Agriculture, Perth, Western Australia.

Averos, X., Knowles, T.G., Brown, S.N., Warriss, P.D. and Gosalvez, L.F. (2008) Factors affecting the mortality of pigs transported to slaughter. *Veterinary Record* 163, 386–390.

Baszczack, J.A., Grandin, T., Gruber, S.L., Engle, T.E., Platter, W.J., Laudert, S.B., Schroeder, A.L. and Tatum, J.D. (2006) Effects of ractopamine supplementation on behaviour of British, Continental, and Brahman crossbred steers during routine handling. *Journal of Animal Science* 12, 3410–3414.

Bayliss, P.A. and Hinton, P.A. (1990) Transportation of broilers with specific reference to mortality rates. *Applied Animal Behaviour Science* 28, 93–118.

Beatty, D.T., Barnes, A., Taylor, E., Pethick, D., McCarthy, M. and Maloney, S.K. (2006) Physiological responses of *Bos taurus* and *Bos indicus* cattle to prolonged, continuous heat and humidity. *Journal of Animal Science* 84, 972–985.

Brown-Brandt, T.M., Eigenberg, R.A., Nienaber, A. and Kachman, S.D. (2001) The thermoregulatory profile of a newer genetic line of pigs. *Livestock Production Science* 71, 253–260.

Chandra, B.S. and Das, N. (2001) The handling and short haul road transportation of spent buffalos in relation to bruising and animal welfare. *Tropical Animal Health Production* 33, 153–163.

Cockram, M.S., Kent, J.E., Jackson, R.E., Goddard, P.J., Doherty, O.M., McGilp, I.M., Fox, A., Studdert-Kennedy, T.C., McLonell, T.J. and O'Riordan, T. (1997) Effect of lairage during 24 hour transport on the behavioural and physiological responses of sheep. *Animal Science* 65, 391–402.

Cockram, M.S., Baxter, E.M., Smith, L.A., Bell, S., Howard, C.M., Prescott, R.J. and Mitchell, M.A. (2004) Effect of driver behaviour, driving events and road type on the stability and resting behaviour of sheep in transit. *Animal Science* 79, 165–176.

Duncan, I.J.H., Slee, G., Kettlewell, P.J., Berry, P. and Carlisle, A.J. (1986) A comparison of the effects of harvesting broiler chickens by machine or by hand. *British Poultry Science* 27, 109–114.

Eldridge, G.A. and Winfield, C.G. (1988) The behavior and bruising of cattle during transport at different space allowances. *Australian Journal of Experimental Agriculture* 28, 695–698.

European Union (EU) (2005) *Official Journal of the European Union* 5.1.2005 L3/19.

Ewbank, R. and Parker, M. (2007) Handling cattle in close association with people. In: Grandin, T. (ed.) *Livestock Handling and Transport.* CAB International, Wallingford, UK, pp. 76–89.

Ferguson, D.M., Niemeyer, D.P.O., Lee, C. and Fisher, A.D. (2008) Behavioral and physiological responses in sheep to 12, 30, and 48 hours of road transport. In: Boyle, L., O'Connell, N. and Hanlon, A. (eds) *Proceedings of the 42nd Congress of International Society for Animal Ethology (ISAE)*, 5–9 August, Dublin, Ireland, p. 136.

Fike, K. and Spire, M.F. (2006) Transportation of cattle. *Veterinary Clinics of North America, Food Animal Practice* 22, 305–320.

Firtzgerald, R.F., Stalder, K.J., Matthews, J.O., Schultz-Kaster, C.M. and Johnson, A.K. (2008) Factors associated with fatigued, injured, and dead pig frequency during transport and lairage at a commercial abattoir. *Journal of Animal Science* 87, 1156–1166.

Garcia, L.G., Nicholson, K.L., Hoffman, T.W., Lawrence, T.E., Hale, D.S., Griffin, D.B., Savell, J.W., Vanoverbeke, D.L., Morgan, J.B., Belk, K.E., Field, T.G., Scanga, J.A., Tatum, J.D. and Smith, G.C. (2008) National Beef Quality Audit – 2005: survey of targeted cattle and carcass characteristics related to quality, quantity, and value of fed steers and heifers. *Journal of Animal Science* 86, 3533–3543.

Gaughan, J.B., Mader, T.L., Holt, S.M., Josey, M.J. and Rowan, K.J. (1999) Heat tolerance of Boran on Tuli crossbred steers. *Journal of Animal Science* 77, 2398–2405.

Gaughan, J.B., Mader, T.L., Holt, S.M. and Lisle, S.M. (2008) A new heat load index for feedlot cattle. *Journal of Animal Science* 86, 226–234.

Gosalvez, L.F., Averos, X., Valdeivira, J.J. and Herranz, A. (2006) Influence of season, distance and mixed loads on the physical and carcass integrity of pigs transported to slaughter. *Meat Science* 73, 553–558.

Grandin, T. (1981) Bruises on Southwestern feedlot cattle. *Journal of Animal Science* 53 (Supplement 1), 213 (abstract).

Grandin, T. (1990) Design of loading facilities and holding pens. *Applied Animal Behaviour Science* 18, 187–201.

Grandin, T. (1994) The welfare of cattle during slaughter and the prevention of downers. *Journal of the American Veterinary Medical Association* 218, 10–15.

Grandin, T. (1997) Handling methods and facilities to reduce stress on cattle. *Veterinary Clinics of North America, Food Animal Practice* 14, 325–341.

Grandin, T. (2007) *Livestock Handling and Transport.* CAB International, Wallingford, UK.

Grandin, T. (2008) Engineering and design of holding yards, loading ramps and handling facilities for land and sea transport of livestock. *Veterinaria Italiana* 44, 235–245.

Grandin, T., McGee, K. and Lanier, J.L. (1999) Prevalence of severe welfare problems in horses that arrived at slaughter plants. *Journal of the American Veterinary Medical Association* 214, 1531–1533.

Hall, S.J.C. and Bradshaw, R.H. (1998) Welfare aspects of the transport by road of sheep and pigs. *Journal of Applied Welfare Science* 1, 235–254.

Hoffman, D.E., Spire, M.F., Schwenke, J.R. and Unruh, J.A. (1998) Effect of source of cattle and distance transported to a commercial slaughter facility on carcass bruises in mature beef cows. *Journal of the American Veterinary Medical Association* 212, 668–672.

Holtcamp, A. (2000) Gut edema: clinical signs, diagnosis and control. In: *Proceedings of the American Association of Swine Practitioners*, 11–14 March, Indianapolis, Indiana, pp. 337–340.

Knowles, T.G. (1999) A review of the road transport cattle. *Veterinary Record* 144, 197–201.

Knowles, T. and Warriss, P. (2007) Stress physiology during transport. In: Grandin, T. (ed.) *Livestock Handling and Transport.* CAB International, Wallingford, UK, pp. 312–328.

Knowles, T.G., Warriss, P.D., Brown, .N. and Edwards, J.E. (1998) The effects of stocking density during the road transport of lambs. *Veterinary Record* 142, 503–509.

Lalman, D. and Smith, R. (2005) *Effects of Preconditioning on Health, Performance, and Prices of Weaned Calves.* Oklahoma Cooperative Extension, Oklahoma State University, Stillwater, Oklahoma.

Lay, D.C., Friend, T.H., Randel, R.D., Bowers, C.C., Grissom, K.K. and Jenkins, O.C. (1992) Behavioral and physiological effects of freeze and hot iron branding on crossbred cattle. *Journal of Animal Science* 70, 330–336.

Lewis, N. (2007) Transport of early weaned piglets. *Applied Animal Behaviour Science* 110, 126–135.

Loneragen, C.H., Dargartz, D.A., Morley, P.S. and Smith, M.A. (2003) Trends in mortality ratios among cattle in US feedlots. *Journal of the American Veterinary Medical Association* 219, 1122–1127.

Mader, T.L., Davis, M.S. and Brown-Brandl, T. (2005) Environmental factors influencing heat stress in feedlot cattle. *Journal of Animal Science* 84, 712–719.

Marchant-Forde, J.N., Lay, D.C., Pajor, J.A., Richert, B.T. and Schinckel, A.P. (2003) The effects of ractopamine on the behavior and physiology of finishing pigs. *Journal of Animal Science* 81, 416–422.

Marai, I.F.M., El-Darawany, A.A., Fadiel, A. and Abdel-Hafez, M.A.M. (2006) Physiological traits as affected by heat stress in sheep: a review. *Small Ruminant Research* 71, 1–12.

McGlone, J. (2006) Fatigued pigs: the transportation link. *Pork*, 1 February. Available at: www.porkmag.com/directories.asp?pgID=728Sed_id=3951 (accessed 21 March 2009).

Meischke, H.R.C. and Horder, J.C. (1976) Knocking box effect on bruising in cattle. *Food Technology, Australia* 28, 369–371.

Minka, N.S. and Ayo, J.O. (2006) Effects of loading behaviour and road transport stress on traumatic injuries in cattle transported by road during the hot dry season. *Livestock Science* 107, 91–95.

Murray, A.C. and Johnson, C.P. (1998) Importance of halothane gene on muscle quality and preslaughter death in western Canadian pigs. *Canadian Journal of Animal Science* 78, 543–548.

National Hog Farmer (2008) Poster. Available at: www.NationalHogFarmer.com/posters (accessed 30 November 2008).

Nijdam, E., Arens, P., Lambooij, E., Decuypere, E. and Stegeman, J.A. (2005) Comparison of bruises and mortality stress parameters, and meat quality in manually and mechanically caught broilers. *British Poultry Science* 83, 1610–1615.

Norriss, R.T., Richards, R.B., Creeper, J.H., Jubb, T.F., Madin, B. and Ken, J.W. (2003) Cattle deaths during sea transport from Australia. *Australian Journal of Veterinary Research* 8, 156–161.

OIE (2009a) *Transport of Animals by Land, Terrestrial Animal Health Code.* World Organization for Animal Health, Paris, France.

OIE (2009b) *Guidelines for the Sea Transport of Animals, Terrestrial Animal Health Code.* World Organization for Animal Health, Paris, France.

Powell, J. (2003) Preconditioning Beef Calves. Cooperative Extension Service, University of Arkansas. Available at: www.uaex.edu (accessed 1 May 2009).

Rademacher, C. and Davis, P. (2005) Factors associated with the incidence of mortality during transport of market hogs. In: Allen, D. (ed.) *Proceedings of the Leman Swine Conference.* University of Minnesota, Minneapolis, Minnesota, USA, pp. 186–191.

Ramsey, W.R., Meischke, H.R.C. and Anderson, B. (1976) The effect of tipping horns and interruption of the journey on bruising cattle. *Australian Veterinary Journal* 52, 285–286.

Ritter, M.J., Ellis, M., Bertelson, C.R., Bowman, R., Brinkman, J., Dedecker, J.M., Keffaber, K.K., Murphy, C.M., Peterson, B.A., Schlipf, J.M. and Wolter, B.F. (2007) Effects of distance moved during loading and floor space on the trailer during transport on losses of market weight pigs on arrival at the packing plant. *Journal of Animal Science* 85, 3454–3461.

Ritter, M.J., Ellis, M., Bowman, R., Brinkman, J., Curtis, S.E., DeDecker, J.M., Mendoza, O., Murphy, C.M., Orelleno, D.G., Peterson, B.A., Rojo, A., Schliph, J.M. and Woltzer, B.F. (2008) Effect of season and distance moved during loading in transport losses of market weight pigs in two commercially available types of trailer. *Journal of Animal Science* 86, 3137–3145.

Ritter, M.J., Ellis, M., Hollis, G.R., McKeith, F.K., Orellana, D.G., Van Genugten, P., Curtis, S.E. and Schilph, J.M. (2009a) Frequency of HAL-1843 mutation of the nanodine receptor gene in dead and non-ambulatory, non-injured pigs on arrival at the packing plant. *Journal of Animal Science* 86, 511–514.

Ritter, M.J., Ellis, M., Berry, N.L., Curtis, S.E. *et al.* (2009b) Review: transport losses in market weight pigs. A review of definitions, incidence and economic impact. *The Professional Animal Scientist* 25, 404–414.

Shaw, F.D., Baxter, R.I. and Ramsey, W.R. (1976) The contribution of horned cattle to carcass bruising. *Veterinary Record* 98, 256–257.

Sutherland, M.A., Bryer, P.G., Davis, B.L. and McGlone, J. (2009) Space requirements of weaned pigs during a sixty minute transport in summer. *Journal of Animal Science* 87, 363–370.

Swanson, J.C. and Morrow-Tesh, J. (2001) Cattle transport: historical, research, and future perspectives. *Journal of Animal Science* 79 (E Supplement), E102–E109.

Tarrant, P.W., Kenny, F.J. and Harrington, D. (1988) The effect of stocking density during 4 h transport to

slaughter, on behaviour, blood constituents and carcass bruising in Friesian steers. *Meat Science* 24, 209–222.
Valdes, A. (2002) *Efectos de los Densidades de Carga y dos Tiempo de Transporte Sobre el Peso Vivo, Rendimiento de la Canal y Presencia de Contusiones s en Novillos Destinados al Faenamiento.* Memoria de Titulo para Optar al Titulo de Medico Veterinario Fac. Ciencias Veerianarias, Universidad Austral del Chile Valdivia, Chile.
Vecerek, V., Grbalova, S., Voslarova, E., Janackova, B. and Maiena, M. (2006) Effects of travel distance and the season of the year on death rates of broilers transported to poultry processing plants. *Poultry Science* 85, 1881–1884.
Warner, A.C.I. (1962) Some factors influencing the rumen microbial population. *Journal of General Microbiology* 28, 129–146.
Warriss, P. (1998) Choosing appropriate space allowances for slaughter pigs transported by road: a review. *Veterinary Record* 142, 449–454.
Warriss, P.D., Bevis, E.A., Brown, N. and Edwards, J.E. (1992) Longer journeys to processing plants are associated with higher mortality in broiler chickens. *British Poultry Science* 33, 201–206.
Webster, A.B. (2004) Welfare implications of avian osteoporosis. *Poultry Science* 83, 184–192.
Weeks, C.A. (2000) Transport of deer: a review with particular relevance to red deer (*Cervus elaphus*). *Animal Welfare* 9, 63–74.
Weeks, C.A., McNally, P.W. and Warriss, P.D. (2002) Influence of the design of facilities at auction markets and animal handling procedures on bruising in cattle. *Veterinary Record* 150, 743–748.
Winks, L.A., Holmes, E. and O'Rourke, P.K. (1977) Effect of dehorning and tipping on live weight gain of mature Brahman crossbred steers. *Australian Journal of Experimental Agriculture and Animal Husbandry* 17, 16–19.
Woods, J. and Grandin, T. (2008) Fatigue is a major cause of commercial livestock truck accidents. *Veterinaria Italiana* 44, 259–262.

8 Animal Well-being and Behavioural Needs on the Farm

Lily N. Edwards*

Colorado State University, Fort Collins, Colorado, USA

Animal welfare is such an important concept in today's society yet it is a term that is difficult for animal scientists, producers, veterinarians and consumers to adequately and completely define. Scientists have provided statements defining good animal welfare (Duncan and Dawkins, 1983; Moberg, 1985; Broom, 1986; Duncan and Petherick, 1991; Mendl, 1991; Mason and Mendl, 1993; Fraser, 1995; Ng, 1995; Sandoe, 1996; Dawkins, 2006). There is no consensus as to how to precisely define animal welfare. The inability of specialists to agree on a complete definition of welfare is in part due to the fact that there are so many different factors that could be used to determine an animal's welfare. Scientists have not been able to create a mathematical formula into which we plug all measures of welfare (biological measures, behavioural measures and psychological components, etc.) to be given a quantitative, all-encompassing measure on a simple scale of 1–10. That would be easy. Unfortunately, we have not even been able to achieve this daunting feat even with humans, perhaps an easier task than with animals considering our first-hand insight into human nature.

Different Definitions of Welfare

Each welfare specialist has their own definition of what 'good' welfare means. Some specialists maintain that an animal's welfare is primarily based on health (Moberg, 1985). Can the animal reproduce? Is it free from disease? Looking deeper at specific physiological mechanisms, levels of various hormones (e.g. cortisol, lactate, epinephrine, etc.) are measured to assess welfare status (Moberg, 1985). Scientists use their knowledge of stress physiology and immune response to help identify an animal's welfare status. For example, calves respond to dehorning with physiological and immunological responses that indicate pain and stress (Wohlt *et al.*, 1994; Petrie *et al.*, 1996; Sylvester *et al.*, 1998; Stafford and Mellor, 2005) (see Chapter 6). It is safe to assume that most people would agree that health, prevention of injuries, reducing death losses and lameness are primary components of an animal's well-being.

Many welfare specialists extend the concept of animal welfare beyond what may be considered the more objective physiological stress data and health status. They say that animal welfare must also include emotional (affective) states, feelings and wants (Duncan and Dawkins, 1983; Dawkins, 1988; Duncan and Petherick, 1991; Sandoe, 1996). Are the cows 'happy'? Are all their needs being met? Are they distressed? In discussion of mental states, some scientists state that the performance of natural behaviours is a necessary component for good welfare. Is the sow in the gestation crate able to perform nesting behaviour? If she cannot nest build, how is this affecting her well-being? The definition of good animal welfare will change depending who you ask. The plethora of definitions regarding animal well-being not only sets the platform for extreme debate, but additionally, it calls attention to the enormous task at hand for all those involved with assessing, determining, ensuring

* Present address: Kansas State University, Manhattan, Kansas, USA

and simply just understanding animal welfare. Despite debates on what animal welfare is, many guidelines and regulations have been created to protect an animal's well-being. It is a little peculiar to imagine rules governing a concept that we, in general, have so much trouble understanding and defining. Some animal welfare concepts and concerns are purely ethical and cannot be completely explained scientifically (see Chapter 2). It is important to explore the vocabulary used in policies and legislation to which we must subscribe to understand which components of welfare should be included as top priorities.

Welfare Concepts in Different Countries

In the USA, the Animal Welfare Act (Office of the Federal Register, 1989) is the federal law, first initiated in 1960, that provides minimum protection to certain species of animals, primarily companion animals and those used in research, for handling, care, housing and treatment. Unfortunately, the law's definition of an animal does not include livestock species raised for food or fibre. Farm animals were exempt from this law due to political pressure from livestock groups. The provisions set forth in the Animal Welfare Act apply mostly to the housing of each animal species. There is one amendment, officially made in 1985, that requires facilities housing non-human primates to provide an 'environment adequate to promote the psychological well-being of the animals'. Although clearly stated, it is unclear exactly what the definition of psychological well-being is and the amendment provides no suggestion as to how to satisfy this requirement. In this law the mental component of animal well-being applies only to non-human primates.

In the UK, the Brambell Commission (Brambell, 1965), a governmental committee, was created to assess the welfare of animals raised in intensive husbandry systems. They wrote a report proposing the conditions that farm animals should experience. Since the revision by the UK Farm Animal Welfare Council (FAWC) in 1979, these animal conditions are now universally referred to as the 'five freedoms' and are as follows: (i) freedom from thirst, hunger and malnutrition; (ii) freedom from discomfort; (iii) freedom from pain, injury and disease; (iv) freedom to express normal behaviour; and (v) freedom from fear and distress (FAWC, 1979). The 'five freedoms' are more inclusive of all aspects of animal welfare and have been the basis for welfare assessment in the UK and many other countries.

Animal Welfare Categories

Dividing the different components of welfare into categories will make it easier to implement effective welfare programmes. Several decades after the five freedoms had been composed, Dr David Fraser (2008) identified the three main approaches to animal welfare assessment and/or basis for definition: (i) biological functioning – physical health, growth, reproduction; (ii) affective states – fear, anxiety, frustration, pain, hunger and thirst; and (iii) natural living – a lifestyle that approximates natural living conditions and the fulfilment of behavioural needs. More recently, at the 2008 American Society for Animal Science meeting, Fraser presented his reorganization of the components of animal welfare into four categories. They are: (i) maintain basic health; (ii) reduce pain and distress; (iii) accommodate natural behaviours and affective states; and (iv) provide natural elements in the environment (see Chapter 1). These four categories include all concepts of animal welfare ranging from health and lack of pathological conditions to emotional states and feelings. These categories will also be easier to implement. Animal welfare may be best defined as a combination of all of these categories, however the exact weighting of each component is unclear (i.e. is there one component that is more important, if two are satisfied can the third be forgotten? etc.). The Brambell Commission (Brambell, 1965) additionally stated that animal welfare 'embraces both the physical and the mental well-being of the animal', similarly drawing on the mental status of the animal as seen in the Animal Welfare Act amendment.

Animal Abuse Versus Animal Welfare

One thing to note in this discussion is the difference between poor animal welfare and animal abuse. Animal abuse is a deliberate act that inflicts obvious, unnecessary harm and suffering. Some examples are starvation, torturing, beating, poking out eyes and depriving of survival needs such as water. When people talk about animal abuse cases they are usually referring to the deliberate acts of abuse that either 'disturbed' or ignorant people commit on animals. In many countries animal cruelty laws were created as regulations against these deliberate acts of abuse. In some countries there are even special humane officers with police powers who enforce these statutes. (I am sure some of you are familiar with the animal TV police shows that feature different humane officers saving cats, dogs, horses and other family pets from atrocious abuse and neglect.) To have good welfare an animal must not experience obvious acts of abuse, but the concept of good animal welfare extends much further. Animal welfare also includes good health, pain relief, reduction of fear and environments that provide for an animal's behavioural needs. As concerns with animal abuse have transformed into concerns for an animal's total welfare, many more areas in which we use animals (research, agriculture, etc.) have come under scrutiny.

The Animal's Mental State is an Important Part of Welfare

In an examination of all the statements made by the various committees and law makers there seems to be a common theme: the emphasis on the emotional (affective) component of the animal's life in the consideration of animal well-being and welfare. Some researchers have clearly emphasized this mental component indicating that an animal's welfare is solely dependent on the psychological needs of the animal, the physical needs being met once the mental ones are met (Duncan and Petherick, 1991). However, most welfare specialists agree that mental (affective) states are an important component of welfare but basic health and prevention of injury should never be neglected. The difficult question arising from these terms is: what is psychological well-being? How do you determine that an animal has good psychological well-being and more importantly how do we measure it? Let us consider a human's psychological well-being. I think it is safe to assume that we believe a human is in a state of mental well-being when he is free from suffering. A psychologically well human, would exhibit contentment with his life and have all of his needs, both for survival and happiness, met. Who has a more desirable mental state? The person who gets to come home from work at five o'clock everyday to have an enjoyable dinner with their family or the person who works until ten, never sees their family and has high blood pressure? Because of our life experiences it is easy for us to identify with both of these situations and choose which one we would prefer. Many people would choose the first alternative, but some high-achieving people who get bored easily may choose the second alternative. Once basic needs for food, safety and shelter are met, the concept of mental welfare becomes more complex.

To reach a state of well-being, humans want to minimize the amount of suffering in their lives. Suffering includes a multitude of negative feelings and emotions such as pain, anxiety, frustration, fear and any other unpleasant sensations associated with physical and/or mental harm to the individual. A human can suffer from chronic, intense pain from physical trauma. Some chronic pain victims feel that pain is unendurable and contemplate ending their lives to end the suffering. A human can also suffer from fear, fear of the economic consequences of losing her job. Or from frustration, the frustration of being stuck in a marriage he is unhappy with but feels he cannot escape from. (As a side note, the American population seems to be experiencing widespread mental suffering when looking at the number of people taking prescription antidepressants and anxiety medications. The National Center for Health Statistics (2007) reported that the use of antidepressants more than tripled between 1988–1994 and 1999–2002.) There are many ways that a human being can suffer, all of which have a negative impact on his overall psychological well-being. Humans try to avoid these situations, looking for situations that are more pleasing, situations that play to the expression of positive emotions.

The Negative Core Emotions

Neuroscience research clearly shows that animals and people have the same core emotions. The emotions of rage, panic (separation anxiety) and fear, are three of the negative core emotions (Panksepp, 1998). These emotional systems have been fully mapped. Electrical stimulation of the thalamus in cats and rats will make them go into a rage (Panksepp, 1971; Olds, 1977; Heath, 1996; Siegel and Shaikh, 1997; Siegel, 2005). The second negative core emotion is separation anxiety. It occurs when a baby animal is separated from its mother or an adult animal is separated from its herd mates (Semitelou *et al.*, 1998; Panksepp, 2005a). The third negative core emotion is fear. The circuit in the fear emotional system motivates animals to avoid predators and they have been extensively researched and studied (LeDoux, 1992, 2000; Rogan and LeDoux, 1995, 1996).

Breeding studies with quail clearly indicate that the fear and panic (separation anxiety) systems are separate systems governed by different genetics. The natural wild quail shows both high fear and high separation anxiety. French scientists were able to breed four separate genetic lines of quail. They were: high fear and high separation anxiety; high fear and low separation anxiety; low fear and high separation anxiety; low fear and low separation anxiety (Faure and Mills, 1998). In their experiment, fear was measured by the tonic immobility test (see Chapter 1). The time the bird remained immobile was used as an index of fearfulness. Separation anxiety (called social reinstatement by Faure) was measured by placing the bird on a moving conveyor which moved it away from a cage full of flock mates. The strength of separation anxiety was measured by the length of time the bird would 'treadmill' on the conveyor to stay near its flock mates.

People do not like feelings of fear or panic, and so they avoid situations that stimulate these negative emotional systems. When a child in a grocery store suddenly finds she is no longer with her mother she panics and begins to search for her. When I am walking my dog in the park alone at night and hear footsteps behind me, I get scared and I walk a little bit faster to get home. People try to escape from the negative emotions and we embrace the positive ones (Grandin and Johnson, 2009).

Would the definition of suffering change if the subjects were animals? It should not. Animals also have the same core emotions. Of course, the behaviours would not be the same. My dog experiences a mild form of rage when he is frustrated that he cannot find the little piece of meat I dropped under the kitchen counter. Instead of communicating his frustration with an annoyed tone of voice as a mildly frustrated human would do, he whines a little bit, letting me know how he is feeling. Fear in animals is easy to identify. Some examples of fear-motivated behaviour are: a dog that crouches down and cowers when a person approaches it with a stick because he is afraid of being beaten or an animal that struggles when it is restrained for veterinary procedures. When cattle are scared of an object they move away from it or baulk and won't walk forward (Grandin, 1997).

The Positive Core Emotions

The brain also has core positive emotions. The most important is seeking or goal-directed behaviour (Panksepp, 1998, 2005a). A pig that gets a fresh flake of straw will vigorously root and chew it up, or a herd of cattle is motivated to walk into a new pasture full of fresh green grass. Manipulating and rooting straw or other fibrous material is a highly motivated behaviour in pigs (Berlyne, 1960; Wood-Gush and Vestergaard, 1989; Day *et al.*, 1995; Studnitz *et al.*, 2007). The circuits that control positive goal-directed behaviours are located in a part of the brain called the nucleus accumbens (Faure *et al.*, 2008). Other positive core emotions are play, sex and mother–young nurturing behaviour/pair bonding (Panksepp, 2005a). Both the positive and the negative core emotional circuits are located in the subcortical more-primitive parts of the brain that are the same in all mammals. Play will occur in rats where the brain's upper cortex has been removed (Panksepp *et al.*, 1994). The emotional systems are the drivers of behaviour that motivate an animal to engage in different behaviours.

Is Anthropomorphism Bad?

It is commonly said that it is not wise to compare human and animal emotions (i.e. human and animal suffering). This attribution of human characteristics to non-human objects (e.g. animals) is called anthropomorphism. The use of anthropomorphism is seen by some as dangerous because people make assumptions about things that are too foreign and subjective to ever truly be understood. A contrary view is that anthropomorphism may not be the sin that some scientists feel it is (Panksepp, 2005a), and actually it may be helpful for predicting an animal's psychological state. One example of how we have used anthropomorphism wisely is in the analysis of opioid control of separation distress in animals. The human psychological responses to opioids led to precise predictions of behavioural responses of animals (i.e. psychological response) to these pharmacological agents (Panksepp, 2005a). What some people forget is that animal welfare is no trickier an entity to define than human welfare (Dawkins, 1998; Broom, 2001). We make assumptions about the emotions of other humans on a daily basis. Although we think we understand the mental capacities of other humans there is no way of knowing if we are correct in our assumptions. Humans interpret and perceive the same situation in different ways. An example would be giving two different people a chocolate sundae. One person would really like to be given the chocolate ice-cream sundae and would find a lot of pleasure eating it. Another person who is diabetic would be frustrated if she was given a sundae because she would be looking at something that she could not eat. Suffering is a private subjective state that only the experiencer can truly understand (Barnard, 2007). If we can understand this caveat for humans yet still make assumptions about human mental states, why can we not do the same for animals and give them the benefit of the doubt?

Anthropomorphism can be especially risky when hard-wired instinctual behaviours (fixed action patterns) are misinterpreted. In humans, eye contact elicits positive emotions but many animals perceive eye contact and staring as a threat. The emotional system that is turned on in people and many animals is different. Instinctual behaviour patterns will be discussed later in this chapter.

Some People Resist the Idea of Animal Feelings

Many people feel more comfortable thinking about what goes on in the minds of other scientists and humans than what goes on in the minds of farm animals. Scientists work to discover the unknown by collecting data, numbers and information that are easily measurable (i.e. objective measurements). An animal's feelings about its surroundings are harder to define with an easily interpretable number. When it becomes necessary to use subjective measurements (i.e. feelings) there is often hesitation to evaluate animals in the same manner (Yoerg, 1991; Ng, 1995). Some scientists are not willing to admit that animals have emotional experiences. It is an interesting phenomenon though that some of these scientists are more willing to attribute feelings to particular animals and not to others. Fear of animal feelings faces many in the workplace, whether it is at the farm, in the lab or at the clinic, but it disappears when observing family pets wagging their tails when we come home from work. The animals we use in research and agriculture are perceived differently from pets. The beagle kennelled at the research facility is no different from the family pet kennelled in the house so he won't chew the carpet to bits while the owners are at work. The difference is our relationship with the animals and how we perceive them. This does not change the fact that dogs, pigs, rats, cats and cows are all animals. If we think our dog is happy when we let him out of his kennel, why can't we think the sow is happy when we let her out of the gestation crate?

Mapping of Emotion Brain Circuits

Animals do have the capacity to experience emotions. As stated earlier, fear of something so spooky and nebulous as animal feelings has prevented many scientists from acknowledging the existence of animal emotions. Emotions are scary because their subjectivity makes them difficult to explain with scientific (objective) measures. This is often translated into a reason for denouncing their existence. Ironically, there is a lot of science that can be used to prove not only the existence of animal emotions, but also the similarities that exist between humans and animals.

Beginning with simple brain structure and neural networks, it has been demonstrated that, although in different proportions, the basic structure and regional function in human and animal brains are highly similar (Jerison, 1997). MacLean was the first to discuss the concept of the triune brain (1990). This is a simplified conceptualization of the brain indicating the adaptive change between reptilian, mammalian and human brain regions. Through this tri-layer concept it can be seen that humans and animals share these brain regions, and in particular the limbic brain region. The limbic system is the home of human emotions and because this region is shared in both humans and animals it is also the seat of animal emotions (Rinn, 1984; Heath, 1996; Panksepp, 1998, 2003; Damasio, 1999; Liotti and Panksepp, 2004). Various emotional circuits (i.e. seeking, rage, fear, panic) are found in this common area (Panksepp, 1998). Anger (Siegel, 2005), fearfulness (Panksepp, 1990), sexuality (Pfaff, 1999) and maternal nurturance (Numan and Insel, 2003) are modulated in this subcortical region; they are all identifiable human and mammalian emotions. It is likely that the main difference between human and animal emotions is the complexity of emotional expression. Humans have a larger cortex than animals to process emotions. It is a misconception that the neocortex in the human brain provides the capacity for emotional experience. In fact, emotional states cannot be generated with neocortical electric or chemical stimulation (Panksepp, 1998). Additionally, when there is injury to the neocortex, certain emotional states, manifested by particular behaviours, such as play, reward seeking and self-stimulation (pleasure) are still present in animals (Huston and Borbély, 1973; Kolb and Tees, 1990; Panksepp *et al.*, 1994).

Similar Neurotransmitters in Humans and Animals

Human and animal brains have similar neurochemical networks. The chemicals, neurotransmitters, that transmit signals between neurons, are the same. Researchers have demonstrated that the neurochemistry of addiction, specifically of opiate and psychostimulants, is highly similar between all mammals (Knutson *et al.*, 2002; Panksepp *et al.*, 2002). The brain networks that respond to addictive drugs are found in the sub-neocortical brain regions, the ancestral regions of the brain that we share with animals (McBride *et al.*, 1999). One way to study the way drugs interact with the brain is through place preference studies (Bardo and Bevins, 2000). In these types of experiments, particular drugs are injected into subcortical areas of the brain when the animal is in particular locations. If the drug produces positive and euphoric feelings, the place that the drug was applied becomes a conditioned place preference. If the drug is aversive, the animal also develops an aversion (avoidance) to that location. Many studies have been done indicating that drugs we consider euphorogenic (or recreational) are also rewarding to animals, producing place preferences, and stimulating the same brain systems (Panksepp, 2004, 2005c). Rats respond with positive vocalizations to places where they have received 'good' drugs and negative vocalizations where they have received 'bad' drugs (Burgdorf *et al.*, 2001; Knutson *et al.*, 2002). The fact that animals exhibit learned place preferences indicates that they are most likely experiencing positive emotional states.

Humans search to make their lives more enjoyable and pleasant by seeking situations that promote positive feelings; animals do the same. In addition to drug addiction, animals respond to anti-anxiety and antidepression drugs. Anti-anxiety medications have become commonplace in many small animal practices (Overall, 1997); the family pet comes in for some Prozac® (fluoxetine) to be relieved of separation anxiety while the owners are at work. There are many scientific papers that detail the effects various psychoactive drugs have on animals. There is a recent review article that alone cites close to 100 studies that examined the effects of antidepressants on animal (mostly rat and mice) behaviour (Borsini *et al.*, 2002). Animals have also been shown to self-medicate and increase consumption of painkillers when experiencing pain from an injured joint (Colpaert *et al.*, 1980). Both antidepressants and anti-anxiety medications are used to treat pets (Seksel and Lindeman, 1998; King *et al.*, 2000; Romich, 2005).

Emotions Required for Survival

Moving away from brain anatomy and neurochemistry, the presence of animal emotions can also be supported from an evolutionary standpoint. Animals, like us, use emotions to determine how to act in their environments. If an animal is fearful of entering a certain place in the forest, it will use that perception of its environment to avoid a potentially threatening situation. Likewise, if my car had broken down on the side of the road while driving home at night and a man approached me offering to help in an insincere manner, my feeling of uneasiness would deter me from accepting his offer. Why would we, humans and animals, have emotions if we did not need them? If emotions were useless they would have probably been eliminated by natural selection a long time ago (Baxter, 1983). It is these 'gut' ancestral feelings that drive an animal's survival. Animals seldom hesitate before determining the appropriate action to take in a life-threatening situation. It would be an anomaly to see a gazelle locking eyes with a lion, pacing back and forth deciding whether or not to run away or wait a couple of minutes and see if the lion leaves. That gazelle instantly runs away from the predator's reach, driven by an intense feeling of fear, an emotion that enables it to survive. The feeling of fear is coming from the animal's fear centre, the amygdala, found in the limbic region of the brain with the other emotional circuitry (Davis, 1992; LeDoux, 1992). The amygdala has also been shown to affect fear conditioning in humans (Bechara *et al.*, 1995; Büchel *et al.*, 1998; LaBar *et al.*, 1998). In animals, damage to the amygdala blocks both learned and unlearned fear responses (Davis, 1992). Some examples of unlearned fear responses would be a horse rearing and becoming agitated when it sees a balloon, or a dog running away from a firecracker. Examples of learned fear responses would be an animal that becomes afraid of a person who beat it or cattle that learn to avoid an electric fence after they get shocked.

Another example of how emotions assist animal survival is distress caused by separation anxiety. We see this distress in both humans and animals. An animal example that immediately comes to mind is weaning in cattle when both the cows and the calves bellow after they are separated. A parallel example in humans would be the 4 year old who is hesitant to leave her mother's side on the first day of kindergarten. The strong social bond between an offspring and its parent is developed as a means of survival and when that bond is broken (at an unnatural, earlier time in particular) emotional reactions and separation vocalizations are triggered (Panksepp, 2005a).

Human Versus Animal Emotional Processing

It seems evident that animals can experience emotions but animals probably do not process these basic emotions in the same manner as humans. This does not mean that they feel less and could perhaps even mean that they feel more. Humans have a greater ability to regulate and inhibit their emotions with use of higher brain functions (i.e. a highly developed neocortex). As higher brain function developed, perhaps lower brain function (i.e. emotional systems) became more and more regulated and inhibited (Liotti and Panksepp, 2004). This hypothesis suggests that animals, with their lower cognitive abilities, may in fact experience more raw unprocessed emotions than humans. Panksepp (2005b) identifies this distinction in emotional experience as emotional awareness, experienced by humans, versus emotional affect, experienced by animals. If one compares animals and human children this theory gains more strength. Children, who have not reached full cognitive and forebrain development, often are more emotional than their parents and experience things such as fear, play and panic, with less inhibition and moderation (Burgdorf and Panksepp, 2006). Humans suffering from traumatic prefrontal brain damage also lack the ability to control their 'gut feelings' the way normal adult humans can (Damasio, 1994). People with frontal cortex head injuries are more prone to going into a rage with a minor provocation (Mason, 2008). In these patients, the brain damage prevented the control and inhibition of basic emotions and the result was lowered inhibition, perhaps similar to children and animals, both of whom have limited forebrain development. It is not that animals do not have subjective states of minds; they just do not have the capacity to inhibit emotional reactions like humans do.

Looking back to the brief mention of calf weaning and kindergarteners, we can see the difference between animal and human interpretations of feelings. (For the sake of discussion, let us assume that after a day of weaning, the calf is returned to its mother.) The kindergartener is nervous and scared to go to school; he is going to be all alone, away from his mother, in a novel environment. The calf is similarly put into a novel situation with other young calves with whom he is unfamiliar. As both mothers are 'taken away' there may be some crying, from both the mother and the son. The human mother says 'Remember, you will only be here for a couple of hours and then I will come to pick you up'. Despite her sadness and concern, the human mother knows that her son will be fine and that she will see him in no time at all. The bovine mother cannot comfort her calf the same way because, like her calf, she does not understand the reason for the separation, how long it will last or any further reasoning other than that her offspring is being taken away. The kindergartener can remind himself that his mother will be back to get him soon. The calf will cry constantly for his mother and his thoughts will bring him no solace because he cannot rationalize the fear of separation that he is feeling. Although that example is not perfect, it begins to illustrate that an animal's inability to foresee the end of a painful and fearful experience may make the experience more distressing.

Looking at it from an alternative perspective, there are some instances in which an animal's limited ability to process information could make him suffer less than humans. Consider the case of castration. When producers castrate calves, the animal suffers from fear of restraint, handling, a novel environment and the pain from the procedure. If someone were to castrate an unwilling young male human there would be an additional aspect of suffering. The male human would not only suffer from the listed events, he would also suffer from knowing that he was about to lead the rest of his life without testicles, a concept likely to be unknowable by the young calf. I think many human men would probably argue that this situation would be a source of extreme suffering. Sometimes ignorance is bliss, or at least a little less psychologically distressing. Unlike an animal, a man knows the pleasures he will never experience.

Behaviour Provides Indications of Emotion

In the expression of human emotions, we not only communicate how we feel by how we act and our particular behaviours but we also have the ability to verbally express how we feel when asked (and have the questioner understand what we mean). Unfortunately, not all of us have the talents of Dr Doolittle and we cannot talk to animals. Therefore, we must rely heavily on an animal's behaviour to determine how it possibly perceives its environment. Behaviour is extremely significant in human–animal communication. A veterinarian, whose job it is to cure disease, cannot simply ask an animal how it is feeling. She must use clues, such as alterations in behaviour (e.g. decreased eating, listlessness, increased water consumption) to determine what is wrong with the animal. Behaviour is our key to understanding an animal's experience in the environments in which we raise them (Darwin, 1872; Dantzer and Mormede, 1983). People use behavioural cues from animals on a daily basis, whether it is the feedlot manager moving cattle through a chute or the pet owner teaching her puppy to sit and stay. It is important to recognize that animal behaviour plays a large part in how we interact and communicate with animals.

One of the five freedoms for captive animals is the freedom to express normal behaviours. As many of us are well aware, it is often difficult to provide the animals we raise and live with, with this opportunity. Whether it is the space restrictions in zoos, the leash laws in our towns or the nature of the confinement of livestock operations, an animal's ability to express normal behaviours has become somewhat limited in today's society. Although we may not be able to allow animals to act as they do in the wild on a daily basis (letting large cats in zoos stalk and kill bunnies would not be accepted by the public), we must look for ways to substitute these behaviours with activities that accomplish the same goal. Hal Markowitz, a founder of enrichment for zoo animals, demonstrates this concept of substitution. One example of his endeavours was the development of acoustic 'prey' as a form of environmental enrichment for captive African leopards (Markowitz *et al.*, 1995). He was able to develop computer-controlled hunting simulation and exercise opportunities for captive felines, ultimately altering their behaviour and enhancing overall well-being. Markowitz was able to replace hunting with a similar experience that achieved the same goal.

Positive Goal-directed Behaviours

Animals have different levels of motivation or drive to perform a variety of goal-directed behaviours. These types of behaviours are directed at achieving a 'goal', the goal being the situation that terminates the behaviour (Manning, 1979). For Markowitz's leopard, the 'goal' was catching prey. The 'goal' of a dog's begging behaviour is to get food. Once he gets the food, the begging behaviour ceases. Some behaviours are more highly motivated than others; the more drive there is, the harder the animal will work to perform them. Animals are highly motivated to eat, drink, sleep, copulate, locomote, play, explore and find sensory stimulation and interact socially (Harlow *et al.*, 1950; Brownlee, 1954; Panksepp and Beatty, 1980; Dellmeier *et al.*, 1985; Dellmeier, 1989). These goal-directed behaviours are motivated by the positive core emotions of lust, seeking, care and play (Panksepp, 1998).

Behaviour Motivated by Negative Emotions

It is especially important to provide environments that reduce stimulation of the core emotions of fear, rage and panic (separation anxiety). Most of our common farm animals are prey species such as cattle, sheep and chickens. They may become more fearful than predators such as wolves, dogs or lions, because fear motivates them to avoid being eaten. Hens are highly motivated to have a hidden secluded place to lay their eggs (Appleby and McRae, 1986; Duncan and Kite, 1987; Cooper and Appleby, 1995, 1996b). When the jungle fowl, the ancestor of the modern hen, lived in the wild, the hens that hid in the bushes to lay their eggs survived and the ones that laid their eggs out in an open clearing got eaten by predators. The fear emotion motivates hens to hide during egg laying. To prevent activation of the fear system, egg-laying hens should be provided with a nest box. This will be discussed further later in the chapter.

Just like humans, animals are driven to perform by their feelings and emotions (Manning, 1979), that is animals too are motivated by their core emotions. Temple Grandin emphasizes the concept of designing animal environments and putting animals in situations that reinforce the positive emotional circuits and avoid the negative emotional circuits to improve an animal's well-being (Grandin and Johnson, 2009). Animals will stimulate these positive emotional circuits through their behaviour because these behaviours are highly motivated. When an animal can no longer express these highly motivated behaviours, some of them seemingly basic behaviours that most of us would expect to be able to do on a daily basis, their environment is no longer ideal.

Hard-wired Behaviours

The behaviours that we observe in animals are either learned or instinctual hard-wired behaviours. We observe many learned behaviours in animals. Dairy cows learn to line-up at milking time. Many zoos have taught primates and various other species to present limbs or stand still for husbandry and veterinary procedures (Grandin, 2000; Savastano *et al.*, 2003). Dogs learn to do a multitude of behaviours during obedience training. Animals learn without the help of humans as well, an example being that of cheetah cubs learning how to efficiently and effectively kill prey. Cheetahs can kill at a young age, but take some time and lessons from their mother to develop their skills (Caro, 1994).

Hard-wired behaviours are different. These are innate behaviours that the animal does not have to learn. There are many examples of these hard-wired behaviours. Display behaviours performed by birds during courtship and mating are instinctual hard-wired behaviours. The nesting behaviour we see sows perform before farrowing is hard wired (Stolba and Wood-Gush, 1984; Jensen, 1986). The prey-catching behaviour we see when dogs chase squirrels in the park is an example of an instinctual behaviour. We observe many hard-wired behaviours in livestock: the point of balance and flight zone responses in cattle during handling (see Chapter 5), rooting behaviour in pigs, dust-bathing behaviour in chickens and using a nest box in hens (Vestergaard, 1982; Appleby and McRae, 1986; Newberry and Wood-Gush, 1988; Stolba and Wood-Gush, 1989; Studnitz *et al.*, 2007). When hens are housed outdoors, they make greater use of an outdoor pasture if trees or bushes are provided for cover (Fig. 8.1).

Fig. 8.1. Trees have been planted to provide these free-range hens with cover. The birds will forage more widely if trees or bushes provide cover from aerial predators.

The birds have a hard-wired instinctual fear of aerial predators. These behaviours are referred to as fixed action patterns (FAPs) because they are behavioural sequences that are always performed in the same manner. A classic example of an FAP is egg-rolling behaviour in Greylag geese (Lorenz and Tinbergen, 1938). When an egg rolls out of a goose's nest, the goose performs a highly predictable pattern of movement to bring the egg back towards the nest. These FAPs are triggered into action by a sign stimulus. Mating displays are triggered by potential mates, prey chasing in dogs is triggered by rapid movement and egg pulling is triggered by an egg (or egg-like object) that is out of the nest. FAPs are hard wired but the particular sign stimulus that triggers the behavioural response can be influenced by learning and emotional experiences. For example, a cow's instinctual response to a human moving into its flight zone is to turn and face the person but keep a safe distance. The size of the flight zone can be modified by learning and experience with people. The cow's flight zone may shrink after becoming habituated to and therefore less fearful of a human's presence (see Chapter 5 for more information). The instinct to turn and face a person is a hard-wired instinct motivated by fear. The turn-and-face behaviour will disappear when the cow has become so tame that she no longer has any fear.

Emotions Motivate Behaviour

Both animal instincts and learned behaviour are driven by animal emotions. A cow's instinct to move away from a predator entering its flight zone is driven by fear; a pig's instinct to root for food is driven by its seeking emotion; a dog's instinct to do a play bow to a fellow puppy is driven by its motivation to play. Behaviour can be examined as patterns of behaviour performed to achieve a specific goal. Positive goal-directed behaviours can be dissected into three phases: (i) searching for the goal; (ii) behaviour directed at the goal; and (iii) quiescence following achievement of the goal (Manning, 1979). The searching phase is called the appetitive phase and the goal-directed phase is called the consummatory phase. A basic principle in animal behaviour is that the searching phase is more flexible and more dependent on learning compared to the consummatory phase. The searching phase has to be flexible to enable the animal to survive in different environments. When wolves and other predators hunt they develop their hunting skills through learning. When the predator kills its prey, the killing bite is a hard-wired instinctual behaviour but what to hunt and eat is learned. In both cattle and sheep the young animal learns from its mother which forages are good to eat and which are not (Provenza *et al.*, 1993). This would be the searching phase. The consumption of the forage, chewing and rumination, is hard wired. This would be the consummatory phase. Feeding behaviour and food-seeking behaviours provide good examples of this staged breakdown. Picture a feral cat hunting in the field. She spends some time stalking her prey and searching for her meal, the appetitive behaviour. At the right time she attacks the field mouse, killing her target (or 'goal'), the consummatory behaviour. After she has eaten the mouse and licked her lips she rests. Feeding behaviour is not as good an example when discussing grazing animals since they are constantly eating, not having as distinct quiescence phase. These goal-directed behaviours are usually highly motivated as there is a reward (the goal) once the behaviours are performed.

Measuring the Strength of Motivation

Scientists can objectively measure an animal's motivation to perform specific behaviours, a useful tool to determine how valuable a resource (e.g. food) or the ability to perform a behaviour (e.g. nest building) is to an animal. Farm-raised mink have been studied to determine how hard they will work, or how much they will 'pay' to gain access to various environments created to stimulate natural (wild) behaviours (Mason *et al.*, 2001). The environments that were created for the mink included novel play items, a water pool, a nest site and a vertical platform. The entrance doors to the various environments were weighted with increasing amounts of weight to determine how much effort the mink were willing to expend to get into the alternative environments. Results indicated that the mink were willing to lift the heaviest door to gain access to the water pool, their motivation being very high to have access to water. Additionally, it was found that when access to these resources was blocked, the animals had higher levels of cortisol, indicative of a stress response.

Many studies have been done measuring motivation for various valuable resources (e.g. food, water, floor substrate, social contact, nest-building material, etc.) for a range of species, including mice

(Sherwin, 2004), rabbits (Seaman *et al.*, 2001), pigs (Pedersen *et al.*, 2002) and hens (Olsson *et al.*, 2002). Several studies have been conducted exploring a hen's motivation to gain access to litter for dust bathing, a behaviour constrained in many commercial systems due to a lack of bedding material to dust bathe with (Vestergaard, 1982; Dawkins and Beardsley, 1986; Petherick *et al.*, 1990, 1991; Matthews *et al.*, 1993; Widowski and Duncan, 2000). Dust bathing is a behaviour that has been observed in wild and domestic fowl and is considered a motivated behaviour. Some studies have found that hens will work, by either pecking a key or pushing a weighted door, to gain access to dust-bathing material such as wood shavings and peat moss indicating a motivation for dust bathing (Matthews *et al.*, 1993; Widowski and Duncan, 2000). Other studies have found results that are not as straightforward. For example, one study found that hens would break a photo beam with their beak to open a door that provided access to litter, after many trials, but they would not learn to peck a key to gain access to the litter (Dawkins and Beardsley, 1986). Other studies have found that hens would peck a key to enlarge their cages but the number of key pecks would not increase substantially when cage enlargement also meant access to litter (Lagadic and Faure, 1987; Faure, 1991). Due to differences in results of these motivation studies, it has been suggested that hens may not have as great a need for litter as was once thought (i.e. dust bathing may not be a highly motivated behaviour). Rather than dismissing dust bathing as a behavioural need, these types of results can actually provide valuable information as to what drives the motivation to dust bathe. For example, certain behaviours could follow a particular biological rhythm (i.e. only occur at certain times of day) making scheduling a necessary component of experimental design. Additionally, visual stimuli are factors affecting motivation for dust bathing in hens, particularly the ability for the hens to see the litter. In the studies indicating that dust bathing was not important, the substrate was not visible (Dawkins and Beardsley, 1986) and in the studies indicating that dust bathing was important, the substrate was visible (Matthews *et al.*, 1993). These examples indicate the complexity of behavioural motivation and thus the importance of thorough assessment of experimental results.

The motivation to dust bathe is probably driven by one of the positive core emotions. Research on motivation to obtain a secluded private nest box shows clearly the hen's need for a nest box. This fear-motivated behaviour has a much stronger motivation, compared to dust bathing. A hen values access to a private nest box prior to egg laying more than gaining access to food following 4 h of feed deprivation (Cooper and Appleby, 2003). Hens will lift weighted doors (Duncan and Kite, 1987; Smith *et al.*, 1990; Cooper and Appleby, 2003) and squeeze through narrow spaces to gain access to a nest box (Cooper and Appleby, 1995, 1996a, 1997; Bubier, 1996). One can conclude from these studies that a nest box is essential for good welfare and access to dust-bathing materials would have a lower priority (see Chapter 15).

In herding animals, motivation to be with a companion, rather than being in isolation, can be measured. It has been demonstrated that calves are motivated to get access to another calf (Holm *et al.*, 2002). The motivation to be with another calf was measured by counting the number of times the calf would press a switch to gain access to its herd mate. Researchers also determined that calves would pay (move switch presses) more for full social contact versus only head contact. Similar to the mink study, the information from this study could be used to determine best-management practices for housing, and individual versus group pen systems in the case of the calf. Motivation studies are useful for determining an animal's behavioural needs, which behaviours satisfy the core emotions and thus which are important to the animal.

Factors that Affect Motivation Strength

There are different clues that help us determine if a particular behaviour is highly motivated in the species we are studying. Some have suggested that if an animal performs a behaviour, even when the materials or stimulus to accomplish the ultimate goal are not present, that behaviour is highly motivated (Black and Hughes, 1974; Van Putten and Dammers, 1976). These are called vacuum behaviours because they are 'set off' for no apparent reason; vacuum behaviours are FAPs carried out even in the absence of the appropriate sign stimulus. An example of this is when sows perform nest-building behaviours (i.e. pawing and rooting for straw and grasses) prior to parturition even when there are no nesting materials present (Vestergaard and Hansen, 1984). Another vacuum behaviour we can observe is the non-nutritive sucking of calves (i.e. suckling on pen mates and other objects) (de Wilt, 1985). Calves can exhibit increased motivation to suckle due to an absence of opportunities to suckle on their mother (Sambraus, 1985). Vacuum behaviours originate from an animal's frustration with not being able to effectively express highly motivated behaviours (Lindsay, 2001). These types of behaviours are important for the animal to be able to perform.

Another indication that an animal has a need to perform a particular behaviour is an increased tendency to perform the behaviour after a period of deprivation. This has been called a 'rebound effect'; when the behaviour is being prevented the drive to perform it grows (Vestergaard, 1980; Nicol, 1987). The behavioural responses of laying hens to space restriction indicated that when given more space after being restricted, hens performed certain behaviours (e.g. leg stretching, wing flapping, feather raising) at a higher rate (Nicol, 1987). The authors acknowledged that although there is potential that this 'rebound' resulted from an increased motivation to perform these behaviours during restriction or it could also simply be a response to a novel environment. Oral deprivation and the result of subsequent feeding behaviour has been studied in dogs (Lawson *et al.*, 1993). Experimental dogs were fed intra-gastrically, receiving no oral stimulation for a set number of days. After the restriction period, dogs were observed to over-eat, or rebound, when finally given food. We can imagine this in our dogs at home too. If we get home late from work and feeding time for our dog is later than usual, he may consume his food at a faster rate than if fed on time.

Abnormal Displacement Behaviours and Stereotypies

When an animal's ability to perform these highly motivated behaviours, such as dust bathing, foraging or nesting, is thwarted, they may begin to develop displacement behaviours, for example bar biting, pacing, rocking, self-narcotizing behaviour, increased aggression, etc. (Table 8.1). Some of these behaviours are stereotypies. A stereotypy is defined as a repetitive behaviour that repeats itself in a pattern that seldom changes and it serves no obvious purpose (Mason *et al.*, 2007; Price, 2008). The development of these repetitive, non-goal-oriented behaviours is a mechanism for the animal to cope with the frustration of not being able to perform certain innate behaviours. Continuing with the food-seeking example, consider a bear in a zoo that rocks back and forth in one area of her pen prior to feeding time. In a zoo the bear does not have to search for her food, she gets fed at a particular time each day. This rocking behaviour has taken the place of long hours of food searching in the wild. Sows in gestation crates sometimes exhibit bar biting (Fig. 8.2) shortly before being fed, perhaps a reaction to the inability to food search and root, innate motivated behaviours (Lawrence and Terlouw, 1993; Day *et al.*, 1995).

Table 8.1. Common stereotypical repetitive abnormal behaviours and other abnormal behaviours. The presence of these abnormal behaviours is an indicator that the animal's needs are not being fulfilled. Steps should be taken to provide environmental enrichment to prevent these behaviours.

Behaviour	Species	Description of the behaviour
Bar biting	Sows	Animal rhythmically bites or mouths a bar or other object
Tongue rolling	Cattle	Tongue is extended and rapidly moved back and forth
Feather pecking	Layer hens	Pecking at another hen, which damages feathers or causes injury
Weaving	Horses	Animal sways back and forth
Cribbing (wind sucking)	Horses	Upper jaw is placed on a fence and the animal rhythmically bites and sucks wind
Pacing	Minks and foxes	Animal circles the cage in a pattern that seldom changes
Wool, hair eating	Sheep, antelope	Pulling wool or hair out of another animal
Belly nosing	Pigs	Animal roots and rubs its nose on other animals
Non-nutritive sucking	Calves	Sucking navels or urine
Tail and ear biting	Pigs	Injures the tail or ears of other animals

Fig. 8.2. Bar biting by a sow housed in a gestation stall. There are large individual differences between sows. Some will be bar biters and others will not.

Some animals may be self-medicating, filling a behavioural need with a maladaptive behaviour. It has been demonstrated that the performance of stereotypical behaviours is related to the release of endorphins, providing the animal with some relief from the stressful environment (Cronin *et al.*, 1986; Dantzer, 1986). Supporting this concept of stereotypical behaviour as a medication is the finding that when anti-addictive pharmacological agents such as nalmefene are given to horses, pigs and mice who perform stereotypies, the stereotypies stop (Cabib *et al.*, 1984; Cronin *et al.*, 1985; Dodman *et al.*, 1988). Animals develop these stereotypical behaviours as a means of coping with a barren environment, an indication that their environment needs to be improved. Animal behaviour can be used as an indicator of an animal's state of well-being (Duncan, 1998).

Behavioural Indicators of Poor Welfare

When animals are observed in confinement captivity, some of these stereotypical and displacement behaviours are often seen: a chicken performing dust-bathing behaviours with no dust; a sow in a gestation crate bar biting; a chained-up dog licking its paw repetitively; or a zoo jaguar pacing the perimeter of its cage. An animal's need to express certain behaviours (e.g. dust bathing, nesting, foraging, locomotion, social interaction, seeking, etc.) is blocked in some confinement livestock production systems. This causes potential problems with animal welfare. The agricultural industry has created some housing systems in which animals are not able to behave naturally. Their drive to perform natural behaviours is thwarted and as a result they develop a higher level of frustration in their housing, manifested in various ways but particularly in their expression of behaviour. Producers can economically raise livestock and poultry in intensive confinement systems. The question is: should we raise them this way? (Bernard Rollin, personal communication, 2008). Chapter 2 by Bernard Rollin provides a further discussion of the ethics of intensive confinement systems. The animals are never going to be able to 'tell' us how they feel and therefore we must try to understand them in a different manner (i.e. through their behaviour). Some people may still fear making assumptions about an animal's well-being through its actions (even though they most likely do this on a daily basis with fellow humans). But, as one researcher profoundly stated, it is important to be 'roughly right on something important than to be accurate but wrong or irrelevant' (Ng, 1995). Animal well-being is that 'something important'.

Environmental Enrichment Helps Prevent Abnormal Behaviour

Providing environmental enrichment can reduce stereotypic abnormal behaviour (Mason *et al.*, 2007). The emphasis has to be on preventing abnormal behaviour from starting. Once it has started, it is difficult to stop. In poultry, feather pecking and cannibalism are hard to stop because the birds teach each other to feather peck. The goal of providing environmental enrichment such as straw for pigs or next boxes for hens is to prevent abnormal behaviour and allow the animals to perform more species-typical behaviours. It should help the biological functioning of the animal (Newberry, 1995). Another advantage of providing environmental enrichment is to help prevent behaviours that may cause injuries or damage to the animals.

In cattle, tongue rolling (Fig. 8.3) can be reduced by providing adequate roughage (Redbo, 1990; Redbo and Nordblad, 1997). Providing hay forages and other roughages will help prevent stereotypic behaviour in horses (McGreevy *et al.*, 1995; Goodwin *et al.*, 2002; Thorne *et al.*, 2005). Horses fed lots of concentrates and pellets may be more prone to stereotypies. Providing litter, straw or materials to forage in may reduce feather pecking and injuries in chickens and turkeys. Providing turkeys with straw and hanging chains reduced injuries from aggression (Sherwin *et al.*, 1999; Martrenchar *et al.*, 2001). Providing ducks with additional foraging materials had no effect (Riber and Mench, 2008). In pigs, providing straw or other fibrous materials for rooting can help prevent tail biting (Day *et al.*, 2002; Bolhuis *et al.*, 2006; Chaloupkova *et al.*, 2007). It is important to provide the straw or other foraging material for the entire life of the animal. If the straw is taken away, tail biting may increase (Day *et al.*, 2002; Bolhuis *et al.*, 2006). Figure 8.4 shows sows bedded on straw.

In systems with liquid manure and slatted floors, it may be difficult to provide straw or forage. Hanging objects such as ropes or chains can also help prevent pigs from directing biting, chewing or nosing towards other pigs. Pigs prefer objects that

Fig. 8.3. Tongue rolling in cattle. Feeding cattle more roughage will help reduce tongue rolling. There are genetic influences on the tendency of different breeds to tongue roll (from Grandin and Deesing, 1998).

Fig. 8.4. Sows housed in a pen bedded with straw. Providing straw will help reduce abnormal behaviour. These systems must be carefully managed to ensure that producers use sufficient straw to keep the animals clean. One of the biggest problems with these systems is failure to use enough straw. This will result in a dirty animal laying in muck.

they can chew up and destroy. Hanging cloth strips and rubber hoses were preferred compared to chains (Grandin, 1989). Jensen and Pedersen (2007) have reported similar results. If non-destructable objects such as chains or balls are used, they MUST be changed frequently to maintain the pig's interest. For caged laying hens, providing hanging strands of white string reduced feather damage (McAdie *et al.*, 2005). The same strings maintained their interest over a period of weeks. In one experiment, they pecked at the strings on the first day for about the same amount of time they pecked them on the 47th day. Poultry may have less need for novelty compared to pigs. Layers in a cage system with access to strands of string had significantly better feather condition at 35 weeks of age. The birds in this experiment were not beak trimmed.

Genetic Effects on Abnormal Behaviour

One must remember that there are strong genetic influences on the occurrence of behaviours that cause damage to other animals. In poultry, genetics plays a big role (Kjaer and Hocking, 2004). Strains of chickens bred for high egg production often have higher rates of cannibalism, aggression and feather pecking (Craig and Muir, 1998). In ducks, Muscovys are more likely to be cannibalistic compared to Peking ducks. In pigs, many producers and lairage managers have observed that certain genetic lines of lean, rapidly growing pigs had a higher incidence of tail biting. Farm managers have also reported higher levels of belly nosing in some lean lines of rapidly growing pigs.

The Dilemma of Maximizing Production

The livestock industry is now faced with a dilemma: continue to maximize production efficiency by maintaining large numbers of animals in small spaces, or provide more space and possibly lower animal production per unit of space to improve animal welfare. Society, producers and scientists are struggling, in different ways, with this concept. The extreme form of confined animal production has 'worked' for several years but is not sustainable, at least not for the animals. Many domestic animal species have been pushed to their biological limits to reach desired levels of production. The biology of the animals will not sustain further manipulation of their biological limits. The agricultural industry was able to develop maize (corn) to reach unimaginable heights of production. Maize has been bred and genetically engineered so that it can withstand tight crowding, growing and thriving with little distance between itself and its identical neighbour. Unlike our livestock species, maize plants do not experience fear, pain or frustration with their environment. We have tried to treat our livestock animals like maize, placing them in crowded living conditions so that we can produce more meat, eggs and milk in less space. By making these environments so limited we have restricted the animal from truly acting like an animal, in the whole sense of the word. Societies around the world realized that housing wild animals in small, sterile environments in zoos was inhumane and bad for animal welfare. Long before the zoo animals received attention, suffering of companion animals was stopped with anti-cruelty laws and legislation. (Companion animals have a different role in society, some pets are treated more like humans than animals because owners easily recognize their dog's capacity for pain, fear and frustration.) The suffering of our beloved companion animals was thwarted at every chance, perhaps even to a fault. Society has also recognized the need to protect laboratory animals from unnecessary suffering. This can be seen in the creation of legislation protecting laboratory animals and the implementation of the animal care and use committees in research institutions around the world. What about livestock? The welfare of farm animals is increasingly becoming a concern in society as seen by the changes in consumer demands and the increase in regulations.

As indicated through the previous discussion, limiting natural behaviour and thus inducing frustration, fear, anxiety and ultimately suffering, is going against the current trend towards an improved production animal welfare status. Ironically, as we limit certain behaviours, other behavioural cues, whether it be lack of behaviour, appearance of abnormal behaviours or exaggerated natural behaviours, begin to tell us that the current system is not working, at least in terms of keeping the animals physically and mentally content. It is necessary that in the next stage of animal agriculture we construct animal-centred outcomes (Hewson, 2003), focus on fitting the farms (now the confinement operations) to the animals, not the animals to the farms (Kilgour, 1978). Manipulating maize to meet our needs for high yields was easier than changing animals because plants do not have the capacity to suffer. We cannot completely breed natural behaviour and thus the capacity for psychological suffering out of animal. Fear-motivated behaviours are needed for basic survival and therefore would be difficult to breed out of an animal. However, the intensity of some behavioural needs can be reduced with breeding. Mother nurturing behaviour has been bred out of some animals. The Holstein cow has less separation anxiety when her calf is removed compared to most beef breeds. Broodiness, the hen's behaviour to sit and incubate her eggs, has been greatly reduced in some chickens (Hays and Sanborn, 1939; Hutt, 1949). A review of research in Chapter 1 indicates that certain behaviours should be accommodated. None the less, we can and should construct environments that enable the animal to live free from aversive situations that cause psychological stress and perform highly motivated behaviours.

References

Appleby, M.C. and McRae, H.E. (1986) The individual nest box as a super-stimulus for domestic hens. *Applied Animal Behaviour Science* 15, 169–176.

Bardo, M.T. and Bevins, R.A. (2000) Conditioned place preference: what does it add to our preclinical understanding of drug reward? *Psychopharmacology* 153, 31.

Barnard, C. (2007) Ethical regulation and animal science: why animal behavior is special. *Animal Behaviour* 74, 5–13.

Baxter, M.R. (1983) Ethology in environmental design for animal production. *Applied Animal Ethology* 9, 207–220.

Bechara, A., Tranel, D., Damasio, H., Adolphs, R., Rockland, C. and Damasio, A.R. (1995) Double dissociation of conditioning and declarative knowledge relative to the amygdala and hippocampus in humans. *Science* 269, 1115–1118.

Berlyne, D.E. (1960) *Conflict, Arousal and Curiosity.* McCraw-Hill Book Company, New York.

Black, A.J. and Hughes, B.O. (1974) Patterns of comfort behaviour and activity in domestic fowls: a comparison between cages and pens. *British Veterinary Journal* 130, 23–33.

Bolhuis, J.E., Schouten, W.G.P., Schrama, J.W. and Wiegant, V.M. (2006) Effects of rearing and housing environment on behaviour and performance of pigs with different coping characteristics. *Applied Animal Behaviour Sciences* 101, 68–85.

Borsini, F., Podhorna, J. and Marazziti, D. (2002) Do animal models of anxiety predict anxiolytic-like effects of antidepressants? *Psychopharmacology* 163, 121.

Brambell, R.W.R. (chairman) (1965) *Report of the Technical Committee to Enquire into the Welfare of Animals Kept Under Intensive Livestock Husbandry Systems*. Command paper 2836. Her Majesty's Stationery Office, London.

Broom, D.M. (1986) Indicators of poor welfare. *British Veterinary Journal* 142, 524–526.

Broom, D.M. (2001) *Coping with Challenge: Welfare in Animals including Humans*. Dahlem University Press, Berlin.

Brownlee, A. (1954) Play in domestic cattle in Britain: an analysis of its nature. *British Veterinary Journal* 110, 48.

Bubier, N.E. (1996) The behavioural priorities of laying hens: the effect of cost/no cost multi-choice tests on time budgets. *Behavioural Processes* 37, 225–238.

Büchel, C., Morris, J., Dolan, R.J. and Friston, K.J. (1998) Brain systems mediating aversive conditioning: an event-related fmri study. *Neuron* 20, 947–957.

Burgdorf, J. and Panksepp, J. (2006) The neurobiology of positive emotions. *Neuroscience and Biobehavioural Reviews* 30, 173–187.

Burgdorf, J., Knutson, B., Panksepp, J. and Ikemoto, S. (2001) Nucleus accumbens amphetamine microinjections unconditionally elicit 50-khz ultrasonic vocalizations in rats. *Behavioural Neuroscience* 115, 940–944.

Cabib, S., Puglisi-Allegra, S. and Oliveria, A. (1984) Chronic stress enhances apomorphine-induced stereotyped behavior in mice: involvement of endogenous opioids. *Brain Research* 298, 138–140.

Caro, T.M. (1994) *Cheetahs of the Serengeti Plains*. Chicago University Press, Chicago.

Chaloupkova, H.G., Illman, I., Bartoš, L. and Spinka, M. (2007) The effect of pre-weaning housing on the play and agonistic behaviour of domestic pigs. *Applied Animal Behaviour Science* 103, 25–34.

Colpaert, F.C., De Witte, P., Maroli, A.N., Awouters, F., Niemegeers, C.J. and Janssen, P.A. (1980) Self-administration of the analgesic suprofen in arthritic rats: evidence of *Mycobacterum butyricum*-induced arthritis as an experimental model of chronic pain. *Life Science* 27, 921–928.

Cooper, J.J. and Appleby, M.C. (1995) Nesting behaviour of hens: effects of experience on motivation. *Applied Animal Behaviour Science* 42, 283–295.

Cooper, J.J. and Appleby, M.C. (1996a) Demand for nest boxes in laying hens. *Behavioural Processes* 36, 171–182.

Cooper, J.J. and Appleby, M.C. (1996b) Individual variation in prelaying behaviour and the incidence of floor eggs. *British Poultry Science* 37, 245–253.

Cooper, J.J. and Appleby, M.C. (1997) Motivational aspects of individual variation in response to nestboxes by laying hens. *Animal Behaviour* 54, 1245–1253.

Cooper, J.J. and Appleby, M.C. (2003) The value of environmental resources to domestic hens: a comparison of the work-rate for food and for nests as a function of time. *Animal Welfare* 12, 39–52.

Craig, J. and Muir, W. (1998) Genetics and the behavior of chickens, welfare, and productivity. In: Grandin, T. (ed.) *Genetics and the Behavior of Domestic Animals*. Academic Press, San Diego, California, pp. 265–297.

Cronin, G.M., Wiepkema, P.R. and van Ree, J.M. (1985) Endogenous opioids are involved in stereotyped behaviour of tethered sows. *Neuropeptides* 6, 527–530.

Cronin, G.M., Wiepkema, P.R. and van Ree, J.M. (1986) Andorphins implicated in stereotypies of tethered sows. *Cellular and Molecular Life Sciences* 42, 198–199.

Damasio, A.R. (1994) *Descartes' Error: Emotion, Reason and the Human Brain*. Grosset/Putnam, New York.

Damasio, A.R. (1999) *The Feeling of What Happens*. Harcourt Brace, New York.

Dantzer, R. (1986) Symposium on 'indices to measure animal well-being': behavioral, physiological and functional aspects of stereotyped behavior: a review and a re-interpretation. *Journal of Animal Science* 62, 1776–1786.

Dantzer, R. and Mormede, P. (1983) Stress in farm animals: a need for re-evaluation. *Journal of Animal Science* 57, 6–18.

Darwin, C. (1872) *The Expression of the Emotions in Man and Animals*. University of Chicago Press, Chicago.

Davis, M. (1992) The role of the amygdala in conditioned fear. In: Aggleton, J.P. (ed.) *The Amygdala: Neurobiological Aspects of Emotion, Memory and Mental Dysfunction*. Wiley-Liss, New York, pp. 255–306.

Dawkins, M.S. (1988) Behavioural deprivation: a central problem to animal welfare. *Applied Animal Behaviour Science* 20, 209–225.

Dawkins, M.S. (1998) Evolution and animal welfare. *Quarterly Review of Biology* 73, 305–328.

Dawkins, M.S. (2006) A user's guide to animal welfare science. *Trends in Ecology and Evolution* 21, 77–82.

Dawkins, M.S. and Beardsley, T. (1986) Reinforcing properties of access to litter in hens. *Applied Animal Behaviour Science* 15, 351–364.

Day, J.E.L., Kyriazakis, I. and Lawrence, A.B. (1995) The effect of food deprivation on the expression of foraging and exploratory behaviour in the growing pig. *Applied Animal Behaviour Science* 42, 193–206.

Day, J.E.L., Spoolder, H.A.M., Burfoot, A., Chamberlain, H.L. and Edwards, S.A. (2002) The separate and interactive effects of handling and environmental enrichment on the behavior and welfare of growing pigs. *Applied Animal Behaviour Science* 75, 177–192.

Dellmeier, G.R. (1989) Motivation in relation to the welfare of enclosed livestock. *Applied Animal Behaviour Science* 22, 129.

Dellmeier, G.R., Friend, T.H. and Gbur, E.E. (1985) Comparison of four methods of calf confinement. II. Behavior. *Journal of Animal Science* 60, 1102–1109.

de Wilt, J.G. (1985) *Behavior and Welfare of Veal Calves in Relation to Husbandry Systems.* Agricultural University, Wageningen, The Netherlands.

Dodman, N.H., Shuster, L., Court, M.H. and Patel, J. (1988) Use of narcotic antagonist (nalmefene) to suppress self-mutilative behaviour in a stallion. *Journal of the American Veterinary Association* 192, 1585–1586.

Duncan, I.J. (1998) Behavior and behavioral needs. *Poultry Science* 77, 1766–1772.

Duncan, I.J.H. and Dawkins, M.S. (1983) The problem of assessing 'well-being' and 'suffering' in farm animals. In: Smidt, D. (ed.) *Indicators Relevant to Farm Animal Welfare.* Martinus Nijhoff, The Hague, The Netherlands, pp. 13–24.

Duncan, I.J.H. and Kite, V.G. (1987) Some investigations into motivation in the domestic fowl. *Applied Animal Behaviour Science* 18, 387–388.

Duncan, I.J.H. and Petherick, J.C. (1991) The implications of cognitive processes for animal welfare. *Journal of Animal Science* 69, 5017–5022.

Farm Animal Welfare Council (FAWC) (1979) Farm animal welfare council press statement. Available at: http://www.fawc.org.uk/pdf/fivefreedoms1979.pdf (accessed 1 July 2009).

Faure, J.M. (1991) Rearing conditions and needs for space and litter in laying hens. *Applied Animal Behaviour Science* 31, 111–117.

Faure, J.M. and Mills, A.D. (1998) Improving the adaptability of animals by selection. In: Grandin, T. (ed.) *Genetics and the Behavior of Domestic Animals.* Academic Press, San Diego, California, pp. 235–264.

Faure, A., Reynolds, S.M., Richard, J.M. and Berridge, K.C. (2008) Mesolimbic dopamine in desire and dread: enabling motivation to be generated by localized glutamate disruptions in nucleus accumbens. *Journal of Neuroscience* 28, 7184–7192.

Fraser, D. (1995) Science, values and animal welfare: exploring the 'inextricable connection'. *Animal Welfare* 4, 103–117.

Fraser, D. (2008) Toward a global perspective on farm animal welfare. *Applied Animal Behaviour Science* 113, 330–339.

Goodwin, D., Davidson, H.P.B. and Harris, P. (2002) Foraging environment for stabled horses: effects on behaviour and selection. *Equine Veterinary Journal* 34, 686–691.

Grandin, T. (1989) Effects of rearing environment and environmental enrichment on the behavior and neural development of young pigs. Dissertation, University of Illinois, Urbana, Illinois.

Grandin, T. (1997) Assessment of stress during handling and transport. *Journal of Animal Science* 75, 249–257.

Grandin, T. (2000) Habituating antelope and bison to cooperate with veterinary procedures. *Journal of Applied Animal Welfare Science* 3, 253–261.

Grandin, T. and Deesing, M.J. (1998) Genetics and animal welfare. In: Grandin, T. (ed.) *Genetics and the Behavior of Domestic Animals.* Academic Press, San Diego, California, pp. 319–346.

Grandin, T. and Johnson, C. (2009) *Animals Make Us Human.* Houghton Mifflin Harcourt, Boston, Massachusetts.

Harlow, H.F., Harlow, M.K. and Meyer, D.R. (1950) Learning motivated by a manipulation drive. *Journal of Experimental Psychology* 40, 228–234.

Hays, F.A. and Sanborn, R. (1939) Breeding for egg production. *Massachusetts Agricultural Experimental Station Bulletin* 307. Agricultural Experimental Station, Amherst, Massachusetts.

Heath, R.G. (1996) *Exploring the Mind–Brain Relationship.* Moran Printing, Baton Rouge, Louisiana.

Hewson, C.J. (2003) Can we assess welfare? *Canadian Veterinary Journal* 44, 749–753.

Holm, L., Jensen, M.B. and Jeppesen, L.L. (2002) Calves' motivation for access to two different types of social contact measured by operant conditioning. *Applied Animal Behaviour Science* 79, 175–194.

Huston, J.P. and Borbély, A.A. (1973) Operant conditioning in forebrain ablated rats by use of rewarding hypothalamic stimulation. *Brain Research* 50, 467–472.

Hutt, F.B. (1949) *Genetics of the Fowl.* McGraw-Hill Book Company, New York.

Jensen, M.B. and Pedersen, L.J. (2007) The value assigned to six different rooting materials by growing pigs. *Applied Animal Behaviour Science* 108, 31–44.

Jensen, P. (1986) Observations on the maternal behaviour of free-ranging domestic pigs. *Applied Animal Behaviour Science* 16, 131–142.

Jerison, H.J. (1997) Evolution of the prefrontal cortex. In: Kragnegor, N.A., Lyon, G.R. and Goldman-Rakic, P. (eds) *Development of the Prefrontal Cortex: Evolution, Neurobiology and Behavior.* Paul H. Brooks Publishing Co., Baltimore, Maryland, pp. 9–27.

Kilgour, R. (1978) The application of animal behavior and the humane care of farm animals. *Journal of Animal Science* 46, 1478–1486.

King, J.N., Simpson, B.S., Overall, K.L. *et al.* (2000) Treatment of separation anxiety in dogs with clomipramine: results from a prospective, randomized, double-blind, placebo-controlled, parallel-group, multicenter clinical trial. *Applied Animal Behaviour Science* 67, 255–275.

Kjaer, J.B. and Hocking, P.M. (2004) The genetics of feather pecking and cannibalism. In: Perry, G.C. (ed.) *Welfare of the Laying Hen.* CAB International, Wallingford, UK, pp. 109–122.

Knutson, B., Burgdorf, J. and Panksepp, J. (2002) Ultrasonic vocalizations as indices of affective states in rats. *Psychological Bulletin* 128, 961–977.

Kolb, B. and Tees, R.C. (1990) *The Cerebral Cortex of the Rat.* MIT Press, Cambridge, Massachusetts.

LaBar, K.S., Gatenby, J.C., Gore, J.C., LeDoux, J.E. and Phelps, E.A. (1998) Human amygdala activation during conditioned fear acquisition and extinction: a mixed-trial fmri study. *Neuron* 20, 937–945.

Lagadic, H. and Faure, J.M. (1987) Preferences of domestic hens for cage size and floor types as measured by operant conditioning. *Applied Animal Behaviour Science* 19, 147–155.

Lawrence, A.B. and Terlouw, E.M. (1993) A review of behavioral factors involved in the development and continued performance of stereotypic behaviors in pigs. *Journal of Animal Science* 71, 2815–2825.

Lawson, D.C., Schiffman, S.S. and Pappas, T.N. (1993) Short-term oral sensory deprivation: possible cause of binge eating in sham-feeding dogs. *Physiology and Behavior* 53, 1231–1234.

LeDoux, J.E. (1992) Emotion and the amygdala. In: Aggleton, J.P. (ed.) *The Amygdala: Neurological Aspects of Emotion, Memory and Mental Dysfunction.* Wiley-Liss, New York, pp. 339–351.

LeDoux, J.E. (2000) Emotion circuits in the brain. *Annual Review of Neuroscience* 23, 155–184.

Lindsay, S.R. (2001) *Handbook of Applied Dog Behavior and Training: Etiology and Assessment of Behavior.* Volume 2. Iowa State University Press, Ames, Iowa.

Liotti, M. and Panksepp, J. (2004) Imaging human emotions and affective feelings. Implications for biological psychiatry. In: Panksepp, J. (ed.) *Textbook of Biological Psychiatry.* Wiley, Hoboken, New Jersey, pp. 33–74.

Lorenz, K. and Tinbergen, N. (1938) Taxis and instinctive behavior pattern in egg-rolling by the greylag goose. In: Lorenz, K. (ed.) *Studies in Animal Behavior and Human Behavior No. 1.* Harvard University Press, Cambridge, Massachusetts, pp. 316–359.

MacLean, P.D. (1990) *The Triune Brain in Evolution.* Plenum Press, New York.

Manning, A. (1979) *An Introduction to Animal Behavior.* Addison-Wesley, Reading, Massachusetts.

Markowitz, H., Aday, C. and Gavazzi, A. (1995) Effectiveness of acoustic prey: environmental enrichment for a captive African leopard (*Panthera pardus*). *Zoo Biology* 14, 371–379.

Martrenchar, A., Huonnie, D. and Cotte, J.P. (2001) Influence of environmental enrichment on injurious pecking and perching behaviour in young turkeys. *British Poultry Science* 62, 161–170.

Mason, G.J. and Mendl, M. (1993) Why is there no simple way of measuring animal welfare? *Animal Welfare* 2, 301–309.

Mason, G.J., Cooper, J. and Clarebrough, C. (2001) Frustrations of fur-farmed mink. *Nature* 410, 35–36.

Mason, G.R., Clubb, R., Latham, N. and Vickery, S. (2007) Why and how should we use environmental enrichment to tackle sterotyped behaviour? *Applied Animal Behavioural Science* 102, 163–188.

Mason, M.P. (2008) *Head Cases: Stories of Brain Injury and its Aftermath.* Farrar, Straus and Giroux, New York.

Matthews, L.R., Temple, W., Foster, T.M. and McAdie, T.M. (1993) Quantifying the environmental requirements of layer hens by behavioural demand functions. In: Nichelmann, M., Wierenga, H.K. and Braun, S. (eds) *Proceedings of the International Congress on Applied Ethology*, Berlin, 26–30 July. Kuratorium fur Technik und Bauwesen in der Landwirtschaft, Berlin, pp. 206–209.

McAdie, T.M., Keeling, L.J., Blohhuis, H.J. and Jones, R.B. (2005) Reduction in feather pecking and improvement in feather condition with presentation of a string device to chickens. *Applied Animal Behavioural Sciences* 93, 67–80.

McBride, W.J., Murphy, J.M. and Ikemoto, S. (1999) Localization of brain reinforcement mechanisms: intracranial self-administration and intra-cranial place-conditioning studies. *Behavioural Brain Research* 101, 129–152.

McGreevy, P.D., Cripps, P.J., French, N.P., Green, L.E. and Nicol, C.J. (1995) Management factors associated with stereotypic and redirected behaviour in the Thoroughbred horse. *Equine Veterinary Journal* 27, 86–91.

Mendl, M. (1991) Some problems with the concept of a cut-off point for determining when an animal's welfare is at risk. *Applied Animal Behaviour Science* 31, 139–146.

Moberg, G.P. (1985) Biological response to stress: key to assessment of animal well-being? In: Moberg, G.P. (ed.) *Animal Stress.* American Physiological Society, Betheseda, Maryland, pp. 27–49.

National Center for Health Statistics (2007) *Health, United States 2007: with Chartbook on Trends in the Health of Americans.* United States Department of Health and Human Services, Centers for Disease Control and Prevention, Hyattsville, Maryland.

Newberry, R.C. (1995) Environmental enrichment: increasing the biological relevance of captive environments. *Applied Animal Behaviour* 44, 229–243.

Newberry, R.C. and Wood-Gush, D.G.M. (1988) Development of some behaviour patterns in piglets under semi-natural conditions. *Animal Production* 46, 103–109.

Ng, Y. (1995) Towards welfare biology: evolutionary economics of animal consciousness and suffering. *Biology and Philosophy* 10, 255–285.

Nicol, C.J. (1987) Behavioural responses of laying hens following a period of spatial restriction. *Animal Behaviour* 35, 1709–1719.

Numan, M. and Insel, T.R. (2003) *The Neurobiology of Parental Behavior.* Springer, New York.

Office of the Federal Register (1989) Final Rules: Animal Welfare Act: 9 Code of Federal Regulations (CFR) Parts 1 and 2. Federal Register Vol. 54 No. 168, 31 August. Office of the Federal Register, Washington, DC, pp. 36112–36163.

Olds, J. (1977) *Drives and Reinforcements: Behavioral Studies of Hypothalamic Function.* Raven Press, New York.

Olsson, I.A., Keeling, L.J. and McAdle, T.M. (2002) The push-door for measuring motivation in hens: an adaptation and a critical discussion of the method. *Animal Welfare* 11, 1–10.

Overall, K. (1997) *Clinical Behavioral Medicine for Small Animals.* Mosby, St Louis, Missouri.

Panksepp, J. (1971) Aggression elicited by electrical stimulation of the hypothalamus in albino rats. *Physiology and Behavior* 6, 321–329.

Panksepp, J. (1990) The psychoneurology of fear: evolutionary perspectives and the role of animal models in understanding human anxiety. In: Burrows, G.D., Roth, M. and Noyes, J.R. (eds) *Handbook of Anxiety No. 3, The Neurobiology of Anxiety.* Elsevier/North-Holland Biomedical Press, Amsterdam, pp. 3–58.

Panksepp, J. (1998) *Affective Neuroscience: the Foundations of Human and Animal Emotions.* Oxford University Press, New York.

Panksepp, J. (2003) At the interface of the affective, behavioral, and cognitive neurosciences: decoding the emotional feelings of the brain. *Brain and Cognition* 52, 4–14.

Panksepp, J. (2004) Affective consciousness and the origins of human mind: a critical role of brain research on animal emotions. *Impuls* 57, 47–60.

Panksepp, J. (2005a) Affective consciousness: core emotional feelings in animals and humans. *Consciousness and Cognition* 14, 30–80.

Panksepp, J. (2005b) Emotional experience. *Journal of Consciousness Studies* 12, 158–184.

Panksepp, J. (2005c) Affective-social neuroscience approaches to understanding core emotional feelings in animals. In: McMillan, F.D. (ed.) *Animal Mental Health and Well-being.* Iowa University Press, Ames, Iowa, pp. 57–76.

Panksepp, J. and Beatty, W.W. (1980) Social deprivation and play in rats. *Behavioural and Neural Biology* 30, 197–206.

Panksepp, J., Normansell, L., Cox, J.F. and Siviy, S.M. (1994) Effects of neonatal decortication on the social play of juvenile rats. *Physiology and Behavior* 56, 429–443.

Panksepp, J., Knutson, B. and Burgdorf, J. (2002) The role of brain emotional systems in addictions: a neuro-evolutionary perspective and new 'self-report' animal model. *Addiction* 97, 459.

Pedersen, L.J., Jensen, M.B., Hansen, S.W., Munksgaard, L., Ladewig, J. and Matthews, L. (2002) Social isolation affects the motivation to work for food and straw in pigs as measured by operant conditioning techniques. *Applied Animal Behaviour Science* 77, 295–309.

Petherick, J.C., Duncan, I.J.H. and Waddington, D. (1990) Previous experience with different floors influences choice of peat in a y-maze by domestic fowl. *Applied Animal Behaviour Science* 27, 177–182.

Petherick, J.C., Waddington, D. and Duncan, I.J.H. (1991) Learning to gain access to a foraging and dustbathing substrate by domestic fowl: is 'out of sight out of mind'? *Behavioural Processes* 22, 213–226.

Petrie, N.J., Mellor, D.J., Stafford, K.J., Bruce, R.A. and Ward, R.N. (1996) Cortisol responses of calves to two methods of disbudding used with or without local anaesthetic. *New Zealand Veterinary Journal* 44, 9–14.

Pfaff, D.W. (1999) *Drive: Neurobiological and Molecular Mechanisms of Sexual Behavior.* MIT Press, Cambridge, Massachusetts.

Price, E.O. (2008) *Principles and Applications of Domestic Animal Behavior.* CAB International, Wallingford, UK.

Provenza, F.D., Lynch, J.J. and Nolan, J.V. (1993) The relative importance of mother and toxicosis in the selection of foods by lambs. *Journal of Chemical Ecology* 19, 313–323.

Redbo, I. (1990) Changes in duration and frequency of stereotypies and their adjoining behaviours in heifers, before, during and after the grazing period. *Applied Animal Behaviour Science* 25, 57–67.

Redbo, I. and Nordblad, A. (1997) Stereotypies in heifers are affected by feeding regime. *Applied Animal Behaviour Sciences* 53, 193–205.

Riber, A.B. and Mench, J.A. (2008) Effects of feed and water based enrichment on the activity and cannibalism in Muscovy ducklings. *Applied Animal Behaviour Science* 114, 429–440.

Rinn, W.E. (1984) The neuropsychology of facial expression: a review of the neurological and psychological mechanisms for producing facial expressions. *Psychological Bulletin* 95, 52–77.

Rogan, M.T. and LeDoux, J.E. (1995) Ltp is accompanied by commensurate enhancement of auditory-evoked responses in a fear conditioning circuit. *Neuron* 15, 127–136.

Rogan, M.T. and LeDoux, J.E. (1996) Emotion: systems, cells, synaptic plasticity. *Cell* 85, 469–475.

Romich, J.A. (2005) *Fundamentals of Pharmacology for Veterinary Technicians.* Thomson Delmar Learning, Clifton Park, New York.

Sambraus, H.H. (1985) Mouth-based anomalous syndromes. In: Fraser, A.F. (ed.) *Ethology of Farm Animals.* Elsevier, Amsterdam, pp. 391–422.

Sandoe, P. (1996) Animal and human welfare: are they the same kind of thing? *Acta Agriculturae Scandinavica* Section A, 11–15.

Savastano, G., Hanson, A. and McCann, C. (2003) The development of an operant conditioning training program for new world primates at the bronx zoo. *Journal of Applied Animal Welfare Science* 6, 247–261.

Seaman, S., Waran, N.K. and Appleby, M.C. (2001) Motivation of laboratory rabbits for social contact. In: Garner, J.P., Mench, J.A. and Heekin, S.P. (eds) *Proceedings of the 35th International Congress of the International Society of Applied Ethology*, 4–8 August. University of California Davis, The Center for Animal Welfare, Davis, California, p. 88.

Seksel, K. and Lindeman, M.J. (1998) Use of clomipramine in the treatment of anxiety-related and obsessive-compulsive disorders in cats. *Australian Veterinary Journal* 76, 317–321.

Semitelou, J.P., Yakinthos, J.K. and Carter, S.C. (1998) Neuroendocrine perspectives on social attachment and love. *Psychoneuroendocrinology* 23, 779–818.

Sherwin, C.M. (2004) The motivation of group-housed laboratory mice, *Mus musculus*, for additional space. *Animal Behaviour* 67, 711–717.

Sherwin, C.M., Lewis, P.D. and Perry, G.C. (1999) The effects of environmental enrichment and intermittent lighting on the behaviour and welfare of male domestic turkeys. *Applied Animal Behaviour Science* 62, 319–333.

Siegel, A. (2005) *The Neurobiology of Aggression and Rage*. CRC Press, Boca Raton, Florida.

Siegel, A. and Shaikh, M.B. (1997) The neural bases of aggression and rage in the cat. *Aggression and Violent Behaviour* 2, 241–271.

Smith, S.F., Appleby, M.C. and Hughes, B.O. (1990) Problem solving by domestic hens: opening doors to reach nest sites. *Applied Animal Behaviour Science* 28, 287–292.

Stafford, K.J. and Mellor, D.J. (2005) Dehorning and disbudding distress and its alleviation in calves. *Veterinary Journal* 169, 337–349.

Stolba, A. and Wood-Gush, D.G.M. (1984) The identification of behavioural key features and their incorporation into a housing design for pigs. *Annales de Recherches Veterinaries* 15, 287–299.

Stolba, A. and Wood-Gush, D.G.M. (1989) The behaviour of pigs in a semi-natural environment. *Animal Production* 48, 419–425.

Studnitz, M., Jensen, M.B. and Pedersen, L.J. (2007) Why do pigs root and in what will they root? A review on the exploratory behaviour of pigs in relation to environmental enrichment. *Applied Animal Behaviour Science* 107, 183–197.

Sylvester, S.P., Mellor, D.J., Stafford, K.J., Bruce, R.A. and Ward, R.N. (1998) Acute cortisol responses of calves to scoop dehorning using local anaesthesia and/or cautery of the wound. *Australian Veterinary Journal* 76, 118–122.

Thorne, J.B.D., Goodwin, D., Kennedy, M.J., Davidson, H.P.B. and Harris, P. (2005) Foraging enrichment for individually housed horses: practicality and effects on behaviour. *Applied Animal Behavioural Science* 94, 149–164.

Van Putten, G. and Dammers, J. (1976) A comparative study of the well-being of piglets reared conventionally and in cages. *Applied Animal Ethology* 2, 339–356.

Vestergaard, K. (1980) The regulation of dustbathing and other patterns in the laying hen: a lorenzian approach. In: Moss, R. (ed.) *The Laying Hen and its Environment*. Martinus Nijhoff, The Hague, The Netherlands, pp. 101–120.

Vestergaard, K. (1982) Dust-bathing in the domestic fowl – diurnal rhythm and dust deprivation. *Applied Animal Ethology* 8, 487–495.

Vestergaard, K. and Hansen, L.L. (1984) Tethered versus loose sows: ethological observations and measures of productivity. I. Ethological observations during pregnancy and farrowing. *Annales de Recherches Veterinaries* 15, 245–256.

Widowski, T.M. and Duncan, I.J.H. (2000) Working for a dustbath: are hens increasing pleasure rather than reducing suffering? *Applied Animal Behaviour Science* 68, 39–53.

Wohlt, J.E., Allyn, M.E., Zajac, P.K. and Katz, L.S. (1994) Cortisol increases in plasma of Holstein heifer calves, from handling and method of electrical dehorning. *Journal of Dairy Science* 77, 3725–3729.

Wood-Gush, D.G.M. and Vestergaard, K. (1989) Exploratory behaviour and the welfare of intensively kept animals. *Journal of Agricultural Ethics* 2, 161–169.

Yoerg, S.L. (1991) Ecological frames of mind: the role of cognition in behavioral ecology. *Quarterly Review of Biology* 66, 287–301.

9 Improving Livestock, Poultry and Fish Welfare in Slaughter Plants with Auditing Programmes

Temple Grandin

Colorado State University, Fort Collins, Colorado, USA

There are four basic types of problems that cause poor animal welfare in a slaughter plant, and to be effective in solving problems that cause animal suffering one must correctly determine the cause of the problem. The four basic types of problems are:

1. **Poorly maintained equipment or facilities.** Examples would be a broken stunner or a slippery floor that causes falling in the stun box or unloading ramp.
2. **Untrained employees or unsupervised employees** who mistreat animals. Examples would be poking a stick in an animal's rectum or slamming an animal down on to the floor with a heavy guillotine gate.
3. **Minor design faults** that can be easily fixed. Examples would be welding steel rods on a poorly designed slippery stun box floor to provide a non-slip surface or eliminating a reflection on shiny metal that makes animals baulk and stop moving.
4. **Major design faults** where a plant will need to purchase a major piece of equipment or rebuild the facility. Examples would be replacing shackling and hoisting of live animals for religious slaughter with a restraining box that holds them in an upright position, or replacing old, undersized equipment with new, larger capacity equipment that can handle a plant's increased line speed. Plants handling wild, extensively raised animals that lack basic equipment such as races will need to build them (see Chapters 5 and 14 for information on design).

Some of the most serious animal welfare problems exist in parts of the world where modern equipment for handling and restraining large animals such as cattle or water buffaloes is not available. Welfare problems become worse if the animal is wild and untamed and not accustomed to being handled by people. Tame animals that are trained to lead are much easier to handle without modern equipment.

Practices that Should be Banned and Severe Welfare Problems that Need Immediate Corrective Action

In many countries, both government regulations and guidelines developed by industry have prohibitions against the worst practices. The World Organization for Animal Health (OIE) has guidelines for handling and stunning of animals which includes lists of practices that cause suffering and should not be used (OIE, 2009). When a slaughter plant is being evaluated, the first step is to eliminate practices or bad employee behaviour that should be prohibited. The OIE (2009) code specifically states that methods of restraint that cause avoidable suffering should not be used in conscious animals because they cause severe pain and stress such as the following:

- Suspending or hoisting animals (other than poultry) by the feet or legs.
- Indiscriminate and inappropriate use of stunning equipment (refers to practices such as using an electric stunner to move animals).
- Mechanical clamping of an animal's leg or feet of the animals (other than shackles used in poultry and ostriches) as the sole method of restraint.
- Breaking legs, cutting leg tendons or blinding animals in order to immobilize them.

- Severing the spinal cord (e.g. using a puntilla or dagger) or using electric currents to immobilize animals, except for proper stunning (see Chapter 5 on the detrimental effects of immobilization). Electrical immobilization is highly stressful and aversive because it paralyses the animal but it remains sensible. Proper electrical stunning induces instantaneous insensibility. Research studies clearly show that immobilization is highly aversive and very detrimental to animal welfare (Lambooij and Van Voorst, 1985; Grandin *et al.*, 1986; Pascoe, 1986; Rushen, 1986).

The OIE also has standards for handling animals. Below are some of the most important parts of the code (2009) of prohibited handling practices:

- Conscious animals should not be thrown, dragged or dropped.
- Animals for slaughter should not be forced to walk over the top of other animals.
- The use of such devices as electric goads should be limited to battery-powered goads on the hindquarters of pigs and large ruminants, and never on sensitive areas such as the eyes, mouth, ears, anogenital region or belly. Such instruments should not be used on horses, sheep and goats of any age, or on calves or piglets. The author adds that no driving aid of any type should ever be applied to a sensitive part of an animal.
- Painful procedures (including whipping, tail twisting, use of nose twitches, pressure on eyes, ears or external genitalia) or the use of goads or other aids that cause pain and suffering (including large sticks, sticks with sharp ends, lengths of metal piping, fencing wire or heavy leather belts) should not be used to move animals.

The OIE (2009) and many countries have regulations that contain detailed specifications for stunning practices used on livestock and poultry. Animals must be rendered insensible and unconscious before dressing procedures begin. Cutting off limbs, skinning or scalding an animal that shows any sign of return to sensibility should NEVER be permitted. The animal must be immediately re-stunned before dressing procedures begin. This is a requirement of both the OIE and the regulations in many countries. If any of these cruel practices occurs, the plant has committed an egregious violation of animal welfare standards and humane slaughter regulations. The EU standards require that a plant has a designated animal welfare officer. United States Department of Agriculture (USDA) standards do not have this requirement.

Assessing Welfare with Numerical Scoring

In Chapters 1 and 3, the advantages of using numerical scoring are covered. Numerical scoring has greatly improved both animal handling and stunning (Grandin, 2003a, 2005). A numerical scoring system developed by Grandin (1998a, 2007a) to score handling and stunning is extensively used in the USA and many other countries. Others have also successfully used numerical scoring to evaluate handling (Maria *et al.*, 2004). The advantage of using numbers to evaluate practices makes it possible to determine if practices are improving or deteriorating. This is the same principle that is used in Hazard Analysis Critical Control Points (HACCP) programmes for monitoring bacteria counts on meat.

The OIE code (2009) supports the use of numerically scored performance standards. They are much more effective than audits that are largely based on paperwork or subjective evaluations. The scoring system developed by Grandin (1998a) for handling and stunning is an animal-based outcome scoring system (i.e. all scores are per animal) where things that can be directly observed can be assessed. To pass the audit, an acceptable numerical score is required on all of the critical control points (CCPs). For large plants 100 animals are scored for each CCP while for small plants 1 h of production is monitored and scored. The five CCPs are:

1. The percentage of animals rendered insensible with one application of the stunner (for all species of animals, poultry and fish). When a captive bolt is used, an acceptable score is 95% and an excellent score is 99%. For electrical stunning of both mammals and poultry, the stunner must be positioned correctly on 99% of the animals. In mammals, tong placement is scored and in poultry, the percentage of effectively stunned birds is scored.
2. The percentage of animals rendered insensible before hoisting, skinning, scalding or plugging of the oesophagus or other dressing procedures. The score must be 100% to pass an audit. For pigs and

poultry, the audit is failed if a pig or bird enters the scalder showing any sign of sensibility. In poultry plants, the audit is failed if a red, discoloured bird is found that does not have its throat cut. It is likely that the bird was scalded alive.

3. The percentage of cattle or pigs that vocalize (moo, bellow or squeal) during handling and stunning. All vocalizing animals that occur in the stun box or restrainer should be scored. In the lead up race, vocalizations that occur during active handling when the animals are being moved into the stun box or restrainer are scored (see Chapter 5 for more information). Each animal is scored as either silent or vocalizing. The goal is 5% or less of the animals vocalizing. Vocalization scoring is not used for sheep at the slaughter plant.

4. The percentage of animals that fall down (i.e. body touches the floor) during handling (use for all mammals). All parts of the facility should be scored including the unloading ramp and stun box. If more than 1% of the animals fall, the plant has a problem that should be corrected.

5. The percentage of animals moved with no electric goads. Use for all mammals – 75% is acceptable, and 95% excellent.

6. For poultry only – the percentage of birds with broken or dislocated wings. The goal is 1% or less on a per bird basis.

There are also acts of abuse that would result in an automatic audit failure. Some of the acts of abuse are listed in the earlier section of the chapter 'Practices that Should be Banned and Severe Welfare Problems that need Immediate Corrective Action'. There would also be an automatic failure if employees deliberately slam gates on animals. Broken bones, bruises, or falls in sensible animals caused by powered devices such as head holders, stunbox gates and other mechanical devices would also automatically result in a failed audit. Repeated use of obviously malfunctioning or broken stunning equipment would also be an act of abuse.

Prestunning Critical Limits

Vocalization critical limits

Cattle

Vocalization (moos or bellows) is scored during handling and in the stunning box or other slaughter point. Vocalization in cattle is an indicator of stress (Dunn, 1990). Grandin (1998b) found that 99% of cattle vocalizations occurred in response to an aversive event such as missed stuns, slipping on the floor, electric-prod use, or excessive pressure from a restraint device. A plant with a restraint device that applied excessive pressure had 35% of the cattle vocalizing (Grandin, 1998b). In another study excessive pressure applied by a head-holding device caused 22% of the cattle to bellow, but when the pressure was reduced, vocalization dropped to 0% of the animals (Grandin, 2001b). Vocalizations are scored during both conventional slaughter with stunning, and ritual slaughter without stunning. In plants with single-file races, vocalizing animals are scored in the race when they are being moved. All vocalizing animals in the stun box or ritual slaughter restraint box or restrainer conveyor are scored. In plants that do not have races or stun boxes, vocalization is scored anytime an animal is being moved by people. All vocalizations are also scored in the pen where animals are stunned or slaughtered. The critical limit is 3% of the cattle vocalizing. If the plant uses a head-holding device for either conventional or ritual slaughter, the standard is 5%. It is yes/no scoring on a per animal basis – an animal is scored as either a quiet animal or a vocalizing animal. Plants can easily achieve this standard (Grandin, 1998b, 2005, 2007a). Data collected by the author in ten beef plants that conduct ritual slaughter indicated that this standard can be achieved when the animal is held with a well-designed head-holding device. For both conventional and ritual slaughter, the audit is failed if any cattle vocalize after hoisting.

Pigs

Squealing in pigs is associated with stressful handling and painful procedures (Warriss *et al.*, 1994; White *et al.*, 1995). Unlike cattle, it is difficult to determine which pig is squealing in a race full of pigs. A sound-level meter could be used by the quality assurance department to monitor sound levels. This works well within a plant, but works poorly for comparing one plant with another plant due to differences in plant design and the number of pigs in a room. For comparisons between plants, squealing of individual pigs can be scored in the following areas: in plants with races, the percentage of pigs squealing in the conveyor, restrainer or stun box are scored; in plants with gas stunning, each individual pig is scored when the gondola is loaded; in plants that stun pigs in groups on the floor, squealing should be scored on each individual pig as it is being handled for stunning. It is yes/no scoring on a per animal basis – a pig is scored as either silent or vocalizing. The limit should be 5% of the pigs vocalizing. The audit is failed if any pig vocalizes after stunning.

Sheep

Do not use vocalization scoring for sheep. As sheep are an ultimate prey animal, they often do not vocalize even when they are stressed and in pain.

Goats

The vocalization standard for goats needs to be developed.

Poultry

Do not use vocalization scoring for poultry prior to stunning. The audit is failed if a bird vocalizes (clucks or cackles) after passing the back-up bleeder person.

Electric-prod (-goad) use critical limits

The critical limit for cattle, sheep, pigs, goats and all other mammals is 75% must be moved with no electric goad. An excellent score is 95% moved with no electric goad. Flags or other non-electrified items should be the primary driving aids. The electric goad should only be picked up when it is needed and then put away after use. If the electric goad is replaced with painful driving methods such as beating, dragging, breaking tails or other abusive practices the audit is failed. Yes/no scoring is used – the animal is either moved with an electric goad or not moved with one. Since it is very difficult to determine if the switch on an electric goad is pushed, electric-goad use is scored if the animal is touched with the electrical device. Plants can easily achieve the standard (Grandin, 2003a, 2005, 2007b).

Falling critical limits

The critical limit for falling in cattle, sheep, goats, pigs and all other mammals is that no more than 1% of animals should fall anywhere in facility from unloading trucks until the animal is stunned or ritually slaughtered. A fall is scored if the animal's body touches the ground. Stun boxes that are designed to make animals fall will result in a failed audit. Plants can easily achieve this standard (Grandin, 2007b). The OIE (2009) codes also use numerical scoring for falling. Corrective action needs to be taken if more than 1% of the animals fall.

Poultry handling – critical limits

Handling of broiler chickens is assessed by counting the percentage of birds that have broken or dislocated wings. Broken wings should be scored with the feathers on to avoid confusing handling damage, which is a welfare issue, with damage caused by the feather-removal equipment. The minimum acceptable score is 1% (of birds with broken wings) in lightweight birds and 3% (of birds with broken wings) in heavy birds over 3 kg. An excellent score for birds weighing over 3 kg is 1% or less. The industry can easily achieve this standard (Grandin, 2007b). Data from 2006 restaurant audits of 22 poultry plants that process heavy chickens weighing 3 kg or more showed that all of the plants had broken wing scores of 3% or less. Six excellent plants processing heavy chickens had 1% or less broken wings. When the audits first started, most plants had 5–6% of birds with broken wings but their scores became better when they improved their handling practices.

Stunning Principles

Principles of captive bolt stunning methods

See Chapter 10 for captive bolt stunning methods.

Principles of electric stunning methods for mammals

When electric stunning is used, sufficient amperage (current) must be passed through the brain to induce instantaneous insensibility. Electricity flows like water. Amperage is the volume of electricity and voltage is analogous to water pressure. Modern electronically controlled stunners can be set to automatically vary the voltage to deliver the required amperage. On older stunning units, the voltage is set and the amperage will vary depending on the resistance of the animal. Wetting the animal will reduce electrical resistance and is strongly recommended. Table 9.1 shows the minimum

Table 9.1. Minimum current levels for head-only stunning (source: OIE, 2009).

Species	Minimum current levels (A)[ab]
Cattle	1.5
Calves (bovines < 6 months of age)	1.0
Pigs	1.25
Sheep and goats	1.0
Lambs	0.7
Ostriches	0.4

[a] In all cases, the correct current level should be attained within 1 s of the initiation of stun and maintained for at least 1–3 s and in accordance with the manufacturer's instructions.
[b] For head-to-body cardiac arrest stunning that kills the animal, higher amperages may be required because the current has to travel a greater distance through the body.

amperage (current) levels for electric stunning (OIE, 2009). To induce instantaneous insensibility, electric stunning must induce a grand mal epileptic seizure (Croft, 1952; Warrington, 1974; Lambooij, 1982; Lambooij and Spanjaard, 1982). When an animal is effectively stunned, there will be a tonic (rigid) phase where the animal is stiff followed by a clonic (kicking) phase.

For pigs, sheep, cattle or goats, there are two basic types of electrical stunning: head-only stunning and head-to-body cardiac arrest. Figure 9.1 shows correct positioning of a head-only electric stunner. Figures 9.2 and 9.3 show the correct positioning of the head electrode, which is critical. It must never be placed on the neck because this will cause the current to bypass the brain. The body electrode placement may be on the back, side of the body, brisket or belly. Passing the current through the feet is not recommended. Head-only stunning is reversible and the lamb or pig must be bled within 15 s to prevent return to sensibility (Blackmore and Newhook, 1981; Lambooij, 1982). When cardiac arrest stunning is used, the stunning-to-bleed interval can be longer because the animal is killed by stopping the heart. Cardiac arrest will mask the tonic and clonic phase of the epileptic seizure. For more information an electrical stunning, refer to Wotton and Gregory (1986), Gregory (2007, 2008) and Weaver and Wotton (2008). Figure 9.4 shows a simple way to achieve cardiac arrest in a plant that has simple equipment. The stunner is applied to the head first to achieve insensibility. After the animal is rendered insensible, the wand or tong is applied to the chest.

To reduce blood spotting (petechial haemorrhages in pork or poultry), some stunners use frequencies higher than the standard 50–60 cycle that comes from the electrical mains. Higher frequencies will reduce the length of time that pigs will remain insensible (Anil and McKinstry, 1994). Very high frequencies of 2000–3000 Hz failed to induce insensibility and should not be used (Croft, 1952; Warrington,

Fig. 9.1. Head-only tong-type electric stunner placed in the correct position on the head. The current must pass through the brain to induce a grand mal epileptic seizure. This will make the animal instantly insensible.

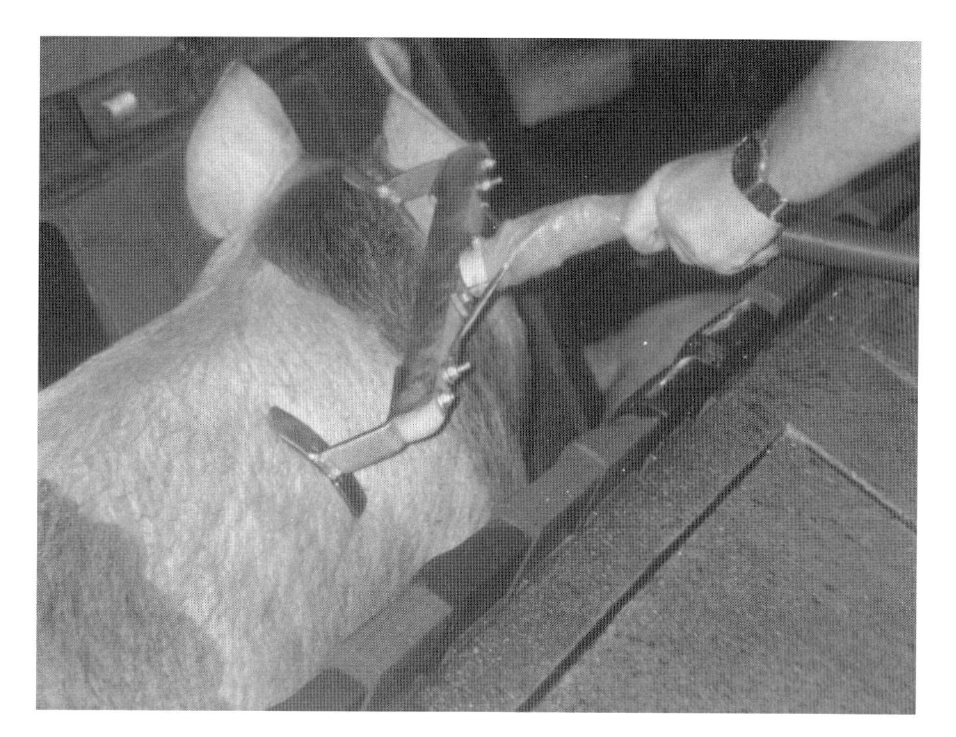

Fig. 9.2. Head-to-body cardiac arrest electric stunner. The head electrode must NEVER be placed on the neck. It must either be in the hollow behind the ears or on the forehead. Placing the body electrode on the side of the body helps reduce blood spotting in the meat. This photo shows correct placement.

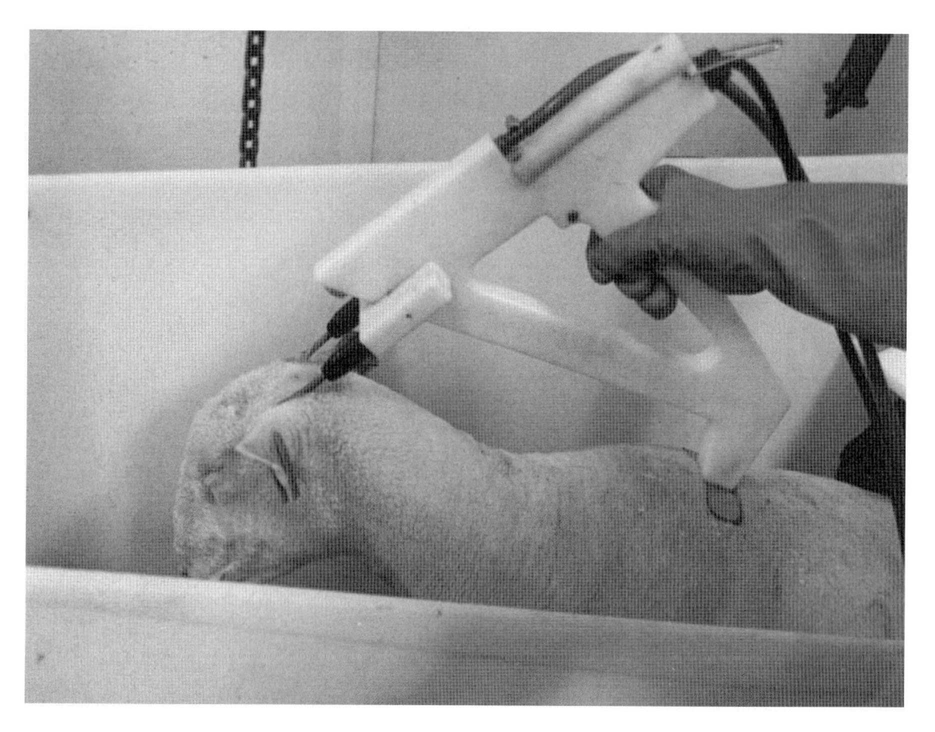

Fig. 9.3. Head-to-body cardiac arrest electric stunner for sheep. To ensure contact through the wool, water is injected through the electrodes. For electrical safety the entire unit is constructed from plastic.

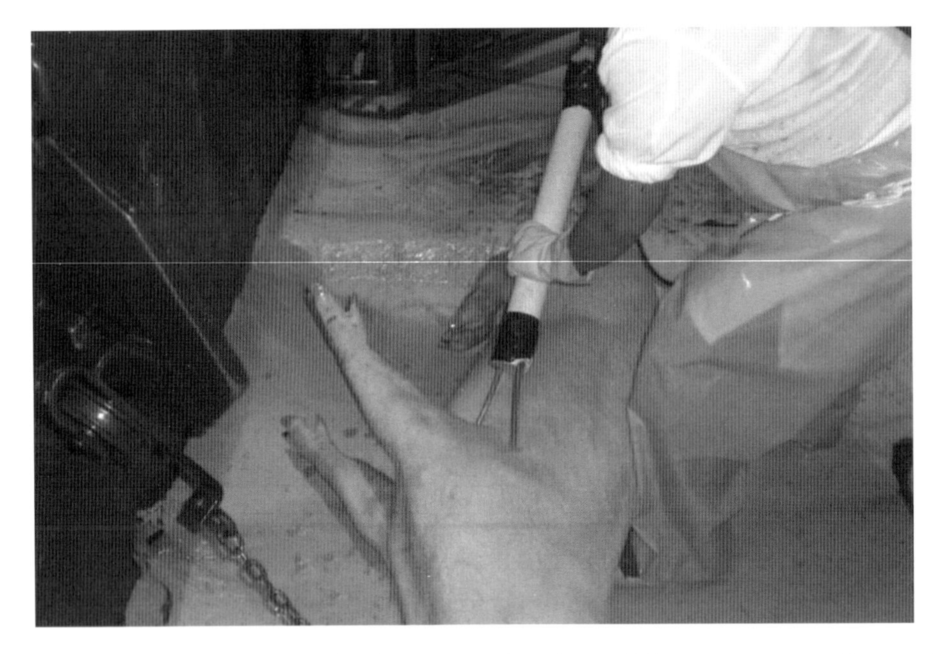

Fig. 9.4. In small plants, many pigs that are head-only stunned regain sensibility because the hoist is very slow. A simple solution to this problem is to apply the stunner to the head first and then apply it a second time to the chest to stop the heart (photograph courtesy of Erika Voogd).

1974; Van der Wal, 1978). Frequencies of 1592 Hz sine wave or 1642 Hz square wave will induce insensibility in small pigs (Anil and McKinstry, 1994). Some effective commercial systems apply an 800 Hz current to the head followed by a 50 Hz current applied to the body. This combination is effective (Lambooij *et al.*, 1996; Berghaus and Troeger, 1998; Wenzlawowicz *et al.*, 1999).

Principles of poultry stunning methods

Chickens and turkeys are stunned and rendered insensible in commercial poultry plants with either electric or gas stunning. To instantaneously render poultry insensible, the current (amperage) levels and electrical frequencies shown in Table 9.2 must be used when electric stunning is used (OIE, 2009).

Table 9.2. Current levels[a] and electrical frequencies for stunning poultry (source: OIE, 2009 – under study).[b]

Species	Current (mA) per bird	
Broilers	100	
Layers (spent hens)	100	
Turkeys	150	
Ducks and geese	130	
Frequency (Hz)	**Chickens (mA)**	**Turkeys (mA)**
<200	100	250
200–400	150	400
400–1500	200	400

[a]All currents are sinewave AC. Other types of currents may require higher amperages.
[b]Additional information on electric stunning of poultry is in Raj (2006).

Table 9.3 shows a comparison of the advantages and disadvantages of electric stunning and gas stunning systems for poultry.

Mechanical design of gas systems for poultry

From an engineering standpoint, there are two basic types of gas stunning systems – the first type is an open system where birds in the transport containers move down a tunnel or pit in a continuous flow on conveyors. The entrance and exit of the system are open to the atmosphere. The second type of system is a closed system (positive pressure). In this system, batches of birds in containers are placed in a chamber that has sealed doors. Carbon dioxide (CO_2) will work well in an open system because it is heavier than air and it will stay in the bottom of a chamber. Argon will also work well in an open system because it is also heavier than air. The main problem with argon is high cost. For best results, open systems should use gases that are heavier than air. This same principle is used for CO_2 stunning of pigs. Pigs in gondolas (elevator cars) are lowered into a deep pit filled with CO_2. In a closed system, the gas is introduced into a closed chamber under positive pressure. The gas moves through a ventilation system that re-circulates it while the chamber is being filled to the specified gas concentration. One advantage of a closed chamber is that the gas mixture can be more precisely controlled compared to an open system. All kinds of gases will work well in a closed system. The disadvantage of a closed system is that it uses a greater volume of gas compared to a CO_2 chamber where CO_2 sits in the bottom of a pit or tunnel. In a closed system, all the gas has to be evacuated before the next batch of birds is put in. Closed systems would work really well with inert nitrogen gas because almost all the oxygen can be flushed out of the chamber. Inert gas systems work best and the bird's reactions to the gas will be reduced if the oxygen level is 2% or less. This is very difficult to achieve in an open chamber. Open systems that use nitrogen have been a commercial failure due to high numbers of broken wings. Further,

Table 9.3. Welfare comparison of electric and gas stunning systems for poultry.

	Stunning	
	Electric[a]	Gas
Advantages	• Instantaneous insensibility (Raj and O'Callaghan, 2004).	• Birds do not have to be removed from the transport containers for stunning. This greatly reduces handling stress. • All birds in the container will be stunned. • Less likely to have problems with employees abusing birds because they do not handle live birds.
Disadvantages	• Birds are hung live inverted on the shackle line. This is stressful for the birds and increases plasma corticosterone (Kannen *et al.*, 1997; Bedanova *et al.*, 2007). • Small birds mixed with large birds miss the water bath and may not be stunned. • Supervising and training employees is more difficult. Problems with people abusing birds are more likely because people handle each individual live bird. • In poorly designed systems, the birds may get small pre-shocks before they enter the water bath.	• Insensibility is not instantaneous. The distress and discomfort to the birds prior to the loss of insensibility will vary depending on the gas mixture. • Requires very careful continuous monitoring of the gas mixtures. Wind around the plant building, changes in plant ventilation and open and closing doors may change gas mixtures in some systems.

[a] Electric stunning takes place in a water bath.

nitrogen works poorly in an open system because it is not heavier than air. Nitrogen is cheap and can be used in a well-designed closed system.

Bird and pig reactions to gas mixtures

There is great controversy in the scientific community on the correct gas mixtures for poultry and it is beyond the scope of this book to fully review all the literature. CO_2 systems use either pure CO_2 or CO_2 mixed with oxygen, argon or nitrogen. Other systems exclusively use inert gases such as nitrogen or argon. A mixture of 90% argon with 2% or less of oxygen is not aversive to pigs or chickens. The animals voluntarily entered a chamber filled with this mixture to obtain food (Raj and Gregory, 1990, 1995). In CO_2-only systems, chickens should never be suddenly introduced into concentrations of CO_2 higher than 30% unless oxygen is added, otherwise this will result in violent flapping and is highly detrimental to bird welfare. Observations by the author of commercial systems that use pure CO_2 indicate that, in order to reduce the bird's reaction to CO_2, the concentration of CO_2 must rise gradually over a period of several minutes. Practical experience in commercial facilities indicates that a smooth gradual increase in CO_2 from 0% to more than 50–55% reduces bird reactions prior to falling over (loss of posture). Chickens require a more gradual slow increase of CO_2 concentration for a longer period of time than turkeys. A slow, gradual increase in CO_2 concentration will prevent wing flapping and panicked attempts to escape from the container.

Another gas system that is used commercially is the biphasic system. In this system, the chickens are initially exposed to an atmosphere of 40% CO_2, 30% oxygen and 30% nitrogen for 60 s. The second stage is the euthanasia phase, which has an atmosphere of 80% CO_2 and air. The biphasic system with added oxygen during the anaesthetic phase has advantages from both a welfare and a carcass-quality standpoint (McKeegan *et al.*, 2007a, b; Coenen *et al.*, 2009). The two most common commercial systems are CO_2 only with a slow induction and the biphasic system.

There is a species difference between birds and pigs. Pigs should be suddenly introduced into a high level of 90% CO_2 (Hartung *et al.*, 2002; Becerril-Herrera *et al.*, 2009). Pigs stunned with 80% CO_2 had more pale soft exudative (PSE) meat, which is a severe quality defect that makes meat pale, soft and watery (Gregory, 2008). Lower levels of CO_2 at 70% were aversive to pigs and are not recommended (Becerril-Herrera *et al.*, 2009). The author has observed pigs in a chamber where they could be easily observed. Pigs free of the porcine stress syndrome (PSS) gene immersed in 90% CO_2 had very little behavioural reaction and remained silent until after they fell over (loss of posture) (Grandin, 2003a). Pigs with the PSS gene had more vigorous reactions (Troeger and Wolsterdorf, 1991). Observations in a commercial plant indicated that pig genetics might have an effect on a pig's reaction to CO_2. Some pigs remained calm and others violently attempted to escape from the container (Grandin, 1988). Studies on the pig's reaction to CO_2 have variable results. Some studies show that CO_2 is not aversive (Forslid, 1987; Jongman *et al.*, 2000). Other studies show that it is aversive (Hoenderken, 1978, 1983). The Forslid (1987) study used pure bred Yorkshire pigs, and in other studies the breed of the pigs is unknown. The author's observations indicate that genetic differences in the pigs may explain some of the differences in the results of different studies.

How to Determine Insensibility

The corneal reflex is the involuntary eyelid-closure response to protect the eyes from foreign objects. It involves two cranial nerves, one sensory and one motor, which converge in the brainstem. When the cornea senses a stimulus, such as the touch of a finger or pen, an impulse is sent through the sensory cranial nerve to a centre in the brainstem. A reflex impulse is then sent from the brainstem to the eyelids to trigger the closure, which is known as the corneal reflex. This test is commonly used to judge brainstem abnormalities in human medicine. However, the corneal reflex only indicates brainstem activity, which is not indicative of perception by the stunned animal. Electrically or gas-stunned animals with a weak corneal reflex triggered by the tip of pen and no other signs of return to sensibility would be in a state of surgical anaesthesia. If the animal has spontaneous, natural blinking that occurs when the eye is not touched, the animal is definitely sensible, and must be re-stunned. People who are assessing insensibility should look at live animals in the lairage so they will know what spontaneous blinking looks like. For animals stunned with either a penetrating or non-penetrating captive bolt, the corneal reflex and all eye movements must be absent (Gregory, 2008). The eyes should open into a wide blank stare and not be rotated (Gregory, 2008). Do NOT use a finger or other thick, blunt object to poke the eyes of animals with small eyes, such as pigs and sheep, when testing small animals for eye reflexes. This causes confusing signs that are difficult to interpret (Grandin, 2001b). A finger may be used on animals with large eyes, such as cattle.

The following signs of insensibility can be used when assessing stunning efficacy (Grandin, 2007a). Additional information is in Gregory (2008). These signs can be used with all types of stunning unless noted otherwise.

1. No rhythmic breathing (rhythmic breathing is when the ribs move in and out at least twice).
2. No response to a pin or knife-tip prick to the nose (apply to nose only).
3. No natural eye blinking (as observed in conscious animals in lairage). Do not confuse with nystagmus (rapid involuntary movements of the eyes). Nystagmus is permissible after electric stunning. It must be absent after captive bolt and gas stunning.
4. No righting reflex when hung on the rail (Fig. 9.5). This can be observed as an arching of the back and sustained backward lifting of the head. This should not be confused with a momentary flop of the head, which occurs when the back legs exhibit involuntary kicking.
5. No vocalization.
6. No rabbit-like nose twitch.
7. No stiff, curled tongue. A limp, flaccid tongue is an indicator of a well-stunned animal.
8. Ignore kicking and other leg movements. People assessing insensibility should look at the head. The head must be loose and floppy (see Fig. 9.5).

If any one of these signs is absent, a return to sensibility is indicated and the animal MUST be stunned again.

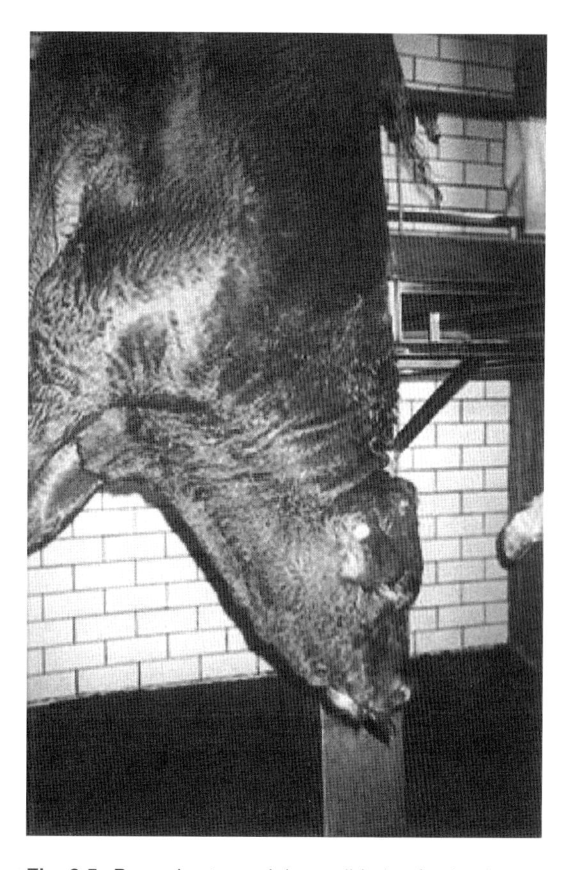

Fig. 9.5. Properly stunned, insensible beef animal. The head hangs straight down. The hind legs or forelegs may move and should be ignored. People assessing insensibility should be trained to look at the head. When the body moves, the head should be loose and floppy.

Evaluation of Stunning

In-plant evaluation of electrical stunning

A person evaluating animal welfare will not know the amperage of a stunning system if it has no amperage gauge or equipment for measuring volts, amperage and frequency. An easy way to evaluate a stunner is to test its ability to induce the tonic (rigid, still phase) followed by the clonic spasms (paddling, kicking phase) of an epileptic seizure. If the plant is using a one-piece head-to-body cardiac arrest stunning wand, the seizure may be masked by the cardiac arrest. The only way to test this system is to connect the electrical leads from the stunner power box to a tong-type stunner that can be applied to the head only. The stunner is acceptable if it induces a seizure. The presence of the tonic and clonic spasm is evaluated AFTER the electric stunner tong is removed from the head. When testing for tonic and clonic spasms, the tong should be held on for 1–3 s. Holding the tong on too long may depolarize the spine and mask the seizure.

Electrical stunner effectiveness may also be difficult to evaluate if a plant uses an immobilizing current to hold the carcass still after stunning. To observe the clonic and tonic spasm, the immobilizer current must be turned off. If the stunner is not capable of inducing an epileptic seizure, it should not be used. The use of low currents or very high frequencies will result in paralysing a sensible animal. Failure to place the stunner on the correct location on the head will also result in failure to induce a seizure. In chickens, the behaviour that indicates a seizure is different. Birds that are properly stunned while hanging on the rail will have vibrating tremors of their wings. Their wings will be held close to the body. When the birds regain sensibility, they may fully extend their wings and do full flapping motions.

Evaluation of gas stunning

To evaluate gas stunning, a system should have either windows or video cameras so that the birds or pigs can be viewed during the induction phase before they fall over (loss of posture) and become insensible. It is the author's opinion that the animal-based outcome that should be measured for auditing and monitoring of commercial systems is the animal's reaction during the induction phase. The author has observed chickens and turkeys in several types of commercial systems. In the best CO_2 systems where the gas is gradually introduced, very few birds flapped their wings and some birds shook their heads and gasped before loss of posture. This is a sign of respiratory distress (Webster and Fletcher, 2001). None of the birds attempted to escape from the container or did continuous violent flapping before loss of posture. One study with turkeys showed that 6.2% of the birds flapped their wings violently prior to falling over (Hansch *et al.*, 2009). Head shaking occurred in 37% of birds and deep breaths occurred in 18% (Hansch *et al.*, 2009). It is the author's opinion that violent flapping prior to loss of sensibility is not acceptable and the gas mixture must be adjusted to avoid this. When evaluating gas stunning systems, the emphasis should be on observing the animal or bird's reaction before it loses posture and becomes insensible. The science for specifying the best gas mixture and methods of induction for poultry are still being researched.

Evaluating the entire poultry handling and stunning system

When a poultry stunning system is being evaluated, you should look at BOTH the stunning method and the handling method as a complete system. Gas stunning greatly improves the handling of the poultry and some discomfort during the induction phase may be a reasonable trade off to eliminate live shackling and greater opportunities for employees to abuse the birds. It is the author's opinion that improving the pre-stun handling of the birds will result in overall better bird welfare even if gas stunning causes gasping and head shaking. If the birds do carry out full, vigorous wing flapping before loss of posture or try to escape from the container, bird welfare would probably be better with electric stunning.

Numerical Scoring of Stunning Practices

A perfect score on every CCP is impossible, so limits for a minimum acceptable score must be set. The following critical limits have been used since 1997 for cattle, pigs and sheep (Grandin, 2005, 2007a) and since 2001 for poultry.

Stunning critical limits

Captive bolt stunning and gunshot

For a minimum acceptable score, 95% of the animals must be rendered insensible with a single shot. This score is easily achieved in well-managed plants (Grandin, 2000, 2002, 2005). An excellent score is 99%. All animals requiring a second shot must be rendered insensible prior to hoisting, bleeding or any other invasive procedure. The penetrating captive bolt is more effective than the non-penetrating captive bolt. Shots that are slightly off target are more likely to fail with a non-penetrating captive bolt. Animals must be bled within 60 s after stunning with a non-penetrating captive bolt. Cattle and pigs should be shot in the middle of the forehead (see Chapter 10). Due to a very thick skull, sheep should be shot on the top of the head (see Chapter 10). Brahman cattle, water buffalo and other animals with a very thick skull may have to be shot behind the poll (hollow behind the horns) (Gregory *et al.*, 2008a). For further information on captive bolt stunning methods, refer to Gregory (2007, 2008) and Grandin (2002) (see Chapter 10).

Electric stunning

Correct placement of the stunning tongs (electrodes) is important to ensure that the current passes through the brain of 99% of the animals. This standard can be easily achieved (Grandin, 2001a, 2003b). Correct placement is essential to induce instantaneous insensibility (Anil and McKinstry, 1998). Figures 9.1, 9.2 and 9.3 show correct placement. The head electrodes must never be placed on the neck because the current will bypass the brain. The stunner wand or tongs also must not be applied to the eyes, ears or rectum of the animal. Acceptable electrode positions on the head are:

- forehead to body (cardiac arrest) – score position of head electrode only;
- hollow behind the ear to body (cardiac arrest) – score position of head electrode only;
- between eye and ear on both sides of the head with a tong-type stunner (head only); and
- top of the head and under the jaw with a tong-type stunner (head only).

For water bath systems for poultry, 99% of the birds must be positioned in the water bath so that the current will pass through the brain.

Hot wanding should occur in only 1% or fewer of the animals. Hot wanding occurs when the electrode is energized before the tong or other device is firmly pressed against the animal. Score a hot wand if the animal vocalizes (squeals, bellows) or pulls away in direct responses to stunner application. Do not score hot wanding vocalizations in sheep. For poultry, score pre-shocks at the water bath entrance as hot wanding. Count each bird as pre-shocked or not pre-shocked.

Carbon dioxide and other controlled-atmosphere stunning methods

Animals must be able to stand or lie down in the gondola or container without being on top of each other. Score an overloaded gondola if there is an animal on top of the others with no space for it to stand or lie down. The standard that should be used for mammals and poultry is that no more than 4% of the gondolas should be overloaded. Animals or birds must never be forced into a gondola on top of other animals. See the sections of this chapter headed 'Mechanical design of gas systems for poultry' and 'Bird and pig reactions to gas mixtures' for further information.

Insensibility critical limits

All animals must show no signs of return to sensibility prior to hoisting to the bleed rail. All poultry that miss the stunner or automatic knife must be cut by the back-up bleeder person. There is a zero tolerance for starting any dressing procedure such as scalding, skinning or leg removal on an animal or bird showing signs of return to sensibility. This standard applies to both standard slaughter with stunning and ritual slaughter (kosher or halal) without preslaughter stunning. For CO_2 and other controlled-atmosphere stunning, there is a need to develop a scoring system for monitoring animal welfare during induction. Windows should be provided so that the induction phase (until the animal or bird loses posture and falls over) can be observed and scored (see the sections headed 'Evaluation of gas stunning' and 'Evaluating the entire poultry handling and stunning system').

Objective Scoring Improves Welfare

Some animal welfare advocates have complained that allowing 1% of the animals to fall or 5% to vocalize is not very good welfare. In slaughter plants that have really good internal and external auditing, the actual numbers are much lower. In 1996 before the restaurant audits started, the worst plant had 32% of the cattle vocalizing and the average was 7.7% (Grandin, 1997a, 1998b). Since the audits started, the average vocalization percentage is under 2% and the worst plant is 6% (Grandin, 2005). Reductions in falling were even more striking. In 2005, data from the most heavily audited plants that were audited by both McDonald's and Wendy's was examined. Over 3000 cattle and pigs were scored in over 30 plants, and 0% fell down. Before the welfare audits were started, 100% of the cattle and pigs were moved with an electric prod in many plants. Today the average electric-goad use is under 20%. In most plants, the only place the electric goad is used is at the entrance of the stun box or restrainer. During the first 4 years of audits, 17.5% of the cattle were moved with an electric goad (Grandin, 2005). Young fed cattle can be moved more easily than old cull cows. The average for young fed beef was 15.2% electric goaded and 29% for old cull Holstein cows (Grandin, 2005). More recent data collected from the 2005 audits of 72 beef and pork plants indicated that only one plant failed on electric-goad use and 23% of the beef plants used electric prods on 5% or less of the cattle. Cattle stunning scores averaged 97.2% on the first shot and 91% (60 out of 66 beef plants) passed the insensibility audit (Grandin, 2005). In 2005, 100% of the beef cattle were rendered insensible prior to hoisting in 42 plants.

When performance is measured, a plant can determine if it is getting better or getting worse. In setting these standards, the critical limit had to be made high enough to force poor plants to improve but not so high that it would be too difficult to improve. The author is often asked by animal activists whether or not the standards should be stricter. The author has resisted changing them because there are differences in the ease of handling different types of animals. Difficult-to-move animals or bad weather would cause too many plants to fail. Plants have been audited with these standards since 1999 and many of the plants are continually improving. Data from a major restaurant company showed that in 2001, 21% of the beef plants and 33% of the pork plants failed on one or more of the five CCPs. In 2005, the failure rate was reduced to 2.5% of the beef plants and 0% of the pork plants.

Cost to Implement Improved Welfare

There are benefits to improved animal handling and stunning. Safety for employees is a sufficient reason for eliminating abusive handling practices of cattle and other large animals. Cruel practices such as shackling and hoisting live cattle prior to ritual slaughter are extremely dangerous. Installation of modern restraint equipment that holds the animal in a comfortable upright position greatly reduced accidents and injuries to employees (Grandin, 1988). Other benefits are reduced bruises and better meat quality. Stressful handling and overuse of electric goads shortly before stunning will greatly increase meat quality problems such as PSE pork (D'Souza *et al.*, 1999; Hambrecht *et al.*, 2005a, b). Multiple shocks from an electric prod will greatly increase lactate levels in the blood and the percentage of downer pigs (Benjamin *et al.*, 2001). The author has worked with plants that were able to export 10% more quality pork to Japan after they improved handling practices.

In the vast majority of plants in the USA, Europe, Australia and South America, major capital costs such as building totally new handling systems were not required. In the USA and Canada, only three plants out of over 75 pork and beef plants had to rebuild the entire animal handling facility. The other plants had to make lots of little improvements in the facility and improve employee training and maintenance of equipment. Improvements in equipment usually cost under US$2000, and in many plants under US$500. The most common simple improvements were installation of non-slip flooring in the stunning box, changes in lighting to improve animal movement and installation of solid panels to prevent approaching animals from seeing people and moving equipment (Grandin, 1982, 1996, 2005) (see Chapter 5). In poultry, the reduction of broken wings was accomplished with better management of catching employees and improved equipment maintenance.

Management of the plant is a big factor. In three plants, audits were failed until the plant manager was changed. Management is a much more important factor than line speed. Electric-goad scores for cattle were very similar at different line speeds ranging from 50 to over 300 cattle/h (Grandin, 2005) (Table 9.4). Good welfare can be maintained at high line speeds if the plant is designed and staffed for the speed. The worst problems occur when equipment is overloaded or the plant is understaffed. One of the worst cases of overloaded equipment observed by the author was a small cattle plant that increased its line speed from 26 cattle/h to 35 cattle/h. This resulted in slamming the stun box door on the cattle. A plant that worked well at 26 cattle/h was terrible at 35 cattle/h. The main

Table 9.4. Percentage of cattle moved with an electric goad at different line speeds (source: Grandin, 2005).

Line speed (cattle/h)	Number of plants	Moved with electric goads (%)
<50	16	20
51–100	13	27
101–200	10	12
201–300	21	24
>300	6	25

reasons that three plants had to be remodelled were either undersized equipment or the wrong type of equipment for their line speed.

Problem Solving Guides

The following are lists of commonly encountered problems at slaughter plants that can cause audit failure and some tried-and-tested solutions to these problems.

Captive bolt stunning (all species)

Problems and corrective actions:

1. Poor maintenance of the gun is a major cause of audit failure. Implement a daily maintenance and cleaning programme (Grandin, 2002). The use of a test stand to measure bolt velocity is strongly recommended (see Chapter 10 for further information on bolt velocity).
2. Air pressure is too low in a pneumatic stunner for effective stunning. Most stunners will require a dedicated air compressor. Air accumulator tanks will only work in very small plants that process four or five cattle per hour.
3. Damp cartridges are a major cause of audit failure for cartridge-fired stunners (Grandin, 2002). Cartridges must be stored in a dry environment such as the office or in a sealed container.
4. Agitated animals make stunning difficult. Two solutions are: (i) improve handling practices so animals will enter the stunning area quietly (e.g. reduce electric-goad use and stop employees from yelling and whistling); and (ii) install a non-slip floor grating to prevent animals slipping on the floor (weld 2 cm diameter steel rods in a 30 × 30 cm square pattern to the floor of the stun box; the rods should be welded flat against the floor and not overlapped).
5. The stunning-to-bleed interval is too long when non-penetrating captive bolt is used.
6. Long hair on cattle may reduce the effectiveness of non-penetrating captive bolt.

Electric stunning (all species)

Problems and corrective actions:

1. Insufficient amperage to induce insensibility. The minimum amperage settings are 1 A for sheep, 1.25 A for pigs and 1.5 A for cattle (see Table 9.1). Old or large animals may need higher settings.
2. Electrodes are in the wrong position to pass the current through the brain. To correct this, retrain the employees on correct electrode positioning. The electrodes (tongs) may have to be redesigned or adjusted to facilitate correct positioning in both manually operated and automated systems.
3. If the animal's body surface is too dry, stunning may not be effective. Wet either the animal or the electrodes to improve conductivity.
4. Dehydrated animals and birds are harder to stun. This is especially a problem in old animals or animals that have travelled a long distance. Providing water during long trips and in the stockyard (lairage) pens at the plant will help prevent dehydration.

Gas stunning (pigs and poultry)

Problems and corrective actions:

1. Gas concentration is either too low or the wrong gas mixture is used (see the section headed 'Evaluating Poultry Stunning Methods' for more information). For pigs 90% CO_2 is recommended (Becerril-Herrera *et al.*, 2009); 70% is aversive for pigs. Another problem that can occur is that the gas is not distributed in the chamber. This may be due to a design fault in the chamber or the ventilation system in the plant may be sucking the gas out of the chamber. Correcting this problem may require the expertise of an engineering professional skilled in the ventilation system. In open gas stunning systems such as the CO_2 systems used for pigs and poultry, the following factors can cause problems: wind blowing around the plant building; changes in the number of ventilation fans turned on in the plant; or opening and closing doors near the chamber. These factors can cause 'stack pressure' that may cause the gas to be sucked out. Stack pressure is a common cause of a sudden appearance of conscious animals emerging from a chamber that has been operating effectively. This may occur when a certain specific sequence of either opening doors or turning on ventilator fans occurs. Doors that slam hard by themselves are moved by differences in air pressure between two rooms or between a room and the outdoors. Differences in pressure in rooms near the gas stunner may cause the stunner to fail. Stack pressure problems will have no effect on positive pressure (closed) systems where the gas is introduced into the chamber with a ventilation system.

2. An overloaded, undersized machine is one of the worst problems with gas stunning. As a plant increases production, the machine may become overloaded. Plant managers who are purchasing a gas stunning machine should purchase a large enough machine to handle future increases in production. An overloaded machine has to be replaced. Specific signs of an overloaded machine are: (i) animals are not rendered insensible because the exposure time has been decreased by speeding up the conveyor; (ii) gondolas or containers are overloaded and pigs or birds do not have enough room to stand or lie down without being on top of each other. Pigs should never be forced to jump on top of other pigs when a gondola is being loaded.

Insensibility

These problems and corrective actions apply to both conventional and religious slaughter:

1. Stunning is not done correctly to induce insensibility – refer to the stunning sections above of the 'Problem Solving Guides'.
2. Stun-to-bleed interval is too long when head-only reversible electric stunning is used. The stun-to-bleed interval for head-only reversible electric stunning should be 15 s or less (Blackmore, 1984; Wotton and Gregory, 1986). For cardiac arrest electrical stunning, the stun-to-bleed interval can be 60 s.
3. Insufficient blood flow after bleeding. This was a major cause of audit failure in many pork plants. The employees should be trained in more effective bleeding methods. In pork, making a larger wound improved blood flow and corrected return-to-sensibility problems (Grandin, 2001a).
4. In plants doing religious slaughter, insensibility will occur more rapidly when the cut is performed on a calm animal (Grandin, 1994). A rapid cut with a very sharp knife is usually more effective. Dull knives that lack a sharp edge should not be used. Excessive pressure applied by a restraint device may cause agitation and delay the onset of insensibility. High vocalization scores are a sign that the restrainer is hurting the animal. A plant with a restraint device that applied excessive pressure had 35% of the cattle vocalizing (Grandin, 1998b). Excessive pressure applied by a head-holding device caused 22% of the cattle to bellow. When the pressure was reduced, vocalization dropped to 0% of the animals (Grandin, 2003b). The knife should be long enough so that the end of the knife remains outside of the neck during the cut. The wound should not be allowed to close back over the knife during the cut. The animal should be restrained in a comfortable upright position and the cut should be performed within 10 s after the head is restrained. If more than 5% of the cattle vocalize, immediate corrective action should be taken. Great care must be taken to avoid excessive pressure from being applied by the rear pusher gate or any other part of the restraint device. Immediately after the cut, the head holder, belly lift, rear pusher gate, and other devices that press against the animal should be immediately released to facilitate bleed out. The animal should not be removed from the box until it has collapsed and has become insensible.

Table 9.5 shows the difference between plants with good and bad procedures for kosher slaughter of cattle (Erika Voogd, personal communication, 2008). The times in Table 9.5 are similar to results in the research literature (Blackmore and Newhook, 1981, 1983; Blackmore *et al.*, 1983; Gregory and Wotton, 1984a). Sheep lose sensibility more quickly after slaughter without stunning compared to cattle. In sheep, the average time to onset of insensibility is 2–14 s (Blackmore, 1984; Gregory and Wotton, 1984b). Time-to-collapse scoring should be used to minimize the time that the animals remain sensible after the throat is cut. Poorly done slaughter without stunning can result in prolonged periods of sensibility (Newhook and Blackmore, 1982; Blackmore, 1984; Gregory and Wotton,

Table 9.5. Time to eye rollback and collapse (loss of posture) during kosher slaughter of cattle (source: Erika Voogd, personal communication, 2008).

	Good technique	Poor technique
Average time to collapse (s)	17	33
Longest time (s)	38	120
Cattle collapsed within 30 s (%)	94	68
Number of cattle	17	19

1984a, b, c). See Grandin (1985/1986) for a review of these studies. Daly *et al.* (1988) reported that the time to loss of sensory responsiveness of the bovine brain varied from 19 to 126 s. To reduce this variability and shorten the time to loss of insensibility will require careful scoring and monitoring of collapse (loss of posture) times.

Vocalization

Problems and corrective actions for both conventional and ritual slaughter:

1. Excessive electric-goad use is a major cause of vocalization. Train employees to use a flag or other non-electric aid as their primary method for moving animals (see Chapter 5). They should only pick up the electric goad when it is needed to move a stubborn animal and then put it back down.
2. Missed stuns cause vocalization – to correct this problem see stunning recommendations under the heading 'Captive bolt stunning (all species)'.
3. Falling or many small rapid slips in the stun box, restrainer or lead up race may cause vocalization. The remedy is the installation of non-slip flooring. Animals panic when they 'jig' and make many repeated small slips in the stun box.
4. Excessive pressure from a head holder or other restraint device that causes pain results in vocalization. Install pressure-limiting devices on hydraulic or pneumatic powered equipment. A head-holding device will need to have its own separate pressure control. It must be set at a lighter pressure than the pressure control for heavy gates and other parts of the apparatus that would require more pressure. There is a problem with the restraint device if animals vocalize in direct response to application of the device.
5. Sharp edges stick into the animal and cause vocalization. Surfaces that contact animals should be smooth. Even a small, sharp edge may stick into the animal. It must be found and removed.

Falling

Problems and corrective actions for both conventional and ritual slaughter:

1. Slick, slippery floors are the primary cause of animals falling. There are a number of ways to provide non-slip flooring:

- (a) Install a non-slip floor grating made from 2 cm diameter metal rods welded in a 30 × 30 cm square pattern to form a flat mat (see Chapter 5). Do not overlap the rods. Gratings work best in high-traffic areas such as stun boxes, races, crowd pens, scales and unloading areas.
- (b) Make grooves in the floor with a concrete grooving machine (can be done by a concrete contractor or the machine can be rented). This method is recommended for large areas of existing slick floors.
- (c) Recommendations for construction of new floors for cattle and other large animals include grooving the floor in pens and races in a 20 × 20 cm square or a diamond pattern. The grooves should be a minimum of 2 cm deep by 2 cm wide. For pigs, sheep and other small animals, smaller grooves can be used that are spaced closer together. Imprinting the pattern of expanded metal mesh into the wet concrete works well. A rough broom finish is not recommended as it wears out too quickly.

Electric-goad use

Problems and corrective actions:

1. Poorly trained employees tend to overuse electric goads when handling animals. Employees should be trained in the behavioural principles of handling animals (Grandin, 2007a, b). Employees should not be allowed to yell or whistle at animals (see Chapter 5).
2. Overloading of the crowd pen that leads up to the single-file race makes handling more difficult. Cattle and pigs should be moved into the crowd pen in small, separate groups and the crowd pen should be filled half full. Sheep can be handled in large groups as one continuous flow (see Chapter 5).
3. Animals baulking and refusing to move is a major cause of excessive use of electric goads. This is a facility problem that must be corrected. To improve the flow of animals, distractions that attract their attention must be removed from the facility. Refer to Table 9.6 to troubleshoot distractions in the race, stun box and restrainers that must be removed. To find the cause of baulking, people should get down in the race and see what the animals are seeing. A calm animal can be used to help locate distractions, as a calm animal will point its eyes and ears towards distractions. A clearly lit entrance to the single-file race or stun box can improve animal movement (Fig. 9.6). Table 9.7

Table 9.6. Troubleshooting guide for finding and removing distractions that cause animals to baulk, turn back or refuse to move.[a]

Distractions that cause baulking	How to improve animal movement
• Race or stun box entrance too dark	• Add indirect lighting that does not shine into the approaching animal's eyes (Fig. 9.6). If sunlight makes the entrance look dark, a shade may need to be installed to block the sunlight.
• Seeing moving people or equipment up ahead	• Install solid sides on races and install shields for people to stand behind. Block the animal's view of moving equipment that causes baulking. Experiment with large pieces of cardboard.
• Air blowing into the face of an approaching animal	• Change ventilation so that air will not be moving at the stun box entrance.
• Excessive noise	• Ensure that there is no yelling, silence hissing air exhausts and install rubber pads on clanging equipment.
• Reflection on metal or wet floors	• Adding or moving lamps will usually eliminate reflections. A person must get in the race and view it from the animal's eye level to determine if the reflection has been removed. Do many experiments with a portable lamp. Existing lamps may need to be moved.
• Small objects that cause baulking	• Remove loose plastic, clothes hung on fences and chains dangling in races.
• High colour contrast	• Paint facilities all the same colour and dress people in clothes that are similar in colour to the walls.
• Floor drains or changes in flooring colour	• Move the drain from the area where animals walk or make the flooring surface look the same.

[a]See Chapter 5 for more information and also Van Putten and Elshof (1978), Grandin (1982, 1996) and Tanida *et al.* (1996). Information from many specialists is in Grandin (2007b).

Fig. 9.6. Placing a lamp at the entrance of the single-file race may improve animal movement by attracting the pigs into the race.

Table 9.7. Solving problems with stun boxes, head holders and conveyor restrainers.

Problem	Possible causes	Remedies
Animal refuses to enter	• See Table 9.6	• See Table 9.6
	• 'Visual cliff' effect in a conveyor restrainer. Animal can see that the conveyor is above the floor	• Install a false floor to provide the illusion of a floor to walk on (Grandin, 2003b). The false floor should be installed so that it is approximately 15 cm below the feet of the largest animal that is riding on the conveyer
	• Hold-down rack touches animal's back while entering	• Raise hold-down rack
Animal becomes agitated or vocalizes	• Slipping on the floor	• Install non-slip flooring
	• Excessive pressure applied by a restraint device	• Reduce pressure. There is an optimal pressure for holding an animal – it must not be too loose or too tight. Install pressure-limiting devices. A head holder will need to have its own separate pressure limiting device – it requires a much lower pressure than heavy gates
	• Held too long in the restrainer	• Stun or ritually slaughter within 10 s
	• Sudden jerky motion of the apparatus	• Install speed-reducing devices so that the device moves with a steady, slow movement. Hydraulic or pneumatic controls should have good throttling control. This enables the operator to move a head holder or other restraint device with a smooth movement like a car's accelerator pedal (Grandin, 1992)
	• One side of a V-conveyor restrainer runs faster than the other	• Both sides should run at the same speed
	• Hold-down rack does not fully block the animal's vision on a conveyor restrainer	• Experiment with cardboard in different positions to block the animal's vision. Lengthen the hold down
	• Animal sees people and other moving things in front of the stun box	• Install a solid shield in front of the stun box

provides assistance for solving problems in the stun box or restrainer such as vocalization or agitated animals.

Welfare Issues with Religious Slaughter

Slaughtering animals without stunning is very controversial from an animal welfare standpoint. It is beyond the scope of this book to do a complete review of the literature on religious slaughter. There are two main welfare issues: the throat cut without stunning; and how the animal is restrained during the throat cut. Some of the worst problems with ritual slaughter are the restraint methods. The author has observed horrible plants where long lines of thrashing live cattle were hung upside down by one leg. The vocalization score was nearly 100% of the cattle.

For the best animal welfare the animal should be held in a comfortable upright position during slaughter without stunning. A poorly designed rotating box that positioned cattle on their backs was more stressful and had more vocalization and higher cortisol levels compared to upright restraint (Dunn, 1990). An improved-design rotating box with an adjustable side and cutting the animal within a few seconds after inversion resulted in vocalization scores that were similar to an upright box. To assess the quality of handling and restraint during religious slaughter, the plant should be scored for vocalization, falling down, and electric-goad use in the same way as conventional slaughter.

The percentage of cattle vocalizing should be 5% or less. People who are implementing animal welfare programmes should concentrate on eliminating painful, stressful methods of restraint (see Chapters 5 and 14).

When slaughter without stunning is done correctly, there was little or no reaction from the cattle during the cut (Grandin, 1994). Barnett *et al.* (2007) reported that only four out of 100 chickens showed a physical response during the cut. When a hand was waved in front of the animal's face, it had a greater behavioural reaction compared to well done Shehita (kosher slaughter) (Grandin, 1994). In most cattle, when the knife first touched the throat, there was a slight twitch. When a hand was rapidly waved within 20 cm of an animal's face, it became agitated because its flight zone had been invaded. The cattle that were observed by the author were extensively raised and they had a large flight zone. Also, the author has observed that when the cut is done poorly, and the wound closes over the knife, the animal may struggle violently. EEG pain measurements indicate that cutting the neck of 109–170-kg calves with a sharp 24.5-cm-long knife causes pain (Gibson *et al.*, 2009a, b). Petty *et al.* (1994) reported that catecholamines were higher after kosher slaughter compared to captive bolt stunning. Another welfare concern after slaughter without stunning is aspiration of blood into the lungs. This can vary from 36 to 69% of the cattle (Gregory *et al.*, 2008b).

Many Muslim religious authorities will permit stunning prior to the throat cut for halal slaughter. In New Zealand, electrical head-only stunning is used for halal slaughter. Other Muslim religious authorities will permit the use of a non-penetrating captive bolt. For poultry, electrical water-bath stunning is often permitted. Gas stunning is not permitted. For kosher slaughter, some rabbis will permit captive bolt stunning immediately after the cut. Some good sources of additional reading are Daly *et al.* (1988), Grandin (1992, 1994, 2006), Levinger (1995), Rosen (2004), Shimshoney and Chaudry (2005) and Gregory (2007, 2008). See the section headed 'How to Determine Insensibility' (p. 168) for additional information on the time required for loss of sensibility. The OIE, the EU, the USA and many other countries allow religious slaughter so that people are free to practise their religious beliefs (Rosen 2004; Shimshoney and Chaudry, 2005).

Careful, Quiet Handling Improves Meat Quality and Food Safety

Careful, quiet handling of pigs shortly before stunning will reduce the incidence of PSE meat. The last 15 min before stunning is critical. Improving handling and reducing electric-goad use in this area may result in 10% more pork that is suitable for lucrative export markets. Electric-goad use and jamming in the lead-up race will raise lactate levels (Edwards, 2009). Hambrecht *et al.* (2005a, b) also reported that careful preslaughter handling will improve pork quality and reduce lactate levels. Benjamin *et al.* (2001) found that multiple shocks with an electric goad will greatly increase lactate levels, open-mouth breathing and body temperature. Measuring of blood lactate on the bleed rail can be used to monitor the quality of handling in the stunning area. Low lactate levels will be associated with careful quiet handling practices. Blood lactate levels can be easily measured with a small lactate analyser that is used by athletes (Lactate Scout, EKF Diagnostic, GmbH, Magdeburg, Germany). Careful, quiet handling produces lactate levels of 4–6 mM and rough handling with high electric-goad use can raise it to 32 mM in pigs that become non-ambulatory (Benjamin *et al.*, 2001; Edwards, 2009). The range of lactate levels in blood collected on the bleed rail ranged from 4.4 mM to 31 mM (Warriss *et al.*, 1994; Hambrecht *et al.*, 2005a, b). When pigs were handled quietly, none of the animals became non-ambulatory, and multiple shocks and rougher handling caused 20% of the animals to go down and become non-ambulatory.

Handling before slaughter also affects the quality of beef. Grain-fed feedlot cattle that became agitated during loading at the feedlot had twice the risk of having their hides contaminated with *Salmonella* spp. (Dewell *et al.*, 2008). Quiet handling will also greatly reduce bruising (see Chapters 1 and 5). Rough handling and excessive use of electric goads increases bruising. Cattle that were handled carefully and quietly had a great reduction in bruising (Grandin, 1981). Repeated use of electric goads 15 min before slaughter resulted in tougher beef (Warner *et al.*, 2007).

Handling Non-ambulatory Animals

Non-ambulatory animals must NEVER be dragged. This is a violation of OIE (2009) codes and EU and USDA regulations for humane slaughter. Pigs, sheep, goats and other small animals can be easily moved on a sledge or other conveyance. Figure 9.7 shows a simple sledge for moving downed non-ambulatory pigs and Fig. 9.8 shows a hand cart that can be easily constructed for moving non-ambulatory pigs, sheep or goats. Large animals such as dairy cows that have become non-ambulatory are very difficult to be moved in a manner that does not cause suffering. One of the best things the USDA did in the USA was to ban the slaughter of non-ambulatory cows in federally inspected plants. Marketing the cows for slaughter when they are in better condition can prevent most non-ambulatory dairy cows. Non-ambulatory pigs and sheep can still be slaughtered in a USDA plant. The emphasis has to be on preventing the occurrence of high numbers of non-ambulatory animals.

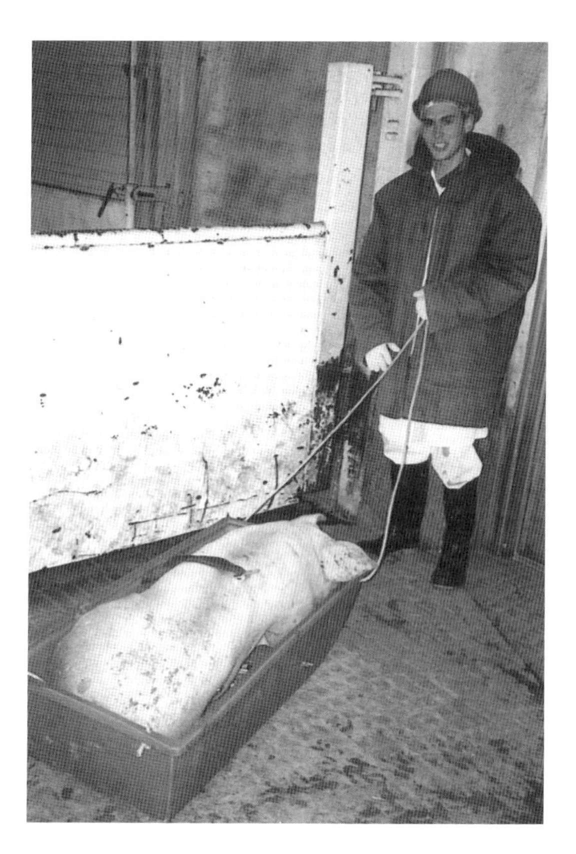

Fig. 9.7. Proper handling of a non-ambulatory pig in a plastic sledge. Downed animals should never be dragged.

Handling and Meat Quality Problems Caused by Beta-agonists

Pigs fed high doses of ractopamine (beta-agonist) may be more difficult to move at the plant (Marchant-Forde *et al.*, 2003). Lairage managers and staff have reported that large market-weight pigs that weigh 125–130 kg (275–285 lb) that have been fed high doses of ractopamine are difficult to move and may become non-ambulatory, as they become fatigued and too weak to move. These pigs do not have the symptoms of PSS. See Chapter 5 and Marchant-Forde *et al.* (2003) for more information. In cattle, a low dose of ractopamine of 200 mg/day for 28 days had only a slight effect on handling (Baszczrak *et al.*, 2006). There were no adverse effects during a short 60-min trip to slaughter.

Cattle fed excessive doses of beta-agonists such as ractopamine or zilpaterol will have tougher meat and less marbling. The author has observed that they may also have symptoms of heat stress and lameness. Their feet will usually look normal but they may be sore footed. Pigs have a higher incidence of hoof lesions when they are fed ractopamine (Poletto *et al.*, 2009). In both cattle and pigs, moderate doses of ractopamine or zilpaterol will increase the toughness of both pork and beef (Carr *et al.*, 2005; Gruber *et al.*, 2008; Hilton *et al.*, 2009). One of the most effective ways to stop the practice of feeding excessive doses of beta-agonists is to charge a fine for moving non-ambulatory animals or for tough meat. When a producer sees money deducted from his pay cheque, he will either stop using beta-agonists or use them in a more responsible manner. These substances are banned in many countries.

Tame Intensively Raised Animals Versus Extensively Raised Animals

A single-file race or stun box is not required if a few halter-broke animals that have been trained to lead are processed. A tame animal can be led on to the slaughter floor and stunned. Small animals such as sheep or goats can be held by a person and either ritually slaughtered or stunned. Simple systems that may be adequate with tame intensively raised animals that are accustomed to people will be highly detrimental to animal welfare if they are used with extensively raised animals. Systems that may be adequate for tame

Fig. 9.8. Hand cart for moving a non-ambulatory animal. It can be easily slid under the animal. This can be constructed in the plant maintenance shop. Non-ambulatory animals must never be dragged. In some countries, animals that arrive downed and non-ambulatory must be shot with a captive bolt on the truck.

Middle Eastern sheep are not adequate for imported Australian sheep that are not accustomed to handling. For these sheep, a single-file race is required. For small plants, a simple race made by a local welder could be used. The sheep could be lined up in the race and removed one at a time by a person. For large Middle Eastern plants, a conveyor restrainer would need to be installed at the end of the race.

Effect of Blood on Animal Behaviour

In some systems, especially in developing countries, the animal can see another animal being bled or is required to walk over a floor covered with blood. The author has observed that cattle will voluntarily enter a kosher restraint box that is covered with blood, provided that the previous animals had remained calm. If a previous animal had been jammed in malfunctioning equipment for 10–15 min, the other cattle would refuse to enter. Research shows that a stressor applied for 15 min, such as multiple shocks from a goad, will trigger the secretion of stress pheromones (Vieville-Thomas and Signoret, 1992; Boissey *et al.*, 1999). If animals remain behaviourally calm, then having blood on the floor may not affect welfare. If an animal becomes agitated for several minutes, the floor may need to be washed before the next animal is dealt with. Observations indicate that stunning an animal in the sight of other animals has almost no effect on behaviour, but the sight of a decapitated head may cause the next animal to panic. They do not appear to understand what has happened provided that the fallen animal's body remains intact.

Answers to Some Questions

Methods for cleaning dirty livestock

The emphasis should be on keeping cattle clean at the farm. Harsh methods such as spraying animals with high-pressure fire hoses should be avoided. Spraying water directly into the face of animals should also be avoided. Some plants have sprayers built into the floor of a holding pen to clean legs and bellies. This is acceptable. Washing should be avoided if the animals are in an environment with below freezing temperatures. The practice of clipping or shaving live cattle should be stopped because it is stressful to the animal and dangerous for people. Some plants that get very dirty cattle wash the carcasses after stunning and bleeding. From a welfare standpoint, this is much better than removing soil from live animals. Another good alternative is clipping the areas where the carcass will be opened with sheep shears. This is done after stunning and bleeding.

Stun-to-bleed interval

Regulations in some countries require that animals are bled within 20 or 60s after stunning. IF an animal shows any sign of return to sensibility, it must be re-stunned BEFORE it is hoisted and hung on the bleed rail. If re-stunning is required, the interval is timed after the second stun.

Procedures during work breaks

DO NOT leave animals in the stun box, restrainer or restraint box during coffee and lunch breaks. During lunch and other long breaks, the lead-up race should also be empty. In poultry plants, do not leave birds hanging on the line during breaks.

How stressful is slaughter?

A review of the literature indicated that cortisol levels in cattle during handling on the farm and after slaughter are similar. Many studies are reviewed in Grandin (1997b, 2007b).

Fish Stunning

Producers and processors of fish have installed systems for stunning fish and rendering them insensible to pain. There is a discussion on research on pain perception in fish in Chapter 1. Research indicates that the process of live chilling of fish may be highly stressful (Lines *et al.*, 2003; Roth *et al.*, 2009). A review of the literature indicates that many fish-farm managers have installed stunning systems. An Internet search revealed many recent patents for fish-stunning equipment. Electrical stunning equipment must induce epileptiform activity in the brain to render fish insensible. Information on electrical systems and electrical parameters can be found in Lambooij *et al.* (2007, 2008a, b) and Branson (2008). Percussion stunning is also used in fish. The technology for stunning fish is rapidly evolving.

Assessing Stunning and Handling of Fish

Stunning of fish can be assessed with numerical scoring the same way that livestock and birds are scored. Fish should be scored on the following variables.

1. Percentage of fish effectively stunned with one application of the stunner.
2. Percentage of fish rendered insensible before processing.
3. Percentage of fish with defects, such as eroded fins, that occurred in the production pens.
4. Percentage of fish that are bruised.
5. Percentage of fish with other carcass defects.

Data will have to be collected to determine the numerical scores for an acceptable level of welfare. When auditing of fish processing and farms first begins, the cut off for an acceptable score should allow the top 25% of the best operations to pass (see Chapter 1).

Conclusions

Implementation of an auditing programme that uses objective numerical scoring will greatly improve animal welfare. Another benefit will be reduced bruises and severe meat quality defects such as PSE and dark cutters. To be effective, the programme should audit CCPs such as the percentage of animals stunned correctly on the first attempt, the percentage of animals rendered insensible, the percentage that fall, the percentage moved with no electric prod and the percentage that vocalize. Numerical scores enable a plant to determine if their practices are improving or deteriorating.

References

Anil, A.M. and McKinstry, J.L. (1994) The effectiveness of high frequency electrical stunning in pigs. *Meat Science* 31, 481–491.

Anil, M.H. and McKinstry, J.L. (1998) Variation in electrical stunning tong placements and relative consequences in slaughter pigs. *Veterinary Journal* 155, 85–90.

Barnett, J.L., Cronin, G.M. and Scott, P.C. (2007) Behavioral responses of poultry during kosher slaughter and their implications for the bird's welfare. *Veterinary Record* 160, 45–49.

Baszczrak, J.A., Grandin, T., Gruber, S.L., Engle, T.E., Platter, W.J., Laudert, S.B., Schroeder, A.L. and Tatum, J.D. (2006) Effects of ractopamine supplementation on behavior of British, Continental, and Brahman crossbred steers during routine handling. *Journal of Animal Science* 84, 3410–3414.

Becerril-Herrera, M., Alonso-Spilsbury, M., Lemus-Flores, C., Guerrero-Legarreta, I., Hernandez, A., Ramirez-Necoechea, R. and Mota-Rojas, D. (2009) CO_2 stunning may compromise swine welfare compared to electrical stunning. *Meat Science* 81, 233–237.

Bedanova, I., Vaslarova, E., Chioupek, P., Pistekova, V., Suchy, P., Blahova, J., Dobsikova, R. and Vecerek, V. (2007) Stress in broilers resulting from shackling. *Poultry Science* 80, 1065–1069.

Benjamin, M.E., Gonyou, H.W., Ivers, D.L., Richardson, L.F., Jones, D.J., Wagner, J.R., Seneriz, R. and Anderson, D.B. (2001) Effect of handling method on the incidence of stress response in market swine in a model system. *Journal of Animal Science* 79 (Supplement 1), 279 (abstract).

Berghaus, A. and Troeger, K. (1998) Electrical stunning of pig's minimum current flow time required to induce epilepsy at various frequencies. *International Congress of Meat Science and Technology* 44, 1070–1073.

Blackmore, D.K. (1984) Differences in behaviour between sheep and cattle during slaughter. *Research of Veterinary Science* 37, 223–226.

Blackmore, D.K. and Newhook, J.C. (1981) Insensibility during slaughter of pigs in comparison to other domestic stock. *New Zealand Veterinary Journal* 29, 219–221.

Blackmore, D.K. and Newhook, J.C. (1983) The assessment of insensibility in sheep, calves, and pigs during slaughter. In: Eikenboom, G. (ed.) *Stunning of Animals for Slaughter.* Martinus Nijhoff, Boston, Masachusetts, pp. 13–25.

Blackmore, D.K., Newhook, J.C. and Grandin, T. (1983) Time of onset of insensibility in four- to six-week-old calves during slaughter. *Meat Science* 9, 145–149.

Boissey, A., Terlow, C. and Le Neindre, P. (1999) Presence of pheromones from stressed conspecifics increases reactivity to aversive events in cattle, evidence for the existence of alarm substances in urine. *Physiology and Behavior* 4, 489–495.

Branson, E. (2008) *Fish Welfare.* Blackwell Publishing, Oxford, UK.

Carr, S.N., Ivers, D.J., Anderson, D.B., Jones, D.J., Mowrey, D.H., England, M.B., Killefer, J., Rincker, P.J. and McKeith, P.K. (2005) The effects of ractopamine hydrochloride on lean carcass yields and pork quality characteristics. *Journal of Animal Science* 83, 2886–2893.

Coenen, A.M., Lankhaar, J., Lowe, J.C. and McKeegan, D.E. (2009) Remote monitoring of electroencephalogram, electrocardiogram, and behavior during controlled atmosphere stunning in broilers: implications for welfare. *Poultry Science* 88, 10–19.

Croft, P.S. (1952) Problems with electrical stunning. *Veterinary Record* 64, 255–258.

Daly, C.C., Kallweit, E. and Ellendorf, F. (1988) Cortical function in cattle during slaughter: conventional captive bolt stunning followed by exsanguination compared to sheehita slaughter. *Veterinary Record* 122, 325–329.

Dewell, G.A., Simpson, G.A., Dewell, R.D., Hyatt, D.R., Belk, K.E., Scanga, J.A., Morley, P.S., Grandin, T., Smith, G.C., Darget, D.A., Wagner, B.A. and Salmon, M.D. (2008) Risks associated with transportation and lairage on hide contamination with *Salmonella enterica* in finished beef cattle at slaughter. *Journal of Food Protection* 71, 2228–2232.

D'Souza, D.N., Dunshea, F.R., Levry, B.J. and Warner, R. (1999) Effect of mixing boars during lairage and pre-slaughter on meat quality. *Australian Journal of Agricultural Research* 50, 109–113.

Dunn, C.S. (1990) Stress reactions of cattle undergoing ritual slaughter using two methods of restraint. *Veterinary Record* 126, 22–25.

Edwards, L.N. (2009) Understanding the relationships between swine behavior, physiology, meat quality, and management to improve animal welfare and reduce transit losses within the swine industry. PhD dissertation, Colorado State University, Fort Collins, Colorado.

Forslid, A. (1987) Transient neocortisol, hippocampal and amygdaloid EEG silence induced by one minute inhalation of high concentration CO_2 in the swine. *Acta Physiologica Scandinavica* 130, 1–10.

Gibson, T.J., Johnson, C.B., Murrell, J.C., Hulls, C.M., Mitchinson, S.L., Stafford, K.J., Johnstone, A.C. and Mellor, D.J. (2009a) Electroencephalographic responses of halothane-anaesthetized calves to slaughter by ventral-neck incision without prior stunning. *New Zealand Veterinary Journal* 57, 77–85.

Gibson, T.J., Johnson, C.B., Murrell, J.C., Chambers, J.P., Stafford, K.J. and Mellor, D.J. (2009b) Components of electroencephalographic responses to slaughter in halothane-anesthetized calves: effects of cutting neck tissues compared to major blood vessels. *New Zealand Veterinary Journal* 57, 84–89.

Grandin, T. (1981) Bruises on Southwestern feedlot cattle. *Journal of Animal Science* 53 (Supplement 1), 213 (abstract).

Grandin, T. (1982) Pig behaviour studies applied to slaughter plant design. *Applied Animal Ethology* 6, 10–31.

Grandin, T. (1985/1986) Cardiac arrest stunning of livestock and poultry. In: Fox, M.W. (ed.) *Advances in Animal Welfare Science*. Martinus Nijhoff, Boston, Masachusetts.

Grandin, T. (1988) Possible genetic effect on pig's reaction to CO_2 stunning. *Proceedings of the 34th International Congress of Meat Science and Technology*, Brisbane, Australia. Commonwealth Scientific and Industrial Research Organization (CSIRO), Brisbane, Australia, pp. 96–97.

Grandin, T. (1992) Observations of cattle restraint devices for stunning and slaughter. *Animal Welfare* 1, 85–91.

Grandin, T. (1994) Euthanasia and slaughter of livestock. *Journal of the American Veterinary Medical Association* 204, 1354–1360.

Grandin, T. (1996) Factors that impede animal movement in slaughter plants. *Journal of American Veterinary Medical Association* 209, 757–759.

Grandin, T. (1997a) *Survey of Stunning and Handling in Federally Inspected Beef, Veal, Pork, and Sheep Slaughter Plants*. United States Department of Agriculture (USDA)/Agricultural Research Service Project 3602-32000-002-08G. USDA, Beltsville, Maryland.

Grandin, T. (1997b) Assessment of stress during handling and transport. *Journal of Animal Science* 75, 249–257.

Grandin, T. (1998a) Objective scoring of animal handling and stunning practices at slaughter plants. *Journal of American Veterinary Medical Association* 212, 6–39.

Grandin, T. (1998b) The feasibility of vocalization scoring as an indicator of poor welfare during slaughter. *Applied Animal Behaviour Science* 56, 121–128.

Grandin, T. (2000) Effect of animal welfare audits of slaughter plants by a major fast food company on cattle handling and stunning practices. *Journal of the American Veterinary Medical Association* 216, 848–851.

Grandin, T. (2001a) Solving return to sensibility problems after electrical stunning in commercial pork slaughter plants. *Journal of the American Veterinary Medical Association* 219, 608–611.

Grandin, T. (2001b) Cattle vocalizations are associated with handling and equipment problems in slaughter plants. *Applied Animal Behaviour Science* 71, 191–201.

Grandin, T. (2002) Return to sensibility problems after penetrating captive bolt stunning of cattle in commercial slaughter plants. *Journal of the American Veterinary Medical Association* 221, 1258–1261.

Grandin, T. (2003a) The welfare of pigs during transport and slaughter. *Pig News and Information* 24, 83N–90N.

Grandin, T. (2003b) Transferring results from behavioral research to industry to improve animal welfare on the farm, ranch, and slaughter plant. *Applied Animal Behaviour Science* 81, 215–228.

Grandin, T. (2005) Maintenance of good animal welfare standards in beef slaughter plants by use of auditory programs. *Journal American Veterinary Medical Association* 226, 370–373.

Grandin, T. (2006) Improving religious slaughter practices in the US. *Anthropology of Food*, 5 May 2006. Available at: http://aof.revues.org/document93.html (accessed 21 March 2009).

Grandin, T. (2007a) *Recommended Animal Handling Guidelines of Audit Guide*, 2007 edn. American Meat Institute Foundation, Washington, DC. Available at: www.animalhandling.org (accessed 5 July 2009).

Grandin, T. (ed.) (2007b) *Livestock Handling and Transport*. CAB International, Wallingford, UK.

Grandin, T., Curtis, S.E. and Widowski, T.M. (1986) Electro-immobilization versus mechanical restraint in an avoid-avoid choice test. *Journal of Animal Science* 62, 1469–1480.

Gregory, N.G. (2007) *Animal Welfare and Meat Production*, 2nd edn. CAB International, Wallingford, UK.

Gregory, N.G. (2008) Animal welfare at markets and during transport and slaughter. *Meat Science* 80, 2–11.

Gregory, N.G. and Wotton, S.B. (1984a) Sheep slaughtering procedures, I. Survey of abattoir practice. *British Veterinary Journal* 140, 281–286.

Gregory, N.G. and Wotton, S.B. (1984b) Sheep slaughtering procedures, II. Time to loss of brain responsiveness after exsanguination or cardiac arrest. *British Veterinary Journal* 140, 354–360.

Gregory, N.G. and Wotton, S.B. (1984c) Time of loss of brain responsiveness following exsanguination in calves. *Resources in Veterinary Science* 37, 141–143.

Gregory, N.G., Spence, J.Y., Mason, C.W., Tinarwo, A. and Heasman, L. (2008a) Effectiveness of poll shooting in water buffalo with captive bolt guns. *Meat Science* 81, 178–182.

Gregory, N.G., von Wenzlawowicz, M. and von Holleben, K. (2008b) Blood in the respiratory tract during slaughter with and without stunning in cattle. *Meat Science* 82, 13–16.

Gruber, S.L., Tatum, J.D., Engle, T.E., Prusa, K.J., Laudert, S.B., Schroeder, A.L. and Platter, W.J. (2008) Effects of ractopamine supplementation and postmortem aging on longissimus muscle palatability of beef steers differing in biological type. *Journal of Animal Science* 86, 205–210.

Hambrecht, E., Eissen, J.J., Newman, D.J., Smits, C.H.M., den Hartog, L.A. and Vestegen, M.W.A. (2005a) Negative effects of stress immediately before slaughter on pork quality are aggravated by suboptimal transport and lairage conditions. *Journal of Animal Science* 83, 440–448.

Hambrecht, E., Eissen, J.J., Newman, D.J., Verstegen, M.W. and Hartog, L.A. (2005b) Preslaughter handling affects pork quality and glycoytic potential of two muscles differing in fiber type organization. *Journal of Animal Science* 83, 900–907.

Hansch, F., Nowak, B. and Hartung, J. (2009) Behavioural and clinical response of turkeys stunned in a V-shaped carbon dioxide tunnel. *Animal Welfare* 18, 81–86.

Hartung, J., Nowak, B., Waldmann, K.H. and Ellerbrock, S. (2002) CO_2 stunning of slaughter pigs: effects of EEG, catecholamines and clinical reflexes. *Deutsche Tierarztliche Wochenschrift* 109, 135–139.

Hilton, G.G., Montgomery, J.L., Krehbiel, C.R., Yates, D.A., Hutcheson, J.P., Nichols, W.T., Streeter, M.N., Blanton, J.R., Jr and Miller, M.F. (2009) Effects of feeding zilpaterol hydrochloride with and without monesin and tylosin on carcass cutability and meat palatability of beef steers. *Journal of Animal Science* 87, 1394–1396.

Hoenderken, R. (1978) Electrical stunning of pigs for slaughter. Why? Hearing on preslaughter stunning. Kavlinge, Sweden, 19 May 1978.

Hoenderken, R. (1983) Electrical and carbon dioxide stunning of pigs for slaughter. In: Eikenboom, G. (ed.) *Stunning of Animals for Slaughter*. Martinus Nijhoff, Boston, Massachusetts, pp. 59–63.

Jongman, E.C., Barnett, J.L. and Hemsworth, P.H. (2000) The aversiveness of carbon dioxide stunning in pigs and a comparison of CO_2 rate vs. the V restrainer. *Applied Animal Behaviour Science* 67, 67–76.

Kannen, G., Heath, J.L., Wabeck, C.J. and Mench, J.A. (1997) Shackling broilers: effects on stress responses and breast meat quality. *British Poultry Science* 38, 323–332.

Lambooij, E. (1982) Electrical stunning of sheep. *Meat Science* 6, 123–135.

Lambooij, E. and Spanjaard, W. (1982) Electrical stunning of veal calves. *Meat Science* 6, 15–25.

Lambooij, E. and Van Voorst, N. (1985) Electroanaesthesia of calves and sheep. In: Eikenboom, G. (ed.) *Stunning Animals for Slaughter*. Martinus Nijhoff, Boston, Masachusetts, pp. 117–122.

Lambooij, B., Merkus, G.S.M., Voorst, N.V. and Pieterse, C. (1996) Effect of low voltage with a high frequency electrical stunning on unconsciousness in slaughter pigs. *Fleischwirtschaft* 76, 1327–1328.

Lambooij, E., Pilarczyk, M., Blaiowas, H., van den Boogaart, J.G.M. and Van de Vie, J.W. (2007) Electrical and persussion stunning of the common carp (*Cyprinus carpio* K.): neurological and behavioral assessment. *Aquacultural Engineering* 37, 117–179.

Lambooij, E., Gerritzan, M.A., Reimert, H., Burggraai, D. and Van de Vie, J.W. (2008a) A humane protocol for electro-stunning and killing of Nile tilapia in fresh water. *Aquaculture* 295, 88–95.

Lambooij, E., Gerritzen, M.A., Reimert, H.G.M., Burggraaf, D., Andre, G. and Van de Vie, J.W. (2008b) Evaluation of electrical stunning of sea bass (*Dicentrarchus labrax*) in seawater and killing by chilling: welfare aspects, product quality, and possibilities for implementation. *Aquaculture Research* 39, 50–68.

Levinger, I.M. (1995) *Shechita in the Light of the Year 2000*. Maskil L. David, Jerusalem, Israel.

Lines, J.A., Robb, D.H., Kestin, S.C., Crook, S.C. and Benson, T. (2003) Electric stunning: a humane slaughter method for trout. *Aquacultural Engineering* 28, 141–154.

Marchant-Forde, J.N., Lay, D.C., Pajor, J.A., Richert, B.T. and Schinkel, A.P. (2003) The effects of ractopamine on the behaviour and physiology of finishing pigs. *Journal of Animal Science* 81, 416–422.

Maria, G.A., Villareol, M., Chacon, G. and Gebresenbet, G. (2004) Scoring system for evaluating stress of cattle during commercial loading and unloading. *Veterinary Record* 154, 818–821.

McKeegan, D.E.F., Abeyesinghe, S.M., McLemen, M.A., Lowe, J.C., Demmeus, T.G.M., White, R.P., Kranen, R.W., Van Bemmel, H., Lankhaar, J.A.C. and Wathes, C.M. (2007a) Controlled atmosphere stunning of broiler chickens, II. Effects of behavior, physiology and meat quality in a commercial processing plant. *British Poultry Science* 48, 430–442.

McKeegan, D.E.F., McIntyre, J.A., Demmers, T.G.M., Lowe, J.C., Wather, C.M., Van den Broek, P.L.L., Coenen, A.M.L. and Gentle, M.J. (2007b) Physiological and behavioral responses of broilers to controlled atmosphere stunning: implications for welfare. *Animal Welfare* 16, 409–426.

Newhook, J.C. and Blackmore, D.K. (1982) Electroencephalographic stunning and slaughter of sheep and calves, Part 1. The onset of permanent insensibility in sheep during slaughter. *Meat Science* 6, 295–300.

OIE (2009) Chapter 7.5. *Slaughter of Animals, Terrestrial Animal Health Code*. World Organization for Animal Health, Paris, France.

Pascoe, P.J. (1986) Humaneness of electro-immobilization unit for cattle. *American Journal of Veterinary Research* 10, 2252–2256.

Petty, D.B., Hattingh, J., Ganhao, M.F. and Bezuidenhout, L. (1994) Factors which affect blood variables of slaughtered cattle. *Journal of the South African Veterinary Medical Association* 65, 41–45.

Poletto, R., Rostagno, M.H., Richert, E.T. and Marchant-Forde, J.N. (2009) Effects of 'step up' ractopamine feeding program, sex and social rank on growth performance, hoof lesions, and Enterobacteriaceae

shedding in finishing pigs. *Journal of Animal Science* 87, 304–313.
Raj, A.B.M. (2006) Recent developments in stunning and slaughter of poultry. *World Poultry Science Journal* 62, 467.
Raj, A.B.M. and Gregory, N.G. (1990) Investigation into the batch stunning/killing of chickens using carbon dioxide or argon induced hypoxia. *Research in Veterinary Science* 49, 364–366.
Raj, A.B.M. and Gregory, N.G. (1995) Welfare implications of gas stunning pigs. Determination of aversion to initial inhalation of carbon dioxide or argon. *Animal Welfare* 4, 273–280.
Raj, A.B.M. and O'Callaghan, M. (2004) Effects of electrical water bath stunning current frequencies on the spontaneous electroencephalogram and somatosensory evoked potentials in hens. *British Poultry Science* 45, 230–236.
Rosen, S.D. (2004) Physiological insights into Shechita. *Veterinary Record* 154, 759–765.
Roth, B., Imsland, A.K. and Foss, A. (2009) Live chilling of turbot and subsequent effect on behavior, muscle stiffness, muscle quality, blood gasses, and chilling. *Animal Welfare* 18, 33–42.
Rushen, J. (1986) Aversion of sheep to electro-immobilization and physical restraint. *Applied Animal Behaviour Science* 15, 315.
Shimshoney, A. and Chaudry, M.N. (2005) Slaughter of animals for human consumption. *Review of Science and Technology* 24, 693–710.
Tanida, H., Miura, A., Tanaka, T. and Yoshimoto, T. (1996) Behavioral responses of piglets to darkness and shadows. *Applied Animal Behaviour Science* 49, 173–183.
Troeger, K. and Wolsterdorf, W. (1991) Gas anesthesia of slaughter pigs. *Fleischwirtschaft International* 4, 43–49.
Van der Wal, P.B. (1978) Chemical and physiological aspects of pig stunning in relation to meat quality, a review. *Meat Science* 2, 19–30.
Van Putten, G. and Elshof, J. (1978) Observations of the effects of transport on the well-being and lean quality of slaughter pigs. *Animal Regulation Studies* 1, 247–271.
Vieville-Thomas, R.K. and Signoret, J.P. (1992) Pheromonal transmission of aversive substances in domestic pigs. *Journal of Chemical Endocrinology* 18, 1551.
Warner, R.D., Ferguson, D.M., Cottrell, J.J. and Knee, B.W. (2007) Acute stress induced by the preslaughter use of electric prodders causes tougher beef meat. *Australian Journal of Experimental Agriculture* 47, 782–788.
Warrington, P.D. (1974) Electrical stunning: a review of literature. *Veterinary Bulletin* 44, 617–633.
Warriss, P.D., Brown, S. and Adams, S.J.M. (1994) Relationship between subjective and objective assessment of stress at slaughter and meat quality in pigs. *Meat Science* 38, 329–340.
Weaver, A.L. and Wotton, S.B. (2008) The Jarvis beef stunner: effects of a prototype chest electrode. *Meat Science* 81, 51–56.
Webster, A.B. and Fletcher, D.L. (2001) Reactions of laying hens and broilers to different gases used for stunning poultry. *British Poultry Science* 80, 1371–1377.
Wenzlawowicz, M.V., Schutte, A., Hollenbon, K.V., Altrock, A.V., Bostelman, N. and Roeb, S. (1999) Field study on the welfare of meat quality aspects of Midas pig stunning device. *Fleischwirtschaft* 2, 8–13.
White, R.G., DeShazer, J.A., Tressler, C.J., Borcher, G.M., Davey, S., Waninge, A., Parkhurst, A.M., Milanuk, M.J. and Clems, E.T. (1995) Vocalizations and physiological response of pigs during castration with and without anesthetic. *Journal of Animal Science* 73, 381–386.
Wotton, S.B. and Gregory, N.G. (1986) Pig slaughtering procedures: time to loss of brain responsiveness after exsanguination or cardiac arrest. *Research in Veterinary Science* 40, 148–151.

10 Recommended On-farm Euthanasia Practices

Jennifer Woods,[1] Jan K. Shearer[2] and Jeff Hill[3]

[1]*J. Woods Livestock Services, Blackie, Alberta, Canada;* [2]*Iowa State University, Ames, Iowa, USA;* [3]*Innovative Livestock Solutions, Blackie, Alberta, Canada*

Introduction

Euthanasia is Greek for 'good death'. It is accomplished when death results in a minimum of pain, fear and distress to the animal. The avoidance of pain and distress requires the use of techniques that induce an immediate loss of consciousness followed by or in conjunction with cardiac and respiratory arrest that ultimately results in loss of brain function. For persons performing euthanasia, a certain degree of technical proficiency, knowledge and appropriate equipment are required.

Livestock owners and others who derive all or a portion of their livelihood from animal agriculture share a moral obligation to ensure the welfare of animals. Part of these obligations is to make certain animals do not experience unnecessary pain and suffering, even in death. Therefore, when disease or injury conditions arise that diminish the quality of life or create pain and suffering that cannot be effectively relieved by veterinary treatment at a reasonable cost euthanasia is indicated.

Reality is that the situations whereby a person may be required to perform euthanasia are frequently on the farm. In some cases it is an emergency procedure perhaps associated with some type of traumatic event. In others it is a decision based upon one's assessment of a sick animal's locomotory status, prognosis for recovery or perceived suffering. Other indications may include loss of productivity, economic restraints of treatment or a dangerous animal. Regardless of the reason, for the sake of the animal and its caretaker, who may also be the person performing this unenviable task, the ability to induce both a rapid and a humane death is of paramount importance.

This chapter provides guidance on the process of euthanasia, beginning with indications for euthanasia and ending with either disposal of the carcass or processing it for meat. Steps in the process that are covered within this chapter include timeliness of application, animal considerations, human concerns, selection of method, application of technique, methods to ensure a humane death and assessing your euthanasia system.

Factors Affecting Timeliness and Effectiveness of Euthanasia

As someone (first author) who is not only a livestock owner but also a professional who practises euthanasia through my work in emergency response to livestock truck accidents, training and research, euthanasia and death is a very large part of my life. It is a job that has to be done and one I have done many times with little hesitation knowing it is best for the animal. One of the most difficult struggles I have had as a livestock owner was when I had to euthanize my 10-year-old daughter's 4H (youth project) lamb named Posie. She was different from the other animals on the farm as, not only was she a pet that I was personally attached to, but I was also dealing with the emotional distress of my daughter.

As sheep often do, she managed to get her head through two fence panels and consumed a large amount of grain. Though we caught her and another ewe shortly after the incident, the grain overload led to acidosis that ended up going systemic. We treated them around the clock for 3 days and the second ewe pulled out of it, but Posie obviously had consumed the most grain and was slowly dying. Acidosis can take several days to kill an animal, but she was getting progressively weaker and showing no signs of recovery. After the third day, it was obvious she was not going to recover and the winter temperature outside was reaching –25°C.

I struggled with being the one who had to euthanize her, as many other owners do. In my head I debated having someone else come out and do it, struggled with continuing treatment though I knew she was terminal and the consideration of taking the easy way and just letting her die a 'natural' death. When I returned from the barn after shooting her, my daughter, in tears, told me it was the right thing to do – 'it was really hard for us, but the best thing for Posie'. The welfare of the animal must always take precedence over our emotions, no matter how hard it may be.

Maintain a caring attitude

People who are really good at caring for animals often have a difficult time making decisions to euthanize them. Blackwell (2004) has found that it was easier for farm stockpeople to euthanize a sick or injured pig if the farm had a written policy that clearly stated the conditions when an animal should be euthanized. Some examples are fractured limbs, emaciated body condition or a specified number of days with no response to veterinary treatment. The policy helps the stockperson because it makes the difficult decision. It is important to make the criteria for euthanasia strict enough to prevent suffering but not so strict that caring employees are not allowed to nurse sick animals back to health (Grandin and Johnson, 2009). If a caring, good stockperson was told to kill almost every sick pig they would probably quit. The livestock industry needs caring, good stockpeople because they improve animal welfare and help make animals more productive (see Chapter 4). Emotional attachment to an animal is just one of the many factors that affect timeliness, efficiency and effectiveness of euthanasia. Other influences include socio-demographic considerations, environmental influences, psychological factors and management practices. When fully understood, each one of these factors can be managed to ensure timely euthanasia and minimize distress.

Cultural factors

Socio-demographic considerations impact the performance of euthanasia to relieve suffering. They include religion, gender, cultural, personality and age. A person's religion can affect their attitude and beliefs towards euthanasia. There are religions that do not condone the taking of life under any circumstances, others may consider certain species sacred. Men tend to have a more positive attitude towards euthanasia than women, even though both are equally willing to perform the task. Gender also influences the preferred mode of euthanasia (Matthis, 2004). Different cultures have contrasting views on animal welfare, ranging from very high awareness to no awareness at all. Research in the US pig industry has shown that most Americans have a more positive attitude towards euthanasia and Mexican employees have a more negative attitude towards euthanasia. Different personalities and temperaments have different attitudes and comfort levels pertaining to euthanasia. Some people have no feelings with regard to the action whereas others are not comfortable with it under any circumstance believing there are always alternatives. This survey also found that the age of the stockperson also influenced attitude – the older the stockperson, the more negative the attitude (Matthis, 2004).

Experience with euthanasia

Other factors that affect a person's attitude include previous experience, management, level of exposure, caring, and social stigma. A stockperson's previous experiences will influence their attitude and approach towards euthanasia. People who are raised around animals and in a farm environment are more likely to be comfortable with euthanasia than those who were not raised around farm animals. Also, people who have had a previous bad experience with poorly done euthanasia are more likely not to be comfortable with it.

The frequency of euthanasia and related tasks can negatively affect a stockperson's attitude. Studies on both pig farms and in animal shelters have shown that the longer a person has been performing euthanasia tasks, the less willing they become to euthanize (*Swine News*, 2000; Matthis, 2004; Reeve *et al.*, 2004). This also applies to the frequency of the performance of euthanasia and related tasks. This phenomenon explains the difference in attitude of swine barn employees – those in farrowing units, where animals are euthanized more frequently, had a more negative attitude towards euthanasia compared to those in a finishing (fattening) barn, where euthanasia is not as frequent (Matthis, 2004).

Caring and killing paradox

The job of a stockperson is to raise and care for animals. When they have to euthanize them, it creates a 'caring and killing paradox' (Arluke, 1994; Reeve *et al.*, 2005). Many people struggle with this, especially in cases where they have put extensive effort into saving a sick or injured animal. Euthanizing animals is also often morally stigmatized in some societies. This 'tainted' image of euthanizing animals can put added strain on the handlers themselves, as they personalize the stigma.

Many studies have documented that euthanasia can have negative psychological effects on the people who perform it. The majority of studies have focused on animal-shelter workers, with some limited research in pigs (Arluke, 1994; *Swine News*, 2000; Matthis, 2004; Reeve *et al.*, 2004; AVMA, 2007). Stockpersons who constantly euthanize large numbers of animals may handle livestock roughly and apply euthanasia carelessly. They may also experience job dissatisfaction, increased missed work days, depression, grief, frustration, sleeplessness, nightmares, isolation, high blood pressure, ulcers and substance abuse.

Psychology of euthanasia

In large slaughter plants, Grandin (1988) reported that the workers who stun (euthanize) the animals fall into four psychological patterns. They are: (i) caring; (ii) box stapler; (iii) sadists; and (iv) those who carry out religious rituals, such as kosher or halal slaughter. The box stapler does his work efficiently and is emotionally detached. This is the most common psychological pattern. One very skilled stunner operator in a huge plant talked about the weather and golf as he shot 250 cattle/h. He was efficient and never abused the animals. The sadist enjoys causing suffering and must be removed. Manette (2004) and Wichert von Holten (2003) also discuss ways to help people who have to kill large numbers of animals deal with the psychological trauma. This is especially a problem after mass kills for disease control purposes. Some people become totally desensitized and others have soothed their grief by performing a funeral ritual for the dead animals (Manette, 2004).

Reaction to euthanasia procedures will vary among individuals with some experiencing no adverse affects, while others are deeply affected. Experience has shown that the more knowledgeable and aware people are made of the correct procedures and reasons for euthanasia, the more comfortable they become with the job. The greater the level of comfort, the less likely they are to experience stress-related symptoms.

On-farm management of euthanasia

Management will often set the tone for the attitude of the stockpeople. Managers must have a proactive approach to all aspects of animal welfare and demand the same attitude from the staff. A stockperson with a good attitude is essential to ensure proper handling and euthanasia procedures on the farm. Research has revealed sequential relationships between the attitudes of the stockperson towards the animals, the behaviour of the stockperson with the animals, the behavioural response of animals to humans (i.e. fear) and the overall performance and welfare of the animal (see Chapter 4).

When hiring or selecting employees for positions that include euthanasia, it must be ensured that the person is comfortable performing the expected task. It is detrimental to employee moral, safety and animal welfare to force people to perform euthanasia who are not comfortable or competent with the procedure. People may be comfortable performing euthanasia, but they may not be comfortable with the modes of euthanasia used within your operation (i.e. blunt force trauma). This must be discussed before they are expected to euthanize animals. Company policy must be clearly posted and there must be an alternative plan for those people uncomfortable with the designated euthanasia process, as euthanasia must be available 24 h a day.

It is also management's responsibility to ensure that stockpeople are trained in euthanasia. Research has shown that the amount and type of euthanasia training a stockperson receives influences their attitude. When people are provided with comprehensive training that covers all aspects of euthanasia, they will often become more comfortable with the procedures and have a better attitude towards euthanasia (Matthis, 2004; Reeve *et al.*, 2004).

Training not only gives them the skill to perform the act, but also the confidence to make timely decisions on when to euthanize an animal.

Management can also alleviate much of the strain on their stockpeople by rotating the euthanasia task throughout the staff, especially in operations where euthanasia is a regular part of the routine. Managers must keep the lines of communication open with employees and note any changes in behaviour, attitude, frequency of sick days, etc. Support should be provided for employees who request it or appear to need it. This support network should include open lines of communication, task rotation when requested and counselling if required. Management of euthanasia requires effective farm management. Grandin (1988) observed that the farms or slaughter plants that had the best stunning and animal handling practices had a manager who was involved enough to care, but not so involved that he becomes desensitized.

Indications for Euthanasia

Indications for euthanasia of livestock include poor health, disease, injury, loss of productivity, economics and safety. There are three possible treatment options for livestock producers when faced with each of these situations: (i) ship the animal for meat processing IF the animal is fit for transport and human consumption; (ii) provide veterinary treatment; or (iii) euthanasia. Just because there is a chance for recovery does not mean treatment is always the best choice for the producer or the animal. When deciding which option is best, there are several questions a person has to ask in order to make a responsible decision.

Whenever the outcome of disease or injury is in doubt, a veterinarian should be consulted for advice. Simply leaving an animal that is suffering to die of natural causes or, in other words, 'letting nature take its course' is *unacceptable*. Furthermore, it is *not* acceptable to prolong an animal's misery by delaying euthanasia for reasons of convenience (i.e. waiting for the veterinarian's weekly visit). It is important that when euthanasia is indicated, it be conducted in timely manner.

Examples of **poor health or injury conditions** that would justify euthanasia include:

- Emaciation and/or debilitation from disease or injury that may result in an animal being too weak to be transported.
- Paralysis from traumatic injuries or disease that result in immobility.
- Cancerous conditions such as lymphoma and squamous cell carcinomas (cancer eye) of cattle. Severe cancer eyes where the eye as ruptured and surrounding tissue is invaded should be immediately euthanized.
- Disease conditions for which cost of treatment is prohibitive.
- Disease conditions where no effective treatment is known (Johne's disease in ruminants), prognosis is poor or time to expected recovery is unusually prolonged.
- Chronic disease where treatment is unlikely to improve outcome (chronic respiratory disease in cattle and sheep).
- Transmittable diseases (zoonotics) where there is significant threat to human health or other animals (rabies, foot and mouth).
- Fractures of the legs, hip or spine that are not repairable and result in immobility or an inability to stand.
- Emergency medical conditions that result in excruciating pain that cannot be relieved by treatment (trauma associated with road accidents).
- A wound significantly impacting a critical biological function (i.e. major organ, muscle and skeletal systems, brain injury) – one example is severe injuries that expose internal organs.
- Profuse bleeding.
- Large umbilical hernia in a pig that has become injured from scraping the floor.
- Animals that are non-ambulatory for more than 24 h. Green *et al.* (2008) found that dairy cows that remain non-ambulatory for more than 24 h are less likely to recover.

Livestock owners are responsible for the welfare of the animals in their care. Although profit is the motivation of most livestock enterprises, the welfare of animals within those operations must always take priority. Whenever it is clear that the cost of therapy outweighs financial return, euthanasia should be considered a potential option. The cost of treating a sick animal includes: cost of medication, expense of increased labour during treatment, veterinary care and input costs to return the animal to production. Other factors that may enter the decision process might include: (i) Will the animal have to endure a painful and lengthy recovery? (ii) Will the animal be likely to return to normal function post recovery (i.e. restoration of quality of life)? (iii) Can the required care be provided during the convalescent period? (iv) Is the animal likely to suffer chronic pain or immobility following recovery? (v) Will weather extremes create inhumane conditions for this animal during and/or after recovery? (Woods, 2009). The answer to these questions is not always clear. None the less, they should be part of the decision-making process whenever livestock owners are faced with the choice of treatment, slaughter or euthanasia.

There are also some indications for euthanasia not related to health, welfare or production. When an animal has become dangerous to handlers, family members, staff or other animals it must be shipped for meat processing or euthanized. No matter how productive the animal is, owners or management are responsible for the safety of all those on the farm or within the facility. It is irresponsible and the owner may be liable if the animal is sold to another producer knowing the animal can hurt someone (see Chapter 5 'Bull Behaviour').

Time for recovery from illness or injury

One of the biggest challenges is determining: *How long should an animal be given to recover?* Industry literature and guidelines would suggest that sick animals should show evidence of significant improvement within 24–48 h from the onset of treatment. Depending upon the specific disorder encountered, most animals will exhibit evidence of a positive response to therapy within 24 h. Rarely do animals recover when they do not show progress within the first 36 h following treatment. Recovery time for animals that experience injuries may be more prolonged and harder to estimate, so they must be evaluated on an individual basis.

In cases where animals are unfit for transport or not acceptable to be processed for meat, it is the responsibility of the producer to euthanize the animals. Examples of this would be animals that are still within the required drug withdrawal period, animals that would experience further pain and suffering during transport, animals that have high fevers or ones in which illness may lead to the condemning of the carcass (e.g. cancer eye).

Most livestock are raised for production purposes whether for raising offspring or providing meat, eggs or dairy products. When the cost of feed, veterinary care, housing and labour exceeds productivity income of an individual animal it is the responsibility of the producer to remove the animal from the production cycle. Producers should cull the animal in a timely manner while the animal is still fit for transport and processing. This will allow for the owner to recover some the cost and maintain good animal welfare. In cases though where the animal is at risk and not fit to transport, on-farm slaughter or euthanasia are the only two options.

Recognition of pain and suffering that may indicate the need for euthanasia

Pain and suffering in livestock are primary indications for euthanasia, but can be easily misinterpreted. Prey animals instinctively avoid expressions of pain in order to evade the notice of predators. Cattle, for example may simply become less responsive or depressed and lame animals will adjust their gait or posture to mask evidence of lameness (Chapters 1 and 6). Predators, on the other hand, freely express signs of pain or discomfort. Consider the reaction of a dog that has inadvertently had its foot stepped on. The response is verbally loud, its flight reaction swift, and on occasion punctuated by aggression.

This is an important distinction, since failure to conduct euthanasia procedures in timely fashion is often related to a misinterpretation of an animal's response to pain and suffering. A survey of respondents on euthanasia of pigs found that sick pigs that appeared to be in pain were easier to euthanize than pigs that showed no signs of pain (Matthis, 2004). The survey also revealed that an overwhelming percentage of people agree that it is more humane to euthanize a sick/injured pig that appears to be in pain than to let it suffer. Experience and education of the stockpeople are important for accurate assessment of animal behaviour, particularly when it is necessary to distinguish distress from normal behaviour.

There are several behavioural signs that can alert you to the fact that an animal may be in pain. Any of these indications may appear alone or in conjunction with other signs and may overlap with signs of distress. If you are unsure if the animal may be experiencing pain or why the animal is in pain, consult an experienced stockperson or professional to assist in assessing the situation.

Indications of pain may include:

- inability or unwillingness to rise;
- inability or unwillingness to walk;
- reluctance to put weight on a limb;
- protection of the painful area;
- vocalization, especially when animal moves or painful area is touched;
- open-mouth breathing;
- arched or hunched back;
- abdomen tucked up;
- drooping head and/or ears;
- tail uncurled (pigs whose tails have not been docked);
- lack of interest in food or water;
- lack of interest in surroundings;
- stays away from other animals;
- does not respond to other animals attempting to interact with it;
- does not respond when touched or prodded;
- standing in a rigid position;
- shivering;
- trembling or profuse sweating;
- animal licks or uses a limb to poke the painful area;
- scratching or shaking affected area;
- kicking or biting at abdomen;
- gets up and lies down frequently;
- walks in circles;
- excessive rolling (horses);
- grunting;
- hiding in bedding;
- licking;
- biting;
- change in personality;
- change in eye brightness;
- glassy eyes, dilated pupils;
- reduced suckling;
- tail flicking;
- unsettled;
- inability to get comfortable;
- their head is turned back;
- lying on their belly when lying down;
- increased blood pressure and heart rate; and
- rough hair or coat that does not lay flat (easier to see in animals with short hair).

This list includes both chronic and acute indicators of pain and distress. See Chapters 1 and 6 for indicators of acute pain after procedures such as dehorning and castration.

According to Moberg (1985) an animal experiences distress when the biological cost of coping with the stressor(s) diverts resources away from other primary biological functions (growth, reproduction, etc.). Distress behaviour includes, but is not limited to, stress vocalization, attempts to flee, aggressive behaviour, rapid or irregular breathing, mouth breathing, salivation, struggling, urination, defecation, rapid heart beat, sweating, dilation of the pupils, tremors/shaking, freezing/standing completely still, grinding teeth and reduced or increased suckling. Just like pain, any of these indicators may appear alone or in combination with others.

It is not uncommon to hear stockpeople say that an animal is 'just not acting like itself' when initially diagnosing a problem. Knowing your animals and being cognizant of their behaviour will assist in early detection of disease or illness, which in turn will allow for an increased chance of recovery or the minimizing of suffering through timely euthanasia. Having employees who can 'read' animals is an asset to the operation.

Pre-euthanasia Considerations and Preparations

It is important that stockpeople do all they can to minimize anxiety, fear, pain and distress for the animal to be euthanized. If the animal to be euthanized is ambulatory and able to be moved without causing distress, discomfort or pain, it may be moved to an area where the carcass would be more easily reached by removal equipment. The use of aggressive means to force an animal to drag itself or dragging of a non-ambulatory animal is not acceptable and it is a violation of both international and many national laws. In cases where movement may increase distress or animal suffering, the animal should be euthanized first, and moved following confirmation of death.

Restraint methods

If restraint is necessary it must occur for the shortest time possible and use the least stressful means available. When choosing a method of restraint it must be safe for handlers and as stress free as possible for animals. If small animals are being restrained by hand for penetrating captive bolt gun (PCBG), extreme caution must be exhibited to avoid shooting the hands or legs of the person performing the restraint. An animal must be restrained so that it cannot strike out or escape if something goes wrong. It must also be possible to remove the carcass from the area following euthanasia.

Proper restraint for pigs depends on the size and condition of the animal. For neonates and early nursery pigs, the animals should be picked up utilizing two points of contact (i.e. leg and flank) either held firmly or placed in an appropriate restraint device (such as a body sling). Under no circumstances should the animals be picked up by one point of contact (i.e. leg) and swung, thrown, etc. (OIE, 2009a). At a minimum, larger animals should be penned. The use of a properly designed restrainer chute minimizes the opportunity for movement and reduces the likelihood of misapplication of the euthanasia method. Snout snares for pigs should be designed for the age and weight of the animal, and be constructed of rope, or smooth, round cable. They must not cut or injure the animal. Snares MUST NOT be used to drag or lead an animal and euthanasia must be performed immediately after the snare is applied. Restrainer chutes (races and crushes) must be designed for age and weight of the animal, and the chute must provide adequate access to the animal to allow for the necessary procedure (i.e. euthanasia). If animals are euthanized in a restrainer chute, the design must allow for animal removal once it is confirmed dead. There are many useful pictures in Sneddon *et al.* (2006).

For cattle, squeeze chutes are acceptable, but animals can be very difficult to remove. A headgate (stanchion, head bale) on the end of a single-file race will often work well. Tame ambulatory animals can be led to the place where euthanasia will be performed with a halter (head collar). Horses and sheep can be restrained in restraint chutes or with halters. If you use a halter, make sure it has enough lead to untie following euthanasia so the animal will not be hanging. For elk/farmed deer a restraint chute (race) designed for their species is recommended. On the farm, poultry can be restrained with the use of cones or by hand.

Euthanasia of fractious animals (such as an aggressive bull or cow) may require capture in a race where the animal can be tranquilized and released to a pen that is accessible by removal equipment. Once the tranquilizer has taken effect and the animal is immobilized, it may be euthanized by an appropriate method that is safe for handlers and assistants.

The safety of other animals in the immediate area must always be considered. If necessary, other animals should be removed from the immediate area to avoid the chance of ricochet, getting hit by a stray bullet or injury due to startling. Biosecurity is also a safety issue for other animals as the sick animal may release contaminated body fluids or blood following euthanasia.

Determination of the Most Appropriate Euthanasia Method

Selection of the best method of euthanasia should include:

- concerns for human safety;
- animal welfare;
- emotional comfort with procedure by the person performing the task as well as bystanders;
- ability to restrain the animal for proper application of the procedure;
- skill of the person performing the procedure;
- cost;
- rendering and carcass disposal considerations; and
- potential need for brain tissue for diagnostic purposes.

Some methods of euthanasia are more costly than others. Tools such as captive bolt guns have a high initial cost, but are relatively inexpensive to use. Anaesthetic overdose has a relatively high cost since a veterinarian must perform the procedure and carcass disposal is complicated by drug residue. The number of animals euthanized within the operation is also a consideration. If you only euthanize an occasional animal, cost is not a significant factor compared with a larger operation that euthanizes animals more frequently or may be in the position of conducting mass killing of livestock.

Each method of euthanasia requires a certain level of skill and training. This is an extremely important consideration as the skill and efficiency of the person performing the task is vital for the proficiency of the task. Improper use of the tool will not only jeopardize the safety of the handler, but also the welfare of the animal. *The vast majority of failed euthanasia is the direct result of human error.*

Individual preference of methods

Individual preference for the method of euthanasia must be considered. People are often more comfortable with one means of euthanasia rather than another. The more comfortable a person is with the chosen acceptable method, the more proficiently they will perform the act. Factors that can determine the level of comfort include religion, background, gender, education/training, previous experience and aesthetics.

Each method of euthanasia has its own degree of aesthetic displeasure. Some means have a more peaceful appearance (i.e. anaesthetic overdose) than others (i.e. blunt force trauma). Aesthetic effectiveness may be affected by blood loss and physical trauma to the animal. Consideration must be given to the comfort level of the person performing the task and that of bystanders (i.e. general public or media). Just because one method is less aesthetically appealing than another does not mean it is less humane. Barbiturate overdose is very effective, peaceful and humane when it is done correctly, but it is not practical on most livestock operations. Animals that are euthanized with anaesthetics cannot be used for food for either people or animals. In many countries, rendering plants may not accept animals that are euthanized with barbiturates.

There are also legal restrictions with certain methods of euthanasia. In many countries, lethal injection can only be delivered by or under the supervision of a licensed veterinarian because barbiturates are a controlled substance and have potential drug abuse. When prescription drugs are used for euthanizing animals, you should find out what the laws are in your country, state or province. In most countries the use of firearms or a captive bolt does not require supervision by a veterinarian. However, some countries require the gun to be registered and the operator must hold a valid firearms licence. In other countries, people are not allowed to own firearms.

Biosecurity considerations include the spread of further disease and clean up of the euthanasia site. Euthanasia techniques that must be followed by exsanguination lead to possible contamination of facilities with blood. Contamination of the area where euthanasia is conducted with brain tissue is a possible risk with gunshot, captive bolt gun and blunt force trauma. Other body fluids may also be released during the euthanasia process, which can contaminate facilities.

When determining which of the various methods of euthanasia should be used, consideration must be given to the type of animal, its size and age. Manual blunt force trauma is only recommended for poultry, piglets and infant lambs and kids. Cervical dislocation is only acceptable for small poultry such as chickens. The skulls of boars, sows, bison and mature bulls are much thicker and more difficult to penetrate with firearms and captive bolt guns, thus limiting the selection of effective tools.

The mode of disposal of the carcass must also be considered in making euthanasia decisions. Such concerns include local and national laws regulating the disposal of dead livestock. If it is likely that a carcass may be consumed by scavengers (buzzards, coyotes, etc.) the animal cannot be euthanized with drugs. Animals that are going to require diagnostic evaluation after death (such as for rabies diagnosis) must be euthanized in a way that will not damage or destroy the brain.

Induction of death

Death is a process and does not occur immediately. Animals are first rendered insensible and then the body begins to die as the brain stops, the heart quits beating, the lungs stop breathing and the blood quits circulating. There is no such thing as immediate death of the entire body. For example, gunshot will instantly destroy the brain, but the heart may beat for several minutes. However, a well-placed gunshot has started the process of death and the animal is instantly rendered insensible to pain and will not recover.

Death may be induced by one or more of the following mechanisms: direct depression of the central nervous system (CNS), hypoxia and physical disruption of brain activity. *Direct depression of the CNS* is achieved through the injection of an overdose of barbiturates. Though not commonly used in a livestock-production setting, inhalant anaesthetics such as ether or halothane will also induce death through depression of the CNS, but they have significant human safety concerns. *Hypoxia*, or lack of oxygen, is achieved by exposing animals to high concentrations of gases such as carbon dioxide (CO_2) or argon, or through rapid blood loss (exsanguination). Gunshot, blunt force trauma or captive bolt stunning devices induce death by *physical damage to the CNS that results in a disruption of brain activity*. Death occurs as a result of respiratory and cardiac failure (AVMA, 2007).

Lethal Injection

For on-farm euthanasia of livestock, one may choose lethal injection with barbiturates. This requires a veterinarian to be available to perform the intravenous injection, which can make its use restrictive. Other restrictions include the need to restrain the animal, safety of the handler especially with large animals, disposal of livestock following death, higher cost and practicality in the euthanasia of several animals. Both the American Veterinary Medical Association (AVMA) and the British Veterinary Medical Association (BVMA) have specific guidelines. The AVMA guides are available free on the Internet and may be accessed by typing 'AVMA Guidelines on Euthanasia' into Google. To access the British guidelines will require either a subscription or membership in the BVMA.

When lethal injection is not available or practical, euthanasia options include gunshot, captive bolt gun, CO_2, electrocution, blunt force trauma and cervical dislocation. Carbon monoxide and maceration are both considered acceptable means of euthanasia. Due to high human safety and animal welfare issues with carbon monoxide and restriction of maceration to only day-old poultry and embryonated eggs, they will not be included in this chapter.

Firearms

Gunshot is an acceptable means of euthanasia for all livestock. In the USA, it was the most popular method for on-farm euthanasia (Fulwider *et al.*, 2008). Gunshot euthanizes by mass destruction of the brain. The degree of brain damage inflicted by the bullet is dependent upon the firearm, nature of the bullet (or shotshell) and accuracy of the shot. For euthanasia purposes, handguns are limited to close-range shooting of less than 5–25 cm (2–10 inches). Shotguns are appropriate for a distance of 1–2 m (1–2 yards) and rifles for long-distance shots.

The correct selection of ammunition is vital for successful euthanasia. The energy of a bullet and its destructive capacity are normally described in terms of muzzle energy (i.e. joules or foot pounds of force). It is recommended that a minimum of 407 J (300 ft lb) of muzzle energy be available for firearms used in the euthanasia of animals up to 180 kg (400 lb), and at 1356 J (1000 ft lb) of muzzle energy for animals heavier than 180 kg (400 lb) (USDA, 2004).

Shotgun 28, 20, 16, 12 gauge or .410

Shotguns are very effective for euthanasia of livestock. They are best used at a distance of 1–2 m (1–2 yards). The 20, 16 and 12 gauge can be used on all weight and species classes. The smaller 28 and .410 gauge shotguns should not be used on larger and mature animals due to skull mass. Number 4, 5 or 6 birdshot is appropriate ammunition for close range only because, as birdshot leaves the barrel of the gun, it begins to disperse or spread, lessening impact and destruction as the distance from the animal increases. Slugs (special shotgun shells) are considered the best choice for euthanasia as a slug is a solid mass that does not disperse as it leaves the barrel. In a situation where animals are loose or where the handler is unable to get close to the animal, a shotgun with solid slugs should be used.

Rifles

The .22 calibre long rifle is one of the most popular firearms, but with an average muzzle energy of only approximately 100 ft lbs, it does not meet the current energy requirements for euthanasia of livestock. Animals up to 180 kg (400 lbs) require a minimum of 300 ft lbs, while animals heavier than 180 kg (400 lbs) will require a minimum of 1000 ft lbs to ensure death (USDA, 2004). The .22 must be followed by either pithing or bleeding. A .22 should NEVER be used on bison, elk or any other large domestic species, because it is not reliable in causing a humane death. If a .22 is utilized, a long rifle, round nose, solid point lead bullet should always be used.

For longer distances of up to 274 m (300 yards) the recommended rifle cartridges would include those shown in Table 10.1. Note the reduction in muzzle energy as it reaches 274 m (300 yards).

Table 10.1. Specifications for rifles (source: USDA, 2004).

Cartridge	Muzzle energy (J) (ft lb in parentheses)	Muzzle energy at 274 m (300 yards) (ft lb in parentheses)
.357 Magnum (rifle)	1593 (1175)	457 (337)
.223 Remington	1757 (1296)	778 (574)[a]
30-30 Winchester	2579 (1902)	883 (651)
.308	3590 (2648)	1617 (1193)
30-06 Springfield	3852 (2841)	1973 (1455)

[a]The .223 Remington is used with a 5.56 mm NATO rifle cartridge.

Larger calibre rifles such as the .308 are not appropriate for close-range shots due to the chance of pass through.

Handguns

Handguns can be used for close-range euthanasia at a distance of 5–25 cm (2–10 inches) with the appropriate round nose, lead bullet ammunition. Common handgun cartridges that exceed the recommended 300 ft lb of muzzle energy include those shown in Table 10.2.

Table 10.2. Specifications for handguns (source: USDA, 2004).[a]

Cartridge	Average muzzle energy (J) (ft lb in parentheses)
.40 Smith and Wesson	553 (408)
.45 automatic	557 (411)
.357 Magnum	755 (557)
.41 Remington Magnum	823 (607)
10 mm automatic	880 (649)
.44 Remington Magnum	988 (729)

[a]When higher muzzle energy is required, it is recommended you use a rifle.

Application of gunshot for euthanasia

FIREARMS MUST NEVER BE HELD FLUSH TO THE ANIMAL'S HEAD OR BODY. The pressure within the barrel when fired may cause the barrel of the gun to explode. Ideally, the gun should be angled so the bullet follows the angle of the neck or the spine. When angled properly the bullet will travel through the brain and ideally end up near the top of the spine or brain stem area. The angle of the firearm is important for safety reasons since it is desired that the bullet remain within the body. Even though the World Organization for Animal Health (OIE) recognizes a shot to the neck as acceptable for killing, it is only in a situation of mass killing for disease control purposes, not for euthanasia of individual animals. The AVMA (2007), the Humane Slaughter Association (2005) and the European Food Safety Authority (EFSA) (2004) does not recommend gunshot to the heart or neck as it does not immediately render the animal unconscious.

Cattle and calves

There are two acceptable points of entry for gunshot in cattle – the front of the head or behind the poll of the animal. If the animal is lying down, has horns or has a thick skull, behind the poll is the better choice over the front of the head (Gregory *et al.*, 2008).

The frontal target area is high up on the head of the animal, NOT BETWEEN THE EYES. An X can be made on the animal's head by drawing imaginary lines between the inside corner of the eye to the horn on the opposite side (or to the top of the opposite ear in an animal without horns). The shot is then placed approximately 2 cm (1 inch) above the intersection on the X – NOT in the centre of the X (Fig. 10.1). The firearm should be positioned so that the muzzle is perpendicular to the skull and the bullet will enter the front of the head and travel towards the tail of the animal. There may be some differences in location of the shot, based on the skull shape and horn mass of an animal.

Pigs

The anatomical site in pigs is slightly more complicated to identify because the pig's brain is smaller relative to the size of its skull. As a pig ages, the skull continues to grow more relative to the brain, which makes the target even more difficult to accurately identify. This is complicated further by the fact that different breeds of pigs can have very different skull shapes and the shape will change as the pig matures. Mature sows have a large sinus cavity located right in front of the brain, setting the location of the brain even further back into the skull, thus requiring deeper penetration with a bullet or penetrating captive bolt. Similarly, boars form a ridge down the front of their skull as they mature, which also complicates penetration to the brain. Due to human safety issues with restraint and ricochet, gunshot is not a practical means of euthanasia for piglets under 5 kg (12 lb).

The ideal location for shooting a market-weight pig (100–135 kg) is approximately 2.5 cm (1 inch) above their eyes, in the middle of their forehead (Fig. 10.2). For older boars and sows, the shot should be located 3–4 cm (1.5–2 inches) above eye level and just to the side of the ridge on their skull. Older, mature animals are best euthanized with shotguns using slugs.

Horses

The brain of the horse is located high on the forehead so the shot will need to be slightly above the

Fig. 10.1. Correct location for shooting cattle with a captive bolt or a firearm. A position slightly above the X is often more effective (from Shearer, 1999).

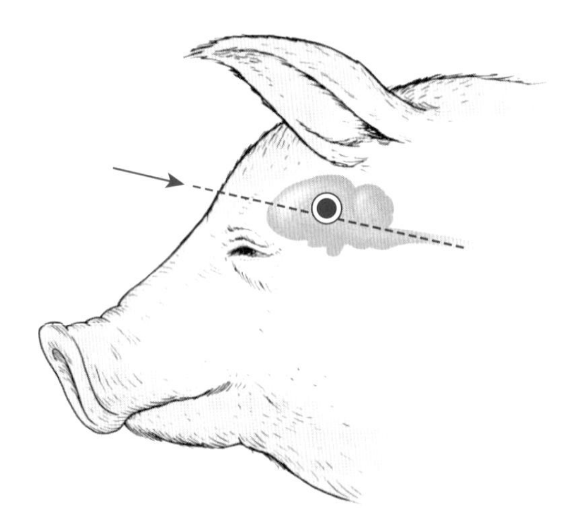

Fig. 10.2. Correct position for shooting swine with a captive bolt or a firearm. Many old diagrams show a position that is too low (diagram by J.K. Shearer).

intersection of an X drawn on the forehead with each line beginning at the inside corner of the eye and progressing to the opposite ear. The specific anatomical site is approximately 2 cm (1 inch) above the centre of the X (Fig. 10.3). It is also important to note that horses may lunge forward or rear up when shot, so the shooter must beware of this when shooting a standing animal and position themselves accordingly.

Sheep and goats

There are three acceptable points of entry on sheep: the front of the head, the top of the head and the back of the poll. When using the landmarks for a forehead shot, the gun should be positioned so that it enters 2 cm above the eyes. Euthanasia of sheep and goats by gunshot can be difficult as different breeds have different skull shapes which will require some alteration to the target area for gunshot.

The ideal shooting position for sheep is a shot to the top of the head. The shot should be placed in the mid-line of the skull, pointing straight down at the throat. This shot will take the projectile through the brain. A shotgun is best to use for this shot to avoid the projectile passing through an animal and injuring other people and animals.

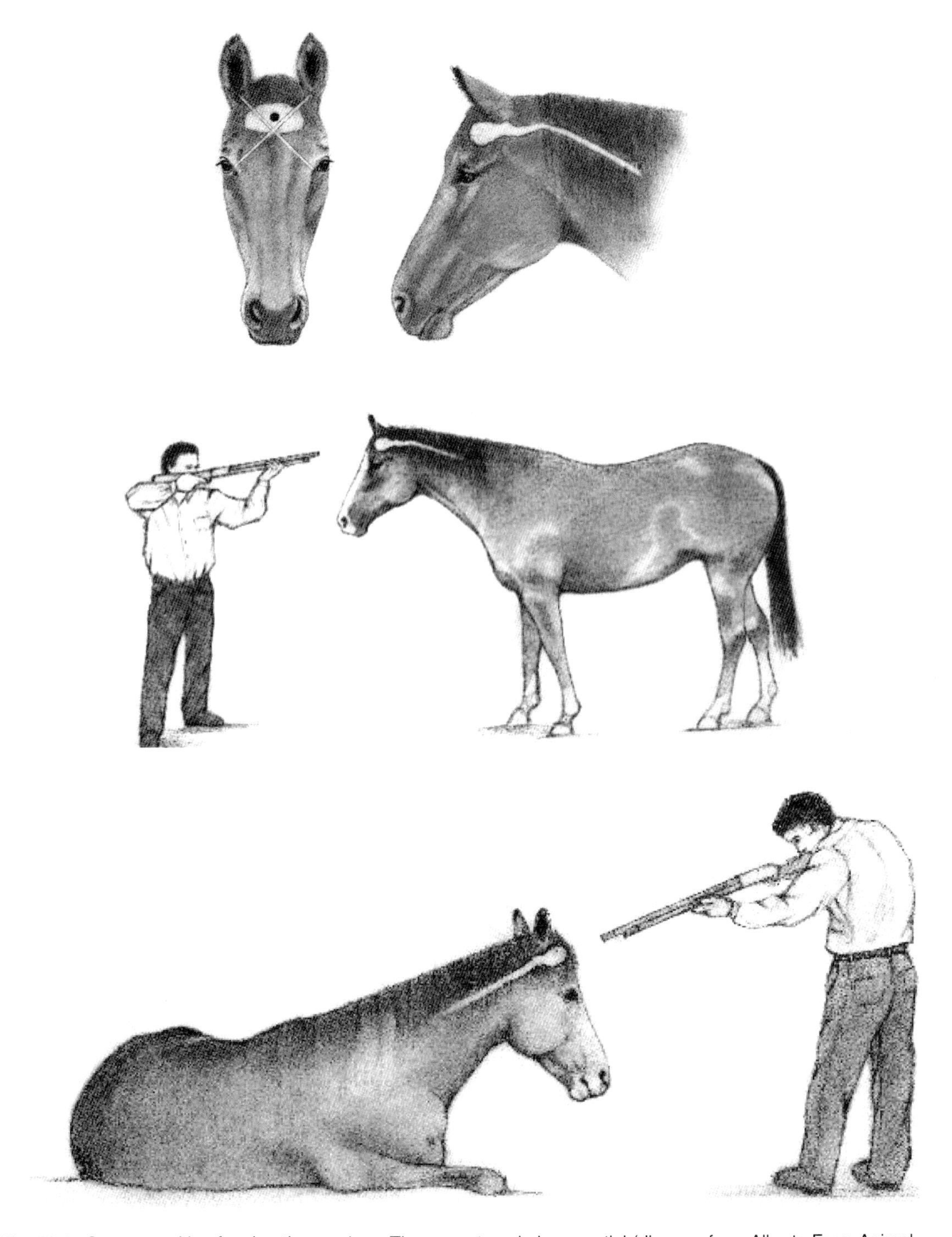

Fig. 10.3. Correct position for shooting equines. The correct angle is essential (diagram from Alberta Farm Animal Care, Alberta, Canada).

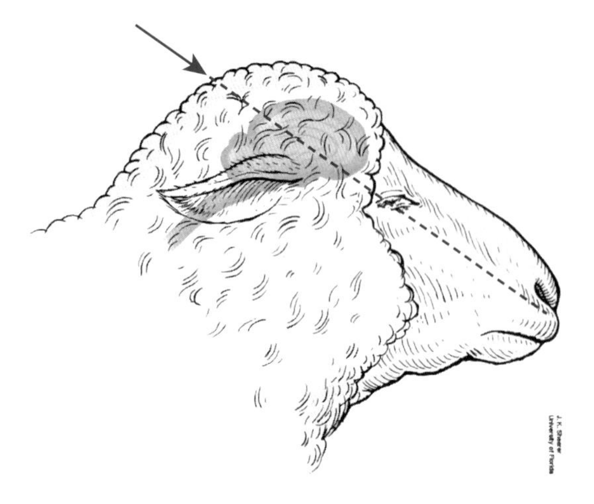

Fig. 10.4. Correct location for shooting a lamb or a goat. Some sheep breeds have extremely thick skulls and shooting from the rear position may be more effective. The correct angle is essential to ensure hitting the brain (diagram by J.K. Shearer).

In horned sheep the most effective shot is behind the poll. The projectile will not have to pass through the extra bone mass created by the horns that cover the area in front of the brain. It is best to make this shot with a shotgun, to eliminate the chance of the bullet passing through and into unintended objects or in the direction of other animals or people. The shot should be aimed towards the throat and mouth (Fig. 10.4).

Poultry

Gunshot is an acceptable means of euthanasia for poultry when the shot is directed perpendicular (at a right angle) to the frontal bone of the bird. Though acceptable, this is not a practical method due to the small size of the bird and the high chance of ricochet.

Farmed/elk deer

A deer's brain is situated high in the skull. The shot will need to be 2 cm above the intersection of an imaginary X drawn on the forehead, with each line starting at the inside corner of the eye and the base of the antler on the opposite side (Fig. 10.5).

Farmed bison

Bison MUST be shot with a high-powered rifle or a handgun that is a .357 Magnum or larger. Due to the massive skull structure of a bison, the best place

Fig. 10.5. Correct location for shooting deer with a captive bolt or firearm (diagram from Canadian Agri-Food Research Council).

to position the firearm is on the side of the head between the base of the horn and the ear. Ideally, the bullet will be delivered on a level line with the base of the horn when the head is carried in a normal walking position. It is best if the shot can be delivered when the bison is standing still to allow the bullet to travel into the skull cavity.

If the shot must be taken from the front, the point of entry should be right above the intersection of the X made between the eye and the base of the opposite horns (Fig. 10.1 same as for cattle). A high-powered rifle is required. It is best to deliver this shot with the bison's head lowered down towards the ground. This can be achieved by offering the bison feed on the ground. Ricochet is a major hazard of shooting a bison in the front of the head.

Penetrating Captive Bolt Gun (PCBG)

Penetrating captive bolt devices consist of a steel bolt, with a flange and piston at one end, which is housed in a barrel. On firing, the expansion of gases propels the piston forwards and forces the bolt out of the muzzle of the barrel. The bolt is retained within the barrel by a series of cushions that absorb the excess energy of the bolt and keep it within the barrel. The bolt is retracted back into the gun either automatically or manually depending upon the design of the PCBG. To be effective, the PCBG must have sufficient bolt velocity. For cattle, the PCBG will be most effective with volt velocities ranging from 55 to 58 m/s. A slower bolt velocity of 41–47 m/s was less effective (Daly *et al.*, 1987). Other important research studies on captive bolt devices are Blackmore (1985), Daly and Whittington (1989) and Gregory *et al.* (2007).

PCBGs are powered by gunpowder or compressed air which must provide sufficient energy to penetrate the skull of the species on which they are being used. Accurate placement, energy of bolt (bolt velocity) and depth of penetration determines effectiveness. Bolt velocity is dependent on maintenance (in particular, cleaning) and storage of the cartridge charges. Many manufacturers of captive bolt guns can also supply a test stand for measuring bolt velocity. Use of this device ensures that the captive bolt is working correctly. Cartridges must never be stored in a damp location as damp cartridges are often less effective. Soft-sounding shots that may be caused by damp cartridges are less effective (Grandin, 2002; Gregory *et al.*, 2007).

PCBGs come in 9 mm, .22 calibre and .25 calibre. Styles include in-line (cylindrical) and pistol grip (resemble a handgun). Pneumatic PCBGs (air powered) must be supplied with sufficient air pressure and air volume to operate and for the most part are limited to slaughter plant environments.

A new era in captive bolt technology

In 2007, a new captive bolt gun was designed to meet the needs of stockpeople for on-farm euthanasia of livestock. Unlike traditional captive bolt guns which were designed for only stunning animals and requiring a secondary step such as pithing or exsanguinations, the euthanizer unit is more effective and a secondary step will not be required. The gun has a longer bolt length and has sufficient force to kill all species and weight classes of livestock. This device is designed to cause sufficient destruction of the brain (cerebral cortex) to cause instantaneous loss of consciousness and destruction of the brainstem (specifically the medulla oblongata), leading to the loss of life functions with no chance of recovery. This gun significantly reduces the cost of the unit and maintenance requirement for farms that house multiple age/weight of animals (i.e. farrow to finish swine farms), farms with mixed species of livestock or for people that deal with a variety of species and weight classes of animals such as veterinarians, animal control, auction marts, transporters and livestock rescue crews. A single gun will be effective on both large and small animals.

PCBGs should only be applied by trained users. The shooter should wear both protective ear and eye gear. The PCBG must never be cocked until the shooter is ready to fire and the shooter should make sure that the safety catch or other safety mechanism which will prevent accidental firing is set until it is ready to be discharged. The muzzle end of the captive bolt should always be kept pointed towards the ground. In-line, cylinder-shaped guns are often the cause of injury due to holding the gun with the bolt end up when the handler confuses the discharge end with the top of the gun.

Animals that are upright and mobile may be difficult to safely euthanize with a captive bolt without restraint. The shooter should wait until the animal stops moving its head before firing and should not chase the animal's head. One well-placed shot should readily cause unconsciousness.

If there is any doubt, a second shot should be applied. If a second shot is required, it should be administered in a different location from the first shot. If the first shot was correctly placed, the second shot should be applied slightly above and to the side of the first hole. If the first shot was incorrectly placed, the second shot should be placed in the correct position.

It is highly recommended that high-powered captive bolt guns designed for on-farm euthanasia be utilized, as traditional PCBGs designed for stunning may only stun the animal and therefore may require a secondary kill step. Secondary kill steps include exsanguination and pithing. When applying the PCBG, it MUST be held firmly against the animal's head. If it is not held flush to the animal's head, the bolt may not penetrate the skull. It may also cause the bolt to slide up the head of the animal. The animal may be approached from the front. However, it is likely to be easier to approach from the rear and off to one side, thus keeping out of the line of sight of the animal. The handler can then reach over the poll of the animal, placing the PCBG on the front of the head. There are two basic types of trigger mechanisms. One type causes the gun to fire when it is tapped on the head and the other type has to be pulled by the operator to fire the gun. The recoil of the PCBG will vary depending upon calibre, buffer configuration, manufacture and the cartridges used. If necessary, the handler should use two hands when firing the PCBG. The style of gun will also make a difference in recoil and noise associated with firing.

Maintenance of captive bolt guns

The most common cause of misfires and ineffective kills is poor maintenance (Grandin, 1998). PCBGs must be cleaned and maintained in order to operate effectively. PCBGs should be cleaned and inspected at the end of each day of use. PCBGs that are not used regularly still need to be cleaned and oiled. All parts must be put back into the PCBG and lost or damaged parts replaced before use. In the case of a misfire, the stunner breech should remain closed for a minimum of 30 s in case the problem is a 'hang fire' due to slow primer ignition.

Refer to the manufacturer's instruction manual for cleaning instructions. Make sure the PCBG is unloaded before cleaning. PCBGs and cartridges must be stored in a dry area. Exposure to humidity will affect both the gun and the effectiveness of the ammunition. Box 10.1 has tips for maintaining captive bolt guns.

Location of shot for captive bolt

The current recommendation for the point of entry on cattle is the same frontal location as gunshot (personal experience of the first author from commercial use suggests that the captive bolt gun needs to be placed approximately 2–5 cm above the intersection of the X due to trajectory differences between a free bullet and a captive bolt) (Fig. 10.1). There is also variation in placement, based on the length and shape of the skull. As an example, Holstein dairy cows have much longer skulls than Hereford beef cows. For animals with heavy skulls such as water buffalo or Brahman cattle, shooting in the hollow behind the poll can be effective (Gregory *et al.*, 2008). The angle must be precise to ensure that the brain is penetrated.

As with cattle, current recommendations for the point of entry on pigs is the same frontal location as gunshot (Dr Lyndi Gilliam, Oklahoma State University, personal communication, 2009). Commercial application, though, has shown that the captive bolt gun needs to be placed approximately 2–5 cm

Box 10.1. Tips for maintaining captive bolt guns (source: Bildstein, 2009).

- It must be cleaned every day if it has been fired. Treat it the same way as a firearm.
- Use gun oil and gun cleaner. Do not use machinery grease, white oil or WD-40.
- Do not get the captive bolt gun wet.
- Use a wire brush to polish the piston end of the bolt every time the stunner is cleaned.
- Replace the bumpers if they are cracked or stiff. This will prevent the bolt from sticking in the head.
- Number the guns and do not mix the parts. Each gun wears differently.
- Have plenty of spare parts for replacement of worn or broken parts.
- Store cartridges in a dry place.

higher then the placement for gunshot due to trajectory differences. PCBG placement is difficult with pigs because, in order to achieve the ideal projection path into the brain, the PCBG must be slightly angled so the bolt travels straight ahead – towards the tail of the pig. If the PCBG is angled too much it may not penetrate the head, but instead ride up the skull. Also, as a pig ages, the skull continues to grow but the brain does not, making the target even more difficult. Mature sows have a large sinus cavity located right in front of the brain, setting the location of the brain even further back into the skull requiring deeper penetration. Boars form a ridge down the front of their skull as they mature, making penetration difficult. Even though the point of entry is the same as a gunshot, the captive bolt gun should be placed slightly to one side of the ridge.

On the farm, horses must be restrained when applying the captive bolt gun, which makes its use very limited. When shot, horses often make a small forward movement (leap into the air) making it dangerous for the handler to stand in front of the horse, holding the gun on the horse's head (Fig. 10.3). The captive bolt gun is best used on horses that are down on the ground and immobile, or restrained in a chute as a human safety precaution. The point of entry will be the same frontal location as with gunshot.

For sheep the point of entry is the highest point/top of the head, the same as with firearms. For goats the captive bolt is best placed behind the poll of the animal. Captive bolt guns are acceptable only on large, mature turkeys with the shot applied to the head while the bird is restrained by the beak. Appropriate placement for farmed elk and deer is the same frontal location as gunshot. Captive bolt guns are not recommended for euthanasia of bison.

A small captive bolt gun is available for euthanizing broiler chickens. To be effective, it has to have a 6 mm bolt driven with an air pressure of 827 kPa. The penetration depth has to be 10 mm (Raj and O'Callaghan, 2001). The gun must be placed on top of the animal's head.

Controlled Blunt Force Trauma

Controlled blunt force trauma induces euthanasia by physical disruption of the brain. A blow to the head can be an effective stand-alone means of euthanasia to small animals with thin craniums. A single, sharp blow must be delivered to the central skull bones with sufficient force to produce immediate depression of the CNS and destruction of brain tissue, without breaking open the skull.

Acceptable tools for controlled blunt force trauma include cartridge and pneumatic non-PCBGs. Non-PCBGs have a metal, blunt, flat or mushroom-shaped bolt and are powered by compressed air or blank cartridges. They are fired from a gun held to the animal's head at a right angle to the skull and result in concussion of the brain resulting in brain haemorrhage, shearing strains and stresses within the brain and tissue deformation.

Non-PCBGs with a convex mushroom-shaped head are classified as controlled blunt force trauma and can be used as a stand-alone means on pigs, lambs and kids under 15 kg (33 lb) and poultry. When used on animals larger than the ones listed above, it must be accompanied by a secondary action such as exsanguination (AVMA, 2007). Effectiveness is also affected by animal size, skull structure and mass, as well as skill of the person using the gun.

When using a non-PCBG on pigs, poultry and lambs, the animal must be restrained and the gun applied to the top of the head. For goats, the gun should be placed on the back of the head behind the poll.

Standard mushroom-head non-PCBGs for on-farm use are designed to only stun young calves and are only acceptable for use when followed immediately by exsanguination. All other means of blunt force are not acceptable for cattle. Controlled blunt force trauma is not an acceptable means of euthanasia for horses, elk and bison. There is a smaller margin of error with most mushroom-head guns. The gun must be placed perfectly perpendicular on the head to be effective. Thick hair on breeds such as Hereford cattle may make a non-PCBG less effective.

Maintenance requirements for the non-PCBGs are the same as the PCBG.

Flat-head bolt for killing small animals

Preliminary tests by the first author with a flat-head captive bolt indicated that it could be very effective for euthanizing small pigs on the farm. A flat head is different from the standard non-penetrating captive bolt. Instead of a mushroom (convex) head, it had a flat head that is 3.8 cm (1½ inches) in diameter. When this was applied to the forehead of 11.4–22.7 kg (25–50 lb) pigs it resulted in more extensive brain damage than a standard penetrating captive bolt. Palpation of the impact site revealed a deep concave depression over 1.3 cm (½ inch) deep. The skin did not break. A piece of skull the size of the bolt diameter was rammed deep into the brain. Most of the brain became mush. The flat-head bolt may be a good alternative for on-farm euthanasia or for killing smaller pigs for disease control. It is also effective for neonatal lambs and day-old calves.

Manual Blunt Force Trauma

Manual blunt force trauma induces death by physical disruption of the brain. A blow to the head can be an effective means of euthanasia for small animals with thin craniums. For livestock, manual blunt force trauma is currently an acceptable means of euthanasia for piglets 5 kg (12 lb) or under, infant lambs under 9 kg (20 lb), infant kids 7 kg (15 lb) and poultry (chicken and turkeys). A single, sharp blow must be delivered to the central skull bones with sufficient force to produce immediate depression of the CNS and destruction of brain tissue, without breaking open the skull.

For manual blunt force trauma, the object must be brought to the animal's head, not the animal to the object. Striking the animal to the object significantly decreases the animal welfare standard. If animals are swung during the application of blunt force trauma, they will experience high stress and a much greater chance of injury with dislocated joints, broken legs, etc. Common acceptable tools used for manual blunt force include ball peen hammers, steel rods, wooden clubs and pipes.

Blunt force trauma to piglets and infant goats must be applied to the top of the animal's head. The top or back of the head is the appropriate application spot for lambs. Poultry will respond violently to blunt force trauma with severe wing flapping and handlers must take precautions to ensure their safety. It is essential that the blow be administered accurately and with absolute determination.

Consistency of delivery is a challenge, therefore manual blunt force trauma is questionable in terms of reliability and effectiveness. Manual blunt force trauma is *not* an acceptable means of euthanasia for cattle, horses, elk or bison.

One of the big problems with blunt force trauma is that caring stockpeople who are good at taking care of infant animals often do not want to use this method. One way to solve the problem is to have another person come in and do it when the primary stockperson is absent or use gas euthanasia. Practical experience has shown that many people would prefer to put an animal in a 'box' (i.e. euthanasia container) instead of hitting it on the head. One person said 'I prefer this, the box does the deed'.

Carbon Dioxide

The time of exposure and concentration level of CO_2 determines its effectiveness and humaneness. CO_2 is approved in many countries for use on pigs, poultry, neonatal sheep and goats. On the farm, CO_2 may only be practical for smaller weight animals because the animal must be placed in an appropriate container. There is currently significant international debate concerning the use of CO_2 for the euthanasia of animals. The use of CO_2 euthanasia for neonate animals (along with domestic waterfowl) requires increased exposure levels and times because they are more resistant to hypoxia.

There are two primary CO_2 techniques utilized in animal production: the high-level precharge system and the gradual-fill system. Each technique must be designed and managed properly to ensure a humane death and minimize stress on the animal. It is recommended that for on-farm euthanasia the animal be exposed to the target minimum level of $\geq 80\%$ for at least 5 min, and all animals evaluated for indicators of death upon removal from the euthanasia container.

Precharged systems

Precharged systems utilize a container in which a high level of CO_2 (>80%, preferably 90%) is injected prior to the animal being placed in the container. This method is recommended for pigs because it will reduce distress, squealing and agitation before loss of sensibility when the pig falls over (see Chapter 9). The animal should be fully immersed within the gas as quickly as possible, in order to minimize the opportunity for aversive reactions to the gas prior to the animal becoming unconscious. The CO_2 must be maintained at the target level until the animal is confirmed dead.

The euthanasia container should be designed to accommodate the behavioural and physical attributes of the target species, including such factors as door design, non-slip flooring, adequate lighting, etc. It is critical that the container is not overloaded with animals. All animals must be provided with adequate space to stand and lie down. It is not acceptable to pile animals into the container as this will jeopardize their welfare and may cause animals to die of suffocation.

For a gas to fill the chamber, the air in the chamber must be displaced by the gas that is being injected. This is the exchange rate. As CO_2 is heavier than air and will stratify within the euthanasia container a multiple-port injection system should be utilized to ensure even levels of CO_2 within the container. Gas flow rates must be regulated and monitored to ensure proper flow. Unregulated release of gas from a canister is not recommended due to the potential to chill the animals, freeze the supply lines or cause lower than desired chamber exchange rate, failed euthanasia, or wasting of CO_2, adding to the cost of the process. Exhaust valves must be designed to avoid pressuring the system during gas injection. Gas flow rates must be slow enough to avoid a high-pitched whistle, which is distressing to livestock. The system should be equipped with a CO_2 monitor or at minimum an alarm that indicates a low CO_2 level within the euthanasia container.

Gradual-fill systems

For a gas to fill the chamber, the air in the chamber must be displaced by the gas that is being injected. This is the exchange rate. As CO_2 is heavier than air and will stratify within the euthanasia container, a multiple-port injection system should be utilized to ensure even levels of CO_2 within the container. Gas flow rates must be regulated and monitored to ensure proper flow. Unregulated release of gas from a canister is not recommended due to the potential to chill the animals, freeze the supply lines or cause lower than desired chamber exchange rate, failed euthanasia or wasting of CO_2, adding to the cost of the process. Exhaust valves must be designed to avoid pressurizing the system during gas injection. Gas flow rates must be slow enough to avoid a high-pitched whistle, which is distressing to livestock. The system should be equipped with a CO_2 monitor or, as a minimum, an alarm that indicates a low CO_2 level within the euthanasia container. Gerritzen *et al.* (2004) contains additional information for poultry (the abstract is available on the Internet and can be downloaded for free).

Adaptations to these methods

Poultry lose consciousness at a much lower level of CO_2 (20–25%); therefore in addition to the traditional precharged and gradual-fill systems, a low-level gradual-fill system can be utilized. A low-level gradual-fill system utilizes a target level of 40% CO_2 concentration but requires a 20 min exposure period.

Due to the aversiveness of CO_2, combinations of CO_2 and various inert gases (argon, nitrogen, etc.) at differing concentrations have also been utilized for euthanasia of livestock and poultry. There is controversy among scientists on the best gas mixtures. Additional information can be found in Raj and Gregory (1995), Raj (1999), Meyer and Morrow (2005), Christensen and Holst (2006) and Hawkins *et al.* (2006). Chapter 9 contains more scientific studies. On the farm, the most practical way to evaluate gas mixtures is to watch the behaviour of the animals. It is absolutely not acceptable to use a gas mixture or methods that cause the animals to attempt to climb out and escape from the container. Gas mixtures and methods should be changed if animals or birds struggle, vocalize or vigorously flap BEFORE they fall down and lose posture. After loss of posture when an animal loses sensibility, vigorous bodily reactions will have no effect on welfare.

Electrocution

Electrocution is considered humane when adequate current passes through the brain to induce a grand malseizure and fibrillation of the heart. Cardiac fibrillation leads to cardiac arrest and ultimately death (see Chapter 9). The animal is either euthanized by an initial passage of current through the brain followed by a second application of the current through the heart of the unconscious animal to induce cardiac fibrillation or through simultaneous induction of unconsciousness and cardiac ventricular fibrillation. Cardiac fibrillation normally leads to cardiac arrest and an interruption in blood flow to the brain and other vital organs.

The amount of current delivered will depend upon the voltage and total impedance in the pathway, which is affected by the species, size of the electrodes, applied pressure, location placement, distance between electrodes and phase of respiration during application of shock. There are a multitude of combinations of voltage, current, frequency and length of application that have been utilized successfully as a euthanasia technique. However, the selection of these factors should be based on scientific evidence and verified in commercial conditions. The absolute minimum of current applied through the brain to induce instantaneous insensibility are shown in Table 10.3.

The correct current should be attained within 1 s of initiation and applied for a minimum of 3 s (OIE, 2009a).

Care must be taken that the electrodes are in full contact with the animal before the electrical shock is applied. When the wand is electrified before placement on the animal it is called hot wanding and it will cause pain. Hot wanding of cattle or pigs will cause the animal to vocalize the instant the electrode touches it. Electrocution of cattle should only be used in a slaughter plant because of the specialized equipment and high amperage that are required (see Chapter 9).

For pigs, electrocution is acceptable on animals over 5 kg (11 lb) when applied properly. The process may be conducted by first applying the electrodes on the opposite sides of the head so that the current travels through the brain. When a commercially available pig stunner is used, the tongs should be placed on the inner base of the ears. After the current has been passed through the head, the tongs are applied a second time across the chest to induce cardiac fibrillation. A commercially available pig stunner is recommended. Homemade devices are often dangerous and may not be effective (see Chapter 9 for further information).

There are no current guidelines or recommendations for on-farm application of electrocution of sheep, horses, elk or bison. Sheep are problematic due to the insulating effects of wool. Specialized equipment that is used in a slaughter plant is not practical or safe for on-farm use.

Table 10.3. Minimum currents[a] necessary for killing (source: OIE, 2009a).

Animal	Current (A)
Cattle	1.5
Calves	1.0
Sheeps/goats	1.0
Lambs/kids	0.6
Piglets	0.5
Pigs	1.3

[a]The frequency of current should be no greater than 100 Hz. Standard 50–60 Hz is recommended.

Cervical Dislocation

Cervical dislocation induces euthanasia by physical disruption of the brain and spinal cord as the procedure dislocates the vertebral column from the skull, which causes damage to the lower brain region and near-immediate unconsciousness. To ensure death, cervical dislocation should be followed by exsanguination.

Cervical dislocation is only acceptable for poultry and requires a higher level of skill than other methods. Proper training is vital to ensure the proper application and is best achieved using a stretching method rather than crushing the vertebrate. Small birds can be dislocated by applying rotational movements to the neck. Adult poultry should be held by the shanks (legs) with one hand, and the head grasped immediately behind the skull with the other hand. The neck is then extended and dislocated using a sharp downward and backward thrust. The bird may react violently with wing flapping. It is best to restrain the wings before euthanasia.

Cervical dislocation should be limited to poultry weighing 3.0 kg (7 lb) or less as it is very difficult on larger, heavily muscled poultry such as broiler breeders and turkeys. However, bovine Burdizzo® castration forceps can be used to kill large poultry through mechanical cervical dislocation. This technique prevents agonal regurgitation and aspiration of crop contents into the respiratory passages if the forceps are left clamped until reflex muscle spasms cease.

Methods that should Never be Used

When intravenous methods are used to euthanize animals, the animal must be rendered insensible with barbiturates or other drugs that are true anaesthetics. Non-anaesthetic agonists must never be used in a sensible animal. If non-anaesthetic drugs are used, to ensure death the animal must be rendered insensible with an anaesthetic drug before the non-anaesthetic agent is administered. Drowning, strangulation or air embolisms are also methods that should never be used. All three authors and the editor of this book have struggled to decide the most practical humane way to euthanize large animals in countries where firearms and cartridge-fired captive bolts cannot be owned by members of the general public. In these countries, the slaughter plants use a pneumatic-powered captive bolt gun that does not use a powder charge. Anaesthetic drugs are not a reasonable alternative due to high cost. The only alternative is to use methods that would not be approved by the professional veterinary associations in many developed countries where there is easy access to captive bolt guns. In this situation, one alternative would be to hit the bovine on the forehead (at X on Fig. 10.1) with a very heavy hammer and then exsanguinate immediately. If exsanguination has to be used as the sole method with no blow to the head, the animal should be cut with a very sharp knife that is twice the width of the neck. The cut should use either the halal or kosher cutting method. The cut should move from the outside of the throat towards the spine and simultaneously cut all four major blood vessels. Do not use the method shown in Fig. 10.6. This method of bleeding should only be used on animals that have been rendered insensible to pain by methods described in this chapter. In conclusion, these methods should only be used if a firearm, captive bolt, lethal injection or other approved method is not available.

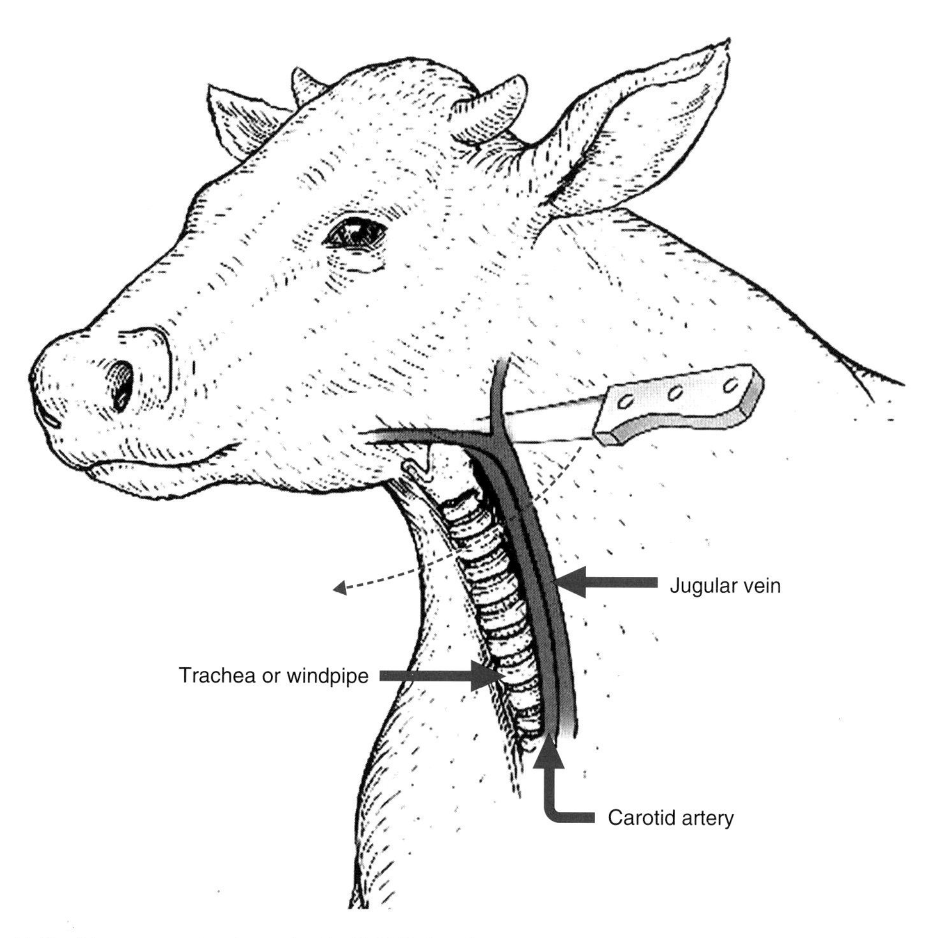

Fig. 10.6. Method for exsanguinating animals AFTER they have been rendered insensible. If a fully sensible animal has to be cut, this method should not be used because stabbing the neck would cause great pain. In parts of the world where no modern euthanasia methods are available, the animal should be cut using the kosher or halal method with a very sharp knife. The length of the knife must be greater than the width of the knife (from Shearer, 1999).

Secondary Kill Steps

There are several reasons why a secondary kill step may be required. In some cases, the euthanasia tool is only intended for or capable of stunning the animal, not killing it. Many captive bolt guns are designed to temporarily stun an animal. If the first shot with a firearm or captive bolt fails, the animal must be immediately shot again. The secondary kill step should occur within 30s after the last shot. A secondary kill step can include either exsanguination or pithing.

Exsanguination

Exsanguination is only recognized by most welfare guidelines as a secondary kill step, not a stand-alone kill method. This is especially true if dull knives (that do not have a sharp edge) or stab methods are used. It is inhumane because it causes the animal great pain, suffering and distress. It has been demonstrated that it may take several MINUTES for the animal to die from exsanguation. During the period when the bleeding animal is still conscious, serious welfare problems are highly likely because the animal can feel anxiety, pain, distress and other suffering (EFSA, 2004). There have been several research projects documenting the length of time it takes an animal to become unconscious, with one project on calves finding a range of 35–680s (Bager *et al.*, 1992). Research shows that sheep become insensible more quickly than bovines (see Chapter 9 for more references).

Exsanguination should be performed using a pointed, very sharp knife with a rigid blade at least 15.2cm (6 inches) in length. The knife should be fully inserted through the skin just behind the point of the jaw and below the neck bones (Fig. 10.6). From this position the knife is drawn forward severing the jugular vein, carotid artery and windpipe. Properly performed, blood should flow freely with death occurring over a period of several minutes. *Exsanguination is not recognized as a sole method of euthanasia and the animal must be stunned prior to bleeding.* This procedure can be very disturbing to observers due to the large volume of blood loss, which also raises biosecurity concerns.

In cattle, the brachial vasculature may also be severed by lifting a front leg and inserting a knife deeply into the axillary area at the point of the elbow and cutting the skin, blood vessels and surrounding tissue until the limb can be laid back away from the thorax of the animal.

For pigs, a sharp knife not less than 12cm (5 inches) should be used. The knife is inserted in the midline of the neck at the depression in front of the breast bone and the skin is raised with the knife point using light pressure and a lifting movement. When penetration has been made, the knife handle should be lowered so that the blade is in a near vertical position, and pushed upward to sever all major blood vessels which arise from the heart. The brachial vasculature can also be severed in pigs as described above for cattle. Great care must be practised while performing exsanguination on all animals because, although they are unconscious, they are still capable of making violent involuntary muscle contractions that cause injury to the handler.

Pithing

Pithing is a technique to cause death by causing significant destruction of the brain and spinal cord tissue. It is performed by inserting a slender pithing rod or tool through the entry site produced in the skull by the penetrating captive bolt stunner or free bullet. The operator manipulates the pithing tool to destroy both brain stem and spinal cord tissue, which results in death. This procedure is sometimes used in advance of exsanguination procedures to reduce involuntary movement in stunned animals.

A pithing rod can be made from a variety of materials such as a discarded cattle insemination gun, high tensile wire, a steel rod or other similar device. Some vendors have disposable pithing rods for purchase. The rod itself must be somewhat rigid, yet flexible. It must be long enough to reach the brain and spinal column through the hole in the skull produced by the projectile from a gun or PCBG. Appelt and Sperry (2007) recommend that pithing should be performed if exanguination is not performed after captive bolt. It will help prevent return to sensibility.

Confirmation of Death in Euthanized Animals

Understanding the process of death is important. Death is a process and does not occur immediately, even though sensibility and the ability to feel pain are lost instantly after correctly applied gun shot, captive bolt or electrocution. The animal is still sensible and may be suffering if it is vocalizing (squealing or bellowing), attempts to get up, lifts up its head or is blinking like a fully alive animal. If any of these signs is present, it must be shot again. Animals are first rendered insensible and then the body begins to die as the brain stops, the heart quits beating, the lungs stop breathing and the blood quits circulating. There is no such thing as immediate death. This process can take up to 3 min to occur when euthanasia is delivered correctly and in the case of barbiturates it may take even longer.

When a person is not knowledgeable about what to expect, he or she may mistakenly interpret any movement as sensibility and a lack of movement as loss of consciousness. Reflex motor activity or muscle spasms may follow the loss of consciousness – this is a normal part of the death process and should not be perceived as the animal being in pain or distress. Certain species and modes of euthanasia will have greater involuntary muscle movement than others (see Chapter 9).

After gunshot or captive bolt, intense muscle spasms in cattle will last on average only 5–10 s, with neonatal lambs lasting for several minutes. Pigs tend to have more violent and longer involuntary muscle spasms lasting 15–20 s, with random, mild convulsions that can last for several minutes. If the animal has an extended period of movement or 'flopping' it may only be stunned and should be reshot. All species may have weak, slow kicking movements for several minutes.

It is also important to recognize that these involuntary movements can be distressful for people inexperienced with these procedures. A forewarning to bystanders ahead of time helps reduce the amount of explanation that may be necessary afterwards. Human safety must also be considered when confirming death. The handler must be aware of the involuntary movements that the animal may make in order to avoid being kicked. It is best to approach the animal along the back/spine in order to avoid contact with the legs or head of the animal.

Confirmation of the onset of death (insensible or unconscious) should occur within the first 30 s following completion of the euthanasia procedure; death must be then be confirmed within 3 min. Confirmation of death can be challenging outside of a veterinarian's office or laboratory. The cessation of a steady, rhythmic heartbeat and absence of respiration are the most reliable modes for determining death, but are extremely challenging to confirm in a field environment due to background noise. If there is any uncertainty over whether an animal is dead or not, the procedure should be repeated or a second method applied.

Absence of rhythmic breathing is determined by visual observation of movement of the chest or palpation for chest movement. However, animals that are unconscious may have very sluggish and irregular breathing making it difficult to determine existence of chest movement. Gasping and random breaths are not considered rhythmic. The lack of chest movement is an indication that the animal is no longer breathing because their respiratory system has shut down (see Chapter 9).

Along with the absence of rhythmic breathing, two of the following five other modes of confirmation must be met before an animal is confirmed technically dead: cessation of rhythmic heartbeat; absence of pulse; lack of palpebral reflex; dilation of the pupils; loss of capillary refill. Animals must not be moved until death is confirmed. It is recommended to wait at least 20 min before moving the animal for disposal.

Cessation of regular heartbeat is accurately evaluated using a stethoscope, electrocardiograph (ECG) or electroencephalogram (EEG), though you may be able to palpate the heartbeat in some circumstances. These measures indicate the animals heart has stopped pumping, ceasing the flow of blood through the body and to the brain. Death is confirmed when you are unable to detect the beating of the heart for more than 3 min. However, this procedure can be very challenging in a field environment. For detection with a stethoscope, you must have a stethoscope and the heartbeat can be difficult to hear because of background noises. In order to utilize an ECG or EEG you must have the proper equipment, which is extremely costly and not commonly used outside a veterinary clinic.

Absence of pulse is detected through palpation of an artery and is an indication that the animal's heart is no longer beating. This technique is very challenging in a field environment requiring operator skill and is primarily utilized on larger species. A pulse is not always palpable, so it cannot be solely relied upon for determining cessation of heartbeat or confirming death.

Palpebral reflex is checked by running a finger along the eyelashes and is to be performed before corneal reflex to avoid pain of touching the eye of a sensible animal. There should be no eye movement or blinking when eyelashes are touched. Lack of palpebral reflex indicates the animal is insensible. This should be done prior to evaluating the corneal reflex by touching the cornea or the surface of the eye. There should be no eye movement or blinking when touched. If the animal is dead, the eye will remain open and the lid does not move.

Visual observation of corneas will determine if there is dilation of the pupils. The pupils will fully dilate when blood is no longer being pumped through the body due to the cessation of heartbeat, stopping the supply of fluid to the eyes. Other tests that can be used are a needle prick on the nose or shining a high intensity flashlight in the eyes. The animal MUST NOT respond to these stimuli.

You can visually observe the mucous membranes in the mouth for loss of colour and capillary refill. When animals die, the mucous membranes become pale and mottled. The loss of capillary refill is indicated when there is no return of colour or refill of blood after pressure is applied. The membranes will become dry and sticky because blood is no longer being pumped through the body due to the cessation of the heartbeat.

An animal is not dead if it lifts its head off the ground or attempts to right itself, vocalizes after application, has the presence of eye movement or blinking, has a palpable pulse, pupils are constricted, responds to painful stimuli (i.e. pinching the nose) or fails at any of the above tests.

No animal should be moved or disposed of for a minimum of 20 min following confirmation of death. This will ensure that the animal is truly dead and will not regain sensibility. Attempts must be made, though, immediately following euthanasia to confirm the method was effective. The onset of rigor mortis and bloating will vary among the species. In pigs, it will normally set in within 4–8 h, sheep 8–12 h and cattle 12–24 h. Rigor mortis should not be used as a means to confirm death due to the extreme delay of onset.

Mass Killing for Disease Control

The OIE (2009b) has guidelines for killing large numbers of animals or birds for disease control purposes. The specifications for gunshot, captive bolt or electrical methods are the same as those specified in Chapters 9 and 10. The following methods should NEVER be used:

- buried alive in a pit;
- thrown in on top of each other in a pit, bag or rubbish bin and allowed to be crushed and suffocated under other animals or birds;
- drowning;
- strangulation; and
- poisoning with substances that would cause a painful death.

The OIE (2009b) states:

> When animals are killed for disease control purposes, methods used should result in immediate death or immediate loss of consciousness lasting until death, when loss of consciousness is not immediate, induction of unconsciousness should be non-aversive and should not cause anxiety, pain, distress, or suffering in animals.

The destruction of large numbers of poultry in a low-stress humane manner is difficult. Filling an entire house with gas may not kill laying hens housed on upper decks of cages. Killing of poultry for disease control purposes is an area that needs further study. Some of the most promising methods for poultry are putting anaesthetics or other agents in food or water and gas stunning in containers (Raj, 2008). Canadian researchers have also successfully used container systems (John Church, personal communication, 2008). US researchers have used firefighting foam to kill broiler chickens (Benson *et al.*, 2007). The foam method needs further research. Discussion within the poultry industry indicates that the bubble size must be carefully controlled to induce a peaceful induction. For all gas or foam methods, the behaviour of the birds prior to loss of posture (falling over) must be evaluated (see Chapter 9). Gerritzen and Sparrey (2008) reported that firefighting foam generated with bubbles filled with CO_2 may work. The bubbles must be large enough so that the birds are killed by CO_2 inhalation and they are not killed by airway occlusion or drowning. Raj *et al.* (2008) have experimented with dry foam filled with nitrogen. As of 2009, foam methods are under study by the OIE.

Carcass Disposal

Euthanasia presents another issue that people frequently fail to consider – disposal of the carcass.

Acceptable means of disposal of dead livestock include scavenging, burial, composting, incineration, rendering and tissue digestion. Scavenging of dead animals is when animals are left out for other animals to eat and for natural decomposition to occur. Burial is a very common method where animals are buried in the ground following euthanasia. When animals are composted, they are mixed with carbon bulking material such as finely ground straw or sawdust and will break down over time into organic material. Tissue digestors digest carcasses by a process of alkaline hydrolysis. Incineration is when carcasses are burned and rendering is a process whereby the carcasses are cooked to destroy pathogens and yield products such as meat, feather, bone or blood meal. Each one of these methods has its advantages and limitations and will be regulated by local laws. Shearer *et al.* (2008) contains information on composting.

Eating Euthanized Animals

In areas of the world where food is scarce, hungry people are going to eat euthanized animals. They may skin and dress the animal on the farm. In places where people are starving, they are not going to throw away food. Common pathogens such as *Salmonella* spp. and *Escherichia coli* and most parasites can be killed by cooking meat until all pink colour has disappeared. People must never be allowed to eat meat from animals or poultry with diseases that are transmissible to people such as tuberculosis, brucellosis, rabies, bovine spongiform encephalopathy (mad cow) or avian influenza (bird flu). In certain areas there may be additional diseases where the carcasses must be destroyed and not consumed due to huge risks to public health.

Personnel Training and Welfare Verification

All farms and ranches are advised to develop personnel training programmes for proper instruction of humane euthanasia techniques. The skill and experience of personnel are of paramount importance when conducting euthanasia procedures. Experience has shown that many people (even those experienced in handling livestock) do not understand the physiological parameters required for proper execution of these techniques. Furthermore, persons should be aware that there is significant danger for the operator (or for bystanders with gunshot) whenever these methods of euthanasia are used. Only those who can demonstrate a working knowledge and proficiency with the techniques should be permitted to perform euthanasia procedures. When these methods are not properly performed, animals may become injured, have varying degrees of consciousness, and experience needless pain and distress.

Experienced persons should assist in the training of inexperienced persons and utilize carcasses to demonstrate anatomical landmarks and application of the various techniques. Carcasses should be used for practice by trainees until they become competent with the procedures. People must also be aware of how to confirm death. In some cases this may require specific training with, and observation of, live animals.

Owners and managers must be acutely aware of all the factors outlined in this section no matter the size of their operation. Whether a farm has a staff of 50 or is owner operated, it is the owner/manager's responsibility to educate themselves on acceptable euthanasia procedures and ensure they appropriately assign tasks to those who are skilled and comfortable with euthanasia procedures.

Training should not only be required for new staff entering the production site, but continual training and assessment for existing staff is essential. The training should be comprehensive and include all aspects of euthanasia with both literature review and hands-on training. A survey of swine barn workers has shown that handlers prefer to be trained on the farm by managers. Stockpeople should not be allowed to euthanize animals on their own until they have been tested and proved to be proficient on skill, efficiency and effectiveness.

It is the responsibility of management to verify that all staff is performing euthanasia properly and in a timely manner. Each employee should be verified on a regular basis and barns assessed for active euthanasia practices, utilizing a standard audit programme or third party verification process. An example of a standard animal welfare assessment for on-farm euthanasia is shown in Box 10.2.

Box 10.2. **Animal welfare assessment for on-farm euthanasia (Woods *et al.*, 2008).**

Animal Welfare Assessment for On-farm Euthanasia
Date of assessment:
Location:
Assessed by:
Employee:
Animal type and weight:

Assessment criteria
Pre-euthanasia

1. Is the employee trained in the process and technique of euthanasia?
Yes or No

Name and date of training programme: ____________________

2. Was the appropriate tool chosen for the species and weight class of the animal?
Yes or No

3. Is the necessary equipment (handling equipment, restraint device, euthanasia tool, etc.) available and in proper working condition?
Yes or No

4. Was the animal exposed to minimal stress, pain and anxiety during handling and restraint?
Yes or No

5. Was the euthanasia decision tree followed to ensure euthanasia was conducted in a timely manner?
Yes or No

Euthanasia

6. Was the procedure applied in accordance to the standards outlined in the training manual?
Yes or No

7. Were indicators of the onset of death confirmed within 30 s following application of the euthanasia technique?
Yes or No

If No – were immediate steps to taken to rectify the situation to ensure death?
Yes or No

8. If a secondary kill step (i.e. exsanguination, pithing, second shot, etc.) was required, did it occur in a timely manner necessary to ensure a humane death?
Yes or No or Not required

Post-euthanasia

9. Was death confirmed within 3 min following application of the euthanasia technique?
Yes or No

10. Was the carcass disposed of in accordance to all applicable rules and regulations?
Yes or No

11. Was the location of euthanasia properly cleaned and disinfected?
Yes or No

12. Was the euthanasia equipment clean and properly re-stored?
Yes or No

If any of these questions were answered NO, it is recommended that the euthanasia training programme be reviewed and the euthanasia process re-evaluated at the next three euthanasia opportunities.

continued

Box 10.2. **Continued.**

Notes or recommendations:

Wilful acts of abuse

Any wilful act of abuse during the euthanasia process is grounds for automatic disciplinary action. Wilful acts of abuse include but are not limited to: (i) intentionally applying goads to sensitive parts of the animal like the eyes, ears, nose, rectum; (ii) deliberate slamming of gates on livestock; (iii) use of euthanasia tools in a manner that violates the manufacturer's recommendations and international standards for use (examples include using an electrocution instrument as an electric goad, shooting an animal in the leg to render it immobile as to ease proper shot to the brain, etc.); (iv) hitting/beating an animal; (v) dragging of a live animal.

Were any wilful acts of abuse observed?
Yes or No

References

American Veterinary Medical Association (AVMA) (2007) *AMVA Guidelines on Euthanasia*. AMVA, Schaumberg, Illinois.

Appelt, M. and Sperry, J. (2007) Stunning and killing cattle humanely and reliably in emergency situations – a comparison between a stunning only and a stunning and pithing protocol. *Canadian Veterinary Journal* 48, 529–534.

Arluke, A. (1994) Managing emotions in an animal shelter. In: Manning, A. and Serpell, J. (eds) *Animals and Human Society*. Routledge, New York, pp. 145–165.

Bager, F., Braggins, T.J., Devine, C.E., Graafhuis, A.E., Meller, D.J., Tavener, A. and Upsdell, M.P. (1992) Onset of insensibility at slaughter in calves: effects of electroplectic seizure and exsanguination on spontaneous electrocortical activity and indices of cerebral metabolism. *Research in Veterinary Science* 52, 162–173.

Benson, E., Malone, G.W., Alphin, R.L., Dawson, M.D., Pope, C.R. and Van Wickler, G.L. (2007) Foam based mass emergency depopulation of floor raised meat type poultry operations. *Poultry Science* 86, 219–224.

Bildstein, C. (2009) Animal Care and Handling Conference, 18–19 March, Kansas City, Missouri.

Blackmore, D.K. (1985) Energy requirements for the penetration of heads of domestic livestock and the development of a multiple projectile. *Veterinary Record* 116, 36–40.

Blackwell, T.E. (2004) Production practices and well-being of swine. In: Benson, G.J. and Rollins, B.E. (eds) *The Well-being of Farm Animals*. Blackwell Publishing, Ames, Iowa, pp. 241–269.

Christensen, L. and Holst, S. (2006) *Group-wise Stunning of Pigs in CO_2 Stunning Systems – Technical and Practical Guidelines for Good Animal Welfare: a Danish Perspective*. Danish Meat Research Institute, Roskilde, Denmark.

Daly, C.C. and Whittington, P.E. (1989) Investigation into the principal determinants of effective captive bolt stunning of sheep. *Research Veterinary Science* 46, 406–408.

Daly, C.C., Gregory, N.G. and Wotton, S.B. (1987) Captive bolt stunning of cattle: effects on brain function of bolt velocity. *British Veterinary Journal* 143, 574.

European Food Safety Authority (EFSA) (2004) *Welfare Aspects of Animal Stunning and Killing Methods*. EFSA AHAW/04-207. EFSA, Parma, Italy.

Fulwider, W.K., Grandin, T., Rollin, B.E., Engle, T.E., Dalsted, N.L. and Lamm, W.D. (2008) Survey of management practices on one hundred and thirteen north central and northeastern United States dairies. *Journal of Dairy Science* 91, 1686–1692.

Gerritzen, M.A. and Sparrey, J. (2008) A pilot study to determine whether high expansion CO_2 enriched foam is acceptable for on-farm emergency killing of poultry. *Animal Welfare* 17, 285–288.

Gerritzen, M.A., Lambooij, B., Reimert, M., Stegeman, A. and Sprujt, B. (2004) On-farm euthanasia of broiler chickens: effects of different gas mixtures on behavior and brain activity. *Poultry Science* 83, 1294–1301.

Grandin, T. (1988) Behavior of slaughter plant and auction employees towards animals. *Anthrozoos* 1, 205–213.

Grandin, T. (1998) Objective scoring of animal handling and stunning practices in slaughter plants. *Journal of the American Veterinary Medical Association* 212, 36–93.

Grandin, T. (2002) Return to sensibility problems after penetrating captive bolt stunning in commercial slaughter plants. *Journal of the American Veterinary Medical Association* 221, 1258–1261.

Grandin, T. and Johnson, C. (2009) *Animals Make Us Human*. Houghton Mifflin Harcourt, Boston, Massachusetts.

Green, A.L., Lombard, J.E., Gerber, L.P., Wagner, B.A. and Hill, G.W. (2008) Factors associated with the recovery of nonambulatory cows in the United States. *Journal of Dairy Science* 91, 2275–2283.

Gregory, N.G., Lee, C.J. and Widdicombe, J.P. (2007) Depth of concussion in cattle shot by penetrating captive bolt. *Meat Science* 77, 499–503.

Gregory, N.C., Spence, J.Y., Mason, C.W., Tinarwo, A. and Heasenan, L. (2008) Effectiveness of poll shooting in water buffalo with captive bolt guns. *Meat Science* 81, 178–182.

Hawkins, P., Playle, L., Golledge, H., Leach, M., Banzett, R., Coenen, A., Cooper, J., Danneman, P., Flecknell, P., Kirkden, R., Niel, L. and Raj, M. (2006) Newcastle Consensus Meeting on Carbon Dioxide Euthanasia of Laboratory Animals, 27–28 February, University of Newcastle-upon-Tyne, Newcastle-upon-Tyne, UK. Available at: www.nc3rs.org.uk/CO2ConsensusReport and www.lal.org.uk/news.html (accessed 30 June 2009).

Humane Slaughter Association (2005) *Humane Killing of Livestock Using Firearms*, 2nd edn. Humane Slaughter Association, Wheathampstead, UK.

Manette, C.S. (2004) A reflection on the ways veterinarians cope with death euthanasia and slaughter of animals. *Journal of the American Veterinary Medical Association* 225, 34–38.

Matthis, J.S. (2004) Selected employee attributes and perceptions regarding methods and animal welfare concerns associated with swine euthanasia. Dissertation, North Carolina State University, Raleigh, North Carolina.

Meyer, R.E. and Morrow, W.E. (2005) Carbon dioxide for emergency on-farm euthanasia of swine. *Journal of Swine Health and Production* 13(4), 210–217.

Moberg, G.P. (1985) Biological response to stress: key to assessment of animal well-being? In: Moberg, G.P. (ed.) *Animal Stress*. American Physiological Society, Bethesda, Maryland, pp. 27–49.

OIE (2009a) Chapter 7.5. *Slaughter of Animals, Terrestrial Animal Health Code*. World Organization for Animal Health, Paris.

OIE (2009b) Chapter 7.6 *Killing of Animals for Disease Control Purposes, Terrestrial Animal Health Code*. World Organization for Animal Health, Paris.

Raj, A.B.M. (1999) Behaviour of pigs exposed to mixtures of gases and the time required to stun and kill them: welfare implications. *Veterinary Record* 144(7), 165–168.

Raj, A.B.M. and Gregory, N.G. (1995) Welfare implications of the gas stunning of pigs I. Determination of aversion to the initial inhalation of carbon dioxide or argon. *Animal Welfare* 4, 273–280.

Raj, A.B.M. and O'Callaghan, M. (2001) Evaluation of a pneumatically operated captive bolt gun for stunning broiler chickens. *British Poultry Science* 42, 295–299.

Raj, A.B.M., Smith, C. and Hickman, C. (2008) Novel method for killing poultry in houses with dry foam created using nitrogen. *Veterinary Record* 162, 722–723.

Raj, M. (2008) Humane killing of non-human animals for disease control purposes. *Journal of Applied Animal Welfare Science* 11, 112–124.

Reeve, C.L., Spitzmüller, C., Rogelberg, S.G., Walker, A., Schultz, L. and Clark, O. (2004) Employee reactions and adjustment to euthanasia-related work: identifying turning-point events through retrospective narratives. *Journal of Applied Animal Welfare Science* 7(1), 1–25.

Reeve, C.L., Rogelberg, S.G., Spitzmüller, C. and Digiacomo, N. (2005) The caring and killing paradox: euthanasia-related strain among animal-shelter workers. *Journal of Applied Psychology* 35(1), 119–143.

Shearer, J.K. (1999) *Practical Euthanasia of Cattle: Considerations for the Producer, Livestock Market Operator, Livestock Transporter and Veterinarian*. Brochure prepared by the Animal Welfare Committee of the American Association of Bovine Practitioners, Opelika, Alabama.

Shearer, J.K., Irsik, M. and Jennings, E. (2008) Methods of Large Animal Carcass Disposal in Florida. University of Florida Extension. Available at: http://edis.ifas,ufl.edu/document_vm133 (accessed 17 April 2009).

Sneddon, C.C., Sonsthagen, T. and Topel, J.A. (2006) *Animal Restraint for Veterinary Professionals*. Mosby/Elsevier, St Louis, Missouri.

Swine News (2000) Euthanasia for hog farms. *Swine News* July.

United States Department of Agriculture (USDA) (2004) *National Animal Health Emergency Management System Guidelines*. USDA, Washington, DC. Available at: www.dem.ri.gov/topics/erp/nahems_euthanasia.pdf (accessed 27 August 2009).

Wichart von Holten, S. (2003) Psycho-social stress in humans of mass slaughter of farm animals. *Deutsche Tierärztliche Wochenschrift* 110, 196–199.

Woods, J. (2009) *Swine Euthanasia Resource*. Blackie, Alberta, Canada.

Woods, J.A., Hill, J.D. and Shearer, J.K. (2008) Animal welfare assessment tool for on-farm euthanasia. Presented at the Welfare and Epidemiology Conference: Across Species, Across Disciplines, Across Borders, 14–16 July, Iowa State University, Ames, Iowa.

Further Reading

American Association of Bovine Practitioners (2009) Practical Euthanasia of Cattle. Available at: www.aabp.org/resources/euth.pdf (accessed 4 July 2009).

American Association of Swine Veterinarians (2009) On Farm Euthanasia Options for the Producer. American Association of Swine Veterinarians, Perry, Iowa. Available at: www.aasv.org/aasv/euthanasia.pdf (accessed 4 July 2009).

Australian Veterinary Association (1987) Guidelines on humane slaughter and euthanasia. *Australian Veterinary Journal* 64, 4–7.

Canadian Food Inspection Agency (2007) *Notifiable Avian Influenza: Hazard Specific Plan.* Canadian Food Inspection Agency, Ottawa, Canada.

Humane Slaughter Association (2004) *Emergency Slaughter.* Humane Slaughter Association, The Old School, Brewhouse Hill, Wheathampstead, UK.

Morrow, W.E.M., Meyer, R.E., Roberts, J. and Lascelles, D. (2006) Financial and welfare implications of immediately euthanizing compromised nursery pigs. *Journal of Swine Health and Production* 14, 34.

University of California Veterinary Medicine (2009) The Emergency Euthanasia of Horses. University of California Davis, California. Available at: www.vetmed.ucdavis.edu/vetext/INF-AN-EMERGUTH-HORSES-HTML (accessed 4 July 2009).

11 The Effect of Economic Factors on the Welfare of Livestock and Poultry

TEMPLE GRANDIN

Colorado State University, Fort Collins, Colorado, USA

During a career spanning over 35 years, the author has learned to understand more and more how economic forces can be used to improve animal welfare. Economic incentives to treat animals better can be very effective. One huge positive force for improving animal welfare is that consumers are demanding that animals be treated better. Corporations, both large and small, can be motivated to improve practices when consumers demand it.

All of the things the author recommends are based on either scientific papers, first-hand experience of implementing welfare auditing programmes, observations during extensive travel to many countries, or research and interviews with other individuals who have implemented effective programmes. In the first section of this chapter, the effects of economic factors on improving animal welfare will be covered. The second half of the chapter will cover economic factors that are detrimental to good animal welfare.

Economic Factors that can Bring About Improvements in Animal Welfare

Alliances between producers and meat companies

In these systems, ranchers and farmers produce animals which must meet specific criteria for animal welfare, food safety and other requirements. The rapidly growing markets in organic and natural meats have created alliance systems where standards can be enforced. Producers are often eager to join these programmes in order to get higher prices. Many of these programmes emphasize local production of the meat, milk or eggs.

Welfare auditing by major meat buyers

The programmes that have been implemented by supermarkets and restaurants to inspect farms and slaughter facilities have resulted in great improvements in how animals are treated (Grandin, 2005, 2007a). These audits have resulted in tremendous improvements in plant facilities. The most noticeable changes are much better repair and maintenance of equipment such as stunners, races and pens. Out of 75 beef and pork plants in the USA that were on the McDonald's approved supplier list, only three had to build totally new systems. Most plants made simple economical improvements that are discussed in Chapters 5 and 9. At three other plants, no improvements occurred until a new plant manager was hired. This shows the importance of management attitude. In the USA, most plants already had at least adequate facilities. In South America and other parts of the world, many new lairages, races and stun boxes have been built to replace poor facilities. Tesco and other supermarket buyers from Europe and McDonald's units within each South American country demanded better animal treatment.

McDonald's auditing programmes are currently operating in the USA, Canada, South America, Australia, Asia and Europe. Large meat buyers such as McDonald's and Tesco have brought about big welfare improvements by using their tremendous purchasing power to enforce standards. When a slaughter plant is removed from their approved supplier list, it may lose huge amounts of money. A single large US plant can lose over a million dollars if it is off the approved supplier list for a year. McDonald's is such a large beef buyer that it purchases beef from 90% of the large or medium-sized US and Canadian plants. Socially responsible buying programmes by big corporations have also brought about environmental and labour improvements. Pressure from activist groups forced the upper management of many big companies to examine the substandard practices of their suppliers.

The author had the opportunity to take upper management people from many different companies on their first trips to farms and slaughter plants. When things were going well, they were happy and when they saw abuses, they became highly motivated to make improvements happen. The executives had to see bad practices with their own eyes to get them to make changes. One executive became really motivated to improve conditions after he saw an emaciated, sick, old dairy cow going into his hamburger. Animal welfare became real and was no longer an abstract concept that was delegated to the public relations or legal department. It is essential to get high-level executives out of the office so they can see bad practices to motivate them to change.

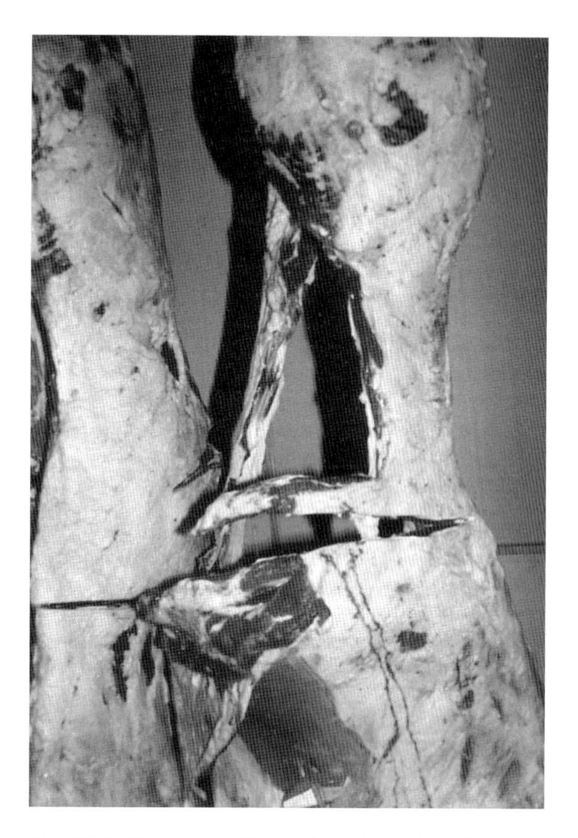

Fig. 11.1. Rough handling and overloaded trucks can cause extensive bruising and damage to the meat. Bruised meat has been trimmed from this carcass. Careful handling will prevent bruises.

Make producers and transporters financially accountable for bruises, poor meat quality, non-ambulatory animals and death losses

Bruises on slaughter cattle were greatly reduced when producers or transporters had to pay for them. In the USA, payment programmes where bruises and other losses are deducted from producer payments greatly improved cattle treatment during transport to the plant. Grandin (1981) found that when producers had to pay for bruises, bruising was reduced by half. Parennas de Costa (personal communication, 2007) in Brazil reported that when supermarkets audited bruises and made deductions from transporters' pay, bruising was reduced from 20% to 1% of the cattle. Bruises cause severe economic losses and large portions of the meat have to be removed from severely bruised carcasses (Fig. 11.1). Carmen Gallo in Chile has also reported that bruises were reduced when transporters were fined for damage to the animals (Grandin and Gallo, 2007). In another case, problems with weak pigs that were too fatigued to walk off the truck or move to the stunner were greatly reduced when producers were fined US$20 for each fatigued pig. Producers reduced non-ambulatory and weak pigs by greatly decreasing the dose of the beta-agonist ractopamine (Paylean®). This feed additive makes pigs big and lean and too high a dose may increase the percentage of non-ambulatory pigs.

Use objective methods for assessing handling and transport losses

Vague guidelines that use terms such as 'adequate space' or 'proper handling' are impossible to implement because one person's interpretation of proper handling will be different from somebody else's. Loading and unloading of trucks and moving animals through vaccination races should be measured with numerical scoring of variables such as the percentage of animals that fall, the percentage of animals that are prodded with an electric goad, and the percentage that move faster than a trot. Moving at a walk is preferable. For more information, see Grandin (1998a, 2007b) and Maria *et al.* (2004) and Chapters 1 and 3. Alvaro Barros-Restano (personal communication, 2006) reported that in auction markets in Uruguay, continuous monitoring has greatly improved handling. Handling practices need to be continuously measured to prevent them from gradually becoming increasingly rough. Measures of death losses, non-ambulatory animals, bruises, injuries, pale soft meat in pigs, and dark cutting, dark firm and dry (DFD) beef should be used to provide either bonuses or deductions from transporter or producer pay. In many countries, there is a large payment deduction for meat quality defects such as DFD in cattle.

Improving meat quality is an incentive for improving animal treatment

When the author first started working with pork slaughter plants in the early 1980s, the handling of pigs was terrible and every pig was shocked multiple times with an electric goad. In the late 1980s, the USA started exporting pork to Japan and the Japanese grader who worked in the plant rejected pale soft exudative (PSE) meat. The author visited many different plants and stopped the excessive use of electric goads in the race that lead up to the stunner. The next day, 10% more pork was suitable for export to Japan because it was not pale, soft and watery due to rough handling. This provided a motivation to handle pigs more carefully. Today several research studies clearly show that how the pigs are handled in the last few minutes in the stunning chute has a big effect on pork quality. Excitement and electric-goad use in the stunning race increases PSE (Grandin, 1985; Hambrecht *et al.*, 2005). One study in feedlot cattle showed that cattle that have a large flight zone and become agitated when people approach have tougher meat (Gruber *et al.*, 2006). An earlier study done by Voisinet *et al.* (1997) also found that excitable cattle have tougher meat and more dark cutters.

Improving animal welfare makes livestock handling safer

The author convinced many slaughter plants, feedlots and farms to improve cattle handling to reduce accidents and injuries to people. Marcos Zapiola in South America has used the same approach to promote quiet handling methods. The author also used safety for people as a major selling tool to eliminate live shackling and hoisting of animals by one back leg in US slaughter plants. Douphrate *et al.* (2009) studied 10 years of worker injury data in the USA. Accidents while handling cattle were a major cause of severe debilitating injuries that required costly medical care. In the USA, shackling and hoisting is still legal due to religious exemptions. When a veal calf slaughter plant replaced shackling and hoisting with an upright restraint for kosher slaughter, there was a huge reduction in accidents. For an 18-month period, before installation of the upright restrainer, there were 126 working days of lost time due to accidents involving employees being kicked or trolleys falling on them. Three workers were absent for more than 3 weeks. After the shackle hoist was replaced with the restrainer, during the next 18-month period one employee was absent for 2 days due to a bruised hand (Grandin, 1988).

Reduce labour requirements with animal-welfare-friendly equipment

The author designed and installed many innovative cattle handling systems that were sold to plant and feedlot management as a way to reduce labour costs. Half the cattle in the USA and Canada are handled in a restrainer system designed by the author. When the author presented her proposal for a totally new handling system at a slaughter plant, many managers bought the system because one or two full-time employees could be removed. The cost savings of reducing labour requirements, improving meat quality, reducing accidents and reducing bruises were all put on the proposal to emphasize how much money they could save.

Increase the economic value of cull animals

Some of the worst abuses the author has observed in transported animals were animals that were not fit for transport. They were treated badly because they were worth very little money. Emaciated, weak, old cows, sow or ewes should be euthanized on the farm and not loaded on to a vehicle. Published materials for assessing body condition, lameness and injuries should be used. The use of pictures and videos to make assessment of the animal's condition more objective is strongly recommended. Livestock quality assurance schemes in many countries have excellent materials for assessing fitness for travel. When programmes are implemented and producers receive more money for cows that are in better condition, they will be motivated to sell their cull animals before they become emaciated (Roeber *et al.*, 2001). In the USA, there are several successful programmes for improving the value of cull cows. They are fed in a feedlot for 60–90 days to improve meat quality to make the meat more valuable.

Promote the use of livestock identification and trace-back

In much of the developed world, animals are required to be identified with either an individual identification number, or the identity of their farm of origin. Animal identification makes it possible to trace animals back to the farm of origin, which enables customers to determine where their meat comes from. Trace-back makes it easier to hold producers and transporters accountable for losses.

Educate consumers about animal welfare

In the developed world, people are becoming more and more concerned about where their food is coming from. Many consumers may stop buying meat from animals transported long distances or that have been badly treated. When consumers are educated, they are willing to buy more socially responsible products. This method can be very effective with affluent consumers. In the USA and Europe, there are expanding markets for local artisan cheese makers and other suppliers of meat, milk or eggs that are raised with high welfare standards. In the UK, the sales of products that were produced under fair-trade agreements rose by 70% in 2007 (Editor in the *Independent*, 2008).

In developing countries, develop local slaughter and dairy-product processing facilities with experienced management

In many countries there is a need for high quality, small, local slaughter plants in areas where animals are raised. Efforts by the government to build local slaughter plants have failed in some countries because the government did not provide funds to hire experienced managers to operate them. There are many rusting hulks throughout the developing world where this mistake has been made. Some of these plants contained equipment that was too expensive and difficult for local people to maintain. Efforts in building producer-owned cooperative plants have had mixed results. The successful cooperatives have clauses in their legal documents to prevent a few producers from buying too big a portion and then selling out, and leaving the others at the mercy of a new plant owner who is no longer legally bound by the original cooperative agreements. The author has witnessed this sad fate for three large cooperatives in the USA. Successful, cooperative plants must have a strong, highly experienced leader and legal documents that will prevent one or two producers from taking over and doing things that are detrimental to other producer members. Fighting between producer board members has destroyed some cooperatives.

Hire and train experienced people who know how to implement practical solutions and pay them well

There is a tremendous need for more educated, experienced people to work on hands-on practical things. They need to have both scientific knowledge and hands-on experience to bridge the gap between making policy and successfully applying it. In the USA there are not enough students who want to become large animal veterinarians (NIAA, 2007). There are similar concerns in Europe. Programmes need to be developed to expose young students to farm animals and economic incentives should be used to encourage students to go into careers that will help animal welfare and sustainable agriculture. Policies and legislation are useless unless there are practical people on the ground who can implement them. Input from practical field researchers will help create policy and legislation that will work. In many fields there is a shortage of people to work in the 'trenches' and make real constructive change. This has happened in many fields ranging from medicine to agriculture. The author urges governments, non-governmental organization (NGO) animal activist groups and livestock companies to support and educate skilled fieldworkers and researchers. These people are essential to make real change and improvements take place.

In some developed countries, there is a shortage of qualified truck drivers. Hands-on jobs such as truck driver and animal handler need to receive more recognition and pay. The author has observed effective programmes in which handling improved when employees received training, better supervision, higher pay, and recognition with a special animal welfare emblem for their hats. For these programmes to work, they must be backed by a firm commitment from upper managers.

Work with government funding agencies to fund practical research

Many government funding agencies in the USA and other countries provide lots of grants for basic research and very little for applied practical research. This bias is motivating university administrations to hire professors of basic research instead of applied research. Researchers who do applied research are more likely not to be replaced when they retire compared to basic research scientists. A recent article in *Nature* discussed problems in human medicine where great discoveries in basic research do not get applied to actual patients (Butler, 2008). The hands-on doctors are doing less and less research and they do not communicate with biomedical scientists. This funding gap between basic and applied research will hurt both human and animal welfare. People interested in improving animal welfare should work with the policy makers of major governmental research funding agencies to direct more money to applied studies.

Implement simple, practical, economical ways to improve handling transport in developing countries

Fancy equipment such as hydraulic tailgate lifts or aluminium trailers is often not appropriate in developing countries. The people do not have the equipment or the money to maintain these items. While travelling to many countries, the author observed that simple improvements make a big difference. Non-slip flooring is essential in vehicles used to haul livestock and on scales, unloading areas and stun boxes. Floors can be made non-slip by welding readily available steel bars on the floor in a grid pattern. A non-slip floor will prevent many serious animal injuries. There is also a need to build ramps for loading and unloading. Many animals in developing countries are injured when they are forced to jump off a vehicle. Training people in animal behaviour and low-stress handling methods is also essential. Further information can be found in Chapter 5 and Grandin (1987, 1998b, 2007a, b), Smith (1998) and Ewbank and Parker (2007). Many people made the mistake of thinking that fancy equipment will solve all the problems. Over the years, the author has learned that good equipment makes good treatment easier, but it is useless without good management.

Private corporate and foundation sources of funding for animal welfare projects and student scholarships

These sources are excellent for funding animal welfare programmes. They often have more vision than government agencies and become willing to fund a programme for a longer period of time. This will help to get it established and keep it running. Programmes in sustainable agriculture and local production of food are activities that many private sources of funding like to support. A good example of an effective project is the mobile slaughter unit that was funded by the Lopez Community Land Trust, a non-profit group that supports sustainable agriculture (Etter, 2008). This system enables small, local sheep and cattle producers to have their animals slaughtered in a United States Department of Agriculture (USDA)-inspected plant and sell their meat without any restrictions. The mobile unit improves animal welfare because the slaughterhouse comes to the farm. To make this type of programme successful, funds must also be provided for hiring competent people to run and operate the system. The project is likely to fail if funds for operating expenses are omitted.

Private sources of money can also be very effective for supporting veterinary and graduate students who go into the animal welfare field. Students will quickly enter the animal welfare field if money is available for projects and scholarships.

Use economic incentives to pay people who handle, load and unload livestock and poultry

Reward animal handlers with extra pay for low levels of bruises, injuries and deaths. In the US and British poultry industries, broken wings were reduced from 5% to 1% by paying a bonus to the chicken loaders when there were 1% or less birds with broken wings. The same system has also worked well for people handling pigs and cattle. The worst way to pay animal handlers is based on how many they can handle per hour. This will result in rough treatment of the animals. The author has observed terrible handling of pigs, cattle and poultry when workers were financially rewarded for high-speed handling. Workers should be rewarded for high-quality handling. It is also essential not to overwork the animal handlers or put them in a situation that is understaffed. Tired people will abuse animals. Internal unpublished data from large pig and poultry companies have shown that death and injuries doubled after the truck loading crews had worked more than 6 h.

Numerous suppliers enable a supermarket or restaurant to effectively enforce high standards

One of the reasons why McDonald's and other hamburger chains were effective in improving the US beef slaughter plants is they had more than 40 plants on their approved supplier list. If one or two plants failed an audit, they could stop buying from them and still have enough beef. Most plants that fail an audit are not permanently 'delisted'. After they make improvements and pass a re-audit they are put back on the approved supplier list. Less progress has been made with improving poultry practices because most US restaurant chains depend on three to five dedicated plants. If one was taken off the approved list, they would not have sufficient product. Wendy's International has been the most effective buyer in bringing about change because they use 27 plants. Even though three large companies own most of the plants, that makes no difference because a single plant is removed from the supplier list, not one of the big corporations. The author has observed that implementing positive change works best when negotiations are kept at the plant level instead of negotiating with a central corporate office. Each individual plant is either financially rewarded with lots of orders or punished by having orders taken away.

Combining animal welfare audits with food safety and quality facilitates implementation

When welfare programmes are first getting started in a country or region, they are easier to implement if they are combined with food safety programmes, health and vaccination programmes. In the USA the same people who do food safety audits do all of the welfare audits in the slaughter plants. This made the programmes easy to implement because no additional people had to be hired, as the food safety auditors were already inspecting the plants. In South America, where the concept of animal welfare is still new, animal welfare is being combined with programmes to prevent injection site damage and programmes to obey medication withdrawal times. The prevention of bruises is a major part of South American programmes. In European countries, where animal welfare is an established concept, there are many people for whom welfare inspection is their only job.

Major Problem Areas where Economic Factors Work Against Animal Welfare

The customer demands live animals

The Australian live sheep trade is a primary example. The welfare of the sheep would be greatly improved if the sheep were slaughtered in Australia and the meat was shipped to the Middle East. The problem is that there are great economic forces working against this. People in the Middle East dislike the taste of chilled meat and they are willing to pay huge prices for live sheep. The religious requirements for halal slaughter could be met in Australia. The main barrier to eliminating this trade is that the customer wants the unchilled (prerigor) meat. The only way to change this would be to increase the customer's awareness of animal welfare issues or convince them that chilled or frozen meat is a good product.

Old cull livestock and poultry of little economic value

Some of the worst abuse during handling, transport and slaughter occurs with old cull breeding stock. In the USA, cull dairy cows and old breeding sows often travel greater distances than young animals that have been fattened on either grain or grass. There is less economic incentive to treat these animals well because they are less valuable than young feed animals. An effective way to reduce abuses is to increase the value of old breeding stock. This provides an economic incentive to treat them better. Producers need to be educated that if they sell animals before they become skinny and emaciated, they will receive more money for them. In the USA and other parts of the developed world, programmes have been implemented in some areas to fatten old breeding stock so that they will become more valuable for meat.

Highly segmented marketing chains with many dealers, agents and middlemen are bad for animal welfare

In the developed world, such as Europe and North America, most high-quality young animals that are fattened for slaughter go directly from the feedlot or farm to a slaughter plant. This makes it much easier to make people accountable for losses. Old breeding stock often passes through a series of auctions or dealers and the origin of the animals may not be able to be traced. In the developing world, all classes of livestock are often sold through middlemen and dealers. In all countries, the sectors of the livestock market where the animals go through a series of auctions, dealers or middlemen will be the most difficult to improve. In a highly segmented market system there is often no accountability for losses. Middlemen and dealers who do not own the animals have little economic incentive to reduce bruises, injuries and sickness because they are not held financially accountable for losses.

In Australia, vaccinating calves and training them to eat from feed troughs reduced sickness (Walker *et al.*, 2006). In the USA, bovine respiratory disease (BRD) is a major problem. Most cases of BRD could be prevented if ranchers pre-weaned and vaccinated their calves before they left the ranch. Half of the ranchers fail to do this because they receive no financial incentives for vaccinating and pre-weaning their calves (Suther, 2006). To prevent sickness, beef calves should be preconditioned 45 or more days before they are shipped to a feedlot. Preconditioning consists of vaccinations, weaning and teaching the calves to eat from a feed bunk and drink from a water trough. Preconditioning will significantly reduce sickness and low animal performance (National Cattlemen's Beef Association, 1994; Arthington *et al.*, 2008). The best way to motivate ranchers to implement these practices is to pay them a premium for preconditioned calves. Cattle that get sick have lower quality grades due to less marbling and are worth less money (Texas A&M University, 1998; Waggoner *et al.*, 2006).

Marketing systems that can be developed to replace segmented marketing chains and improve welfare

Marketing systems that can be developed to replace segmented marketing chains and improve welfare include:

- **Special sales where only certified preconditioned calves are sold.** State or provincial cattle associations work with local producers to develop these programmes. Buyers will pay more for the cattle (Troxel *et al.*, 2006). Verification by an independent third party that preconditioning procedures were done increased premiums buyers paid to calves (Bulot and Lawrence, 2006).
- **Animals are enrolled in a contractual agreement** with a supermarket, restaurant, meat company or other buyers to produce the animals to certain specifications such as cage free, preconditioned calves, etc. Producers are paid a premium to be in these programmes.
- **Producers sell locally grown animals through a cooperative** that has strict standards for welfare. For all types of programmes, auditing is required to ensure that people follow the specifications.

Overloading the animal's biology causes suffering

Broiler poultry, laying hens, dairy cows and pigs that are raised in intensive systems have been genetically selected over the years to provide more and more meat and milk. Some of the welfare problems caused by single-minded selection for production traits have resulted in increased lameness and leg problems in dairy cows and chickens (Knowles *et al.*, 2008). These problems have gradually become worse over the years and some new people entering the industry do not know that high percentages of lame animals are totally abnormal. There are three basic ways that animal biology can be overloaded to the point where welfare will be poor: genetic selection solely for rapid weight gain; large muscle growth; or ever-increasing milk production.

Single-minded over-selection for production traits has many bad side effects such as reduced disease resistance in pigs (Meeker *et al.*, 1987; Johnson *et al.*, 2005). A study in Belgium showed that the slower-growing genetic types of broiler chickens on organic farms had less fluctuating asymmetry compared to conventional fast-growing genetics (Tuyttens *et al.*, 2007). Fluctuating asymmetry occurs when one side of an animal may be either bigger or smaller than the other side. Animals that are not symmetrical often have genetic defects. A high-producing American Holstein gives twice as much milk, but she may burn out after only two lactations. A New Zealand grass-fed Holstein lasts twice as long. The producer wants to get the short-term economic gains, but in the long run, economics may be worse due to a lack of disease resistance and a higher cost of replacement heifers.

Beta-agonists and growth hormone (rBST)

Feed additives such as beta-agonists that make animals grow lean muscle can cause serious welfare problems if they are fed for too many days or at too high a dose.

In one study, ractopamine (Optaflex®) fed at 200 mg/day for 28 days caused a slight increase in the speed cattle moved during handling in British, Continental and Braham crossbred steers (Baszczak *et al.*, 2006). There were no adverse effects on welfare. Anecdotal reports indicate that it may cause hoof problems in Holsteins housed in a muddy lot, when given at the same dose. Stockyard managers at slaughter plants and my own observations indicate that higher doses for longer times cause weakness and more 'downer' non-ambulatory pigs. It can also make pigs more difficult to handle (Marchant-Forde *et al.*, 2003). The manufacturers have lowered the recommended dosage from 18 g/t of feed to 4.5 g/t of feed. In cattle, over-use and too high a dose of zilpaterol (Zilmax®) or ractopamine in cattle has resulted in lameness and heat stress. Stockyard (lairage) managers in two plants have reported that the outer shell of the hoof fell off feedlot cattle fed too high a dose of beta-agonists. Both the author and the managers have observed that the hooves look normal and they are not elongated like the hoof of an animal with laminitis. Data obtained through the Freedom of Information Act from the US Food and Drug Administration (2006) approval trials for zilpaterol indicate that it made beef tougher when fed at 6.8 g/t of feed. The Warner Bratzler shear force test results were 3.29 kg for the control and 4.01 kg for the zilpaterol cattle with a significance level of $P < 0.001$ (US Food and Drug Administration, 2006). Research conducted by Vasconcelos *et al.* (2008) showed that cattle fed zilpaterol had less marbling ($P < 0.01$). In pigs, ractopamine made pork tougher unless it was aged for 10 days (Xiong *et al.*, 2006). Fernandez-Duenas *et al.* (2008) have also reported a significant increase in Warner Bratzler shear force values in pigs fed ractopamine. All of the studies showed that beta-agonists greatly increased muscle mass and the area of the loin. The cost of this increased amount of meat is poorer meat quality and bad effects on animal welfare unless beta-agonists are used very carefully.

Quality and quantity of meat are two opposing goals. Beef cattle fed too many beta-agonists will have less marbling and tougher meat. Meat companies who want high quality meat have banned or greatly restricted these products in their programmes. Unfortunately there are some meat companies who pay a premium for animals with a high percentage of lean meat. This has provided an economic incentive to overuse beta-agonists, which has resulted in many lame cattle. They do this because they are selling beef to low-income consumers and they put all the meat through a needle machine to tenderize it. Beta-agonists can cause lameness, weakness and hoof problems because they are vasoconstrictors.

The growth hormone rBST has caused problems for dairy cows unless it is used in a very carefully managed programme. It can cause excessive loss of body condition and increase mastitis (Willeberg, 1993; Kronfield, 1994; Collier *et al.*, 2001). People who are assessing welfare in feedlots and farms that use these products should carefully evaluate animal-based outcomes such as body condition, lameness and host stress symptoms such as panting.

Poor insurance programmes reward bad practices

Mistreatment of animals and high levels of bruising and death losses will occur if a truck driver or transport company is reimbursed for every dead, injured or bruised animal. The most effective insurance programmes pay for catastrophic losses such as rolling a truck over but they will not pay for the first five dead pigs. There is a deductable amount of losses that are not covered. This motivates drivers and transporters to handle animals well so they will not have to pay for losses caused by rough handling or carelessness.

Fatigued, overworked truck drivers and stockpeople

To save on transport costs, some drivers are doing too many trips and not getting sufficient sleep.

The worst problems often occur with independent truckers who want to earn more money, but some transport companies have drivers on schedules where getting sufficient sleep is impossible. This has resulted in many truck accidents. Jennifer Woods, a livestock specialist in Canada, found that a high percentage of livestock truck accidents were due to fatigue – 49% of the accidents occur between midnight and 9:00 in the morning. Another indicator that fatigue was a major reason for the accidents was that 80% of the accidents involved a single vehicle and 84% were trucks that rolled to the right because a sleeping driver ran off the road. Weather had little effect. During the winter when there is ice on the road, fewer accidents occurred compared to October, when livestock truck traffic was greatest. These data are based on a total of 415 commercial livestock truck accidents in the USA and Canada (Woods and Grandin, 2008). Data on the accidents were obtained from news reports on the Internet and research from industry.

Overworked employees and overloaded equipment are detrimental to animal welfare

Many people who are interested in animal welfare assume that slaughter plants with a high line speed are bad. Data collected by Grandin (2005) indicated that when equipment is designed and staffed with enough people, line speed had no effect on the percentage of cattle stunned with one shot from a captive bolt (Table 11.1). The author has observed that the worst problems occur when either equipment or people are overloaded. This occurs when the plant's meat sales exceed the capacity of the plant's equipment and staffing. One plant that slaughtered 26 cattle/h worked really well, but when its line speed was increased to 35 cattle/h with no increase in employees and no equipment modifications, employees repeatedly slammed the stun box door on cattle. The plant ended up having its USDA inspection suspended for violation of the Humane Slaughter Act.

Table 11.1. Effect of line speed on stunning in beef slaughter plants (source: Grandin, 2005).

Line speed (cattle/h)	Cattle rendered insensible with a single shot from a captive bolt stunner (%)[a]
<50 (16 plants)	96.2
51–100 (13 plants)	98.9
101–200 (10 plants)	97.4
>200 (27 plants)	96.7

[a] Percentages shown are average values for the number of plants.

Conflict of interest during audits

It is important that the person who is evaluating a farm or slaughter plant does not have a financial conflict of interest. Audits for animal welfare and food safety have the greatest credibility when they are conducted by either a third party, independent auditor, a representative from the company that is buying the meat, a government employee, or an employee of the meat company that has a contract with the producer to grow livestock or poultry. A farm's regular veterinarian has a conflict of interest. If he or she is too strict, the farm may fire him/her. This situation would be like a traffic policeman giving his boss a fine for speeding. To prevent bribery, auditors must be paid enough so they will be less likely to accept bribes. The farm veterinarian plays a very important role in helping his/her clients to obey the standards. The farm veterinarians should conduct internal audits on a regular basis so his/her clients will be ready for the external audit. It is also a conflict of interest for an auditor to be profiting from selling equipment, drugs, feed or any other service to the people who are being audited.

Failure to treat sick animals in organic or natural programmes

Most people who raise livestock or poultry for organic or natural programmes do a good job and, if an organic method for treating a sick animal fails, they will use antibiotics. The organic programmes in the USA and some other countries forbid all use of antibiotics. If a sick animal is treated with antibiotics it has to be removed from the organic programme. The author has observed some cases where cattle had hair falling out due to lice or severe coughing and they were not treated so they could maintain their organic status. Successful organic medicine has a major emphasis on good management practices to prevent disease. One good way to prevent disease is to use breeds or genetic lines of animals that are more hardy and disease resistant. Research done by the USDA showed that certain genetic lines of lean, fast-growing pigs were more susceptible to porcine respiratory and reproductive syndrome (PRRS) (Johnson *et al.*, 2005). For an organic grass-based dairy, New Zealand Holstein genetics may be a better choice than American Holsteins because they are more hardy. At one poorly run organic dairy in the USA, 30% of the Holstein calves died due to the failure to give them antibiotics.

Effect of the price of grain on welfare and the destruction of pasture

In the USA, the beef industry started putting cattle in feedlots during the 1960s because grain was cheap. Cheap grain provided an incentive to take cattle off pasture and raise them in feedlots. The pork and poultry industries also greatly expanded due to availability of cheap grain. Since the early 2000s, the use of grain to make ethanol has increased around the world. This has resulted in high grain prices. The author visited Brazil and Uruguay in 2007 and 2008 and learned that the high price of grain is an economic incentive for turning pastureland into crops. Up to 30% of the prime pastureland in Uruguay has already been converted to growing soybeans. In Brazil, more and more cattle are being moved off pasture into feedlots. A similar situation is also occurring in Argentina. Much of Argentina's grain crop goes to Europe and the Argentinean treasury receives huge amounts of income in grain taxes. It has been estimated that taxes on exported grain may produce 80% of the country's tax revenue (Nation, 2008). Ranchers are converting more and more of their land to grain. In the USA, county extension agents in Illinois report that hilly pastureland that is not really suited for crops is being planted with crops.

To have good animal welfare in a feedlot requires dry ground. The author has observed from travelling all over the USA, Canada, Australia, New Zealand and Central and South America that, in areas where the rainfall is greater than 50 cm (20 inches) of rainfall/year, it is very difficult to keep the surface of the feedlot pens dry. This is the reason why so many US feedlots are in the high plains area where rainfall is less. In the 1970s, ranchers tried building dirt feedlots in the rainy south-eastern USA but they gave it up due to the mud. Some people are moving cattle back to the rainy Midwestern grain-growing areas to take advantage of feeding ethanol by-products, which are wet and more expensive to ship than dry grain. This has resulted in some cattle being put in indoor feedlots.

Sacrificing the well-being of the individual for greater profits

When pigs or chickens are jammed too tightly into a house, the welfare and productivity of each individual animal usually declines. Unfortunately, there is an economic incentive to do this because the overall output of eggs or meat per house may be greater. The bad economic incentive to overstock a building is most likely to occur in places where both land and buildings are expensive.

Standards become less strict when demand for the product exceeds the supply

The author has worked with many of the grocery stores that sell either organic or high welfare standard meat, milk or eggs. In the beginning, they start out with strict standards. When the product becomes popular, demand exceeds supply. The grocery store cannot get enough products so it is tempted to either lower its standard or start using less reliable suppliers. The author observed a natural beef company that became lax in enforcing its own standards and the practices became so bad in its overworked little slaughter plant that it got shut down several times by the USDA for humane slaughter violations. Standards can also slip when the original founder of the company leaves and the company is sold to a large corporation. When the founder leaves, the company no longer has the founder's vision.

Customers are very poor and feeding a hungry family has priority over animal welfare

People in this situation will buy the cheapest meat they can get. This is especially a problem in poor, developing countries. Feeding the family is their first priority. The meat company that is encouraging the use of zilpaterol is selling to low-income customers.

Cutting horns off mature animals is a painful, stressful procedure

When cattle are transported, those with horns will have more bruises than cattle with no horns. A recent study in West Africa showed that breeds with massive horns had more bruises and injuries compared to breeds with smaller horns (Minka and Ayo, 2008). Tipping (cutting the tips) of the horns does not reduce bruising (Ramsey *et al.*, 1976). Some transporters and producers cut horns on adult animals to reduce damage during transport. However, this practice of cutting horns on adult animals should be banned as it is a painful and stressful procedure. Cutting horns of large feedlot cattle is currently occurring in the USA and other countries because the slaughter plant tells them to do this, and has no concern for welfare. Instead, horns should be removed from small calves before they grow long, or polled (hornless) cattle breeds should be used. In addition, many injuries on African cattle could probably be reduced by having lower stocking densities on the truck, using non-slip flooring and quiet handling. To encourage producers to dehorn calves, charging a 'horn tax' works well. There should also be an additional financial deduction for cut horns on mature animals.

Welfare problems caused by war and corruption

Unstable or corrupt conditions in a developing country may create economic incentives that motivate people to severely abuse animals. An extension animal scientist, who cannot be named due to concerns for his safety, has worked in a country where corruption and bad decisions by the government have provided an economic incentive to the local people to use atrocious management practices. Water buffalo and local cattle are given so much rBST (growth hormone) and oxytocin that they are milked until they have become emaciated skeletons. The dairies are located on the perimeter of a large city. Due to war and other conditions, there is no economic incentive to rebreed the animals for another cycle of lactation. The animals are just 'used up' and then dealers bring in more animals that they purchase from destitute farmers who live in the surrounding countryside. The drugs used are cheap copies that are obtained from firms in Asia.

Livestock industry in developed countries profits by selling inappropriate breeding animals to developing countries

Breeding animals such as dairy heifers are shipped around the world to many countries. Some of this trade is beneficial and helps improve livestock genetics but there are cases where the trade has caused severe welfare problems. Private industry profits because government agencies in developing countries mistakenly believe that high-producing animals such as Holstein dairy heifers will improve their dairy industry. Sometimes these heifers die quickly because the receiving country does not have the facilities to protect them from heat stress or sufficient feed to support high production. Another disadvantage of this trade is that local hardy breeds of cattle, pigs or chickens may be lost. In the future, the hardy local breeds may be essential for breeding disease resistance. Some of the most innovative programmes cross-breed imported animals with local hardy breeds. In Mexico, the author observed that Holsteins are crossed with local Zebu cattle to provide both hardiness and heat tolerance.

Well-intentioned Legislation with Bad Consequences

Over the years, the author has observed many situations where well-intentioned legislation or activist activities may be detrimental to improving animal welfare. Prime examples of these are the laws in the USA banning horse slaughter for human consumption. The closure of two out of the three US horse slaughter plants has already resulted in unwanted horses being transported even further distances to either Canada or Mexico for slaughter. Some have been transported all the way to Mexico City. Live horses are also being shipped to Japan. When the Humane Society of the USA lobbied the government to pass this law, nobody thought about worse fates that some unwanted horses could suffer. The fates that are worse than slaughter in Texas or Illinois are: (i) longer transport times; (ii) transport under substandard conditions in Mexico; (iii) being neglected and left to starve in the desert (high hay and grain prices have made this problem worse); and (iv) being ridden and worked in Mexico until they become totally debilitated. The author has seen these worse fates and they are awful. Horse slaughter became such an emotional issue that animal advocates chose to ignore the observations of people in the field that indicated that there are worse fates than slaughter in a US plant.

Animal welfare can also become worse when legislation makes standards so strict that an animal industry is shut down in one country and production is transferred to another country that has atrocious standards. Eggs are now being imported from Eastern Europe into Western Europe. The standards for animal welfare in Eastern Europe are poor. Exporters of eggs, milk and meat should be required to adhere to the standards of the importing country.

Conclusions

An understanding of how economic factors affect how farm animals are treated will help policy makers to improve welfare. Holding people financially accountable for losses or providing incentives for low losses will greatly improve animal treatment. The wise use of the tremendous purchasing power of large meat buyers has already brought about some dramatic improvements. Unfortunately, there are also many situations where economics is detrimental to animal welfare. One of the worst problems is pushing animals past their biological limits by either single-minded selection for production or the over use of performance-enhancing substances.

References

Arthington, J.D., Qiu, X., Cooke, R.F., Vendramini, J.M.B., Araujo, D.B., Chase, C.C. and Coleman, S.W. (2008) Effects of preshipping management on measures of stress and performances of beef steers during feedlot receiving. *Journal of Animal Science* 86, 2016–2023.

Baszczak, J.A., Grandin, T., Gruber, S.L., Engle, T.E., Platter, W.J., Laudert, S.B., Schroeder, A.L. and Tatum, J.D. (2006) Effects of ractopamine supplementation on behavior of British, Continental and Brahman crossbred steers during routine handling. *Journal of Animal Science* 84, 3410–3414.

Bulot, H. and Lawrence, J.D. (2006) The value of third party certification of preconditioning claims in Iowa. Department of Economics Working Papers, Iowa State University, Ames, Iowa.

Butler, D. (2008) Crossing the valley of death. *Nature* 453, 840–842.

Collier, R.J., Byatt, J.C., Denham, S.C., Eppard, P.J., Fabellar, A.C., Hintz, R.L., McGrath, M.F., McLaughlin, C.L., Shearer, J.K., Veenhuizen, J.J. and Vicini, J.L. (2001) Effects of sustained release bovine somatotropin (sometribove) on animal health in commercial dairy herds. *Journal of Dairy Science* 84, 1098–1108.

Douphrate, D.L., Rosecrance, J.C., Stallone, L., Reynolds, S.J. and Gilkes, D.P. (2009) Livestock handling injuries in agriculture: an analysis of workers compensation data. *American Journal of Industrial Medicine* 5 February (Epub). Available at: www.pubmedcentral.nih.gov (accessed 7 July 2009).

Editor in the *Independent* (2008) Editorial and opinion. *Independent* (London), 24 May, p. 44.

Etter, L. (2008) Have knife will travel: a slaughterhouse on wheels. *Wall Street Journal*, 5 September, p. 1.

Ewbank, R. and Parker, M. (2007) Handling cattle raised in close association with people. In: Grandin, T. (ed.) *Livestock Handling and Transport*. CAB International, Wallingford, UK, pp. 76–89.

Fernandez-Duenas, D.M., Myer, A.J., Scramlin, C., Parks, C.W., Carr, S.N., Killefer, J. and McKeith, F.K. (2008) Carcass meat quality and sensory characteristics of heavy weight pigs fed ractopamine hydrochloride (Paylean). *Journal of Animal Science* 86, 3544–3550.

Grandin, T. (1981) Bruises on Southwestern feedlot cattle. *Journal of Animal Science* 53 (Supplement 1), 213 (abstract).

Grandin, T. (1985) Improving pork quality through handling systems. In: *Animal Health and Nutrition*. Watt Publishing, Mount Morris, Illinois, pp. 14–26.

Grandin, T. (1987) Animal handling. *Veterinary Clinics of North America, Food Animal Practice* 3, 323–338.

Grandin, T. (1988) Double rail restrainer conveyor for livestock handling. *Journal of Agricultural Engineering Research* 41, 327–338.

Grandin, T. (1998a) Objective scoring of animal handling and stunning practices in slaughter plants. *Journal of the American Veterinary Medical Association* 212, 36–39.

Grandin, T. (1998b) Handling methods and facilities to reduce stress on cattle. *Veterinary Clinics of North America, Food Animal Practice* 14, 325–341.

Grandin, T. (2005) Maintenance of good animal welfare standards in beef slaughter plants by use of auditing programs. *Journal American Veterinary Medical Association* 226, 370–373.

Grandin, T. (2007a) Introduction: effect of customer requirements, international standards and marketing structure on the handling and transport of livestock and poultry. In: Grandin, T. (ed.) *Livestock Handling and Transport*. CAB International, Wallingford, UK, pp. 1–18.

Grandin, T. (2007b) *Recommended Animal Handling Guidelines and Audit Guide*, 2007 edn. American Meat Institute Foundation, Washington, DC. Available at: www.animalhandling.org (accessed 5 July 2009).

Grandin, T. and Gallo, C. (2007) Cattle transport. In: Grandin, T. (ed.) *Livestock Handling and Transport*. CAB International, Wallingford, UK, pp. 134–154.

Gruber, S.L., Tatum, J.D., Grandin, T., Scanga, J.A., Belk, K.E. and Smith, G.C. (2006) Is the difference in tenderness commonly observed between heifers and steers attributable to differences in temperament and reaction to pre-harvest stress? Final Report, submitted to the National Cattlemen's Beef Association, Department of Animal Sciences, Colorado State University, Fort Collins, Colorado, pp. 1–38.

Hambrecht, E., Eissen, J.J., Newman, D.J., Verstegen, M.W. and Hartog, L.A. (2005) Preslaughter handling affects pork quality and glycolytic potential in two muscles differing in fiber type organization. *Journal of Animal Science* 83, 900–907.

Johnson, R., Petry, D. and Lurney, J. (2005) Genetic resistance to PRRS studied. Swine research review. *National Hog Farmer*, 15 December, pp. 19–20.

Knowles, T.G., Kestin, S.C., Hasslam, S.M., Brown, S.N., Green, L.E., Butterworth, A., Pope, S.J., Pfeiffer, D. and Nicol, C.J. (2008) Leg disorders in broiler chickens: prevalance, risk factors and prevention. *PLOS One* 3(2). Available at: www.pubmedcentral.nih.gov/articlerender.fegi?articl=2212134 (accessed 6 July 2009).

Kronfield, D.S. (1994) Health management of dairy herds treated with bovine somatotropin. *Journal of American Veterinary Medical Association* 204, 116–130.

Marchant-Forde, J.N., Lay, D.C., Pajor, J.A., Richert, B.T. and Schinckel, A.P. (2003) The effects of ractopamine on the behavior and physiology of finishing pig. *Journal of Animal Science* 81, 416–422.

Maria, G.A., Villaroel, M., Chacon, C. and Gebresenbet, C. (2004) Scoring system for evaluating stress to cattle during commercial loading and unloading. *Veterinary Record* 154, 818–821.

Meeker, D.L., Rothschild, M.F., Christian, L.L., Warner, C.M. and Hill, H.T. (1987) Genetic control of immune response to pseudorabiens and atrophic rhinitis vaccines. *Journal of Animal Science* 64, 407–413.

Minka, N.S. and Ayo, J.O. (2008) Effect of loading behavior and road transport stress on traumatic injuries in cattle transported by road during the hot, dry season. *Livestock Science* 107, 91–95.

Nation, A. (2008) High grain prices are creating more confinement grain feeding in Argentina. *The Stockman Grass Farmer*, January, pp. 1–4.

National Cattlemen's Beef Association (1994) *Strategic Alliances Field Study*. Coordination with Colorado State University, Texas A&M University and the National Cattlemen's Beef Association, Englewood, Colorado.

National Institute of Animal Agriculture (NIAA) (2007) Shortage of food animal veterinarians: a call for action. In: *Cattle Health Newsletter* Summer. NIAA, Bowling Green, Kentucky, p. 6.

Ramsey, W.R., Meischke, H.R.C. and Anderson, B. (1976) The effect of tipping horns and interruption of the journey on bruising cattle. *Australian Veterinary Journal* 52, 285–286.

Roeber, D.L., Mies, P.D., Smith, C.D., Field, T.G., Tatum, J.D., Scanga, J.A. and Smith, G.C. (2001) National market cow and bull beef quality audit, 1999: a survey of producer-related defects in market cows and bulls. *Journal of Animal Science* 79, 658–665.

Smith, G. (1998) *Moving Em, a Guide to Low Stress Animal Handling*. Graziers, Hui, Kamuela, Hawaii.

Suther, S. (2006) CAB (Certified Angus) drovers survey: benchmarks for producer practices and opinions. *Drover's Journal*, Vance Publishing, Lenexa, Kansas, pp. 18–19.

Texas A&M University (1998) *Ranch-to-Rail Statistics 1992–1998*. Department of Animal Science, Texas A&M University, College Station, Texas.

Troxel, T.R., Barham, B.L., Cline, S., Foley, D., Hardgrave, R., Wiedower, R. and Wiedower, W. (2006) Management factors affecting selling prices of Arkansas beef calves. *Journal of Animal Science* 64 (Supplement 1), 12 (abstract).

Tuyttens, F., Heyndrickx, M., DeBoeck, M., Moreels, A., Nuffel, A.V., Poucke, E.V., Coillie, E.V., van Doogan, S. and Lens, L. (2007) Broiler chicken health, welfare and fluctuating asymmetry in organic versus conventional production systems. *Livestock Science* 113, 123–132.

US Food and Drug Administration (2006) New animal drug application for Zilmax (zilpaterol). Freedom of Information Act Summary. 10 August. US Food and Drug Administration, Washington, DC, pp. 141–258.

Vasconselos, J.T., Rathman, R.J., Reuter, R.R., Leibovich, J., McMeniman, J.P., Hales, K.E., Covey, T.L., Miller, M.F., Nicholas, W.T. and Galyean, M.L. (2008) Effects of duration of zilpaterol hydrochloride feeding and days on the finishing diet on feedlot cattle performance and carcass traits. *Journal of Animal Science* 86(8), 2005–2015.

Voisinet, B.D., Grandin, T., O'Connor, S.F., Tatum, J.D. and Deesing, M.J. (1997) *Bos indicus* cross feedlot cattle with excitable temperaments have tougher meat and a higher incidence of borderline dark cutters. *Meat Science* 46, 367–377.

Waggoner, J.W., Mathis, C.P., Loest, C.A., Sawyer, J.E. and McColum, F.T. (2006) Impact of feedlot morbidity on performance, carcass characteristics and profitability of New Mexico ranch-to-rail steers. *Journal of Animal Science* 84 (Supplement 1), 12 (abstract).

Walker, K.H., Fell, L.R., Reddacliff, L.A., Kilgour, R.J., House, J.R., Wilson, S.C. and Nichols, P.J. (2006) Effects of yard weaning and behavioral adaptation of cattle to a feedlot. *Livestock Science* 106, 210–217.

Willeberg, P. (1993) Bovine somatotropin and clinical mastitis: epidemiological assessment of the welfare risk. *Livestock Production Science* 36, 55–66.

Woods, J. and Grandin, T. (2008) Fatigue is a major cause of commercial livestock truck accidents. *Veterinaria Italiana* 44, 259–262.

Xiong, Y.L., Gower, M.J., Elmore, C.A., Cromwell, G.L. and Lindemann, M.D. (2006) Effect of dietary ractopamine on tenderness and postmortem protein degradation of pork muscle. *Meat Science* 72, 600–604.

12 Improving Animal Welfare: Practical Approaches for Achieving Change

Helen R. Whay and David C.J. Main

University of Bristol, Langford, Bristol, UK

Introduction

This chapter addresses the implementation of knowledge to improve animal welfare. There is a vast wealth of information available about what management practices, activities and processes lead to improvements in animal welfare. Such information comes from the work of animal welfare scientists and agricultural scientists or is captured from the experiences of farmers, livestock hauliers, auctioneers and abattoir staff and is extensively published elsewhere. However, this chapter will not cover the technical nuts and bolts of how to meet animals' requirements. This chapter is about the process of taking this vast wealth of knowledge and translating it into actions by those who have responsibility for animals. In short, this chapter is about changing the behaviour of human beings.

Human-behaviour change is a science in itself, commonly associated with the term 'social medicine'. For years, workers in human medicine have been trying to encourage groups within the population to quit smoking, lose weight, drink less alcohol, practise safe sex or eat more fruit and vegetables. This need to change human behaviour has also extended beyond health to encompass activities such as recycling refuse, saving water, reducing speed on roads, wearing seat belts while driving and so on. The relevance of this for animal welfare is that we have to recognize that the lives of modern domesticated animals, including production animals, are under the total and absolute control of humans. Humans decide what food they have access to, how much and when. Humans decide what environment they should live in, how much bedding should be available, how clean and dry it will be and how much light they will have to see it by. Humans control animal health; they choose management strategies that might promote or prevent transmission of infectious disease, they decide on vaccination programmes and decide when and if treatment is offered. Humans even control how and when an animal will die. Once we accept that animals have virtually no genuine control over their own lives then it becomes obvious that in order to achieve welfare improvement we need to work with the humans responsible for their care. This may come as something of a surprise to those of us who entered agriculture or veterinary science because we thought we wanted to work with animals.

Targets for Behaviour Change: Who Do We Want to Influence?

If we accept that humans are the mediators of animal welfare, it is worth looking at the various players who have a stake in implementing change. Animal carers, in this case farmers, transporters and abattoir staff, are the most obvious targets for animal welfare behaviour change interventions. The knowledge generated by science needs to feed through to and be implemented on the ground by those people who have direct control over animal lives. It is important to recognize that within this group there are people who behave in a competitive, entrepreneurial way and who automatically and regularly implement changes to the management of their livestock. There are also those who want to do the best for the livestock under their care and who do spontaneously make some management changes but find keeping up with and filtering current knowledge a challenge, perhaps finding that translating information so that it is relevant to their own situation is difficult, and who often struggle to divert from existing routines in order to try out new ideas. Finally, there is the group who, again, would like to manage their stock in the best way possible but always seem to be paddling hard just to stay still and find virtually no opportunities to keep up with the latest information or to implement management changes; these are the 'crisis management' folks. Beyond the producers there are, of course, many other groups with an interest in implementing behaviour change: farm advisors and sales representatives, farm assurers and standard-setting bodies, legislators, animal welfare charities and campaigners, veterinary surgeons and animal health technicians, animal welfare scientists, retailers and in some cases the purchasers of the final product. These different interest groups often come with different motivations for wanting to see changes implemented by livestock carers. Most, although possibly to varying degrees, would like to see change implemented to make the lives of animals better. A farm advisor is also looking for the credibility and enhancement of reputation that comes with their advice actually being implemented on farm. Legislators are interested in seeing legislation properly implemented, not only because laws are introduced for a purpose, but because failure to implement legislation undermines its role within society. Animal welfare scientists hope to see their work taken up and used; but further to that, the process of implementation may allow them to see how well their findings actually work at the farm level. Then, as a final example, there are the all-important, but often overlooked, customers. Customers of animal products have tremendous power to influence animal welfare; they just don't know they have it. Unfortunately only a relatively small proportion of food customers purchase primarily on the basis of the animal welfare provenance of their food. However, interest in ethical purchasing is growing and a recent survey carried out by a UK supermarket (Talking Retail, 2008) found that their customers rated animal welfare as a greater purchasing motivation than environmental issues. Many of the interest groups described above can be seen as having a stake in seeing animal carers implement changes on their farms, but it is also worth remembering that these interest groups can also be targets for behaviour-change interventions themselves.

So far this chapter has highlighted that changing human behaviour is the route to improving animal welfare and that many groups have an interest in seeing changes implemented. However, if it were as simple as 'wish and it be done' then it would probably not merit a chapter within this book. Unfortunately there is considerable evidence that humans often find changing their behaviour extremely difficult; one only has to look at the number of people still smoking cigarettes or trying to lose weight to appreciate the magnitude of the problem. This challenge is then magnified by the fact that implementing animal welfare improvements is about asking people to make changes on behalf of a third party: the animal. If people find it difficult to implement changes such as taking daily exercise or reducing alcohol consumption that protect their own health, it is clearly a challenge to ask people to make changes to improve the lives of animals. An example from the agricultural sector illustrating how knowledge and change have failed to be successfully implemented is dairy cattle lameness in the UK. Since the late 1980s information about the levels of lameness among the UK herd has been regularly collected (Clarkson *et al.*, 1996; Whay *et al.*, 2003). These data tell us that on any one day, between 20 and 25% of the national dairy herd are likely to be lame; this figure has remained more or less constant despite the huge wealth of available information about methods for reducing lameness. When asked why new lameness management strategies are not being implemented, farmers most commonly report lack of time, skilled labour or the unpopularity of the task (K.L. Leach, Z.E. Barker, A. Bell, C. Maggs, H.R. Whay and D.C.J. Main, 2009, unpublished paper). Throughout this chapter, examples relating to dairy cattle lameness will feature quite prominently; this is because it is an area where there has been significant investment in behaviour change research and an area in which the authors have first-hand experience.

In order to bring about a change in behaviour to address an animal welfare problem, there must be genuine awareness that a problem exists coupled with knowledge about possible solutions (i.e. actions that can be taken or changes made to improve the situation). Unfortunately, awareness of a problem and knowledge of a solution do not necessarily lead to changes being made. It is the job of those who see their role as facilitating change to create situations and present options for change in a format that will help people to take the step that leads to changes in routines, plans or working practices.

An important point to bear in mind is that not all people are susceptible to the same types of intervention. Figure 12.1 illustrates the concept of dividing up producers into categories according to which type of support they are most likely to respond to. On the extreme left of the diagram are producers whose animals are in the poorest welfare and who are unlikely to access and take up mechanisms of support in order to help themselves. This group is only susceptible to enforcement as a means of promoting change. This might be enforcement of legislation and codes of practice and using tools such as farm assurance schemes to set minimum standards and ensure compliance before products can reach the marketplace. The middle section of the diagram describes the majority of producers, who are likely to make changes with a combination of encouragement and enforcement but are unlikely to actuate change for themselves so require external contact to initiate and, to some extent, sustain the process. The far-right group are those producers who are self-motivated and self-actuating to a high degree; they tend to be competitive, prepared to take risks and are regularly looking for new market opportunities. This group does not require intervention as such, but benefits from receiving access to new knowledge and scientific findings. They are rewarded by being able to produce premium-priced products, cornering market share or accessing niche marketing opportunities. This group has a tendency not to feel the need to either share or receive information from their peers. The broken curve denotes the possibility of achieving improvements in animal welfare among all three groups by matching them with appropriate intervention approaches.

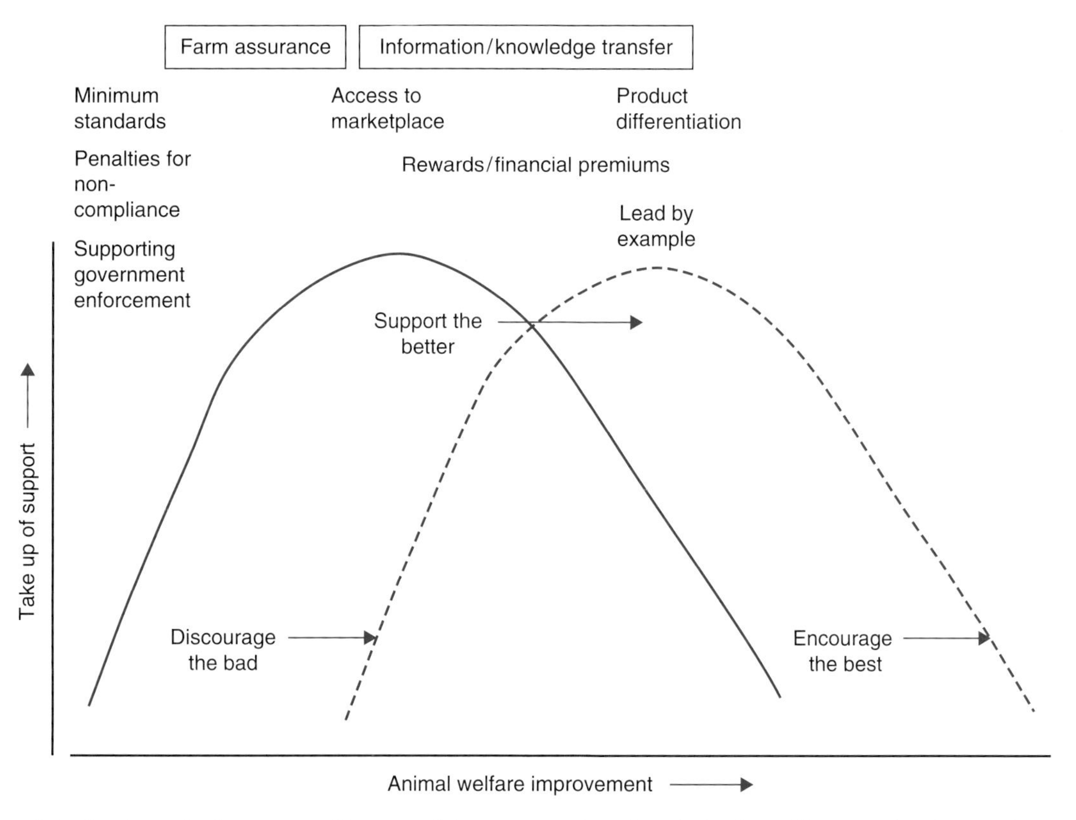

Fig. 12.1. A schematic diagram matching potential categories of producer behaviour with appropriate intervention approaches (adapted from Defra, 2008).

The following sections within this chapter will review some of the currently available approaches for both encouraging and enforcing change.

Encouragement as a Means of Implementing Change

The key principle of encouraging sustainable behaviour change is to transfer ownership of both the problem and the solution to the person, in this case the producer or farmer, responsible for implementing change. The golden rule to remember is that telling people what to do, no matter how encouragingly, simply does not work – no matter how wonderful, strikingly obvious or just plain sensible the information is, no matter how beautifully and enthusiastically presented the solutions may be, and no matter how simple and straightforward the change you are hoping to see implemented. This point is well illustrated in the excellent book *Fostering Sustainable Behaviour* (McKenzie-Mohr and Smith, 1999) where the authors Doug McKenzie-Mohr and William Smith give the example of some work done by Scott Geller (Geller, 1981). Geller and colleagues ran a series of workshops for householders providing information about the importance of home energy conservation and providing advice on simple measures that could be implemented by each householder to conserve energy. The workshops were designed to give the householders the message that it was entirely possible for them to make these changes in their own homes. The attitude of the participating householders was measured before and after the workshops and it was found that following the workshop, attendees reported a much greater awareness of energy conservation issues and an intention to implement energy-saving measures within their own homes. However, this change in attitude was not translated into a change in behaviour. During follow-up visits to 40 of the participating households it transpired that only one person had followed through on the recommendation to lower the thermostat on the hot-water boiler. Two participants had placed blankets around their boilers as advised, but it transpired that they had done this before attending the workshop. Indeed, the only significant change to be implemented was that eight of the workshop participants had installed low-flow showerheads, though this was not precipitated by the advice alone, as each participant who attended the workshop was provided with a free low-flow showerhead.

The concept of ownership of change is much more subtle that simply saying to a producer 'it is you who has to implement these changes so therefore you are the owner of them'. It is also not quite as cynical as trying to make someone think that the changes were their idea in the first place. Giving someone ownership of change is about creating opportunities for them to explore and realize their problem and allowing them to be a partner in generating ideas for possible solutions. Pointing out to a farmer that their animals are suffering from a welfare problem and even going as far as suggesting they may be breaching laws or codes of practice is a challenge to their dignity and professionalism and can easily be seen as a hostile act. To return to the previously used example of dairy cattle lameness: imagine a scenario where a farmer is aware that he or she has lame cows in the herd but has not yet considered the implications of this lameness. By asking a series of questions about what effect it has on the cows, how the cows feel when they are lame, what the time costs are for the farm workers, and in what ways lame cows cost money, the farmer is able to build up a picture of all the potential consequences of having lame cows on the farm and rationalize how important these pitfalls are. So now the problem belongs to the producer and is personal to their farm. Others may not accept that lameness is a problem until they hear about and discuss the problem with other farmers in their area. Likewise, ownership of solutions is important because there are rarely one-size-fits-all solutions to problems on farms, as no two farms operate in exactly the same way. While considering and discussing possible solutions, a farmer needs to run through implementing the change on their own farm in their mind; this process of rehearsing changes is an important step in achieving implementation. In fact, rehearsing change may even extend as far as carrying out a series of tests of the desired change before fully committing to it. Again, it is also often important to hear from others that they too are trying out changes, and often it is more reassuring to hear from a fellow farmer that a particular action worked than to hear it from a farm advisor, veterinary surgeon or welfare scientist.

So the key aims of an encouragement approach are to:

- pass ownership of the problem and solution to the farmer;
- give farmers the opportunity to mentally rehearse any changes they might make and even encourage them to try out changes before fully implementing them; and
- encourage farmers to discuss their problems with colleagues on other farms – it is very valuable for them to know that making changes is a normal behaviour.

In fact, encouraging change seems to be much more about creating the right environment for farmers to realize and address problems for themselves than about being too forceful and hands on in addressing a problem directly. However, it is worth remembering a few cautionary points. It will never be possible to achieve 100% compliance with any intervention; people are individuals and may well have other priorities in their lives that overshadow the purpose of your proposed area of intervention. People are often unpredictable and do not make decisions that we, from our own standpoint, consider logical (bearing in mind that to others we appear equally illogical). In a guide to human-behaviour change Kerr *et al.* (2005) wrote: 'interventions often make the mistake of assuming human behaviours are straight forwardly controlled by reason, attitude and will', which is a good reminder that people do not behave in a particular way just because we want them to. There also seems to be some basic law of intervention returns that dictates that you only achieve success commensurate with the effort you put in, and this appears to be true both for the facilitator and the implementer, so don't expect changing behaviour to be easy and be prepared to commit sufficient resources to get the job done. Finally, the types of intervention described in this chapter are of course a form of manipulation, albeit well intentioned and on behalf of animals that are unable to represent themselves. However, those who develop skills in this type of manipulation can give in to the temptation to manipulate others for their own Machiavellian reasons; this would be unethical and diminish all those who try to promote behaviour change for the benefit of other humans and animals.

The following section describes three broad approaches (social marketing, participatory methods and farmer groups) for encouraging behaviour change that are currently being implemented and tested in the agricultural sector. Each will be considered in terms if its potential strengths and weaknesses.

A social marketing (type) approach

Social marketing is widely used to promote change among groups of people. It is an extension of the principles used in marketing and advertising. We are all familiar with the way advertisers try to persuade us to change our habits and purchase products by implying that owning a particular item will make us glamorous, popular, save us time and so on. Advertisers also attempt to reassure us by letting us know that many others are making a similar purchase and offering us special incentives to encourage us to spend our money. Social marketing is also about trying to persuade people to do something different and uses a range of tools to achieve this, but the two key differences from commercial marketing are that social marketing is: (i) directed towards encouraging change that affords a benefit to society (or in our case to animals and their owners); and (ii) works very specifically to identify and overcome barriers to change.

In an animal welfare and agricultural context, we have found that we have been able to apply many of the tools of social marketing but have had to make some modifications to take into account the isolated nature of farmers. Farmers in the UK often work alone on their farms, they have very limited contact with others and their days involve completing a lot of repetitive, routine tasks. So social marketing for farmers needs to include more contact with individuals than would normally be expected. To understand how social marketing works the following example will illustrate the types of tools available while describing a project currently running in the UK to encourage dairy farmers to take action to reduce lameness in their dairy cattle.

Benefits and barriers

For every desired change in behaviour there will be both perceived benefits and perceived barriers. These may be internal, external or both:

- **Internal benefits** may include a change that aligns with a person's ethical position (i.e. believing that they should do everything possible for the welfare of their animals), offering a feeling of pride or fellowship with others who are working on a similar problem.

- **External benefits** may include believing that the change will save time, offer economic benefit or perhaps contribute to making other tasks on the farm easier as well. For example keeping the feet of cows clean in order to reduce infectious lameness may also result in cleaner udders and faster milking times.
- **Internal barriers** to change may include a fear that the change will be inconvenient, time consuming and difficult to implement within existing routines. A further internal barrier might simply be fear of change itself.
- **External barriers** might include a lack of appropriate equipment, for example the yard scraper may be inefficient and need repair, modification or replacement. Other external barriers might be losing time on other important jobs, having a farm layout that makes the change difficult or hearing from others that making changes is a waste of time and won't work.

It is important for anyone promoting behaviour change to recognize that these perceptions of benefits and barriers exist. Also that other potential routes of behaviour change may exist which may be similar but will not be of equal animal welfare benefit. For example a farmer may perceive that it is indeed important to keep the yards cleaner than they currently are but be reluctant to increase the frequency of yard scraping each day, and instead he or she proposes to flood wash the yards. This will keep the yards very clean and exploit the currently unused wastewater from the milking parlour. However, the reason for cleaning the yards was to help reduce digital dermatitis by creating a cleaner, dryer environment for the cows and the flood washing will regularly soak the cows' feet, increasing their susceptibility to infectious lesions. Ensuring a proper understanding of benefits and barriers from the farmers' point of view is a cornerstone of social marketing and running focus groups is a good way of doing this. It also allows learning of local colloquial terms that are used and finding phrases and quotes that will make sense when speaking to farmers about making changes.

Facilitation

Each farmer receives a visit from a member of the behaviour change project who has a good knowledge of cattle lameness and the types of changes needed to resolve the problem. The visit is not about giving advice but about helping the farmer to generate solutions that are appropriate to his or her own farm. The facilitator will walk around the farm with the farmer and ask questions about particular aspects of the farm that are likely to be risk factors for lameness. The facilitator will address barriers to change presented by the farmer by encouraging him or her to weigh them against potential benefits. The facilitator will also share experiences of other farmers by describing actions they have taken, and offer contact details of other farmers (with their permission) that have found ways of tackling a similar problem. At the end of the facilitated visit, before leaving the farm, the facilitator will compile an action list of the changes the farmer has identified, including notes on who will be responsible for implementing each change (the farm manager, herdsman, tractor driver, etc.) and when the change is going to be implemented, along with a space to tick when the change has been introduced. This list is then left with the farmer for the coming year.

'Norms'

Norms describe the process of reassuring farmers that others are also making changes (i.e. that it is normal behaviour to make changes to reduce lameness). Despite the fact that we mostly perceive ourselves as being individuals who plough our own furrow, social research shows that we are actually more comfortable and reassured when we know that others are doing the same thing. A good example that illustrates this point was an American advertisement run in the 1970s that showed a Native American shedding a tear over the widespread corruption of the environment. This was a powerful and effective campaign. Later, to capitalize on this success, a second advertisement was designed that showed people leaving their rubbish and waste at a bus stop while the tear-shedding Native American looked on. Much to the consternation of the advertisers, this poster precipitated an increase in littering at bus stops. What happened was that this second poster showed people that littering at bus stops was normal and something that everyone else does; in effect, it gave them permission to drop litter at the bus stop.

The way norms have been used within the cattle lameness project was first to create an image and name for the project (illustrated in Fig. 12.2) so

Fig. 12.2. The lameness project logo, used as a 'norm' to create a group identity and to let farmers know they are working together with other farmers to reduce lameness.

that all the participants are aware they belong to a larger project in which others are involved and that they have a group identity they can be proud of.

As previously described, norms are also created through describing what changes other farmers have made on their farms. This helps to address perceived barriers, but also acts to reassure each farmer that others are also making changes and overcoming problems. The activities of other farmers are relayed using verbal descriptions, photographs of what they have changed (with their permission) and using quotes (e.g. 'You can fix things around the farm but you can't fix cows' feet if you damage them'). The project also produces a regular newsletter which features case examples of farms where changes have been implemented.

Commitment

Commitment on the part of project members is critical for sustaining behaviour change. Although farmers who join a project such as the lameness project being described here are showing some degree of commitment simply by joining up, this is often more in the form of a spectator than a fully signed-up member. There are various techniques to encourage more positive commitment. People tend to like to think of themselves as behaving consistently, so by displaying signs of commitment to a project they are more likely to be loyal to the project and continue with any changes they have implemented. Within the lameness project, all participating farmers are given a jacket lapel badge and a car sticker of the project logo (shown in Fig. 12.2) and are encouraged to display them. Although this is a relatively small act, by showing others that they are part of the project they are more likely to go on to take on the more challenging behaviour change. To use an example from a different context (Goldstein *et al.*, 2007), it was found that a group of people who were asked to wear a lapel pin for a charitable cancer society were subsequently nearly twice as likely to donate money to the charity as people who were not asked to wear a lapel pin.

In the lameness project, pictures of changes that have been made are shown to other farmers also considering similar changes. This is not only useful in creating norms and overcoming barriers among the farmers considering change, but the farmer who has already implemented the change is very clearly told that others will be shown the photograph of the change he or she has made. This serves multiple purposes: by agreeing to allow the photograph to be shown to others, the farmer is agreeing to display a further outward sign of commitment to the project; knowing that others are seeing the photograph also encourages the farmer to sustain the change that has been made. This is further reinforced by asking permission to give their contact details to other farmers who might wish to come and view the changes.

A further area where commitment is promoted is through asking farmers to put their signature on the action plan which is drawn up during the facilitation visit. Permission is also sought to compile a list of farmers who have signed up to their action plan so that it can be posted on the project web site. Again, this is to encourage as much outward display of commitment to the project as possible.

Prompts

Prompts are aides-memoire to help remind people of their intended changes in behaviour. Although people's intentions to change a particular practice or habit are generally good, new activities can easily be forgotten or slip the mind, especially when they involve making changes to existing routines or when people find themselves under time pressure. Prompts need to be targeted as closely as possible to the desired behaviour change; very generic prompts tend to be less effective. Within the lameness project, extremely targeted prompts have so far proved difficult to create because there is such a huge diversity of different behaviour changes being initiated. To date, the prompts that have been used (besides the logo and strap line shown in Fig. 12.2) are cartoons targeting specific activities (an example is shown in Fig. 12.3).

A catalogue of suppliers of equipment, services and materials that are commonly needed when making lameness-reducing changes is presented to the farmer at the time when the facilitated action list is generated. The catalogue is intended to prompt picking up the telephone and placing an order or booking a service, etc., as a common stalling point for action is farmers saying they don't know where to buy a material (e.g. wood shavings to spread on cows' beds to increase their lying comfort); the catalogue overcomes this. Farmers whose action list involves introducing or changing foot-bathing routines are given a small laminated card that gives the appropriate dilutions for different foot-bath chemicals and explains in very straightforward measures how to achieve the dilution for a range of foot-bath sizes. Finally, a poster has been made that highlights the financial losses associated with lameness (Fig. 12.4). This is only used once farmers have begun to initiate behaviour changes. Contrary to popular opinion, especially in the agricultural sector, promoting the financial benefit of behaviour change rarely precipitates action. However, once behaviours are changed and people start to notice the financial benefit for themselves, it is a useful tool in sustaining behaviour change.

Incentives

Incentives can be very powerful tools for behaviour change; they may take the form of financial rewards or penalties, offers of approval or acceptance or small 'drug-company type' gifts. It is remarkable how motivated people can become to earn relatively small 'gifts', which in reality they could purchase for themselves for much less effort. This phenomenon is regularly observed in the relationship between pharmaceutical companies and veterinary surgeons who will work hard to promote a particular drug to their clients in order to receive a mug or padded waistcoat. To date, the lameness project has not fully exploited the 'drug-company type' gift, although occasionally small logoed gifts such as a woolly hat or an insulated mug are handed out to farmers who have implemented a change, to reinforce the approval for the change being voiced by the facilitator. As part of the lameness project, there is a positive control group that is able to access modest but not insignificant financial rewards for having implemented changes outlined in its lameness action plan. Although the final outcome of this incentive scheme has not yet been fully evaluated, the highest level of behaviour change appears to be in this group.

Fig. 12.3. A cartoon prompt to remind farmers to work on early detection and treatment of lameness (cartoon by Steve Long).

Other models of financial incentive already exist within the agricultural sector. One example that is widely used around the world is the payment scheme for bulk somatic cell counts in milk. The milk payment that farmers receive is adjusted according to their levels of bulk milk somatic cell counts. Dekkers and colleagues (1996) reported that dairy farmers in Ontario, Canada could achieve substantial additional milk revenues through this scheme. An incentive-based intervention approach to improve the welfare of commercially produced broiler chickens in Sweden has been described by Algers and Berg (2001). An animal welfare programme for broilers set up by the Swedish Poultry Meat Association monitored the standards of the buildings and equipment used in the rearing of the birds and a score was formulated on the basis of the assessment. This score then determined the future maximum stocking density at which a producer was allowed to keep broilers. The stocking densities at which the birds could be kept ranged from 20 kg/m^2 to 36 kg/m^2; this had obvious economic implications for the producers and as such was promoted as a reward system for those producers who chose to improve their broiler rearing facilities. Contained within the animal welfare programme was the foot health programme that monitored and scored the prevalence of foot-pad dermatitis in the birds at the time of slaughter. As well as being part of the stocking-density reward programme the foot-health monitoring programme was linked to an advisory

COSTS OF LAMENESS

Cows produce less milk when they are lame

Milk loss facts for lame cows

Loss per 305 day lactation associated with:

- Medium lameness score = –442.8kg
- High lameness score = –745.6kg
- Claw lesions = –360kg

Milk loss greatest in early lactation
Milk losses for 2 months before a sole ulcer
Milk losses for 5 months after sole ulcer and white line disease

Lame cows are less likely to get in calf

Fertility facts for lame cows

Increased intervals for lame cows

- Calving to conception = 14–50 days longer
- Calving to first service = 4 days longer
- First service to conception = 8 days longer

10% fewer pregnancies/first service
0.42 more services per conception
1.16 times more likely to be treated for anoestrus

Lame cows are more likely to be culled early

Culling facts for lame cows

8.4 times more likely to be culled early

Many cows that become lame have higher yields before they become lame than cows that never become lame **SO** lame cows are culled later in lactation than cows culled for mastitis or fertility.

Cows that become lame make less money for the farmer

Overall costs of a case of lameness

Reduced fertility = £46.14

Reduced milk yield = £55.05

Culling/replacements = £53.72

Treatment costs = £23.32 (1.4 treatments/cow)

TOTAL COST = £178.23 per lame cow

The poster shows the results of numerous studies that have measured the effects of lameness. The overall costs of lameness are the average cost of lameness taken from the DAISY Research Report No. 5 in 2002. These figures are intended as a guide but will vary from farm to farm.

Fig. 12.4. Poster highlighting the financial losses associated with lameness (compiled by Dr Zoe Barker, cartoons by Steve Long).

service that was available to the farmers. Within the first 2 years of running the foot health programme the prevalence of severe foot-pad lesions had decreased from 11 to 6%.

As demonstrated by the examples above, incentives have a potentially powerful role in encouraging behaviour change. The limitation of some of the described approaches is that they can require considerable external investment through government, retailers and food processors or charities looking for routes to promote animal welfare improvement. Incentives also have a role within enforcement models of behaviour change, such as restricting access to markets unless specified animal welfare criteria are met. This aspect of incentive use will be discussed in more detail in the enforcement section of this chapter.

Summary of social marketing approach

Social marketing is a well-known and widely used tool for promoting human-behaviour change. As can be seen from the description above, it requires careful planning and needs to contain many strategies within a single approach. This approach has sustaining change integrated within the plan, which is clearly a strength, and the only limitation of this intervention is the innovativeness and imagination of those designing it. The approach described here includes one-to-one facilitation visits to all those farmers participating in the project. Despite this proving a cornerstone for this project, it is clearly labour intensive, which translates into a relatively high-cost intervention, although of course intervention cost has to be balanced against the level of change achieved. It will be interesting in future projects to apply social marketing principles without the one-to-one facilitation element.

Recommended further reading

Hastings, G. (2007) *Social Marketing: Why Should the Devil Have All the Best Tunes?* Elsevier, Oxford, UK.

McKenzie-Mohr, D. and Smith, W. (1999) *Fostering Sustainable Behaviour. An Introduction to Community-based Social Marketing.* New Society Publishers, Gabriola Island, British Columbia, Canada.

Participatory approaches

The participatory approaches described in this section were created for use in community development, primarily in developing countries. The approaches described here are powerful and based around empowering the most marginalized and vulnerable people to make changes to improve their own lives. Previously we have used these approaches in Asia to help the owners of equids used for traction work to make improvements to the welfare of their animals without risking their income. More recently we have brought some of these approaches on to farms in the UK in order to help farmers become aware of, plan and own behaviour changes to improve the lives of their animals. The approach described here is commonly referred to as either PRA (Participatory Rural Appraisal) or PLA (Participatory Learning and Action), although many variations on these names exist. Within this section we will discuss two main areas of participatory approaches: (i) the principles that guides their use; and (ii) the types of tools available. These have been 'reinterpreted' to be relevant to a commercial Western agricultural context, so apologies in advance to any development workers who may be reading this chapter.

Principles of participatory approaches

The key to the participatory approach is to develop interactive tools or exercises that allow a single farmer or a group of farmers to share, present, analyse and enhance their knowledge of a problem they may be facing; in this case our interest is animal welfare. The first principle to be aware of is that of participation. Put most purely, this means that 'all people have a right to play an active part in shaping the decisions which affect their lives' (International HIV/AIDS Alliance, 2006). So for us this is about getting farmers involved in and wanting to make changes to the lives of their animals because these changes will affect their lives as well. It is also important to recognize that on a farm there may be many voices that need to be included in discussions about change, not just the farm owner or manager. Some of these voices are rarely heard and often not respected, for example the relief milker, the farmer's wife or the son who feeds the calves when he gets home from school. So our goal is to include as many participants in the process as possible, the more people who have a stake in and support a change the more likely it is to happen. The second principle is that this approach respects the fact that local people (in this case farmers) hold much more knowledge about their own situation than we as external parties could ever hope to do. It would be arrogant and presumptuous of us to simply give advice or instruction without recognizing that our understanding of how a particular farm works and why it developed the routines and practices that it currently uses is very limited. We generally rush to bring new knowledge to farmers without stopping to consider and use the knowledge they already hold. So a participatory approach aims to bring out this knowledge in a way that is visual, engaging and organized for the farm team to consider and learn from. The third, and often overlooked, principle of using a participatory approach is about the need for proper reflection and analysis of information. There is a common temptation to use a tool and complete an exercise as rapidly and efficiently as possible without pausing to review, think about and discuss what can be learned from having completed the exercise. In fact, there would be no point conducting the exercise at all if no subsequent reflection or review takes place.

As may be becoming apparent, this process relies on someone bringing these principles and tools to a farm or group of farmers and guiding them through the process. This person is usually the external agent who is interested in helping a farm to implement behaviour change and is usually called a facilitator. The success of the process is heavily dependent on the competence of the facilitator. The term facilitator roughly translates into 'making it easy' and can be interpreted as someone whose role is to find ways to help others through the principles outlined above without bringing their own values or prejudices into the discussion. Good facilitators encourage others to speak, usually by asking questions without offering opinions, and spend the majority of their time listening rather than talking.

Participatory tools

As already mentioned, these tools came about for use among marginalized communities in developing countries. They are designed to be as accessible as possible so exploit local materials such as dried beans, sticks, stones, etc. and are usually carried out on the ground. As far as possible written words are avoided so as not to exclude the illiterate, and pens and paper are not often used as many people are uncomfortable with this medium. However, in accordance with the principles of participation, it is perfectly acceptable to modify the materials used so that Western farmers feel more comfortable using them. We tend to work sitting around a table, use pens and large sheets of paper and brightly coloured sticky Post-it® notes. We do still try to use pictures and not rely solely on written words. The reason for this is that according to official figures, one in six adults in the UK is functionally illiterate, to the extent that they cannot use a telephone directory to find a plumber. So drawings avoid exclusion of the less literate and act as prompts to help remind people what a particular question or category referred to. This is also helpful in situations where we would like to include migrant farm labourers from Eastern Europe but we do not share a common language with them.

The following three tools or exercises are illustrated pictorially and their purpose explained. When facilitating exercises such as these, it is important to break them down into steps and work on completing one step at a time.

EXAMPLE TOOL 1: SEASONAL LAMENESS CALENDAR The seasonal lameness calendar illustrated in Fig. 12.5 is compiled by the facilitator from the farm records. Ideally this would be done in collaboration with members of the farm team, but we often find that presenting the final product for analysis and discussion is still a useful exercise when the farm personnel have only limited time. This chart allows farmers to consider management changes in relation to peaks of lameness, it helps them to identify potential risks that are specific to a particular type of lameness and to consider how particular management activities could be timed to pre-empt a rise in lameness levels.

EXAMPLE TOOL 2: MATRIX FOR PRIORITIZING LAMENESS MANAGEMENT ACTIVITIES The matrix illustrated in Fig. 12.6 is built up in stages. Initially the participant(s) are asked to make a list of potential changes they are considering that might contribute to reducing lameness levels among their dairy cattle; these are illustrated on separate cards and laid out in a column. They are then asked to consider each of these possible changes with reference to how much welfare benefit each one would confer to their cattle. Each one is scored out of ten with ten indicating the greatest benefit. Next they are asked to score how much each of the changes will be likely to cost. The question is phrased so that again a score of ten represents the most positive outcome (i.e. the least cost). When considering cost, the participant(s) can consider the potential cost of implementing a permanent change to a daily routine versus a one-off

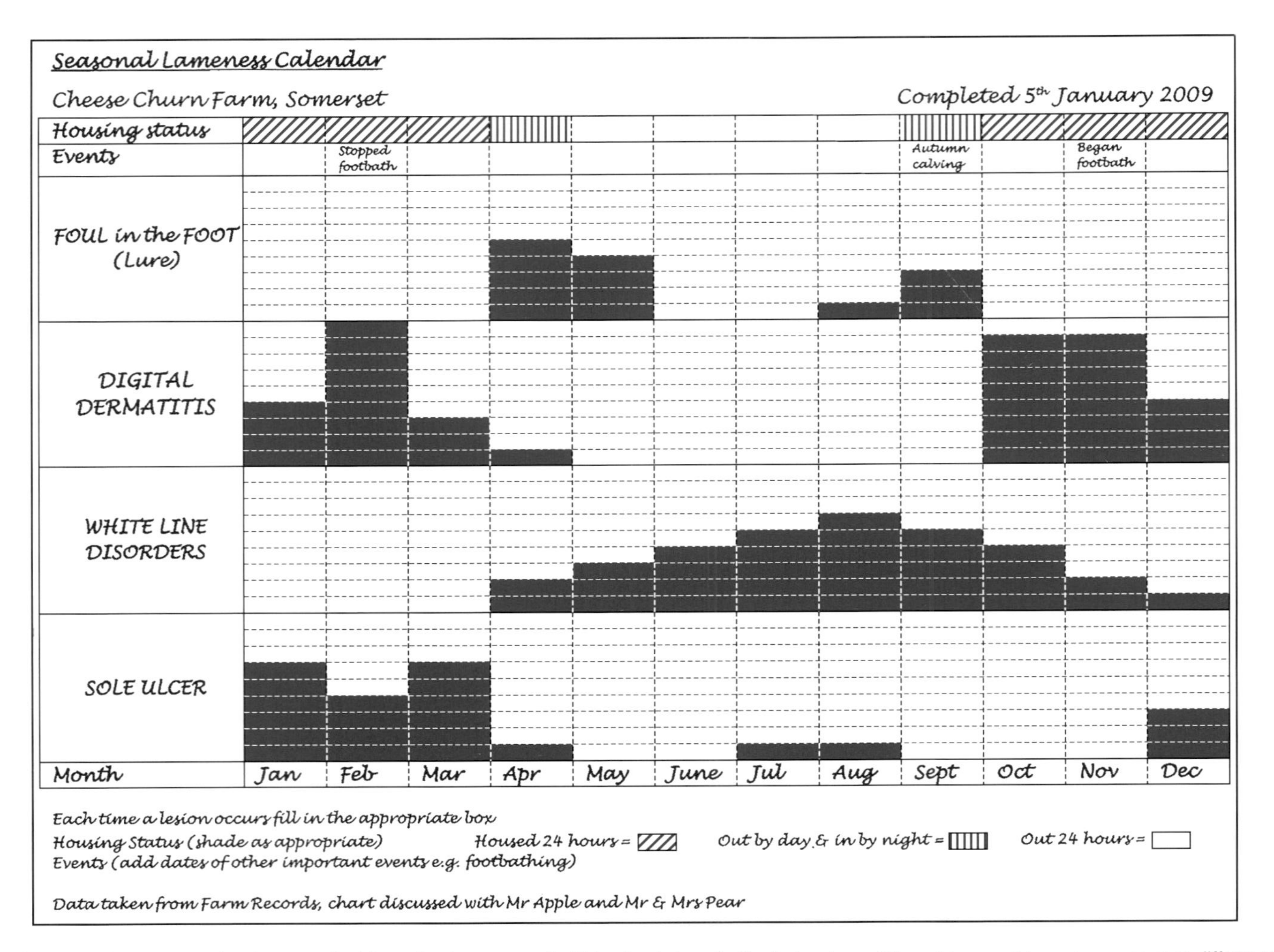

Fig. 12.5. Seasonal lameness calendar compiled from the farm records. This chart visually illustrates how different types of lameness occur at different times and can be used to promote discussion about potential risk factors for different types of lameness and the timing of lameness management activities (compiled with the assistance of Dr Zoe Barker).

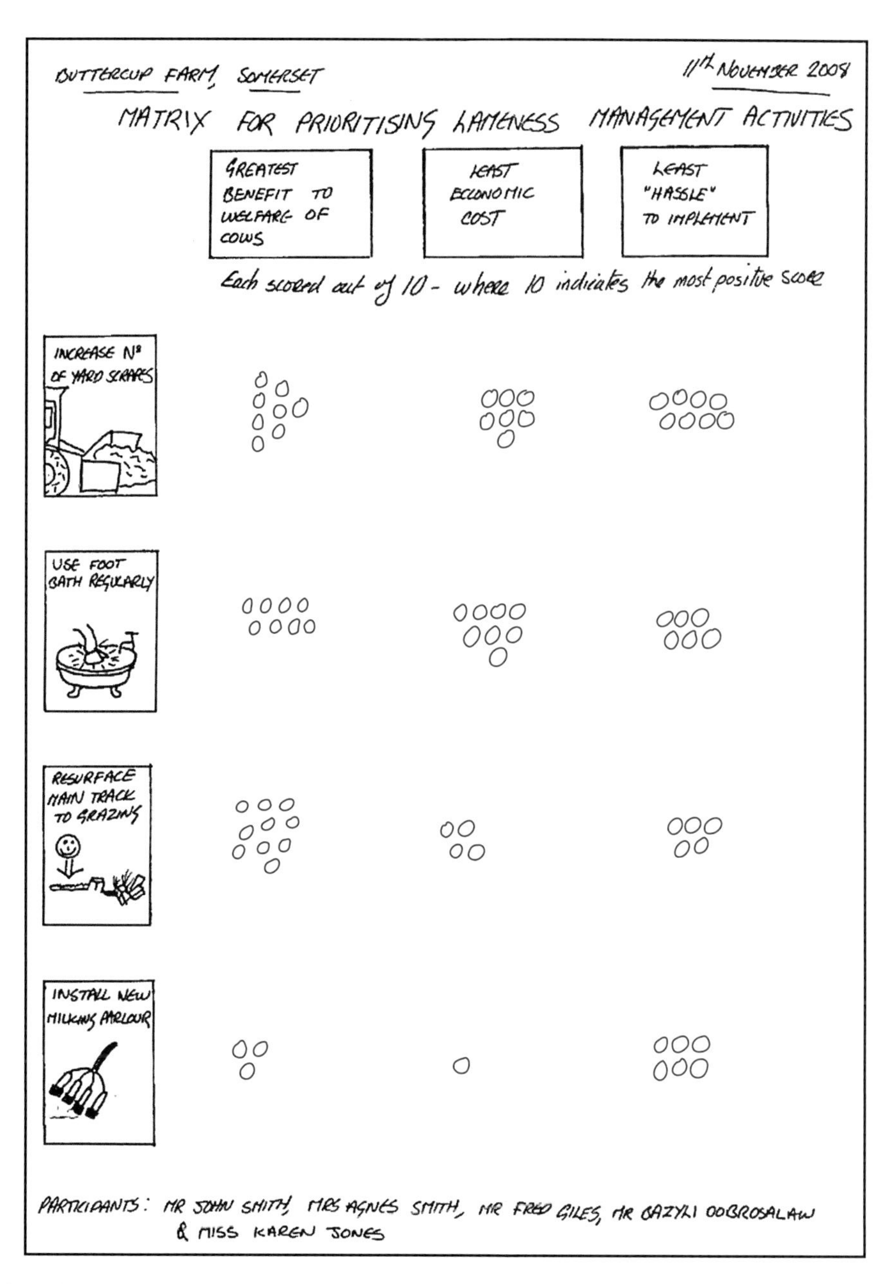

Fig. 12.6. Matrix for prioritizing changes. This matrix is built up in stages and helps participants to view the potential positive and negatives aspects of a change in manageable chunks.

capital investment. Finally, the participant(s) are asked to score the 'hassle' (extra daily work) factor of implementing the change, once again a score of ten represents the least hassle. Once the matrix is completed the starting point for the review is to look where the greatest scores appear. This exercise is useful for prioritizing actions and breaking down considerations of the implications of implementing a change into bite-sized chunks.

EXAMPLE TOOL 3: LAMENESS TRANSECT WALK The lameness transect walk (Fig. 12.7) is useful for giving

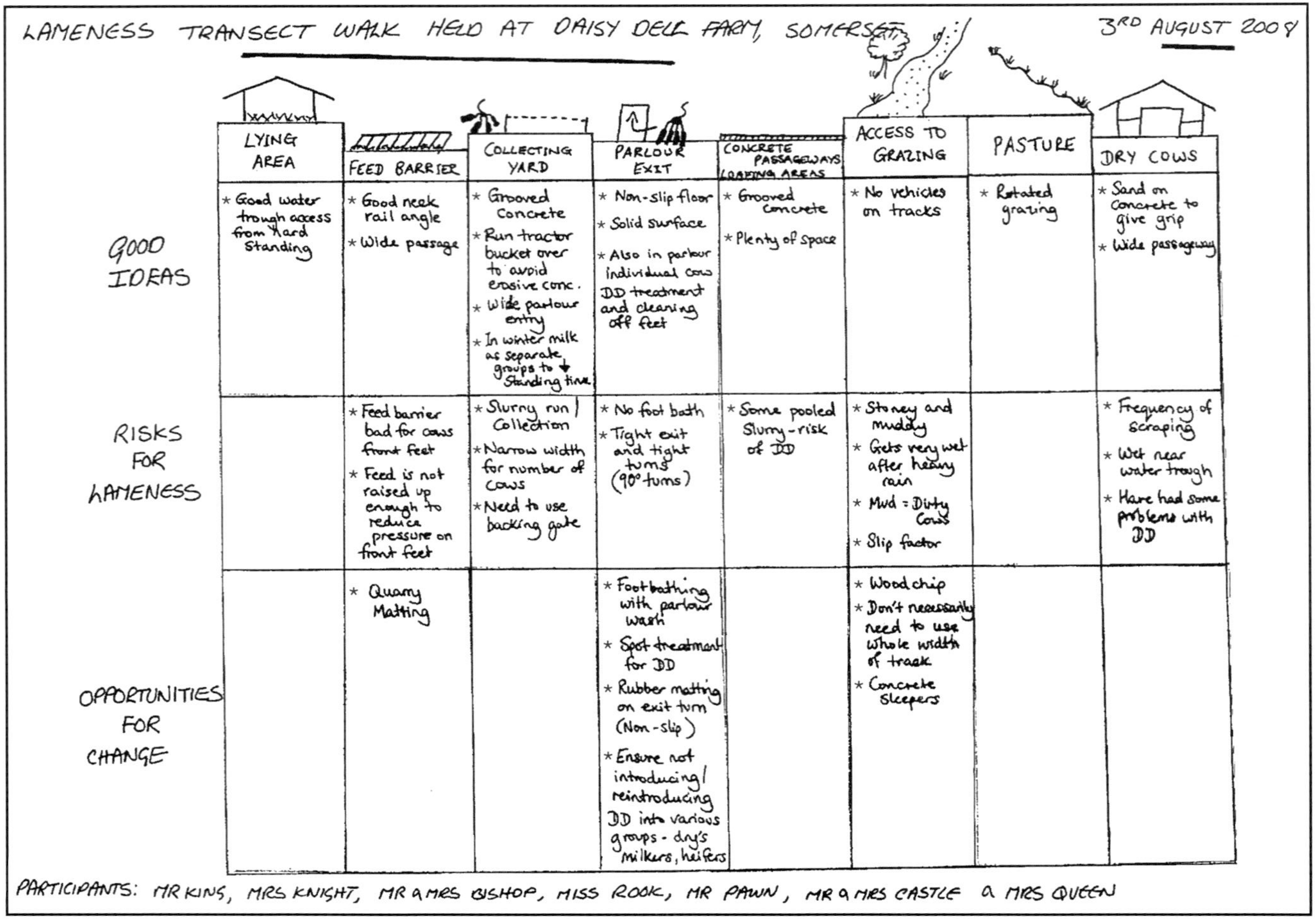

LAMENESS TRANSECT WALK HELD AT DAISY DELL FARM, SOMERSET — 3RD AUGUST 2008

	LYING AREA	FEED BARRIER	COLLECTING YARD	PARLOUR EXIT	CONCRETE PASSAGEWAYS LOAFING AREAS	ACCESS TO GRAZING	PASTURE	DRY COWS
GOOD IDEAS	* Good water trough access from hard standing	* Good neck rail angle * Wide passage	* Grooved concrete * Run tractor bucket over to avoid erosive conc. * Wide parlour entry * In winter milk as separate groups to ↓ standing time	* Non-slip floor * Solid surface * Also in parlour individual cows DD treatment and cleaning off feet	* Grooved concrete * Plenty of space	* No vehicles on tracks	* Rotated grazing	* Sand on concrete to give grip * Wide passageway
RISKS FOR LAMENESS		* Feed barrier bad for cows front feet * Feed is not raised up enough to reduce pressure on front feet	* Slurry run/collection * Narrow width for number of cows * Need to use backing gate	* No foot bath * Tight exit and tight turns (90° turns)	* Some pooled slurry - risk of DD	* Stoney and muddy * Gets very wet after heavy rain * Mud = Dirty cows * Slip factor		* Frequency of scraping * Wet near water trough * Have had some problems with DD
OPPORTUNITIES FOR CHANGE		* Quarry matting		* Footbathing with parlour wash * Spot treatment for DD * Rubber matting on exit turn (Non-slip) * Ensure not introducing/reintroducing DD into various groups - dry's, milkers, heifers		* Wood chip * Don't necessarily need to use whole width of track * Concrete sleepers		

PARTICIPANTS: MR KING, MRS KNIGHT, MR & MRS BISHOP, MISS ROOK, MR PAWN, MR & MRS CASTLE & MRS QUEEN

Fig. 12.7. This chart produced from a farm walk is transcribed from a very large chart built up by a group of farmers who wrote their comments on to brightly coloured Post-it ® notes. The giant chart formed the basis for a facilitated discussion following the walk. A copy of the transcribed chart was then sent to all the farmers who came to the meeting (transcribed with the assistance of Miss Clare Maggs).

structure to a meeting of a group of farmers. At the beginning of the meeting the group is asked to select the route 'following in the footsteps of the cows' for a walk around the host's farm. Each participant is given a farm map on a clipboard and asked to note down any good ideas they see, any potential risks for lameness and any opportunities for change. After the walk each member of the group is invited to transcribe their observations on to brightly coloured Post-it® notes and stick them on to a giant pre-prepared chart. A facilitator then promotes a discussion around areas that have been highlighted on the chart, often by a high density of stickers. The facilitator's role is to keep the discussion moving along and to bring out as many potential opportunities for change as possible. Although the discussion is based around the farm being visited, all participants benefit from hearing the ideas, suggestions and experiences of others and often bring particular problems they are facing into the discussion.

Summary of participatory approaches

These participatory tools and the participatory approach described in this section are very useful for helping farmers through the process of recognizing problems, considering potential solutions and prioritizing activities. If they are introduced sensibly, the unusualness of the approach is easily overcome as the participants recognize the usefulness of the results. It is, however, important to recognize that simply using exercises such as these alone is unlikely to stimulate behaviour change, the outputs and realizations from the exercises need to be formulated into an action plan and it might be useful to bring in some of the commitment tools from the social marketing section to help secure behaviour change. The sustainability of the change being implemented is dependent of the level of farmer ownership and the quality of the analysis carried out when working with these tools. This method is intended to help people see that they are in control of the changes to be made on their farm, but this positive must be balanced against the question of how likely continuation of implementation of change and introduction of new innovations will be if contact with the facilitator is not maintained.

Recommended further reading

International HIV/AIDS Alliance (2006) *Tools Together Now! 100 Participatory Tools to Mobilise Communities for HIV/AIDS*. International HIV/AIDS Alliance, Hove, UK. Available at: www.aidsalliance.org/sw44872_asp (accessed 6 July 2009).

Kumar, S. (2002) *Methods for Community Participation: a Complete Guide for Practitioners*. ITDG Publishing, London.

Farmer groups

The concept of farmers forming self-help groups or, more commonly, discussion groups is not new, and these groups are often useful for breaking down the isolation associated with farming and for sharing new ideas. This section looks at two approaches to how farmer groups can be managed to achieve optimum likelihood of changing behaviour and addressing areas of animal welfare concern. The two models of farmer group structures to look at in this section are the 'stable school' and the 'monitor farm'. The actual formation of groups is often not easy, but is a discussion for another time. For the purposes of this section we will assume that we have a group of farmers looking for a structured group format.

Stable schools

The stable school system is adapted from a system known as Farmer Field Schools running in Uganda and has been tested as the Stable School approach in Denmark by Mette Varst and colleagues (Vaarst *et al*., 2007). In Denmark, six organic farmers formed a group with the stated goal of using health promotion to reduce and, where possible, phase out antibiotic usage in their cows. To participate in the stable school each farmer agreed to host a visit from the group and to be open to receiving management recommendations from the group. The visit consisted of a farm walk where the host was asked to present an aspect of the farm he or she was proud of as well as two problem areas. Once the walk was completed the group returned to the farmhouse or office and examined the farm's accounts. Following this a facilitated discussion took place in which the group worked to develop recommendations on changes that could be made to improve the host farm's management and help reduce the need for antibiotics. The goal was to prevent problems arising requiring antibiotic treatment, not to withhold treatment from animals in need. The facilitator's role was to encourage participation by all group members and act as the recorder and reporter of the outcome of the discussions. After all farms had been visited, each then received a follow-up visit so that in total each farm was visited twice. This approach produced some notable successes with reference to the common goal of reducing antibiotic usage. However, as well as this common goal each farmer also formed his or her own 'local' goals, which in some cases were animal-health oriented and in others were based on improving family or social environments. These were important markers of the success of the project and many participants felt they benefited very much from joining a group.

Monitor farms

The monitor farm approach was originally developed in New Zealand and has more recently been imported to Scotland. The purpose of the farmer groups currently engaged in the monitor farm approach is to improve farm performance and profitability and it is likely that animal welfare may be implicit within these goals. In the context of this chapter we will have to extrapolate this approach to consider how it might work when explicitly focused on animal welfare improvement.

The concept of the monitor farm is that one farmer agrees to open up his farm business to a group of farmers and possibly members of allied industries such as veterinary surgeons and feed specialists. The monitor farm hosts regular visits from the group, during which aspects of the farm's management and potential changes to be made are discussed. The regular meetings then allow the group to observe the progress of the farm business and to see the effect of implementing suggested changes (or not). The monitor farm also hosts an open day once a year for visitors outside the group to view the farm and learn about the process. Once again, this process is supported by a facilitator who helps with setting the objectives of the meetings, managing the logistics and, where needed, analysis of data.

Clearly, the monitor farm itself is likely to be implementing changes, although it is not obliged to follow the recommendations of the group. The group members themselves are also likely to form the habit of trying to implement changes on their own farms, benefiting from hearing others' experiences and learning from the implementation of changes on the monitor farm for which they already feel some level of ownership. In some cases, members of the group have been reported to also develop cooperative activity. It is also suggested that the 'trickle through' effect can lead to changes being implemented in the wider community. Trickle through is a commonly promoted concept within development, but can be very difficult to demonstrate.

Summary of farmer groups

The two examples of very structured farm groups given here are both potentially very powerful tools for stimulating behaviour change. Within the stable school, all participants join with the expectation of making changes to their farms. To some extent these farmers pre-select themselves as being willing to implement change with a little encouragement and support. The monitor farm system allows farmers who may not initially be open to change to join as viewers rather than implementers, although it would be hoped that with time they would begin to see themselves as implementers of change as well. A critical difference between the two approaches is that within the stable school system all participants must be prepared to open up their farms and businesses to external scrutiny and dissection. Not all farmers would be comfortable with this and would exclude themselves from this type of group. The monitor farm system, though not so directly influential on each member, is likely to bring in a wider group of participants who may grow in confidence as a result of membership.

Recommended further reading

ADAS (2008) *An Investigation into the Role and Effectiveness of Scottish Monitor Farms.* A report prepared for the Scottish Government by ADAS (Agricultural Development and Advisory Service) UK Ltd. Scottish Government Publications, Wolverhampton, UK. Available at: www.scotland.gov.uk/publications/2008/10/29093936/0 (accessed 7 July 2009).

Vaarst, M. (2007) Participatory common learning in groups of dairy farmers in Uganda (FFS approach) and Danish stable schools. *DJF Animal Sciences* 78 (special edition). Available at: http://orgprints/13731/ (accessed 7 July 2009).

Enforcement as a Means of Implementing Change

Legislation, policy instruments and farm assurance

Insisting that farmers take particular actions to improve the welfare of animals under their care is a well-established route to compel behaviour change. Traditionally, this approach has centred on setting legal requirements that a farmer must comply with. Failure to meet these legal requirements could result in prosecutions and/or penalties depending upon the local regulatory requirements.

Legislation is clearly a well-recognized mechanism for governments to bring about changes in behaviour. However, legislation is just one of a set of tools also known as 'policy instruments' used by government to implement policy. The Farm Animal Welfare Council (FAWC) (2008) has reviewed the policy instruments that the UK government could use to improve animal welfare. A total of 11 categories of instruments were identified, including direct action, education, incentives, research and promoting private markets (see Table 12.1). One example of a policy instrument has been the government's role in producing guidance in the form of species-specific animal welfare codes. The welfare codes promote good animal husbandry and animal care as well as presenting relevant pieces of animal welfare legislation in a clear and easily accessible format. The animal welfare codes have been influential in guiding both farmers and farm inspectors and advisors. While legislation may remain the most obvious tool available to governments, the FAWC suggested that: 'To achieve the levels of animal welfare that people want requires a coordinated approach to the use of policy instruments to achieve desired behavioural change'.

However, more recently it is the marketplace that is insisting producers maintain minimum animal welfare standards. Failure to meet the criteria of the marketplace could result in a farmer finding him or herself unable to sell livestock products; this is clearly a very powerful motivator for change. In many countries, these marketplace requirements are achieved by insisting that farmers become members of farm assurance schemes. The farm assurance schemes aim to provide the purchaser of a product an assurance that animal welfare (and food safety) standards are being met. Often these schemes are driven by retailers that are keen to ensure minimum standards for their products. The retailers' primary motivation here is to avoid negative media publicity associated with problems on the farms that supply them. The minimum these schemes normally aim to achieve is to verify compliance with national legislative requirements. However, some retailers and schemes may formulate additional requirements either as part of the company's corporate social responsibility (CSR) policy or to obtain a marketing differential from other products, and animal welfare may be included within their remit. The examples describing the influence that the fast-food chain McDonald's has had on animal welfare through the requirements they set for slaughterhouses, presented in Chapters 1 and 9, are based on this principle.

Although the requirement for farmers to comply with specific standards may originate from different sources, for example politicians representing citizen interests or the marketplace representing consumer interests, the net result is the same: a powerful statement to producers that they have no choice other than to comply. Both 'enforcement' approaches (legislation and marketplace requirements), therefore, have the same inherent strengths and weaknesses

The **strengths of the enforcement approach** are:

1. It can be effective in ensuring the worst extremes of animal welfare problems are avoided.
2. A clear set of requirements can be transparent and communicated to all farmers.
3. Provided a careful development process is followed that includes incorporation of relevant practical experience and scientific knowledge, then the standards can be fair to producers.
4. Considering the penalties involved (prosecution or inability to access the marketplace), farmers will be highly motivated to comply.
5. Once an inspection and certification system has been established, the minimum standards can be supplemented to include additional requirements that can be used in marketing initiatives and, thereby, lead to a price premium (i.e. a positive financial incentive to be members of such schemes).

The **weaknesses of the enforcement approach** are:

1. The inherent 'negative-ness' of this approach can generate bad feeling with farmers as it conveys the message: 'We do not trust you to care for your animals'.
2. Farmers may be encouraged to comply up to a minimum standard, but it does not recognize excellence, continuous improvement or innovation beyond the minimum required.
3. Written requirements are difficult to construct, so are often poorly defined and, therefore, ambiguous. Standards are also often developed in a haphazard manner with insufficient consultation.

Table 12.1. Types of government intervention and their relative strengths and weaknesses (source: Webster *et al.*, 2006 reproduced in FAWC, 2008).

Type of policy instrument	General example	Applied to animal welfare and health[a]	Strengths	Weaknesses
1. Legal rights and liabilities	Rules of tort law	Animal Welfare Act (England and Wales) 2006. Animal Health and Welfare (Scotland) Act 2006	Self-help	May not prevent events resulting from accidents and/or irrational behaviour
2. Command and control	Secondary legislation. Health and safety at work	Minimum space rules for poultry	Force of law. Forceful. Minimum standards set. Immediate. Transparent	Intervention in management. Incentive to meet, not exceed standard. Costly. Inflexible
3. Direct action (by government)	Armed forces	Welfare inspections by state veterinarians and local authorities. Border controls	Can separate infrastructure from operation	Danger of being perceived as 'heavy handed'
4. Public compensation/ social insurance	Unemployment benefit	Compensation for animals slaughtered for welfare reasons during 2001 FMD outbreak. Cross-compliance. Pillar II monies[b] for farm animal welfare improvements	Insurance provides economic incentives	May provide adverse incentives. Can be costly to taxpayers
5. Incentives and taxes	Car fuel tax	Cross-compliance. Pillar II monies for farm animal welfare improvements	Low regulator discretion. Low cost application. Economic pressure to behave acceptably	Rules required. Predicting outcomes from incentives difficult. Can be inflexible
6. Institutional arrangements	Departmental agencies, levy boards, local government	Animal Health, Meat Hygiene Service, Veterinary Laboratories Agency, local authorities	Specialist function. Accountability	Potential for narrow focus of responsibility
7. Disclosure of information	Mandatory disclosure in food/drink sector	Reporting of notifiable diseases. Labelling	Low intervention	Information users may make mistakes
8. Education and training	National Curriculum	Animal welfare in veterinary education, national school curriculum	Ensures education and skills required by society	Can be too prescriptive and inflexible

9. Research	Research councils	Funding for animal welfare research through BBSRC, Defra, charities, etc.	Provide information to policy	May duplicate or displace private sector activities
10. Promoting private markets (a) Competition laws	Office of Fair Trading. Airline industry. Telecommunications	Market power of companies in the food supply chain and prices to farmers to meet production costs	Economies of scale in use of general rules. Low level of intervention	No expert agency to solve technical/commercial problems in the industry. Impact of global commodity costs. Uncertainties and transaction costs
(b) Franchising and licensing	Rail, television, radio	Veterinary drugs/treatments. Animal husbandry equipment	Low cost (to public) of enforcement	May create monopoly of power
(c) Contracting	Local authority refuse services	Hire of private vets to provide public services	Combines control with service provision	Confusion of regulatory and service roles
(d) Tradable permits	Environmental emissions. Milk quotas	Permits for intensive livestock production systems (e.g. The Netherlands)	Permits allocated to greatest wealth creators	Require administration and monitoring
11. Self-regulation (a) private (b) enforced	(a) Insurance industry. (b) Income tax	(a) Farm assurance schemes, veterinary profession, industry codes of practice. (b) Defra 'welfare codes'	High commitment. Low cost to government. Flexible	(a) Self-serving. Monitoring and enforcement may be weak

[a] Abbreviations used: BBSRC, Biotechnology and Biological Sciences Research Council; Defra, Department for Environment, Food and Rural Affairs; FMD, foot and mouth disease.
[b] A level of subsidy available through the Common Agricultural Policy of the EU.

4. If no marketing claims arise from the marketplace requirements, then farmers required to comply with a standard are at a commercial disadvantage to producers (perhaps from a developing country) that do not have such an approach.

It seems reasonable that farmers are required to achieve a minimum standard. Indeed it is in all farmers' interests that unacceptable welfare conditions are regulated, as the actions of a few can have a very significant adverse effect in the media and on public opinion. So the enforcement approach is certainly useful for regulating the extremes.

The experience of Temple Grandin's approach based on setting requirements to achieve a set of welfare outcomes within the slaughterhouse environment is very effective (see Chapters 1 and 9). The key here seems to be focusing on a few parameters and demonstrating to slaughterhouse managers that achieving change is possible. The challenge will be to replicate this success on the farm where the welfare problems are often more chronic in nature and directly related to an inherent component of the production system, for example resolving the problem of tail biting in pigs, which seems to be intrinsic to some types of production system. In a slaughterhouse environment, welfare improvements due to modifications in handling systems can be demonstrated in the next batch of animals, whereas on the farm it can be a considerable time before the outcome and positive effect of changes can be seen. Once a dairy cow becomes lame with a sole ulcer, her foot is damaged for life; it will not be until the next generation of heifers joins the herd that the outcome of implementing new preventive measures will be seen.

The use of standards in an enforcement approach

It is beyond the scope of this chapter to review all the possible legislation and standard options that might influence animal welfare. Most countries will have legislation that aims to avoid cruelty or 'unnecessary suffering'. This is a useful absolute baseline standard for animal welfare. However, beyond this baseline the legal requirements will, of course, vary between countries although a common minimum standard for farm animal welfare is applied within several countries of the European Community (Directive 98/58/EC). Similarly, penalties for failure to comply with standards will also vary between countries.

In order to understand the potential power of using standards to improve animal welfare it is useful to examine what types of standards are available. The examples given below cover standards for health planning, treatment of both individuals and groups of animals with welfare problems and the provision of resources, and Box 12.1 illustrates how these standards can be applied to the problem of cattle lameness. The potential impact of these standards will of course vary. For example, it is likely that standards that require specific actions directly associated with lameness will be more beneficial than actions associated with records and protocols, although of course these can have a role. As can be seen from the variety of approaches shown in this chapter, it can be very valuable to link the standards with animal-based outcomes on the farms. Animal-based outcomes are measures of welfare made by looking directly at the animal itself, for example how many cows are limping, what body condition score animals are in or how animals respond behaviourally to an approaching human. Animal-based outcome measures are distinct from resource-based measures that define the environment in which the animal should live but do not describe what the welfare of the animal should be. Animal-based outcome measures may take the form of farmers self-assessing their own animals or involve external observers, inspectors or assessors. The findings from an animal outcome assessment may lead to a requirement for a farmer to take specific actions once an inspector observing animals with a particular condition such as lameness finds that the prevalence of the problem has gone above a pre-defined threshold.

The following four examples illustrate how standards may be used to enforce changes intended to improve animal welfare (these have been developed in association with Miss Siobhan Mullan).

Standards associated with health planning activities

Health planning is an important tool for farmers to achieve early recognition of welfare problems including disease, to identify potential risks for their introduction and spread and to implement management of identified risks (Defra, 2004). These standards can ensure that farmers undertake preventative or corrective health planning activities

Box 12.1. An illustration of how enforcement standards may be applied to the problem of lameness in cattle.

Examples of UK Standards Relevant for Dairy Cattle Lameness

The following are examples from either UK legislation or from the UK welfare codes that are relevant for dairy cattle lameness. Although UK welfare codes are recommendations, farm assurance schemes often require adherence with them as well as UK legislation.

Requirements associated with health planning activities

- The stock-keeper should draw up a written health and welfare plan with the herd's veterinary surgeon and, where necessary, other technical advisors, which should be reviewed and updated each year.*
- The written health and welfare plan should…look at…lameness monitoring and foot care.*
- A record must be maintained of…any medicinal treatment given to animals.**
- …you may find it useful, as part of the health and welfare plan, to note specific cases of…lameness…and where appropriate the relevant treatment given.*

Requirements associated with individual affected animals

- Any animals which appear to be ill or injured must be cared for appropriately and without delay. Where they do not respond to such care, veterinary advice must be obtained as soon as possible.**
- If a lame animal does not respond to the veterinary surgeon's treatment, you should have it culled rather than leave it to suffer.*
- …you must not transport any cattle off-farm that cannot stand up unaided or cannot bear their weight on all four legs when standing or walking.*

Requirements associated with herd/group level outcomes

- …all keepers of animals must…ensure that the needs of an animal for which he is responsible are met to the extent required by good practice and those needs shall be taken to include…its need to be protected from pain, suffering, injury and disease.***

While not currently explicit, standard guidance notes for inspectors could state that prevalence of lameness above a certain percentage could be used as evidence of failure to provide this need.

Requirements associated with provision of specific resources

- The floor should not slope too steeply…as steeper slopes can cause leg problems, slipping and falling.*
- [For slatted accommodation] the gaps between the slats should not be wide enough to cause foot injuries.*
- [For cubicles] you need to have enough bedding to…prevent them from getting contact or pressure sores.*

*Defra Welfare Code of Recommendation: Cattle 2003 PB7949

**Welfare of Farmed Animals (England) Regulation 2007 SI 2078

***Animal Welfare Act 2006 Chapter 45

relevant to a specific parameter. This may involve changes to housing, facilities, routines or planned use of medication. These standards can be required of the producer at all times. Examples of standards that should be implemented at all times are as follows:

- Farmers (and/or their veterinary surgeon) must inspect animals for presence of 'a condition' on a regular (e.g. weekly) basis.
- Those inspecting animals for the presence of 'this condition' must record such information in a timely manner.
- The records associated with 'this condition' must be reviewed by the producer and their veterinary surgeon on a regular basis (e.g. every 6 months).
- A written action plan must be developed that includes preventative husbandry and medicine related actions that minimize the prevalence of 'this condition'.
- The written action plan should also include predefined farm-specific intervention levels that determine when investigation activities need to be conducted.

Requirements associated with individual affected animals

Standards can be directed at ensuring that a farmer suitably manages the most severely affected individuals. This is applicable for all individual animals with a welfare problem above a pre-defined severity threshold; this is relevant for more severely affected animals, for example severely lame animals. Requirements could include one or more of the following components:

- All animals affected with 'this condition' must receive prompt treatment and nursing care.
- Veterinary advice must be obtained for affected animals that fail to respond to treatment within 3 days.
- Animals affected with 'this condition' must not be transported for slaughter.
- Animals that are severely affected by 'this condition' must be humanely euthanized/culled.

Requirements associated with herd/group level outcomes

Standards can be directed towards ensuring that the farmer appropriately manages welfare problems that occur at the herd or group level. This would be applicable where the prevalence of animals moderately affected by a particular problem exceeds a pre-defined intervention guideline, for example the combined prevalence of moderate and severely lame cows exceeds 20% on the day of assessment. Requirements could include one or more of the following components:

- Producers must manage 'the condition' so that the overall prevalence does not exceed the pre-defined intervention guideline.
- Producers must develop a farm-specific written action plan in collaboration with their veterinary surgeon when the prevalence of 'the condition' exceeds the intervention guideline.
- Producers must demonstrate that appropriate preventative actions have been taken when the herd level exceeds the intervention guideline.
- Additional verification visits will be required if the prevalence of 'the condition' exceeds the intervention guideline. Such visits will require an additional fee.

Requirements associated with provision of specific resources

Farm assurance and marketplace requirements include many resource-based standards: animals must be provided with defined facilities, space, bedding, etc. These are sometimes referred to as engineering standards. Animal outcomes can also be used to assist in the interpretation of such requirements. If a standard requires an 'appropriate' or 'sufficient' resource, for example bedding substrate, then animal outcome measures of cleanliness and body lesions and swellings can help define what is 'sufficient' in terms of delivering adequate animal welfare. For example:

- Animals must be provided with sufficient resource (e.g. bedding) to ensure that no animals develop the relevant outcome (e.g. swollen hocks).

Criteria for use of animal-based outcome measures in enforcement approaches

It is clear that animal-based outcome measures are not only a useful tool for farmers but that they should form part of welfare enforcement mechanisms. In order to implement standards that incorporate animal-based outcome assessments the following conditions will also need to be considered:

1. **Competency** of inspectors is obviously an important criterion. For most of these standards it would be important for the inspector to observe the condition among the animals. The inspector, therefore, needs the relevant knowledge, skills and attitude to conduct these tasks. The standard may also require observation of written records or protocols and verification that appropriate actions have been taken on the farm. To assess these standards the inspector would need to have sufficient knowledge of the system to make reasonable judgements.
2. **Guidance notes** are crucial for all standards as there are inevitably opportunities for variable interpretation.
3. **Training** inspectors, producers or veterinary surgeons in how to carry out the assessment of relevant animal-related outcomes is important and ensuring consistency of assessment is a key challenge.
4. **Benchmarking** information is essential for setting intervention guidelines and advising farmers. The farmer would need to know how their assessment results compare to other units of a similar type. Also the product retailer would be keen to have this information in order to describe to interested consumers the welfare status of goods they are selling.
5. **Sampling protocols** are also needed to define how many animals need to be observed. This will depend on the variability and prevalence of parameters within normal units. For example, on pig units, if a parameter is normally relatively consistent

within pens or groups, then fewer pens would need to be assessed to give reasonable certainty that the prevalence has been properly counted. If the parameter is generally known to have a relatively low prevalence, then more animals in more pens will need to be assessed to produce a definitive assertion of the on-farm prevalence. In addition, if the purpose of the assessment is to produce an overall farm average then sampling from a representative proportion of pens will be satisfactory. However, if a farmer needs to know where the problem originated then more pens will need to be sampled.

6. **Advisory information** that allows farmers to manage these animal outcomes needs to be available. Although farm assurance schemes are usually not permitted to provide advice alongside an impartial assessment, there does need to be some mechanism for getting advice to farmers. Often the veterinary surgeon should be best placed to do this.

Recommended further reading

Webster, A.J.F. and Main, D.C.J. (2003) Proceedings of the 2nd International Workshop on the Assessment of Animal Welfare at Farm and Group Level. *Animal Welfare* 12(4), 429–713.

The link between enforcement and incentives

Most farm assurance schemes and marketplace standards are usually perceived by farmers to be a negative enforcement approach because failure to comply can result in inability to access the market. However, some purchasers of livestock products may be prepared to pay an additional premium in recognition that farmers are complying with additional standards. This may, therefore, be perceived as a type of incentive. So, for example, the Royal Society for the Prevention of Cruelty to Animals (RSPCA) Freedom Food scheme aims to deliver a high level of assurance with respect to livestock welfare, and this may be valued by consumers. Any premium generated by these claims would be an incentive for farmers to participate in such schemes. Additional potential incentives for participation in such schemes might be the creation of a niche market for a specialist produced product, receiving contracts with a particular retailer (especially if they have a good track record for dealing fairly with their suppliers) or even ensuring continuation of a strong generic product market.

Another example was the Sustainable Dairy Project, which provided a premium for producers supplying Tesco, a major UK retailer, in return for following some additional welfare requirements including the requirement for self-evaluation of lameness. Tesco received a Good Business Award for Innovation from the RSPCA in 2007 for recognition of its efforts.

In the same way that incentives may be perceived positively, some financial instruments are quite clearly disincentives. Two useful examples exist here, again for lameness in dairy cattle. First, there is variation within Europe over the interpretation of hygiene regulation for dairy production. Some authorities have interpreted that lame cows have been defined as 'unhealthy' and, as such, milk from these animals, even if free from relevant medication, cannot enter the food chain. This means that farmers are directly affected by a loss in milk sales for every lame cow; it provides a very strong disincentive for having lame cows on the dairy farm.

Secondly, in order to gain subsidies under the Single Farm Payment scheme, European farmers are required to demonstrate cross-compliance (i.e. adherence to legal standards for aspects of their farm management such as environmental protection and animal welfare). These requirements are verified by government agencies and failure to comply can result in withdrawal or reduction in scheme payments. Even though the welfare standards are non-specific, the guidance provided to veterinary inspectors may be much more specific. For example the European Community Directive 98/58/EC requires that: 'Any animals which appear to be ill or injured (a) shall be cared for appropriately without delay; and (b) where they do not respond to such care, veterinary advice shall be obtained as soon as possible'. The guidance provided in the UK to the veterinary inspector explicitly refers to lame animals as it states that:

> If a significant number (over 5% is the current suggested intervention point) of dairy cows or dairy goats are lame at the time of inspection, the Inspector should make enquiries to establish if the cause of the lameness has been recognized and whether appropriate action has been taken and veterinary advice or FHWP [Farm Health and Welfare Plan] protocol has been followed.

Summary of enforcement

The enforcement approach, whether driven by government or by the marketplace, has a role to play particularly in ensuring that the worst standards of welfare are addressed. However, linking this approach with incentives may further enhance the potential for behaviour change. This may be achieved by providing incentives such as premiums or disincentives such as withdrawal of subsidy.

Concluding Notes

This chapter has highlighted several different approaches to promoting behaviour change among farmers that should lead to animal welfare improvements. Several practical tools for achieving this have been proposed. It is likely that more than one method applied in a coordinated manner will be most effective as different farmers respond to different approaches. It is also important to highlight that behaviour change, whether through the route of encouragement, enforcement or both, rests on the farmers having a good knowledge of how to farm in a caring and humane manner, i.e. 'professional knowledge'. Without this knowledge, farmers will be unaware that problems exist and ill equipped to find solutions. In this chapter we have argued that the majority of farmers do hold a great wealth of professional knowledge both general and specific to their own situations. In fact, there currently seems to be a danger that farmers often receive such a barrage of information that they find it difficult to sort and filter. It is also important to reiterate that holding this knowledge (although vitally important) is not sufficient on its own to precipitate behaviour change. All of the tools described here offer mechanisms to help farmers deal with and act upon the knowledge and skills they already have.

As a final note, we want to point out how crucial it is that any intervention is carefully monitored so the success or failure can be identified and appropriate modifications made to the intervention programme. Responsive monitoring not only protects the animals from unintentional harms that may arise from introducing changes but allows us all to learn more rapidly and efficiently from the process of making change.

References

Algers, B. and Berg, C. (2001) Monitoring animal welfare on commercial broiler farms in Sweden. *Acta Agriculturæ Scandinavica Section A, Animal Science* Supplement 30, 88–92.

Animal Health and Welfare (Scotland) Act (2006) HMSO, London.

Animal Welfare Act (England and Wales) (2006) HMSO, London.

Clarkson, M.J., Faull, W.B., Hughes, J.W., Manson, F.J., Merritt, J.B., Murray, R.D., Sutherst, J.E., Ward, W.R., Downham, D.Y. and Russell, W.B. (1996) Incidence and prevalence of lameness in dairy cattle. *Veterinary Record* 138, 563–567.

Dekkers, J.C.M., Van Erp, T. and Schukken, Y.H. (1996) Economic benefits of reducing somatic cell count under the milk quality program of Ontario. *Journal of Dairy Science* 79, 396–401.

Department for Environment, Food and Rural Affairs (Defra) (2003) *Welfare Code of Recommendation: Cattle.* PB7949. Defra, London.

Department for Environment, Food and Rural Affairs (Defra) (2004) *Animal Health and Welfare Strategy.* Defra, London.

Department for Environment, Food and Rural Affairs (Defra) (2008) *A Roadmap for Environmental Behaviours. In a Framework for Pro-environmental Behaviours.* Defra, London.

European Community (Directive 98/58/EC) Concerning the protection of animals kept for farming purposes. *Official Journal of the European Community.*

Farm Animal Welfare Council (FAWC) (2008) *Opinion on Policy Instruments for Protecting and Improving Farm Animal Welfare.* FAWC, London.

Geller, E.S. (1981) Evaluating energy conservation programs: is verbal reporting enough? *Journal of Consumer Research* 8, 331–335.

Goldstein, N.J., Martin, S.J. and Cialdini, R.B. (2007) *Yes! 50 Secrets From the Science of Persuasion.* Profile Books, London, pp. 18–21.

International HIV/AIDS Alliance (2006) *Tools Together Now! 100 Participatory Tools to Mobilise Communities for HIV/AIDS.* International HIV/AIDS Alliance, Hove, UK. Available at: www.aidsalliance.org/sw44872_asp (accessed 6 July 2009).

Kerr, J., Weitkunat, R. and Moretti, M. (2005) *ABC of Behaviour Change. A Guide to Successful Disease Prevention and Health Promotion.* Elsevier Churchill Livingstone, Amsterdam.

McKenzie-Mohr, D. and Smith, W. (1999) *Fostering Sustainable Behaviour. An Introduction to Community-based Social Marketing.* New Society Publishers, Gabriola Island, British Columbia, Canada.

Talking Retail (2008) Co-op Shoppers Put Animal Welfare Above Climate Change. Available at: www.talkingretail.com (accessed 1 December 2008).

Vaarst, M., Nissen, T.B., Østergaard, S., IKlaas, I.C., Bennedsgaard, T.W. and Christensen, J. (2007) Danish stable schools for experimental common learning in groups of organic dairy farmers. *Journal of Dairy Science* 90, 2543–2554.

Webster, S., Brigstocke, T., Bennett, R., Upton, M. and Blowey, R. (2006) An economic framework for developing and appraising animal health and welfare policy. Report to the Department for Environment, Food and Rural Affairs. In: Farm Animal Welfare Council (FAWC) (2008) *Opinion on Policy Instruments for Protecting and Improving Farm Animal Welfare.* Farm Animal Welfare Council, London, p. 4.

Welfare of Farmed Animals (England) Regulations (2007) HMSO, London.

Whay, H.R., Main, D.C.J., Green, L.E. and Webster, A.J.F. (2003) Assessment of the welfare of dairy cattle using animal-based measurements: direct observations and investigation of farm records. *Veterinary Record* 153, 197–202.

13 Practical Methods for Improving the Welfare of Horses, Donkeys and Other Working Draught Animals in Developing Areas

Camie R. Heleski, Amy K. McLean and Janice C. Swanson

Michigan State University, East Lansing, Michigan, USA

Preface

I (first author) have spent the majority of my academic career as Coordinator of the Michigan State University (MSU) Horse Management Program. Among my primary goals has been to teach students strategies to optimally manage horses to: maximize competitive performance; enhance diet formulation; optimize profitability; enhance breeding programmes; and select horses based largely on aesthetic parameters.

What a rude awakening for me to become passionately involved with the welfare of working horses and donkeys beginning in 2000 with my first of three trips to Brazil. I now think in terms of what are five to 15 essential things these owners can do to enhance the welfare of their animals, realizing that in enhancing the welfare of their cart-horse or donkey, they, in turn, enhance the well-being of their family (Starkey, 1997; Kitalyi *et al.*, 2005).

Instead of worrying about amino acid profiles to enhance the growth of young stock, I try to teach owners of working equids the importance of letting their animal graze or forage whenever it is not actively working; instead of worrying about choosing the perfectly shaped dressage girth so as to not pinch a horse behind its elbows, I encourage people to provide enough clean padding on harnesses so the animals do not have to work with open sores; instead of trying to train a student how to perform a ridden manoeuvre with the slightest of leg pressure, I encourage owners not to beat their animals with sticks or whips; and rather than focusing on how to optimally wrap an injury to minimize scarring, I simply try to ensure the working equid owner knows that packing a wound with traditional 'medicines' such as mud or manure or diesel fuel will only contribute to infection.

Most of the owners/handlers of these horses, donkeys and mules are not intentionally cruel; rather they live in suboptimal environments and equine husbandry information and outreach are rare commodities. In some cases, it is the limitation of their financial resources that prevents them from doing more for their animals, but in many cases, it is simply a lack of knowledge on their parts.

If you are a professional/paraprofessional working in a developing part of the world, your work is to be commended. You are serving a very important purpose and one that will greatly improve the welfare of many animals. The suggestions that follow are designed to involve minimal financial inputs. We hope to bring about a transition in the husbandry of working equids from reactive health care (treating wounds or rehabilitating animals too ill or undernourished to continue working) to proactive husbandry so that animals can work more productively, have enhanced longevity and well-being. This, in turn, contributes to the well-being of families who depend on these animals and, as we have seen first hand, contributes to the pride they have in their working animals.

Introduction

The importance of working draught animals in developing areas of the world is often overlooked (e.g. Pearson *et al.*, 1999). Research on these animals in refereed publications is comparatively scarce (though in preparation to write this chapter, it was impressive how much information on draught animals is available if one knows where to look). Based on world equid population statistics (e.g. Pollock, 2009), of the approximately 55 million horses, nearly 84% are used for work in developing countries; of the 41 million donkeys, 98% are used for this purpose; and of the 13 million mules and hinnies, 96% are used for performing work in developing countries. According to some estimates, 98% of the world's equine veterinarians work on a mere 10% of the world's equine populations (Pollock, 2009). It is understandable that most equine veterinarians would work in developed areas of the world; however, there is tremendous working equid suffering taking place that merits our attention. For people who have not yet worked with livestock in developing parts of the world, the inevitable question comes...why help the animals when the people are often suffering? It has been our experience that when working animal welfare is enhanced, it subsequently enhances the well-being of their respective families. Furthermore, there are environmentally friendly aspects to supporting the sustenance, possibly even the growth, of draught animal usage in developing countries where efficiency may be best captured through use of lower rather than higher forms of technology (FAO, 2002).

With this in mind, this chapter is written to provide equine professionals/paraprofessionals working in developing regions of the world with fundamental knowledge important to enhancing the husbandry and subsequent welfare of working animals, with a particular emphasis on working equids. Direct experience working with carthorses in southern Brazil and with draught donkeys in Mali, West Africa is combined with a review of literature to form the basis for insight and recommendations presented in this chapter. We have tried to make this chapter as realistic and applicable as possible. We recognize that most of the owners/users of working draught animals have very few financial resources to enhance the well-being of the animals in their care. Our suggestions are based on those enhancements that can provide maximal benefit to the animal's comfort and longevity, with minimum inputs, other than time.

The remainder of the chapter discusses 15 recommendations that can greatly enhance the welfare and longevity of working equids and can be implemented with very little financial cost:

1. Check harness fit and modify when necessary.
2. Check shaft height and cart balance.
3. Clean the horse or donkey in between work sessions (washing/brushing).
4. Provide clean harness padding where needed.
5. Clean and treat lesions (and other wounds) with antibiotic salve as needed.
6. Provide forage as frequently as possible.
7. Add high quality forage/concentrate when animal is working especially hard or starting to get thin.
8. Try to get the animal into shade during work breaks.
9. Provide salt, preferably trace mineral salt.
10. Provide fresh, clean water multiple times per day. Horses require significantly more than donkeys.
11. Deworm the animal for parasites that are found in your region.
12. If at all possible, vaccinate for critical diseases, based upon veterinarian advice.
13. Keep the hooves maintained adequately.
14. If at all possible, do not work lame animals or those with serious injuries/sickness.
15. Treat the animal fairly – no beating! Very few equids need beating to go forward unless they are completely exhausted (in which case they should be rested) or are completely overburdened.

Harness Fit, Harnessing and Loading Practices

One of the least expensive ways to enhance working equid welfare is by improving the harness fit, to add padding where needed, to keep the animal clean where the harness is likely to rub, to keep the harness itself clean, and to be certain the cart's load is as balanced as possible. All of these factors will help to minimize the harness lesions that tend to be prevalent in working equids (e.g. Diarra *et al.*, 2007; Sevilla and León, 2007) (Fig. 13.1). In addition, animals that are of moderate body condition score (i.e. not thin – body condition scoring will be discussed further in the 'Nutrition' section) will have less friction between body contact points and the harness, generally leading to fewer lesions. If the animal has sweated during the work day, which is more likely in horses than donkeys (Bullard

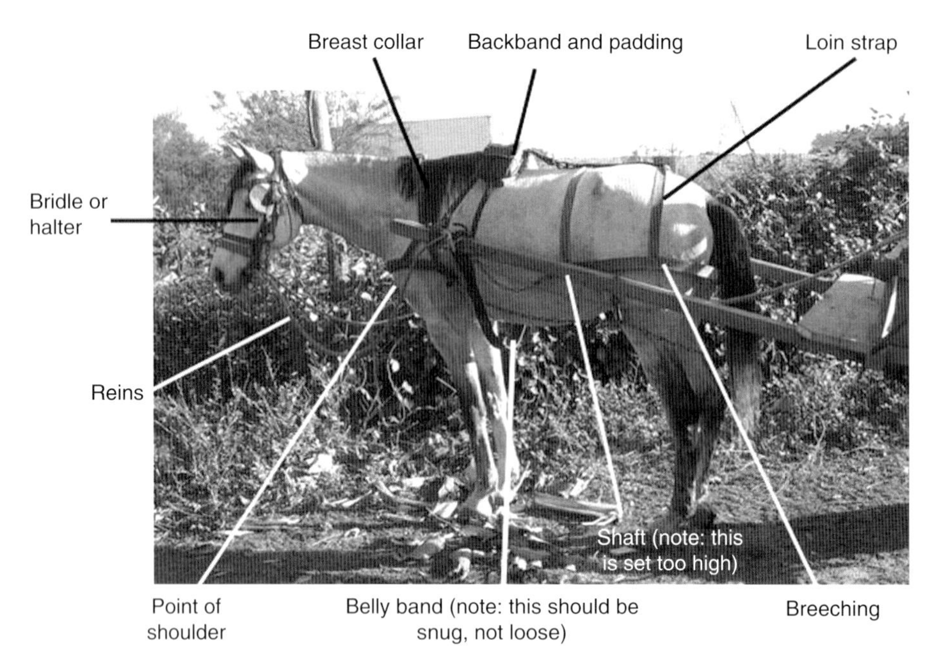

Fig. 13.1. Diagram showing well-positioned harness, shafts and all appropriate parts of a harness to help distribute loading.

et al., 1970), with mules typically being intermediate, the animal should be rinsed off or sponged off at the end of the work day. If water is a rare commodity, the animal can be brushed off the following day prior to having its harness put on. In this way, the sweat does not build up under the harnessing area and cause additional friction on subsequent work days. When dirt or sweat accumulates, it increases the friction of the harness rubbing on the animal. When the animals are thin, there is less natural padding on the body and lesions are even more likely to occur.

In one study on Brazilian carthorses (Zanella *et al.*, 2003), we initially found that 96% of the horses had harness lesions, most frequently on their withers or behind their elbows. After an educational workshop for the horse owners that emphasized the previously discussed recommendations, the percentage of lesions dropped to 62% when re-checked 4 months later. A 34% reduction in harness lesions of working horses, as shown in this study, is a significant welfare improvement.

A survey in Mali by Diarra *et al.* (2007) found that more than 60% of the donkeys had significant harness lesions. It should be noted that with both the horses and the donkeys, many of the lesions were still open sores. Working a full day with a harness rubbing against an open sore is very painful. Improvements in harness fit and harness padding will greatly enhance the welfare of working equids (e.g. Ramswamy, 1998).

Figures 13.2, 13.3, 13.4 and 13.5 show horses and donkeys with generally well-placed harness adjustments. Each country has slightly different modifications to standard harness designs and commonly used carts (see Figs 13.6 and 13.7). A carefully designed harness for all owners of working equids is currently financially unfeasible. In the mean time, persons working with these populations should develop efficient, reasonably comfortable harnesses for the animals they are engaged in helping if deficiency in harness design is responsible for decreased welfare. Readers are encouraged to visit the Draught Animal News (DAN) web site for downloadable documents containing numerous suggestions, pictures and drawings for enhancing draught animal harnesses and implements in developing countries. Type 'Draught Animal News' into a search engine such as Google. See Table 13.1 for tips on harness padding and cleaning.

Our observations of horses in Brazil show many of the horses have the shafts placed too high

Fig. 13.2. Typical harness set up for working carthorse in Brazil. The shafts are slightly high and ideally should be longer. The shafts should be level with the point of shoulder in most cases. Notice the harness in the photo. This is a good example of a harness for working horses. The breast collar or strap placed over the shoulders of the horse and around the chest is properly adjusted and made of material that is sturdy but will not rub the horse. Also, the horse is equipped with a belly band and backband that have adequate padding placed above the shoulders and withers. Both of these straps, the backband and the belly band, help to stabilize the load and keep the shafts in place. This horse is also equipped with breeching, which is needed to help stop the cart and load. Often primitive harnesses will not include breeching (horizontal band around the rear) but this is needed when animals are pulling heavy loads, for stopping and especially when working in hilly or mountainous terrain. Traces or trace chains, the metal chains attached to the harness at the breast collar and then to the cart, are also needed so the animal's power can better reach the cart and be implemented. Again, primitive or crude harnesses often lack traces and these should be added to assist the animal.

(Fig. 13.8), which places considerable pressure on to the withers – sometimes referred to as vertical weight (Jones, 2008a). Many of the horses are observed to have lesions in the wither area (see Fig. 13.9) that may cause considerable discomfort. Our experience with Malian donkeys has shown that many of them also have wither lesions and/or fistulous withers (Doumbia, 2008). It appears this can be due to shaft placement being too high, and/or the harness padding material is too abrasive and contributes to these lesions (see Fig. 13.10 for appropriate padding). In West Africa, the Society for the Protection of Animals Abroad (SPANA) has been working hard to distribute appropriate harness back band padding to donkey owners (see Fig. 13.11). A recent study by Burn and others (2008) found that many of the ridden pack donkeys in Jordan had under-tail lesions caused by poorly made/poorly cleaned cruppers under the tails (see Fig. 13.12).

Load

As a general recommendation, working equids should not be asked to pull more than their own body weight for any considerable length of time (Jones, 2008a). Setting a specific recommendation for length of time is difficult because it also depends on: whether the terrain is flat or hilly; whether the road is hard packed, sandy or rocky; whether the wheels have good bearings and turn smoothly or are themselves a large contributor to friction; the

Fig. 13.3. Another typical harness set-up for a working horse in Brazil. Again, shafts are somewhat too high and should be longer, extending out to the point of shoulder. Notice there is no breeching on this horse. Notice, though, that this horse has a leather strap around his tail called a crupper. This too can help hold a load off the horse, or at least more evenly distribute the load.

Fig. 13.4. Another typical donkey-cart set-up observed in western Africa. Note the appropriate placement of shafts as compared to the point of the shoulder. Also note that, in this case, the donkey would be primarily guided by its halter and reins. We observed that when the back padding was made from cotton-type materials, it served its purpose considerably better than when abrasive plastic burlap padding was used.

Fig. 13.5. Example of a donkey being driven with a halter and reins. Notice the length of shafts, which is good, and they are at the ideal height, even with the point of shoulder. Also, the cart has higher wheels and a lower axle than some shown in earlier pictures. This makes it easier for the donkey to pull the cart forward because the axle and wheels are properly balanced. Also, a synthetic material, probably foam, has been used for backband padding for the donkey. This material is soft, not abrasive, easy to clean and probably recycled. This donkey is also in good physical condition. Donkeys and horses that receive proper nutrition can be more productive when working.

Fig. 13.6. A typical cart used for transporting people and picking up small loads of goods or recyclables (southern Brazil).

speed the donkey or horse is being asked to travel at; how many hours the draught animal needs to pull the load; whether breaks will be provided; and the fitness/condition of the equid. Kilo for kilo, donkeys and mules are generally considered to be stronger than the typical light horses found in most developing parts of the world.

Regarding carried loads, the numbers can be even more difficult to calculate. In developed parts of the world, 20–25% of the horse's weight has become the recommendation for how much a horse can carry, assuming average condition of the horse, average terrain to be covered, etc. (Wickler *et al.*, 2001). Many working donkeys are observed to carry more than 50% of their own body weight over considerable distances and often over rough terrain. This does not mean it represents an ideal

Fig. 13.7. A typical working-cart donkey harness set-up in western Africa. Note the back padding to help minimize harness lesions. A donkey such as this would be primarily guided with a stick. Also, notice that the harness is very crude. There is no halter or bridle for guiding and communicating with the donkey, the belly band is made of a thin rope, which could lead to lesions. It should instead be made of wider and thicker material. Also, this harness has no traces. The traces could help transfer the power of the donkey pulling to moving the cart. Also, if the wheels were higher on the cart then the load would roll more easily.

Table 13.1. Harness padding and repair materials to help prevent sores and lesions.

Good materials	Bad materials
• Cotton canvas • Soft leather • Sheepskins • Woven wool blanket • Foam padding • Twine or string for quick repairs	• Woven plastic bags – the WORST material • Burlap • Lightweight cloth that wrinkles and forms ridges that rub • Polyethylene sheets – these are too hot • Inner tube rubber – this is too hot • Wire for quick repairs – wire should never come in contact with the animal

Cleaning tips for harnesses
• Scrape off crusted dirt and sweat • Pick debris out of woven wool • Replace foam monthly, if possible • Use soap and water to clean but ensure animal contact surfaces are allowed to dry completely before using • If saddle soap or oil for leather is available, it should be used to keep leather soft

Fig. 13.8. These shafts are set significantly too high. You can see the horse trying to brace the load by putting its front legs well under its body. The distribution of weight on the cart could also contribute to the shafts being placed so high. One positive aspect about this photo is the use of breeching and trace chains. However, if the horse is improperly hitched, then the breeching and traces cannot be used to their fullest to alleviate extra weight being placed on the forelimbs of the horse. Unevenly distributed loads can lead to lameness and create lesions from rubbing the horse's body.

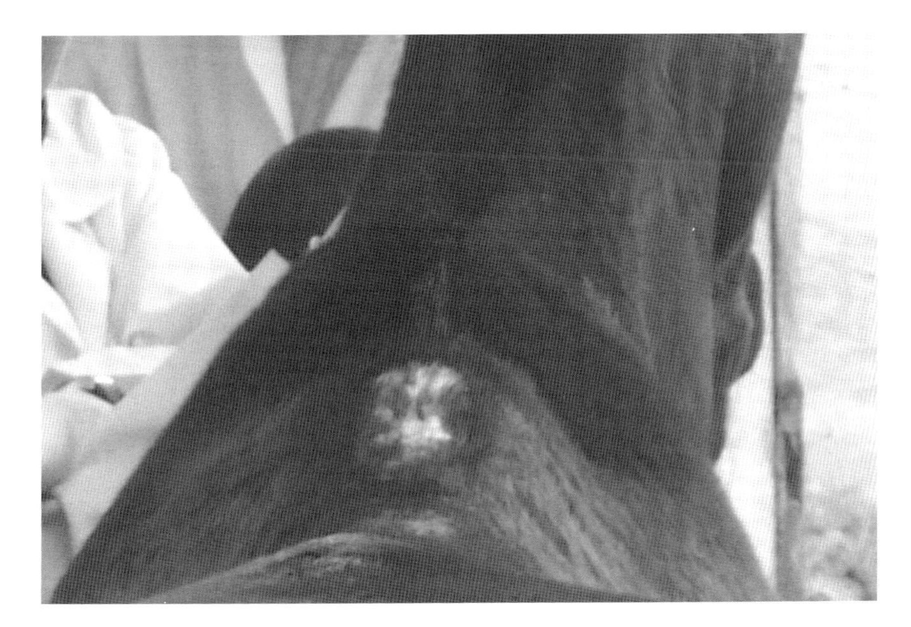

Fig. 13.9. Wither lesions are very common in working carthorses in Brazil, often because the harness does not fit well, because the padding is insufficient or because the horse is quite thin. Additionally, traditional beliefs often mean the horses do not get bathed, so dirty harnesses are often put back on sweat-caked horses, adding to the friction problem.

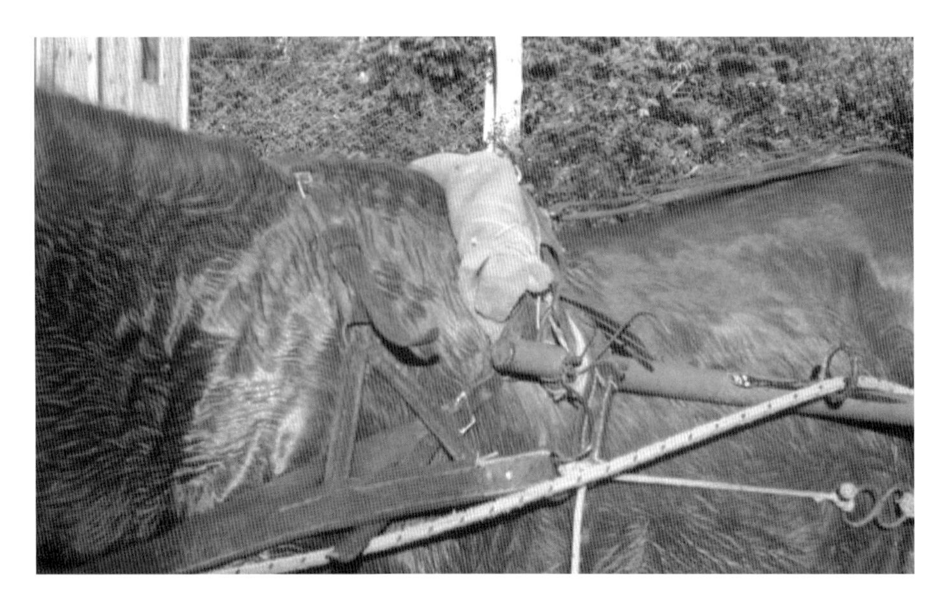

Fig. 13.10. One type of padding (foam) sometimes used to help protect the horse's withers from harness lesions. If the padding is kept clean, it can greatly reduce the number of lesions that a horse might get. Do not use woven plastic bags or padding. This material is abrasive and will cause sores.

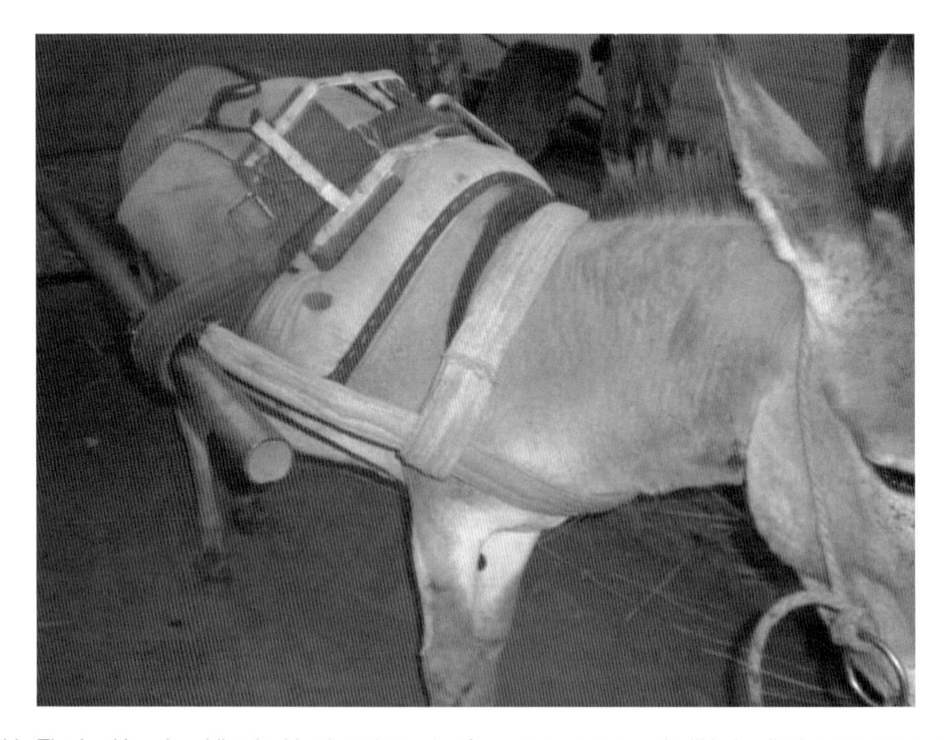

Fig. 13.11. The backband padding in this photo is made of a cotton canvas and will help alleviate pressure on the donkey's withers and back. On many occasions the material used for padding is abrasive and causes lesions. To prevent lesions, use cotton, animal skins or foam padding. The padding should also be kept clean, and the donkey should be groomed in between work sessions to reduce friction.

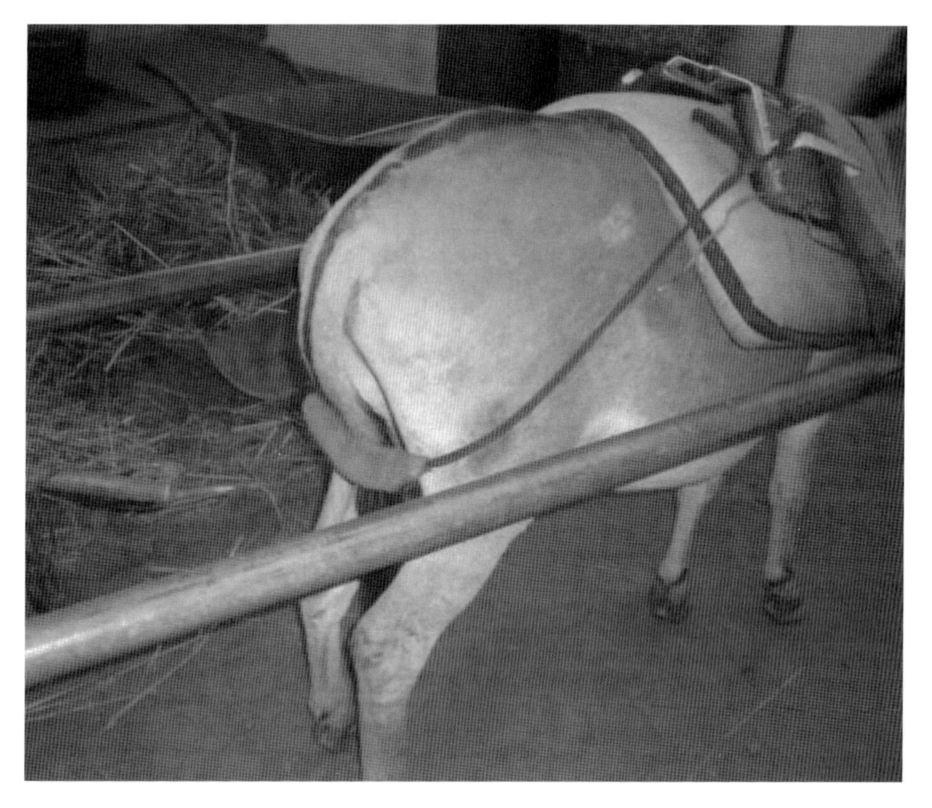

Fig. 13.12. This photo displays a crude form of a tail crupper. Breeching (Fig. 13.2) would be ideal in terms of keeping weight off the donkey and in terms of helping to stop the cart. If a tail crupper is being used, make sure the material will not rub under the tail. Often the material used for cruppers will rub and cause severe lesions under the tail, causing the animal to not want to work. The tail crupper should not be too tight or too loose – it should fit snugly.

situation, and may be partially why many of these animals have significant lesions on their body.

Nutrition

Most of the working equids we have observed are thinner than would be desirable (see Figs 13.13 and 13.14). Many of them are dangerously thin and their body condition is likely to reduce the length of their working life and contribute to significantly reduced welfare. Thin animals often develop more harness lesions and may have subsequently reduced immune systems that make them more likely to become ill. A few different body condition scoring systems have been implemented by professionals/paraprofessionals working in the various countries (e.g. Henneke *et al.*, 1983; Svendsen, 1997), however, the bottom line is usually something like: overweight (almost never observed in working equids), moderate (see Figs 13.15 and 13.16), thin, and various levels of extremely thin (see Figs 13.13 and 13.14) (Pritchard *et al.*, 2005).

Water

The most important nutrient to the well-being of working equids is water. Many owners are unaware how much water is required for their animal, especially if it is a horse. Horses may sweat up to 15 l/h of trotting work, especially when working in hot, humid climates (Clayton, 1991). It could be expected that a working horse in a warm climate might need 40–60 l/day (10–15 gallons/day) of clean, fresh water (Fig. 13.17). However, a donkey may need only 20 l/day (5 gallons/day) (Fig. 13.18). Donkeys are the best choice for really hot climates because they sweat less than horses. Water would ideally be offered intermittently throughout the

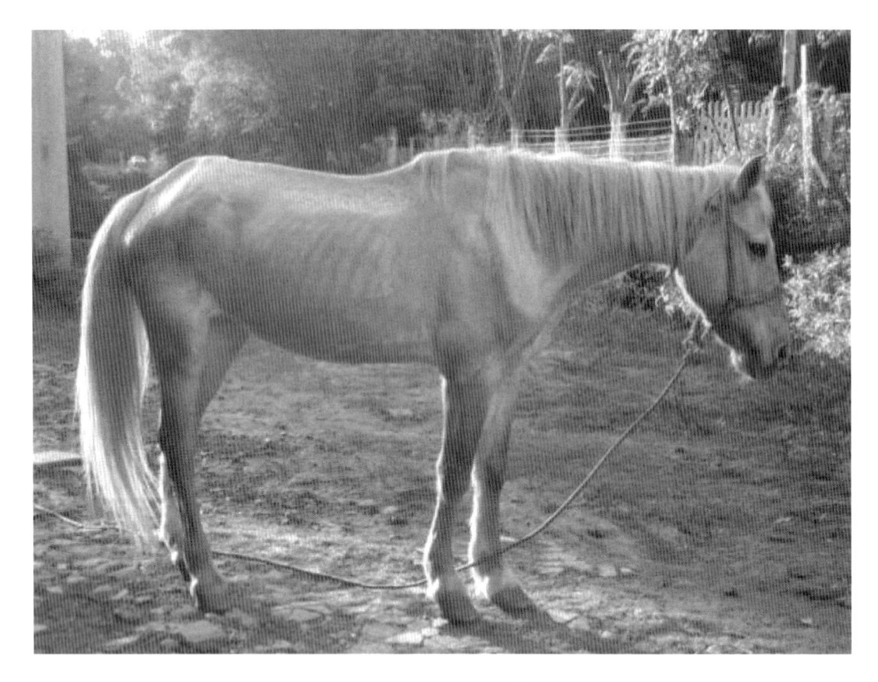

Fig. 13.13. This horse is too thin. Note that her ribs, her spine and her hip bones are all quite noticeable; also her neck is very thin. She would be approximately a '2' on the 1–9 Henneke (1983) scoring system. An animal this thin will have trouble working as hard as her owner might want her to. She may also have more trouble resisting disease.

Fig. 13.14. This donkey is too thin. Note that the ribs and hip bones are showing and the thinness of the neck. Also note the various lesions on the donkey – some from harness wear, some from abusive treatment.

Fig. 13.15. This horse is in good, moderate shape. He would be a body condition score of '5' on the Henneke (1983) system. Note the smooth covering of flesh over the ribs, hip bones and spine. Also notice that the coat is shiny from receiving better nutrition and grooming.

Fig. 13.16. This donkey is in good, moderate body condition. Note the smoothness of flesh covering the ribs and the smoothness of the neck to wither junction. Also note the shine of the hair coat due to receiving better nutrition and better grooming.

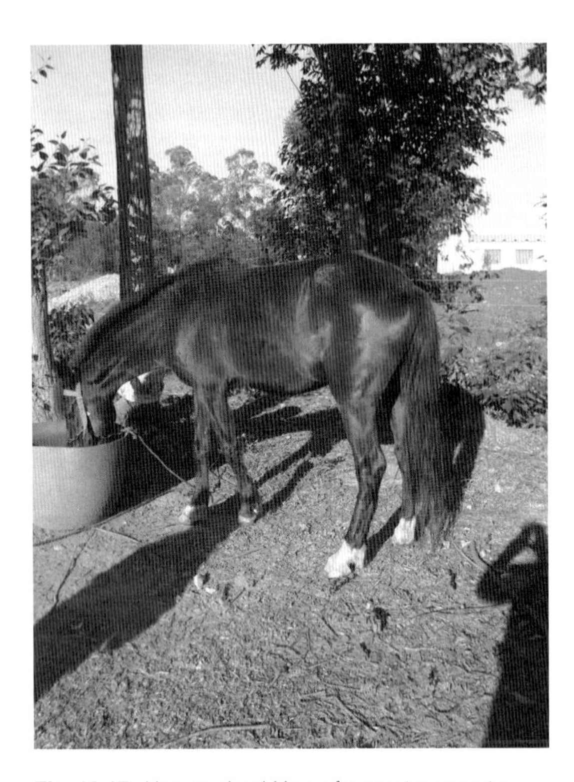

Fig. 13.17. Horses should have frequent access to fresh, clean water. This is especially important for working horses and those who live in hot, humid environments. A horse needs 40–60 l/day (10–15 gallons/day). This is really important in hot climates. An easy way to teach the correct amount of water is to tell people how many buckets are required each day.

day. In southern Brazil (Zanella *et al.*, 2003) most working horses were only offered water at night, and often just one bucket was made available. When skin hydration was checked in nearly 5000 working equids (Pritchard *et al.*, 2007), 50% of horses and 37% of donkeys showed significant signs of dehydration. Another problem, at least in southern Brazil, is the presence of leptospirosis in still (stagnant) water (R. Zanella, personal communication). Often horses were turned loose in small paddocks containing small, stagnant ponds as the only water source. Sadly, the seroprevalence of leptospirosis was quite high. In addition to the potential damage this can cause to horses (e.g. moon blindness/uveitis, abortions and depression), it is also a zoonotic disease, which creates a further concern to the humans living nearby.

Fortunately donkeys are far more drought resistant than horses (NRC, 2007). It is not uncommon for wild donkeys to go to water only once every 2–3 days. However, when they do gain access to water, they can drink a large amount in a very short amount of time (Bullard *et al.*, 1970). Working donkeys should be provided with water daily. In hot weather, they should be offered water several times per day. As previously discussed, the quality and health status of the water is important.

A major problem with insufficient hydration in working equids is that food intake becomes depressed (NRC, 2007) and leads to an apathetic animal incapable of performing its usual work duties at full capacity. Decreased work performance could lead to owner frustration and increases the likelihood of whipping or beating of the animal.

Despite traditional beliefs to the contrary, repeated studies with heavily worked horses show no ill effects in allowing the horses, mules or donkeys to drink their fill whenever water is made available and returning to work (Jones, 2004; Pearson, 2005).

Forage

Few working equids are able to consume enough calories to keep up with their work load, thus many are observed as being thin (e.g. Pearson, 2005). Owners should allow the animal to forage whenever possible. Also, work breaks should be provisioned to allow the animals to eat. In southern Brazil, grass quality is generally good, however, the owners of working horse often live in small villages on the edge of the city (peri-urban) where dozens of horses have been staked out and have over-grazed the grass to the ground. The horse may spend the evening grazing, but under these conditions his calorie intake will be minimal. The more-informed horse owners observed on this study sent their children to cut bags of grass from neighbouring ditch banks. Horses receiving the cut grass were in better condition than the average horses observed (Fig. 13.19). (Note: cut grass should never be stored in bags for any length of time before being given to the horses as it will go mouldy and lead to sick horses.)

In addition to the observations noted above, draught horses identified as being in good condition were supplemented with leftover melons from local fruit stands, leftover bread from local bakeries and

Fig. 13.18. Donkeys are more drought resistant than horses, but they still require frequent access to clean, fresh water. When donkeys do not receive enough water, their feed intake goes down, which in turn reduces their work output. Donkeys are adapted for living in the desert and they require less water. They will need 20 l/day (5 gallons/day).

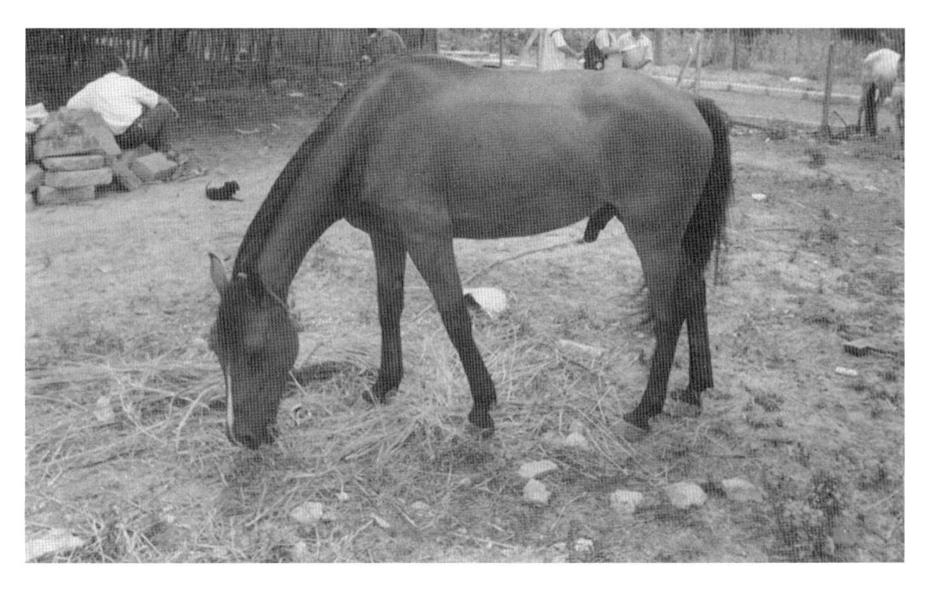

Fig. 13.19. This horse is eating fresh forage. In this case, the grass was cut from a nearby ditch bank.

wheat bran and oatmeal obtained from the local market. We discovered that extra calories from even the most unorthodox sources were utilized by the horses. Obviously the potential for toxic plants should be considered. Most horses, even very thin ones, are good at avoiding toxic plants. This is not an endorsement of unorthodox feed resources, but there are more working equids suffering from lack of food than the rare one that might be lost to toxicity. In these developing parts of the world, everything becomes a risk–benefit assessment. The utilization of locally available food resources, even those that deviate from what is considered 'normal' feed, should be considered.

Working donkeys require fewer daily calories than working horses (NRC, 2007). The donkey is truly amazing with respect to what minimal resources are required for survival. Many of the donkeys we have observed subsisted on seasonal opportunity forage. For example, when cowpea bean hay was occasionally available during the growing season, this was a higher quality food source for the donkeys that helped them regain body condition lost during the droughty season (Fig. 13.20). At other times of the year, donkeys had to be content with only straw or maize stover (stalks and leaves left after the maize has been harvested), both of which are very low in nutritional content. Somehow the donkey is able to effectively utilize every possible nutrient from feedstuffs (e.g. straw) that would be useless to many animals (Pearson, 2005; NRC, 2007). On rare occasions, donkeys were offered supplemental grain, but this practice was unusual. Donkeys, and other equids, do need some of their nutrient needs met with fresh, green forage in order to accommodate their vitamin requirements.

Coauthor Amy McLean has considerable experience working with mules, though not in developing areas of the world. In general, the nutrition needs of the mule (or hinny) are intermediate between horses and donkeys.

Salt

The need for salt is a well-established nutritional requirement for horses (NRC, 2007). Horses should have access to salt, preferably trace mineralized

Fig. 13.20. Donkey eating forage. This is cowpea bean hay, which is a reasonably high-nutrition forage available to donkeys in western Africa. Donkey forage is often purchased at the market, especially during the dry season. Also, notice that the donkey's owner has chosen to unharness his donkey and let him eat while he is not being worked. When donkeys and horses are not working their harness should be removed and they should be allowed to graze or offered food and especially water. If possible, they should be moved into the shade during breaks.

(TM) salt, developed for the region of the world that the animal is living in. For example, there are many parts of the world that are selenium deficient. In those areas, TM salt will often have the addition of selenium. Horses that are not receiving salt are likely to become dehydrated and depressed more quickly than horses with access to salt. Their water intake, in some cases, will become depressed. Salt is a cheap nutrient to supply in most parts of the world. We would like to see all mobile veterinary clinics servicing working equids provide TM salt to their client animals.

Less is known about the salt requirements of the donkey (Pearson, 2005; NRC, 2007). Donkey husbandry guides typically recommend supplying TM salt to donkeys. If TM salt is not available, even a handful of regular salt or access to rock salt will benefit the donkey in replacing the major electrolytes, such as sodium and chloride.

Mules tend to sweat an intermediate amount between horses and donkeys but little documentation exists that verifies this anecdotal observation. It is assumed their salt requirements are likely to be intermediate between that of horses and donkeys.

Parasite Control

In developed countries, deworming a minimum of four times per year is considered typical and a rotation of products is usually recommended. (The recommendations for deworming horses in the developed countries are currently under debate and are likely to be revised in the near future.)

In developing parts of the world, any enhancement in parasite control would be considered an improvement. During our owner interviews in southern Brazil (Zanella *et al.*, 2003), we found that less than 5% of horse owners were deworming – despite the animals being staked out in close proximity and eating in areas of high concentration of manure. Twice per year deworming with a low-cost Ivermectin product could dramatically reduce parasite loads. Unfortunately there was no research conducted to verify the recommended strategy due to funding limitations. An advantage of Ivermectin-based products that is sometimes forgotten by people in developed countries is that Ivermectin helps get rid of external parasites that may plague these working equids. Many skin conditions have been observed in both working horses and donkeys. It is likely that at least some of these skin conditions are exacerbated, if not caused, by external parasites.

Health Management (Wound Care, Vaccines)

Perhaps nowhere is deficiency in owner knowledge more obvious than in wound care and administration of vaccines. Horses are often staked out amid junk and debris, much of which has sharp edges. Not surprisingly, this can lead to significant injuries (see Fig. 13.21). Unfortunately there has been little communication providing information on first aid for working equids (Pearson and Krecek, 2006). Owners have frequently been observed putting mud, manure and diesel fuel, among other traditional remedies, on to open wounds. While there would be a chance that mud might provide some benefit (e.g. by keeping flies out of the wound) many of the substances will simply prolong healing and add insult to the injury. Furthermore,

Fig. 13.21. Working equids are frequently kept in areas that have not been checked for junk and debris lying around. This animal has a cut on its leg, and wounds such as this wire cut are common. All owners should have antibiotic salve on hand to help treat wounds after cleaning with soapy water. Ideally, working equids should have up-to-date tetanus immunizations.

contracting tetanus is a real concern in developing areas of the world. Tetanus vaccinations would be a relatively inexpensive way to address this problem (Colorado State University, 2008).

Basic wound cleaning, ideally with soap and clean water, is a significant first step towards enhancing wound care. This, followed by the administration of antibiotic wound ointments could dramatically reduce the number of badly infected wounds. Bandaging the wounds should be suggested with careful forethought. Inappropriate bandage care is likely to create more problems than it solves. It is important to remember that in the developing world you are basically working in a 'triage' scenario. It is rare that we think about optimal care; rather we need to think of ways to minimize animal suffering and enhance the chances of animal longevity.

Lesions from poorly fitting harnesses are another common source of injuries (see Fig. 13.22). Ways to minimize these injuries are mentioned in the section on harnessing. With respect to the wound, it should be cleaned and dressed. Under the best conditions, the animal should have time off until the wound has healed. This is often not possible; at minimum, sufficient padding needs to be applied to the harness where the trouble spot has been identified (see Figs 13.10 and 13.11).

Another commonly observed injury in working equids is the rope burn (see Fig. 13.23). In horses, this is typically caused by staking the horse out without allowing it sufficient time to understand the concept of being tied. The horse then gets the rope wrapped around its ankle, panics and fights until there is a deep, burn-type injury around the pastern area. In developed countries the practice of staking out a horse is discouraged because of the high risk of rope injuries. However, our experience in Brazil indicates the only chance the animals have to forage adequately is if they are staked out and preferably often. Surprisingly, we observed many horses with no evidence of rope burns. It is possible that horses under these work conditions are so fatigued at the end of their work day that they do not have the strength to panic and fight the rope. If the opportunity is available, having the rope or chain wrapped within a length of old watering hose is one way to make it safer. Also, if the horse has been slowly

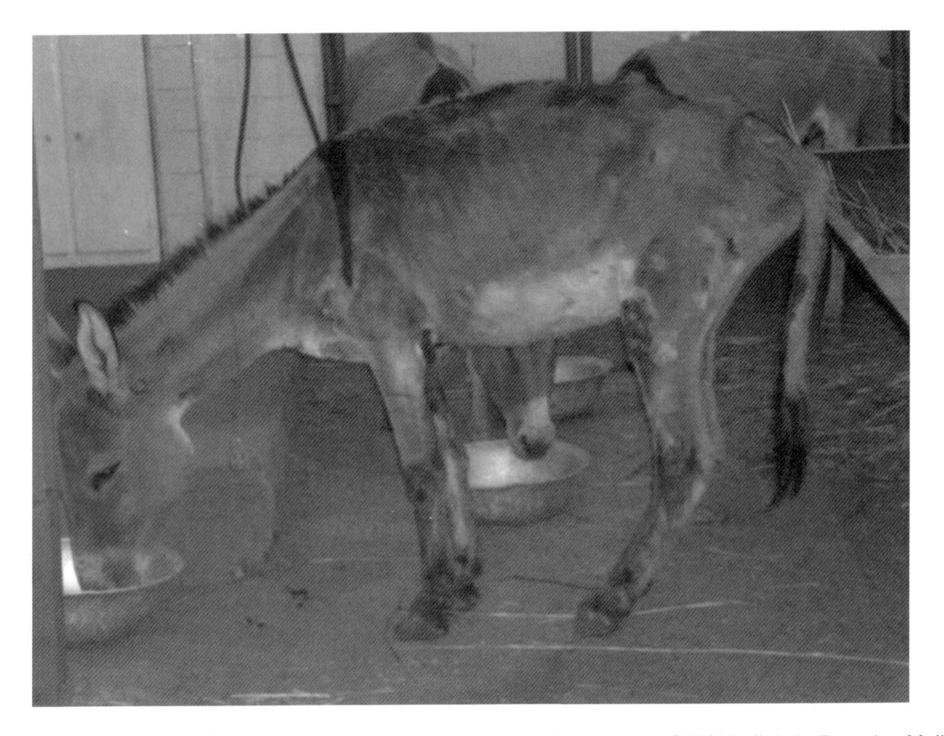

Fig. 13.22. This thin donkey with numerous lesions is being rehabilitated by the SPANA clinic in Bamako, Mali. Eventually this donkey will be healthy enough to return to work.

Fig. 13.23. Rope burns are also commonly observed in the carthorse population. On the positive side, it is good for horses to be tied out so they can eat grass whenever they are not working. On the negative side, some horses panic when they first get tangled in their tie-out rope and this can cause rope burns such as this.

trained to accept the concept of giving to rope pressure, this will also assist his learning curve.

One other unfortunate cause of injuries for working equids is leaving on certain items of tack/equipment. For example, we have observed donkeys whose halters had been left on so long that the skin was attempting to grow around the rope. We have seen horses wearing crudely made permanent splint boots (pieces of rubber tied on to the legs with binder twine) in which the twine has begun cutting into the leg. Others (Jones, 2008b) have observed harnesses and bridles being fixed with wire, which then often cuts into the animal, especially when left on too long.

Vaccines

In developed parts of the world, a great deal of money is spent on vaccinating equids for nearly every disease they might encounter. For the most part, this is a good practice. When it comes to the working equid in the developing world, a risk–benefit assessment needs to be considered, and takes into account the unlikelihood that the resource-poor owner can afford vaccines.

When minimal-cost or donated vaccines can be obtained, it is important to recommend those vaccines that have a high efficacy rate for preventing zoonotic diseases. For example, in southern Brazil, we were able to procure vaccines for many horses to be vaccinated and boostered against tetanus, eastern equine encephalomyelitis (EEE, otherwise known as sleeping sickness) and western equine encephalomyelitis (WEE, also called sleeping sickness). Workers operating in a geographic area need to consider the endemic disease risks for the region. For example, some regions may need to consider rabies as a potentially important vaccine. Other regions may need to consider West Nile virus (WNV) or Venezuelan equine encephalomyelitis (VEE) as potentially important vaccines. Typically, cheap vaccines with high efficacy rates that prevent diseases with significant human risk are the only ones that can reasonably be considered. It is likely that Coggins testing for equine infectious anaemia (EIA) should be implemented, but unless replacement equids can be provided for those families who own a positive animal, this process will make no sense to the draught-animal owner.

Hoof care

Surprisingly, during our research work in southern Brazil, we saw few hoof-care problems. The owners were spending a reasonably high percentage of their limited income to have their horses trimmed and reshod every 6–10 weeks. The workmanship of the farriery was quite good. As a consequence, we rarely saw lame horses being worked. At least in southern Brazil, owners had identified the relationship between proper hoof-care practice and the impact on their economic well-being through reduced lameness of their horses.

This has not been the finding in other locations. In one study in Mexico (Aluja, 1998), a high percentage of donkeys and horses showed evidence of poor hoof care, with overgrown hooves being the most frequently observed problem. Furthermore, even in the stoic donkey, many were observed to be mildly to severely lame.

Lame equids need to be rested. Due to the difficulty in implementing this practice, SPANA clinics in Africa frequently provide replacement donkeys for owners who are willing to relinquish their lame donkey. In this scheme, owners do not have to lose income, but their lame animal is provided with a chance to rest, which is often the only therapy that is required.

Animals with high maintenance and chronic lameness need to be culled from working. There are no additional inputs for tolerating these types of lameness. The animal suffering is not only debilitating to the animal, but it is highly likely that family well-being is also affected.

Fair Treatment

Our experience in southern Brazil was that, while the horses might be sadly neglected in terms of calorie inputs and water needs, it was extremely rare to see them purposely mistreated. In our weeks of observation, with hundreds of horses observed, we saw only two horses being goaded with the whip (and this was when driven by teenagers who appeared to be going for a joy ride in their 'car').

Our experiences in Mali were different. Many of the donkeys were guided with no bridle or halter, instead they were directed with a stick carried by the donkey handler. The stick was often used roughly and repeatedly. Numerous croup and loin lesions were observed that appeared to be a direct result of being struck with the stick (see Fig. 13.24). Some researchers have reported owners in parts of Africa use sticks with nails protruding from them (Herbert, 2006). Obviously this contributes to the serious, and occasionally infected, lesions sometimes observed on the animals. We are currently engaged in a study examining the differences in using halters with reins to guide donkeys (see Fig. 13.5) versus the stick method (see Fig. 13.7) of guiding donkeys. Theoretically, the halter guiding method will result in less beating of the donkeys and less painful lesions that are difficult to heal. If owners can be trained to recognize the importance and benefits of compassionate handling of their animals, it will greatly enhance the welfare of these hard-working animals (Swann, 2006; McLean *et al.*, 2008).

Reproduction and Raising Young Stock

In truth, very few smallholder farmers are set up to raise young stock appropriately. However, replacement animals must be raised. In Brazil, we rarely observed young stock at any of the villages visited. When owners were asked about this, it appeared

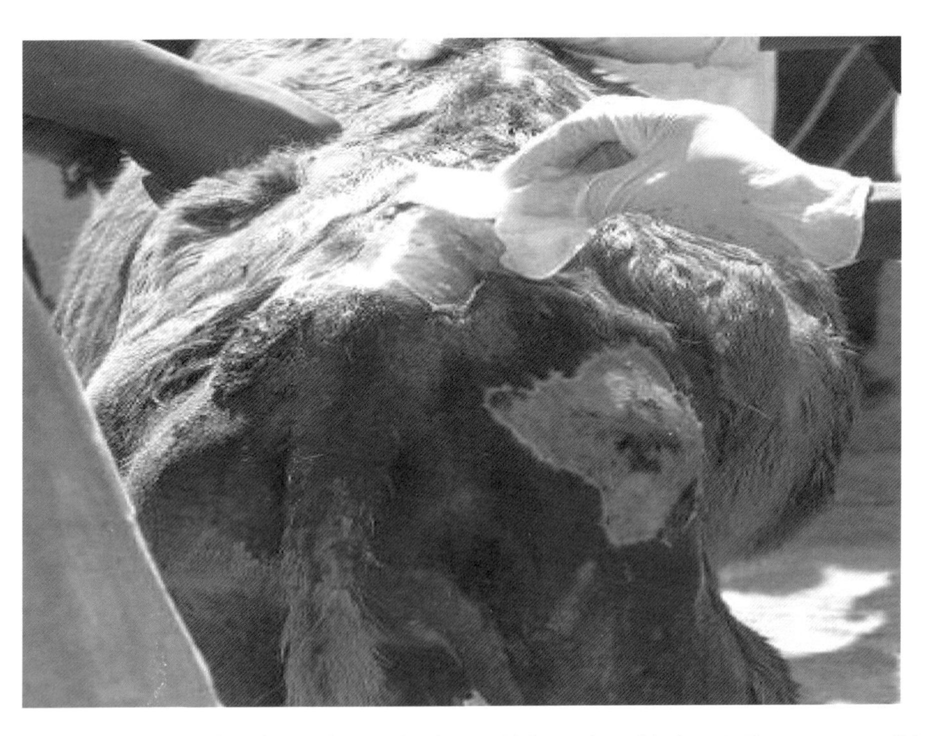

Fig. 13.24. The Malian donkeys have been observed to have a high number of lesions in the croup area. It is believed this is because they get lacerations and deep tissue bruising from being hit with a stick. Some lesions ultimately become infected.

that they typically bought young stock from farmers who were raising horses for other purposes. We did observe a few young animals being raised, and in most instances it was obvious that the animals were receiving inadequate protein and inappropriate calcium:phosphorous ratios (NRC, 2007) (i.e. young stock were stunted and often had heads that were inappropriately large for their bodies.)

In Mali, young donkeys were observed tagging along with their working dams fairly often. Again, as an attribute to the resiliency of the donkey, these young stock often looked surprisingly healthy.

As a side note, because castration is often hard to perform in resource-poor parts of the world, it is not uncommon for stallions to be the working animals nearly as often as geldings and mares. This is not without its problems, in terms of the potential for aggressive behaviour or unplanned breeding. As with other behaviours, often these animals work so hard that there is little interest in causing some of the behavioural problems that would be considered typical to stallions. Another reason that these stallions may have fewer behaviour problems compared to stallions in the developed world is due to better socialization with their own kind. In developed countries, the practice of keeping young stallion colts alone in stalls prevents them from learning how to be socialized with older, experienced horses. Horses that are reared alone, regardless of sex, are more likely to attack other animals (see Chapter 5).

Conclusion

More resources, in terms of time and money, are needed to enhance the welfare of working equids in resource-poor areas of the world. In addition, encouraging all opportunities for information dissemination between organizations and experts with the husbandry information and those who could derive benefit needs to take place (Dijkman *et al.*, 1999; Pearson and Krecek, 2006) (see Fig. 13.25). We assume this is true for all species of draught animals. Although we have concentrated on relating our experience with working equids, other working livestock can benefit from some of these recommendations when applied in a species-specific manner (Fig. 13.26). Helping working equids sometimes feels like the story of the boy and the starfish[1] but it is that boy's words that encourage us to keep going.

Note

[1] **The Starfish Story – an original story by Loren Eisley**

One day a man was walking along the beach when he noticed a boy picking something up and gently throwing it into the ocean.

Approaching the boy, he asked, 'What are you doing?'

The youth replied, 'Throwing starfish back into the ocean. The surf is up and the tide is going out. If I don't throw them back, they'll die.'

(a)

(b)

Fig. 13.25. Many times, it is the children who bring the draught animals to the veterinary clinics for treatment and/or immunizations as here in Brazil (a) and Mali (b). We have begun to implement educational programmes on working equid husbandry that are specifically geared to a juvenile audience.

Fig. 13.26. A water buffalo in the Philippines pulls a plough with a very simple harness. The most common problem is sores on the shoulders. When this type of harness is used on Brahman cattle, sores on the hump are common. This animal is in very good body condition, which will help reduce sores. Yokes and other components that contact the animal must be smooth and have no rough areas or sharp edges (photograph courtesy of Temple Grandin).

'Son,' the man said, 'don't you realize there are miles and miles of beach and hundreds of starfish? You can't make a difference!'

After listening politely, the boy bent down, picked up another starfish, and threw it back into the surf. Then, smiling at the man, he said... 'I made a difference for that one.'

Acknowledgement

We would like to thank Dr Temple Grandin for inviting us to contribute this chapter to her book. We deeply appreciate her effort to broaden the awareness in the welfare science community about the plight of a high percentage of draught animals in the developing world.

References

Aluja, A. (1998) The welfare of working equids in Mexico. *Applied Animal Behaviour Science* 59, 19–29.

Bullard, R.W., Dill, D.B. and Yousef, M.K. (1970) Responses of the burro to desert heat stress. *Journal of Applied Physiology* 29(2), 159–167.

Burn, C.C., Pritchard, J.C., Farajat, M., Twaissi, A. and Whay, H.R. (2008) Risk factors for strap-related lesions in working donkeys at the World Heritage Site of Petra in Jordan. *Veterinary Journal* 178, 261–269.

Clayton, H.M. (1991) Thermoregulation. In: Clayton, H.M. *Conditioning Sport Horses.* Sport Horse Publications, Mason, Michigan, p. 66.

Colorado State University (2008) Working Equids in Ethiopia Receive Donated Vaccines. *theHorse.com* Article #13295. Available at: www.thehorse.com/PrintArticle.aspx?ID=13295 (accessed 19 December 2008).

Diarra, M.M., Doumbia, A. and McLean, A.K. (2007) Survey of working conditions and management of donkeys in Niono and Segou, Mali. *Journal of Animal Science* 85 (Supplement 1), 139 (abstract #59).

Dijkman, J.T., Sims, B.G. and Zambrana, L. (1999) Availability and use of work animals in the middle Andean hill farming systems of Bolivia. *Livestock Research for Rural Development* 11(2), 1–12.

Doumbia, A. (2008) Fistulous withers: a major cause of morbidity and loss of use amongst working equines in West Africa. An evaluation of the aetiology and treatment of 33 cases in Mali. In: *Proceedings of the 10th International Congress of World Equine Veterinary Association*, Moscow, Russia, pp. 581–583. Available at the International Veterinary Information Service: http://www.ivis.org/proceedings/weva/2008/shortcom9/1.pdf?LA=1 (accessed 7 July 2009).

Food and Agriculture Organization (FAO) (2002) Draught Animals Plough On. Available at: www.fao.org/ag/magazine/009sp1.htm (accessed 24 September 2002).

Henneke, D., Potter, G., Kreider, J. and Yates, B. (1983) Relationship between condition score, physical measurements and body fat percentage in mares. *Equine Veterinary Journal* 15, 371–372.

Herbert, K. (2006) At Work in Morocco. *theHorse.com* Article #7001. Available at: www.thehorse.com/printarticle.aspx?ID=7001 (accessed 1 June 2006).

Jones, P. (2008a) Animal tolerance. *Draught Animal News* 46(1), 17–19. Available at: http://www.link.vet.ed.ac.uk/ctvm/Research/DAPR/draught%20animal%20news/danindex.htm (accessed 7 July 2009).

Jones, P. (2008b) Thoughts on harnessing donkeys for work, based on practical experiences in southern Africa. *Draught Animal News* 46(2), 64–70. Available at: http://www.link.vet.ed.ac.uk/ctvm/Research/DAPR/draught%20animal%20news/danindex.htm (accessed 7 July 2009).

Jones, W.E. (2004) Water, dehydration, and drinking. *Journal of Equine Veterinary Science* 24(1), 43–44.

Kitalyi, A., Mtenga, L., Morton, J., McLeod, A., Thornton, P., Dorward, A. and Saadullah, M. (2005) Why keep livestock if you are poor? In: Owen, E., Kitalyi, A., Jayasuriya, N. and Smith, T. (eds) *Livestock and Wealth Creation, Improving the Husbandry of Animals Kept by Resource-poor People in Developing Countries.* Nottingham University Press, Nottingham, UK, pp. 13–27.

McLean, A., Heleski, C.R. and Bauson, L. (2008) Donkeys and bribes...maybe more than just a cartoon! In: *Proceedings of the 42nd Congress of the International Society for Applied Ethology*, Dublin, Ireland. International Society for Applied Ethology, Dublin, Ireland, p. 72.

National Research Council (NRC) (2007) Donkeys and other equids. In: *Nutrient Requirements of Horses*, 6th revised edn. The National Academies Press, Washington, DC, pp. 268–279.

Pearson, R.A. (2005) Nutrition and feeding of donkeys. In: Matthews, N.S. and Taylor, T.S. (eds) *Veterinary Care of Donkeys.* Available at the International Veterinary Information Service: http://www.ivis.org/advances/Matthews/pearson/chapter.asp?LA=1 (accessed 7 July 2009).

Pearson, R.A. and Krecek, R.C. (2006) Delivery of health and husbandry improvements to working animals in Africa. *Tropical Animal Health Production* 38, 93–101.

Pearson, R.A., Nengomasha, E. and Krecek, R. (1999) The challenges in using donkeys for work in Africa. In: Starkey, P. and Kaumbutho, A. (eds) *Meeting the Challenges of Animal Traction.* Intermediate Technology Publications, London, pp. 190–198. Available at: http://www.atnesa.org/challenges/challenges-pearson-donkeys.pdf (accessed 7 July 2009).

Pollock, P.J. (2009) Helping working equids in Egypt. *theHorse.com* Article #13537. Available at: www.thehorse.com/Print.aspx?ID=13537 (accessed 29 January 2009).

Pritchard, J.C., Lindberg, A.C., Mail, D.C.J. and Whay, H.R. (2005) Assessment of the welfare of working horses, mules and donkeys, using health and behavior parameters. *Preventative Veterinary Medicine* 69, 265–283.

Pritchard, J.C., Barr, A.R.S. and Whay, H.R. (2007) Repeatability of a skin tent test for dehydration in working horses and donkeys. *Animal Welfare* 16, 181–183.

Ramswamy, N.S. (1998) Draught animal welfare. *Applied Animal Behaviour Science* 59, 73–84.

Sevilla, H.C. and León, A.C. (2007) Harnesses and equipment communly used by donkeys (*Equus asinus*) in Mexico. In: Matthews, N.S. and Taylor, T.S. (eds) *Veterinary Care of Donkeys.* Available at the International Veterinary Information Service: http://www.ivis.org/advances/Matthews/chavira/chapter.asp?LA=1 (accessed 7 July 2009).

Starkey, P. (1997) Donkey work. In: Svensen, E.D. (compiler) *The Professional Handbook of the Donkey.* Whittet Books, Stowmarket, Suffolk, UK, pp. 183–206.

Svendsen, E. (1997) Parasites abroad. In: Svensen, E.D. (compiler) *The Professional Handbook of the Donkey.* Whittet Books, Stowmarket, Suffolk, UK, pp. 227–238.

Swann, W.J. (2006) Improving the welfare of working equine animals in developing countries. *Applied Animal Behaviour Science* 100, 148–151.

Wickler, S.J., Hoyt, D.F., Cogger, E.A. and Hall, K.M. (2001) Effect of load on preferred speed and cost of transport. *Journal of Applied Physiology* 90(4), 1548–1551.

Zanella, R., Heleski, C. and Zanella, A.J. (2003) Assessment of the Michigan State University equine welfare intervention strategy (MSU EQWIS-ACTION) using Brazilian draught horses as a case study. In: *Proceedings of the 37th International Congress of the International Society for Applied Ethology*, Abano Terme, Italy, p. 192. Available at: http://www.applied-ethology.org/isaemeetings_files/2003%20ISAE%20in%20Abano%20Therme,%20Italy.pdf (accessed 7 July 2009).

14 Successful Technology Transfer of Behavioural and Animal Welfare Research to the Farm and Slaughter Plant

Temple Grandin

Colorado State University, Fort Collins, Colorado, USA

In the fields of both human medicine and animal welfare, there is a great disconnect between the researchers who do the research and the practical application of the research findings. This problem is greatest in the developed countries. Deelan Butler (2008) wrote in the journal *Nature* that 'a chasm has opened up between biomedical researchers and the patients who need their discoveries'. She calls it the 'valley of death'. There is a great need in human medicine for translational research to move basic scientific research discoveries from the laboratory bench to the bedside (Begley, 2008).

Similar problems exist in the animal welfare field. Recently the author consulted with a large commercial pig producer. Most of the managers were not aware that ethology and behaviour were actual fields of scientific study. They knew that sow gestation stalls have to be phased out, but they had no idea that hundreds of behavioural research studies existed. They did not know that many scientific papers were available that could help them make a successful transition to group housing. The author has shown many managers how to access the scientific databases. For more information, refer to Chapter 7 on transport. When the author showed the managers how to access www.scirus.com, Google Scholar and PubMed, they were amazed at all the information they could find.

Failure to Get Behaviour Research Out of the Lab

In many developed countries, the people who do the research, both in medicine and in animal welfare, are usually highly specialized researchers. These people are not practising doctors, veterinarians or farm managers. The problem is that some people are getting too specialized and research has become separated from practice. The reason why this has happened is that people who go into research often have a completely different career path compared to people who become practising veterinarians, doctors or managers. In many countries, a researcher is rewarded for having lots of publications and bringing grant money into his/her institution. They are not financially rewarded for transferring research results to the field. As soon as one research project is finished, they are writing grants for the next project. Often the practical research that can be applied in the field brings in much less grant money compared to large basic research studies. Ian Taylor chairs an important UK task force on technology. He states that 'applied science should be raised to the status of basic science' (Taylor, 2009). He visited universities and found that scientists were frustrated that they did not receive financial incentives to continue projects after publication of the scientific papers.

Good Innovations Remain In the Lab

There are many innovative designs that have been developed by researchers but they were never transferred to the industry. In the 1980s, a talented researcher, also named Ian Taylor, discovered that a sow will waste up to 10–20% of her feed in a poorly designed feeder. To prevent feed wastage, a sow feeder should have a large, deep bowl. This allows a pig to eat like a slob and not toss feed out on the floor. Twenty-five years later, many pork producers still use the same poorly designed feeders and tonnes of feed is wasted. These poorly designed feeders were prominently featured in a 2009 catalogue from a major swine equipment company. There has also been lots of research on better farrowing pens, which provide the sow with greater freedom of movement and still protect her piglets. Many of these innovative designs never went to the next stage of research, which would be installation on farms where they could be tested against conventional farrowing stalls. Figure 14.1 shows a simple turn-around farrowing stall that worked well in a research station, but it was not widely adopted by the North American pig industry.

One of the reasons why many of these excellent research results did not get transferred to the farm is researchers being given the wrong financial incentives. In many developed countries, academic promotion is solely based upon publishing papers and getting grants, not on successful technology transfer. One way to solve this problem is for the livestock industry and governments to provide additional grant funding for activities which will transfer research results into commercial practice. A lot of the money will be needed for travel expenses, which are essential for successful implementation of new methods.

Lack of vision may also stop technology transfer

Transferring new knowledge and technology from the university to industry often takes more work than doing the actual research in the first place (Grandin, 2003). The field of diffusion research has many examples of good technologies that failed at some stage of the transfer to the market. These include the personal computer and the mouse, which were invented by the Xerox Corporation between 1970 and 1975 but failed to transfer from the research team to the manufacturing and marketing department (Rogers, 2003). The personal computer did not finally make it to the market until 1984, and then only because Steve Jobs at Apple hired a lot of Xerox employees who had worked on the original personal computer that had died in the Xerox research labs. Other superior technologies have either never made it to market at all or failed when they did. It is likely that Xerox's failure to market the personal computer was the result of a lack of vision. Upper management may have dismissed the personal computer as an interesting novelty that consumers would never buy (Grandin and Johnson, 2009).

Fig. 14.1. Turn-around farrowing stall that worked really well on a research farm. The sow can be closed in when the piglets are newborn like in a conventional farrowing stall. The photo shows the stall in the open turn-around position. Dimensions are critical. In one study these stalls performed poorly because the welding shop changed some critical dimensions. Details matter.

Principles of Successful Technology Transfer

Ethologists, veterinarians and animal scientists need to spend more time transferring the results of their research to industry. The author will now outline four steps that are required for successful transfer of research results from the lab to producers. These recommendations are based on years of work on improving livestock handling, inventing new equipment and implementing animal welfare auditing programmes.

Communicate your results outside the research community

It is important to publish your research in peer-reviewed scientific journals so knowledge does not get lost. Publishing in refereed journals also helps to ensure that the research results are verified by the use of good scientific methods. Researchers also need to publicize their work by giving talks and lectures, writing articles for industry magazines, and creating and maintaining web sites. One of the reasons the author was able to transfer cattle handling designs to the industry is that she wrote over 300 articles on her work for the livestock industry press. Every research project had several articles published about it. She also gave talks at producer meetings, and she posted her designs on her web site where anyone could download them for free. People are often too reluctant to give information away. The author discovered that when she gave lots of information away she got more consulting jobs than she could handle. She gave the standard designs away for free and made a living by charging for custom designs and consulting.

Make sure your early adopters do not fail

For a technology transfer to work, the first people who adopt it have to succeed. Researchers and developers need to choose companies with management that believes in what they are doing, and they need to inspect every detail. Early adopters must be supervised during every step of the installation and start-up to make sure that the new method works correctly. Choosing the right farm or plant to test new technology is critical. The place should have a manager who loves the project and is dedicated to making it work. A poor manager with a bad attitude can make a good system fail. The author had a promising technology fail because the plant manager was not interested in the project. The people at the corporate office wanted the project but the local plant manager did not like having the hassle of working on a new project.

Supervise other early adopters to prevent mistakes that could cause the method or technology to fail

Transferring the author's centre-track technology out of the first plant that used it successfully was very time consuming (Fig. 14.2). Lots of time was

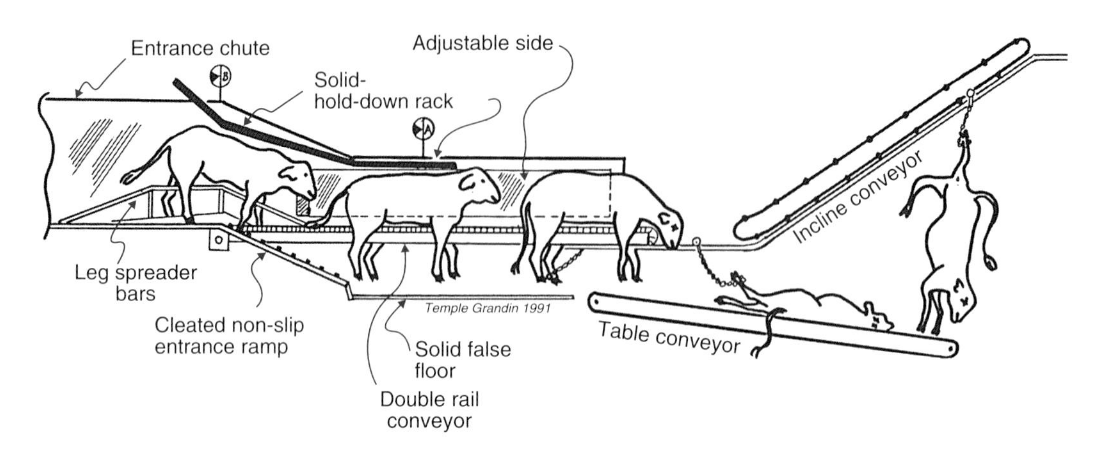

Fig. 14.2. Side view of the centre-track conveyor system that is used in large beef slaughter plants. The cattle straddle a moving conveyor built from metal slats that are 26 cm wide. The outside width of the moving conveyor and the frame must not exceed 30 cm.

spent on site at the next nine veal and beef plants to make sure the equipment had been installed correctly. The steel welding companies often made many poorly designed modifications. They thought they were improving the system but their modifications made it work worse. In half the plants the author visited, the author found either installation mistakes or modifications that would have caused the system to fail. Transferring the technology from the first plant to the next nine plants required more time and travel than installation in the first plant.

Do not allow your method or technology to get tied up in patent disputes

There have been cases of companies buying patent rights to good technologies to prevent a new design from being adopted. That happened in the pig industry in the 1970s, when a designer in Ireland developed a humane, low-cost electric stunner for small pork plants. His design was very clever. He made it out of bicycle parts, which were cheap, and he designed it so it ran automatically and you didn't have to pay an employee to operate it. That made it even cheaper to run, so small companies could afford it. But they never got a chance to buy it because one of the big equipment companies that manufactured and sold an expensive stunner for large plants bought the patent rights. They put the new invention on the shelf and removed it from the market.

That was a terrible waste because the new design was more humane than the equipment the small plants were using and it was too small to work in the larger plants. The company that bought the rights was not going to lose any money if the cheaper design for small plants went on the market. They just wanted to get rid of anything that had any possibility of competing with their system. (Grandin and Johnson, 2009).

Case Study of Successful Transfer of Technology From the Research Lab To the Industry

The original idea for the centre-track restrainer came from work done at the University of Connecticut in the 1970s (Westervelt *et al.*, 1976; Giger *et al.*, 1977). In the early 1970s, the Council for Livestock Protection, a consortium of US animal welfare groups, granted US$60,000 to researchers at the university to develop an alternative method to cruel shackling and hoisting of conscious calves and sheep by one rear leg for slaughter in kosher plants. The Humane Society of the USA and other major animal non-governmental organization (NGO) groups contributed to this project. The researchers developed and built a plywood prototype, but many more components needed to be invented to make a commercially usable system (Fig. 14.3). With the prototype the researchers determined that straddling a conveyor was a low stress, comfortable system for restraining an animal (Westervelt *et al.*, 1976). Before the invention of the straddle (centre-track) design, all the plants were using a V-restrainer that had two angled conveyor belts on either side of the animal, squeezing the animal as it carried it along. The V-restrainer was invented for pigs and it worked well for fat, plump pigs. It was uncomfortable for a more angular, lean animal like a calf or cow.

In 1985 the Council for Livestock Protection awarded another US$100,000 grant to build an operational system inside a commercial veal

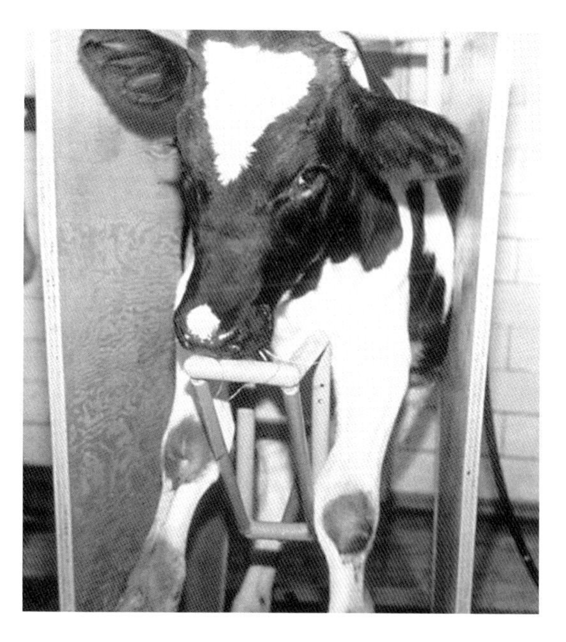

Fig. 14.3. Prototype calf restrainer that was used to determine that a straddle-type restrainer was a low-stress method for holding calves and sheep. A space between the bars enables the animal to rest on its armpits and prevents excessive pressure on the sternum (brisket).

slaughter plant. A major factor that made the transfer of the research prototype to a commercial plant successful was that the same funding agency funded both the scientific research and the development of the first commercial system. The author was hired to design the commercial system and supervise its installation. The council also helped find a plant to build it in – a plant run by an innovator named Frank Broccoli. Frank was very interested in helping to develop a new system. His support and enthusiasm helped make the project successful. The council patented the original straddle idea in the 1970s. They did this so nobody else could patent it and the design was put in the public domain for everybody to use.

Invent critical components in the field

The author had to invent an entrance design so the cattle or calves would walk on to the conveyor belt and straddle it with their legs in the correct position. The original plywood prototype had no usable entrance design and no way to adjust it for different sized animals. The author also had to design adjustable sides to allow for different sized animals (Fig. 14.4). To visualize this you can think of the adjustable sides in a printer feed tray that you can narrow for envelopes and expand for typing paper. Without those two new parts the system would not work (Grandin, 1988, 1991) (Fig. 14.4).

The author destroyed all the world patent rights and put her entrance and adjustable-side designs in the public domain. This was done by publishing diagrams of the two inventions in a meat magazine to provide a verifiable proof of date of the invention. That prevented companies outside the USA from patenting the designs. Placing the designs in the public domain was done to ensure that the technology would be used.

In 1990, the author received a second large grant from a non-profit group to make a larger centre-track conveyor system for adult cattle (Figs 14.2 and 14.4). In 2008, centre-track conveyor restrainer systems were used in over 25 plants in the USA, Canada and Australia. Half of all cattle in the USA and Canada are handled in this system when they go to slaughter.

Fig. 14.4. A large steer is held in the centre-track restrainer system. He remains calm because his body is fully supported by the conveyor. Adjustable sides can be easily adjusted for different sized animals. A gap has been left between the adjustable side and the framework to prevent fingers from being pinched.

Industry may use the technology differently

Another interesting twist in this successful technology transfer is that most of the centre-track systems are being used for regular slaughter with stunning. Even though the original grants were for kosher calf and sheep slaughter equipment, the conventional beef slaughter plants made the greatest use of the system and were the early adopters. Today it is being used for kosher slaughter in a few plants (Fig. 14.5). This shows that sometimes a technology gets used successfully in a different manner from how it was originally intended. Some people may not understand why the author destroyed the patent rights when she could have made money from the design. It is extremely difficult to make money from an invention where a low number of units are sold. If the price of the units gets too high, the plants would not buy them. The cost of worldwide patents would exceed the value of the profit from a device that is used in only 25–30 plants worldwide. From an animal welfare standpoint, the device was a success. The beef plants that use the centre-track restrainer slaughter over 40,000 cattle/day.

Transfer to industry is more costly than original research

The grants for transferring this technology out of the research lab were more than double the original grant for the basic scientific research. Additional grant money was used to fund one commercial calf system and one commercial cattle system. These grants covered the cost of the equipment and its installation. Both plants had to pay for a building to house the equipment, plumbing and electrical work. Equipment should never be totally free. People are usually more motivated to make something work if they have to share some of the costs. After these first two systems were successfully installed, the meat industry paid for all of the other units, and had them built by several different equipment companies. Another hassle was teaching each new equipment company how to build the system correctly. One company built the adjustable sides backwards and another company used lighter gauge steel and many parts broke. Over 20 systems were already being used, but new shops still had to be trained for construction in other countries. The author spent hours on the

Fig. 14.5. Head holder mounted as the end of the centre-track restrainer for kosher or halal slaughter. The round 7-cm-diameter pipe fits behind the animal's poll. See Grandin (2007) and www.grandin.com for further information.

telephone going over pictures and drawings. It was usually possible to fix the problems without travelling to the plant. Today live video feeds over the Internet would be very helpful in solving problems with equipment.

People need to see it work

Even though the calf system had been operating for 4 years, and had been publicized widely, none of the beef plant managers was able to visualize how this system could be scaled up for large cattle. Nobody was willing to invest in a large cattle system until they had seen a large cattle system in operation. After a large cattle system had been built and people were able to see it, the technology spread rapidly through the industry. For a 2-year period, the author went to the equipment start-ups at the next seven plants to correct and prevent installation mistakes. Both the calf plant and the beef plant had to sign contracts which enabled people who were interested in buying the equipment access to look at it.

Avoid Premature Transfer of Technology

A common mistake is to attempt to transfer a new invention or method to the entire industry before it is fully perfected. Before any attempt was made to install the centre-track technology in a second plant, the author made sure that EVERYTHING was working perfectly on the first system. Below is an example of a failure of a new technology due to launching it before it was completely perfected.

The USA had a really bad problem in the 1980s with the premature sale of electronic sow feeders for group housing of sows. In these systems, each sow has an electronic key that allows her to enter a feeding stall and get her precise portion of feed. In the early systems, a sow had to back out of the feeding stall and the next sow who was waiting often bit her. Many farmers installed the systems and then ripped them out because they worked poorly. The marketing department at the equipment company was selling systems before the research and development team had finished inventing them. In the 1990s many of the problems with electronic sow feeders were solved. The innovation that really improved electronic sow feeders was a gate that allowed a sow to exit through the FRONT of the feeding stall (Fig. 14.6). Today they work well. When a technology fails, it sometimes takes 10–20 years for people to be willing to try it again.

The disastrous launch of electronic sow feeders for group housing was probably a major contributor to the spread of individual gestation stalls throughout the USA. When the US pork industry underwent a huge expansion in the 1990s, sow stalls were chosen over group systems. Previous bad experiences with electronic sow feeders probably played a role in the decision of the three largest pork companies to build stalls (Grandin and Johnson, 2009).

(a)

(b)

Fig. 14.6. Electronic sow feeders are now perfected and they work well. (a) An overall view of the feeder. (b) A sow's eye view of the rear entry gate. The gate is closed because a sow is in the feeder. The angled race on the left-hand side of the photo is the front exit gate. The front exit gate was a critical innovation that helps prevent the next sow in line from biting a sow in front of her (photographs courtesy of University of Minnesota).

Successfully Switching to Group Sow Housing

Many farm managers will be supervising the change over to group housing systems for sows. There are many different systems that are available. A simple very effective system is to lock up the sows in individual feeding stalls for feeding. This system is low tech and can be easily built in any country. No electricity is required and old sow gestation stalls can be easily converted to feeding stalls. Electronic feeders work really well in developed countries and the computerized records enable a farmer to determine how much a sow is eating.

Genetics

Some genetic lines of lean, rapidly growing pigs are very aggressive. The author observed that the incidence of tail biting more than doubled when a popular lean genetic line was introduced. Many producers who have switched from gestation stalls to group housing found that they needed to change their pig genetics to lines with less aggressive animals. Sows that bite vulvas and inflict serious injuries to other sows should be immediately removed from group housing. One must remember that for over 25 years, there has been no culling for temperament. Because the sows lived in stalls, producers stopped culling aggressive animals. Aggression does not occur when animals are housed individually.

Group size

Some of the most successful systems have sows housed in large groups of over 60 sows/pen. Large groups fight less. When pigs are mixed, small groups of five or six in a small pen will often get in severe fights. One of the reasons this happens is that there is no way for an animal that is being attacked to get away. Sows can be housed in either small 'static' groups of five or six where they are never mixed again or large 'dynamic' groups where animals are constantly being added and removed. Dynamic groups MUST be LARGE groups of 60 or more sows per pen. Box 14.1 provides suggestions to help with a successful transition to group housing.

Providing hiding places

Another option that may help reduce fighting is providing 'hides'. McGlone and Curtis (1985) built small half stalls along one wall of a pen. When a pig entered the 'hide', its head and shoulders were protected from another pig by a solid partition.

Box 14.1. Tips for successful conversion from sow gestation stalls to group housing.

- Pig genetics may have to be changed. Some lines of lean pigs are very aggressive. Pigs can be genetically selected to be less aggressive. Never mix gentle genetic lines of pigs with aggressive genetic lines. The pigs from the gentle genetic line may be seriously injured by the pigs from the more aggressive genetic line.
- IMMEDIATELY remove sows that severely injure other sows from group housing. Quick removal is essential to prevent other sows from learning her aggressive behaviour. The aggressive sow and her offspring must be removed from the future gene pool. Cull the aggressive sow after she has her piglets and feed out all her pigs as market pigs. A few gestation stalls may need to be kept for holding the really aggressive sows until they farrow.
- When sows are mixed, put them in a new pen to prevent the problem of newcomers invading another sow's territory.
- Providing sows with straw or other fibrous plant material to manipulate will help reduce fighting during mixing. In addition, try adding more bulk and fibre to sow diets. Adding fibre such as soybean hulls, wheat bran or sugarbeet pulp to sow diets resulted in faster growing piglets (Goihl, 2009).
- Do not constantly mix sows housed in small groups. Groups of five or six sows should be kept together as a 'static' group. Groups of over 60 sows can be 'dynamic', where sows are constantly added or subtracted. Big groups fight less because an animal that is attacked can escape.
- The farm manager must be enthusiastic about changing the system. A manager who is against the change may cause it to fail.
- If possible, use new gilts (young females) instead of sows that have lived their entire life in stalls.

Providing 'hides' reduced injuries from fighting. When pigs fight they attack the shoulder using an instinctual hard-wired behaviour pattern. The little half stall covers up the area where pigs instinctually attack. Injuries were reduced in pens equipped with 'hides'. Innovative managers and scientists need to explore the option of providing protected space. The paper can be downloaded for free from the Internet and it has clear photos and diagrams that could easily be enlarged.

Welfare Problems Caused by Building Contractors

The author has worked with many building contractors for over 35 years. Building contractors like to build buildings that are labour efficient for them to build. Contractors can precisely estimate costs for steel fabrication they do in their shop because they know their exact labour costs. When a contractor has to build something complicated on the farm, his costs may vary because bad weather can delay construction.

The weather problem provides an economic motivation to design buildings that can be quickly erected on the farm which minimizes costly delays caused by rain and storms. These designs are often not best for the animals. A well-designed, naturally ventilated dairy cow barn or covered lairage should have a high pitched roof and a large ridge vent. Contractors may prefer to build a flatter roof with a smaller ridge vent because less labour on the farm is required to build it. A flatter roof with a small 30 cm ridge vent will result in a hotter barn in the summer. During cold weather, in a naturally ventilated building that lacks a large ridge vent there can be greater condensation and water dripping from the inside of the roof. The author has visited lairages, stockyards and dairies in many parts of the world. When a large space has to be covered with a roof and naturally ventilated, the best buildings have a high-pitched roof and a large ridge vent that is 2 m or more wide (Fig. 14.7). This creates a chimney effect that pulls heat and moisture out of the building.

Fig. 14.7. A roof with a high pitch and a large ridge vent will keep lairages, dairies and other covered livestock facilities cooler in the summer and well ventilated. The use of this design will reduce or eliminate the need for mechanical ventilation.

Contractors may also try to persuade farmers to move away from naturally ventilated buildings to more intensively managed fan-ventilated buildings. The fan-ventilated buildings are faster and easier for a contractor to build because the high-pitched roof and large ridge vent are not required. They also have an economic incentive to make money selling fans and other expensive ventilation components.

In the developing countries in warm regions, complicated mechanized ventilation systems should be avoided. They have high energy costs and are more difficult to maintain. During travels to Brazil, Chile, the Philippines, Mexico and China, the author has observed locally designed, naturally ventilated buildings that worked extremely well. In China, a naturally ventilated broiler chicken house was kept cool by evaporation by sprinkling water on the roof. The building has no mechanical components.

The Importance of Behaviour in Design

One of the most difficult concepts to transfer to the livestock and poultry industry is the importance of design features on equipment that are there to influence the animal's behaviour. On both dipping vats and the centre-track restrainer system, the author designed an entrance ramp which provided non-slip footing (Grandin, 1980, 2007). The principle of providing non-slip footing is hard for many people to understand. The author has had to return to five or six beef plants and several dip vats to put the non-slip ramps back in after a user replaced it with a slick, slippery ramp. If an animal refused to enter the system, the user just assumed that he needed to change the ramp instead of finding the distractions that have been described in Grandin (1996) and Chapter 5.

The mistake that people make is they often attempt to use force instead of using behavioural principles to get an animal to move through a facility. The first step in fixing an existing facility is to remove distractions. People should use behavioural principles first before resorting to force.

It is hard for some people to understand the importance of a non-slip floor in a veterinary race or a stun box. Animals will remain calmer if they have a good footing. Almost every time the author has visited a new beef slaughter plant, problems were observed with a slippery floor in the stun box. In 2008, the author visited four beef and veal plants for the first time. All four places needed to install non-slip flooring in the stun box. After it was installed, the managers were amazed at the difference it made. Another technology that has been difficult to transfer is the use of solid shields to control what animals see during handling. On the centre-track restrainer systems, there are two large metal panels that are visual shields to control animal behaviour (Grandin, 2003).

The first solid panel is located over the entrance and the first part of the conveyor (Fig. 14.4). The purpose of this panel or hold down is to block the animal's vision until it is all of the way into the conveyor and restrainer (Grandin, 2003). If this panel is too short, the animals often become highly agitated. A 0.5 m difference in the length of this panel had a huge effect on cattle behaviour. Well-intentioned welding shops shortened this panel on restrainers in two beef plants. This resulted in the cattle with large flight zones going crazy. To prove that this piece of metal was required, the author laid a piece of cardboard across the entrance to lengthen the hold down. The plant was able to run for the rest of the day with cardboard holding in 560 kg wild cattle. That is the power of behaviour. At the end of the shift, the cardboard was replaced with metal.

The second component that some people have removed is a false floor that prevents the animals from seeing the 'visual cliff' effect. The restrainer conveyor is 2.2 m off the floor and cattle refuse to enter if they see the 2.2 m drop-off. A false floor has to be installed to provide the optical illusion that there is a floor to walk on. The false floor is positioned so that it is 15 cm below the hooves of the largest animal. Unfortunately, people often removed the false floor because they did not see the purpose of it. They thought it was extra metal that they did not want to clean and maintain. On many occasions the author has had to return to a plant to solve problems with animals refusing to enter the restrainer. The problem was fixed by replacing the false floor that they had removed.

Why is it Hard to Understand Behavioural Features of Design?

A person would never remove a mechanical component on a piece of equipment such as a drive unit because they know that a conveyor has to have a drive unit in order for it to move. The problem with parts that are there for behavioural reasons is that some people do not see the purpose of them, so they remove them. This is most likely to occur when the people the author trained have left and new people have replaced them. Animals notice small sensory details that people often fail to notice. Since animals do not have verbal language, they are much more aware of tiny visual details in their environment that people often fail to notice (Grandin and Johnson, 2005). Figure 14.8 shows a white curtain made from lightweight conveyor belting that prevents the next animal in line from seeing movement of the stunner. Animals will often stop and refuse to move when they see rapid movement or high contrasts of light and dark (see Chapter 5).

Technology is Easier to Transfer than Management Know-how

During a long career of working with many clients, the author has observed that people are more willing to adopt a new technology to solve a problem than to improve their management methods. For example, they are more willing to purchase new equipment compared to spending the time to train and supervise the employees who will be handling animals. The author's consulting business sells twice as many books on the design of livestock handling facilities compared to materials on behavioural principles of handling cattle. Good equipment is essential but it is useless unless good management is combined with it.

The research presented in Chapter 4 shows very clearly that good management and good stockmanship will produce healthier animals that have better welfare and production. Learning good stockmanship takes time, dedication and hard work. A common mistake is to assume that all the problems on a livestock or poultry farm can be fixed by buying new equipment. Good equipment can fix only half of the problems. The other half of the problems have to be fixed by good management practices. Well-designed equipment provides the tools that make good treatment of animals easier, but it is useless if management fails to train and supervise employees.

Curved Handling Systems

During a long career, the author has designed many curved handling systems for ranches (Figs 14.9, 14.10 and 14.11), farms and slaughter plants. The behavioural principles of these systems are explained in Chapter 5. A curved race works well

Fig. 14.8. A curtain prevents the next animal in line from seeing the movement of the large pneumatic stunner. This curtain should be replaced if it becomes ripped or torn. It is important for managers to pay attention to these design details.

Fig. 14.9. Round curved pen and curved race system. Correct layout is essential. There is a straight section before the single-file race turns. This facilitates cattle movement so that the cattle see a place to go. If animals refuse to move past the one-way backstop, it should be equipped with a remote control rope so it can be held open when cattle enter.

because it takes advantage of the animal's natural tendency to circle back in the same direction it came from. Another reason why curved races are efficient is that animals entering the race cannot see the people at the other end when they first enter. The systems must be laid out correctly (Figs 14.10 and 14.11). The most common mistake is bending the race too sharply at the junction between the single-file race and the crowd pen. An animal entering the race must be able to see two to three body lengths up the race before it curves. Even though the principles of layout are clearly stated on the web site www.grandin.com, some people still lay the systems out wrong. Layout mistakes can wreck the system. The junction between the crowd pen and the single-file race is very critical. Some people may have a difficult time seeing the importance of behavioural principles of layout.

Determine the Real Causes of Problems

When a problem occurs, it is important to determine the real cause of the problem. Over the years, poultry and pigs have become more fragile and difficult to handle and transport. Instead of building more and more technological fixes, maybe a better solution to weak animals would be to breed a hardier animal. A slight decrease in productivity would be more than paid back by improved welfare, less sickness and fewer losses to death.

The author learned this lesson about the importance of determining the true cause of problems back in 1980. She was asked by a large pork slaughter plant to design a conveyor system to transport pigs that were too weak to walk up a big, long ramp. This total engineering approach to solving this problem was all wrong. A conveyor was installed in the bottom of the single-file pig race and it caused the pigs to jam up and flip over backwards. The entire expensive mess had to be torn out.

When the author started tracking the weak pigs, it was discovered that they all came from a single large farm. Instead of changing all the equipment at the plant, changes should have been made on this one farm. The weak pig problem could have been

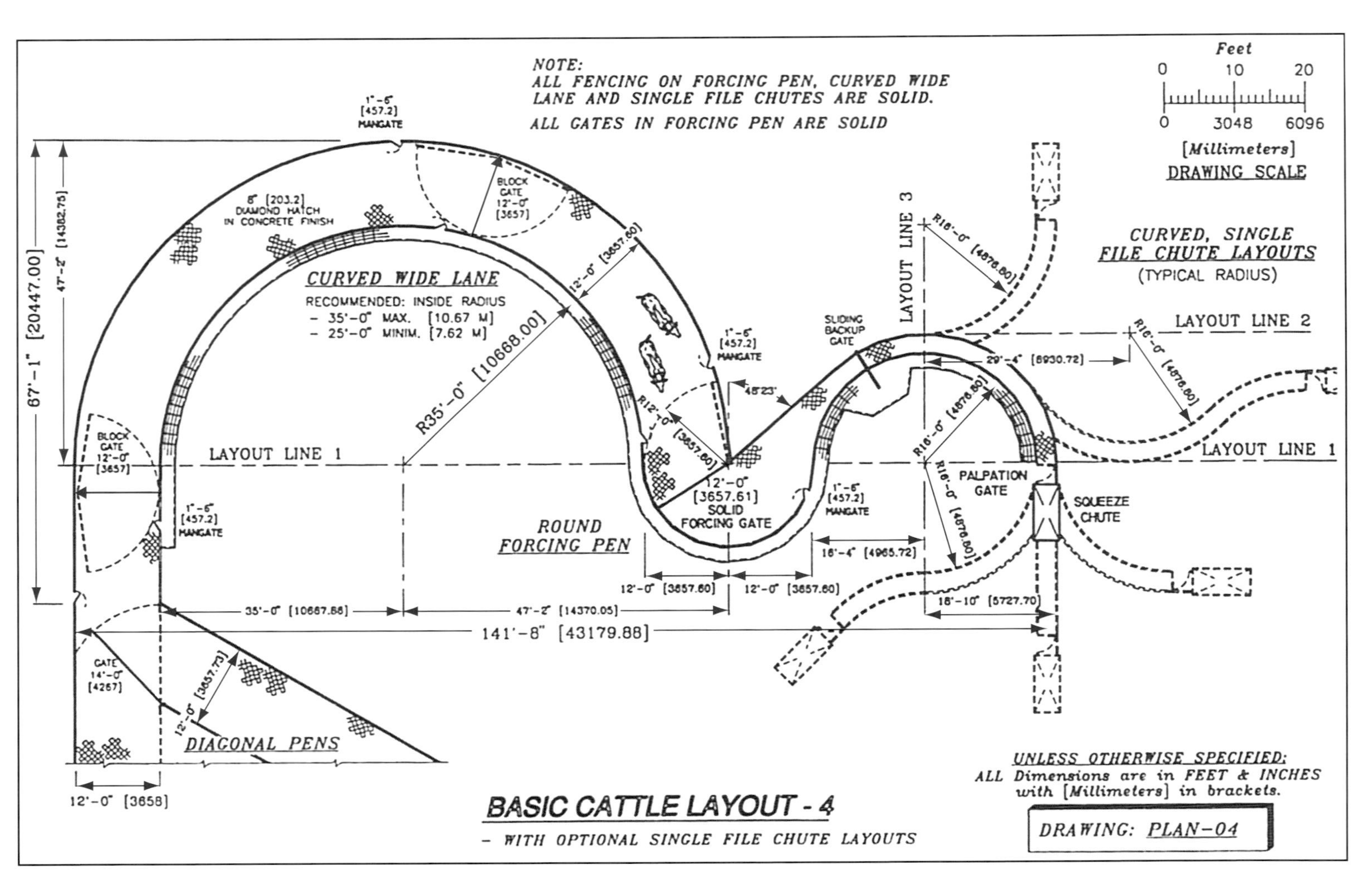

Fig. 14.10. Curved layout showing all the different ways that the curved single-file race can be laid out. The layout of the junction between the single-file race and the crowd pen is critical. Lay this part out exactly as shown.

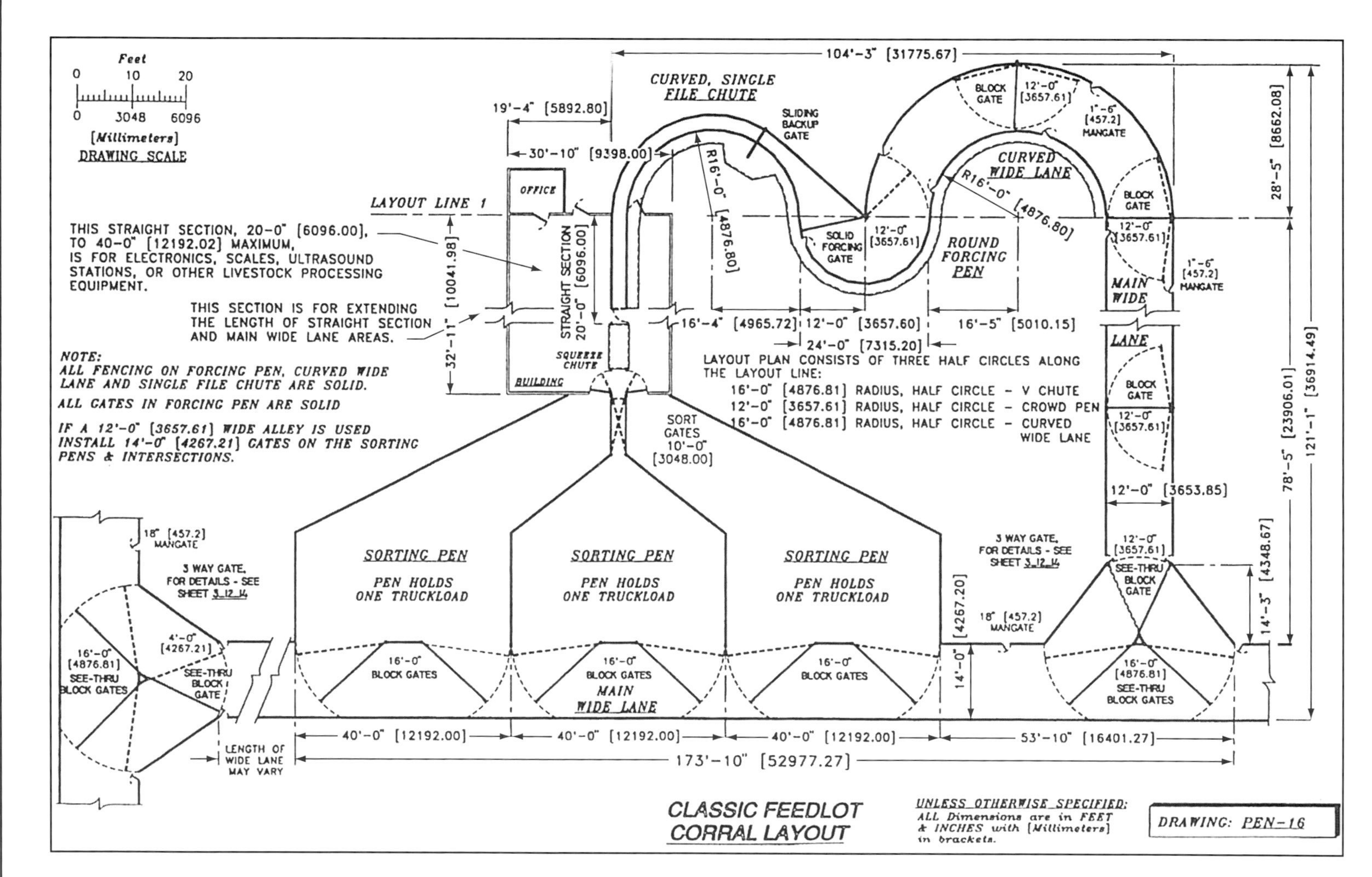

Fig. 14.11. Curved layout for handling cattle on a large ranch or feedlot. After the animals are vaccinated and treated, they can be sorted three ways.

easily corrected by changing boars and breeding out a genetic defect called 'spraddle leg'. The farm also had smooth metal flooring that caused the pigs to grow excessively long hooves. Correcting the problems at this one farm would have been much easier and cheaper than rebuilding plant equipment to accommodate animals with genetic defects and excessive hoof growth. The lesson to be learned from this case is that managers, veterinarians and engineers must find out what the true cause of a problem is before they initiate expensive methods to correct it.

A Lot of Little Changes Equals Big Improvements

Many older facilities can often be greatly improved with many small improvements. You would never want to copy the old, bad designs for a new facility, but a high standard of animal welfare can usually be attained in many older facilities. Below is a case history from one meat plant.

Step 1. Train the employees in basic behavioural principles (see Chapter 5).
Step 2. Repair broken gates and other equipment that directly contacted the animals. Small repairs such as fixing broken or dragging gates was much more important than fixing a leaking lairage roof.
Step 3. Identify and correct all distractions that cause animals to back up or baulk. Below is a list of distractions that had to be eliminated in this particular facility:

- Block sunbeams shining in the main alley with curtains.
- Install a shield to prevent approaching cattle from seeing a person who kept track of cattle ID tags.
- Add a lamp to illuminate a dark race entrance.
- Install a curtain to prevent cattle from seeing people walking by.
- Close in the side of the single-file race to block vision.

All of these distractions had to be located and fixed before the cattle would move easily.

A Summary of Principles for Successful Technology Transfer

1. Knowledge of BOTH the practical and the scientific aspects of your area of speciality in animal welfare is essential. One must keep one's knowledge up-to-date by continually visiting farms and plants. Continuous reading of the scientific literature is also required.
2. Continually writing and communicating ideas and methods – similar lectures and articles must be presented many times to many different audiences. Build a web site full of practical information with free access.
3. Steady, sustained effort over a long period of years will bring about more lasting improvements than short intense bursts of activity.
4. The methods or technologies you implement must be practical and they must work on the farm or in the plant. You need to get out into the farms and plants. Talk to the people who will be using the equipment. These people are often really good at making incremental improvements. However, they often do not have the vision for trying something totally new. They need to see really good videos to be convinced that a new technology will work.
5. Be willing to be criticized, especially when doing new, pioneering things. Accept constructive criticism and be willing to modify what you have been doing.
6. Have the wisdom to work on changing things that can be changed and do not bash your head against the things that cannot be changed.
7. Be positive in your communications and always keep your word. Present the SAME message to the livestock industry, the public and animal advocacy groups. Never be nasty, mean or rude.
8. Find innovative farmers and plant managers who want to make your methods work to be the early adopters. To be successful, the early adopters must believe in your methods and research.
9. The time required to successfully transfer your research results to the livestock producers may require more work and time than doing the scientific research.
10. Maintain confidentiality so you can maintain access to farms and slaughter plants. When bad practices were observed, the author would write and talk about them without revealing the location. If the plant or farm was identified, it would have caused the industry to shut its doors on my efforts to reform it.

References

Begley, S. (2008) On science: where are the cures? *Newsweek*, 10 November, p. 56.
Butler, D. (2008) Crossing the valley of death. *Nature* 453, 840–842.

Giger, W., Prince, R.P., Westervelt, R. and Kinsman, D.M. (1977) Equipment for low stress animal slaughter. *Transactions of the American Society of Agricultural Engineering* 20, 571–578.

Goihl, J. (2009) Bulky gestation diets help piglets grow faster. *Feedstuffs* 23 March, pp. 12–13.

Grandin, T. (1980) Observations of cattle behavior applied to the design of cattle handling facilities. *Applied Animal Ethology* 6, 19–31.

Grandin, T. (1988) Double rail restrainer for livestock handling. *International Journal of Agricultural Engineering* 41, 327–338.

Grandin, T. (1991) *Double Rail Restrainer for Handling Beef Cattle.* Paper No. 91–5004. American Society of Agricultural Engineers, St Joseph, Michigan.

Grandin, T. (1996) Factors that impede animal movement at slaughter plants. *Journal of the American Veterinary Medical Association* 209, 757–759.

Grandin, T. (2003) Transferring results of behavioral research to industry to improve animal welfare on the farm, ranch and the slaughter plants. *Applied Animal Behaviour Science* 81, 215–228.

Grandin, T. (2007) Handling and welfare in slaughter plants. In: Grandin, T. (ed.) *Livestock Handling and Transport.* CAB International, Wallingford, UK, pp. 329–353.

Grandin, T. and Johnson, C. (2005) *Animals in Translation.* Scribner (Simon and Schuster), New York.

Grandin, T. and Johnson, C. (2009) *Animals Make Us Human.* Houghton Mifflin Harcourt, Boston, Massachusetts.

McGlone, J.J. and Curtis, S.E. (1985) Behavior and performance of weanling pigs in pens equipped with hide areas. *Journal of Animal Science* 60, 20–24.

Rogers, E.M. (2003) *Diffusion of Innovations*, 5th edn. Free Press, New York.

Taylor, I. (2009) Learn to convince politicians. *Nature* 457, 958–959.

Westervelt, R.G., Kinsman, D., Prince, R.P. and Giger, W. (1976) Physiological stress measurement during slaughter of calves and lambs. *Journal of Animal Science* 42, 831–834.

15 Why Are Behavioural Needs Important?

Tina Widowski

University of Guelph, Guelph, Ontario, Canada

Introduction

For a number of reasons, animal behaviour has always been a key component in discussions about animal welfare. One reason stems from the viewpoint held by many that good welfare means that animals should be able to lead relatively natural lives, or at least to behave in ways that are consistent with the nature of their species (Fraser, 2003). Although performing all natural behaviour patterns is not necessarily a requirement for good welfare, from a scientific perspective, a species' behavioural biology – its sensory capacities and its general behavioural traits – do determine how the animals perceive and adapt to the ways we house and handle them (Spinka, 2006). Therefore, it is important that the care we provide matches up with their behavioural traits. Further to this is the notion that the performance of some specific behaviour patterns, for example nesting by hens or rooting by pigs, may be important to animals and that they may suffer if they do not have the opportunity to perform them (Dawkins, 1990; Duncan, 1998). This view has led to some of the most difficult and contentious issues regarding the welfare of farm animals.

Importance of feelings

Another reason that behaviour plays an important role comes from the viewpoint that animal welfare is mainly to do with how an animal feels (Duncan, 1996; Rushen, 1996). According to this viewpoint, animals should be housed and handled in ways that prevent negative feelings such as pain, fear and frustration and that may even promote positive feelings such as pleasure or contentment. Chapter 8 contains information on the biology of emotions. Although animals' feelings cannot be measured directly, a variety of experimental techniques have been developed so that we now have a number of well-established ways for scientifically gauging animals' perceptions and emotions (Kirkden and Pajor, 2006). Some techniques rely on animal preferences and the fact that an animal will work very hard to obtain something that it finds highly rewarding and will avoid or try to escape from something that it finds unpleasant. Other behavioural techniques aimed at assessing welfare quantify postures or vocalizations that animals have evolved to use as signals to communicate emotions such as fear, pain or distress (Dawkins, 2004).

Problem behaviour patterns

A final reason that behaviour plays a key role in animal welfare is that in many commercial settings, animals perform behaviour patterns that are aggressive or that cause injury to other animals, which directly reduces the welfare of the victims. Tail biting in pigs and feather pecking in poultry are two examples of behaviour problems that can have devastating effects on the welfare of the target animals. In addition, many animals develop behaviour patterns that appear to be abnormal and these are often interpreted as indicators of poor welfare (Mason and Latham, 2004). These include repetitive abnormal behaviours such as sham chewing and bar biting in sows and tongue rolling in cattle. Substantial research has been aimed at identifying the environmental, genetic and neurophysiological mechanisms underlying these types of behaviour problems in order to help us understand why they develop, what they mean for welfare and how we can prevent them (see Mason and Rushen, 2006).

Use outcome-based measures

In practical settings, brief snapshots of animal behaviour can be difficult to interpret, and therefore most behavioural measures used to assess welfare are often better suited to experimental studies. Animals' natural patterns of behaviour, preferences and aversions are usually determined from controlled experiments, and this information can then be used to make recommendations on input-based indicators of welfare for housing or management practices. Some behavioural indicators of fear, pain or thermal discomfort such as vocalizations, gait or postures have been validated by laboratory studies and can be used as outcome-based measures of welfare (see Chapters 1 and 3). Many of these are readily recognized and used in day-to-day practice by knowledgeable stockpeople, and some can be applied to welfare audits. Behaviour that results in either poor condition or injury to either the animal performing them or to group-mates can be indirectly assessed by using outcome-based measures such as wound scores (Turner *et al.*, 2006), feather scores (Bilcík and Keeling, 1999; LayWel, 2009) or body condition scores (see Chapter 3). However, understanding the underlying causes of those conditions is important for interpreting them in relation to an animal's welfare and for troubleshooting problems.

In this chapter, some of the scientific concepts and approaches used to measure animal behaviour in relation to animal welfare will be discussed. This will include how an understanding of behaviour and motivation can provide us with insight as to how animals perceive the environments we provide for them and help us to develop best practices for managing their physical and social environment. It will also include some discussion of indirect measures of behaviour and how those can be used to identify and solve behaviour problems.

The Importance of Understanding the Behavioural Biology of the Animal

Behaviour is a central part of the mechanisms that every animal has for maintaining its health, survival and reproduction. Throughout its natural history, each species has evolved complex strategies and coordinated sets of responses for obtaining nutrients, protecting itself from predators and other harm, finding mates and taking care of its young. To a large degree, behaviour is under genetic control since genes determine the development of sensory organs, the muscles and the neural systems that coordinate them. All of our livestock species have been domesticated for thousands of years, and the process of domestication as well as intensive genetic selection for production traits has led to some behavioural changes in modern breeds compared to their wild ancestors. However, most of the changes in behaviour that have occurred are quantitative rather than qualitative (Price, 2003). What this means is that domesticated breeds retain many (if not most) of the behavioural traits of their wild ancestors, but that those traits vary in the degree that they are expressed.

Comparison of domestic and wild animal behaviour

Jensen and colleagues recently conducted a comprehensive set of studies comparing the behaviour of White Leghorn layers selected for egg production with that of jungle fowl, which is the wild ancestor of domestic fowl (Jensen, 2006). The birds were incubated, hatched and reared under identical conditions and then observed in a semi-natural environment (outdoors but with food and some shelter provided) and during a series of behaviour tests. The modern egg-laying breed showed all of the same motor patterns and the same social and sexual signals as their wild ancestors, but generally they were less active, less fearful of humans and novel objects, and performed less exploratory and less anti-predator behaviour (Shütz *et al.*, 2001). The modern strain of birds fed more intensively – they were less apt to search for food and ate more from a localized food source, and they engaged in more sexual behaviour. Although they tended to space themselves more closely together, the modern strain was surprisingly more aggressive than the wild type after new groups were formed. There were no differences between wild and domestic stock in either the nature or the amount of what the authors called 'comfort behaviours', which included preening and dust bathing (Shütz and Jensen, 2001). Using modern genomic techniques, Jensen and colleagues (see Jensen, 2006) also investigated the underlying genetic mechanisms for changes in behaviour that might be associated with changes in the genome due to selection for production traits. Their results using quantitative trait loci (QTL) analysis suggest that changes in just a few regulatory genes may be responsible for many of the behavioural effects considered to be part of the domestic phenotype (Jensen, 2006).

These recent results from the studies on laying hens and jungle fowl are in agreement with the idea that domestication and artificial selection for production traits generally tend to increase the frequency of behaviour that supports greater feed efficiencies, higher reproductive rates, and makes the domesticated animal easier for humans to handle than wild animals, but that it has not changed many of their basic behavioural predispositions. A number of studies have also been conducted on feral (populations of domesticated animals that have gone back to the wild) and on domesticated chickens (Wood-Gush and Duncan, 1976), pigs (Jensen and Recén, 1989) and cattle (Rushen *et al.*, 2008) held in wild or semi-natural environments in order to determine how animals spend their time, organize their social groups and what resources or features of the physical environment that seem important for supporting their 'natural' behaviour patterns. These types of studies do not necessarily imply that keeping animals in a natural environment is a requirement for their welfare. 'Natural behaviour' is actually quite flexible and variation in natural behaviour allows animals to adapt to variations in geography, environmental conditions and food supplies. In addition, natural conditions (and natural behaviour) are not always good for welfare because they can cause animals to suffer from stress or injury (Spinka, 2006). However, some consideration of species typical behaviour can provide insight into their physical and social requirements and help us to understand some of the welfare problems that can occur in commercial agricultural settings (Rushen *et al.*, 2008).

Lack of foraging opportunities in intensive production systems

Within intensive production systems there are several areas of housing and management that deviate substantially from livestock species-typical behavioural biology. These include the ways we feed them, the ways we manage maternal behaviour and mother–young relationships, the sizes and composition of the groups that we keep them in and the lack of physical resources (e.g. vegetation, soil or substrate) that support some behaviour patterns. With regard to feeding, individual species use a wide variety of strategies for searching, finding, preparing and finally ingesting their daily nutrient requirements. Examples include grazing and ruminating in cattle and sheep, scratching and pecking in poultry and rooting and chewing in pigs. In natural feeding systems, foraging and feeding behaviour often occupy a large part of an animal's daily time budget. For example, the observed time spent grazing for pastured dairy cattle ranged from 8.6 to 10 h/day depending on the strain of dairy cow and level of concentrate feeding (McCarthy *et al.*, 2007), and time spent rooting and foraging ranged from 12–51% of daylight hours in gestating sows managed outdoors, even though the sows were fed a standard daily allotment of feed in the morning (Buckner *et al.*, 1998). This is in contrast to the grazing and foraging opportunities provided for dairy cows in zero-grazing systems and sows in most indoor housing systems. In young mammals, ingestion of milk usually involves contacting the mother in ways that stimulate milk let down, like nuzzling or butting, in addition to sucking behaviour. In free-living systems calves have been observed suckling cows until 7–14 months (see Rushen *et al.*, 2008) and piglets will nurse for anywhere between 10 and 17 weeks (see Widowski *et al.*, 2008). Most dairy calves are fed from buckets from the first day of age, while commercial piglets are commonly weaned from sows to dry diets at around 21–28 days of age, or even earlier in some systems. These wide departures in feeding system can lead to behaviour problems such as cross-sucking in calves and belly nosing and belly sucking in piglets and may contribute to the development of oral stereotypies in adult ungulates. These topics will be addressed in more detail in a later section.

Nesting behaviours are highly motivated

Because of their size and general lack of physical development, many newborn or newly hatched animals have very different physical requirements from their older counterparts. In nature, neonates are vulnerable to hypothermia and predation. Therefore most species of birds and some species of mammals construct a nest for incubation of eggs and early care of offspring. Even though maternal nests are no longer necessary for survival of chicks or piglets, hens and sows are still highly motivated to perform nesting behaviour before egg laying and farrowing, respectively. Both hens and sows perform elements of nest building in the confines of cages and crates and show behaviour changes that have been interpreted as signs of frustration when not given the resources to nest. In addition, many animals engage in bathing or grooming as part of their natural behavioural repertoire. For example, domestic fowl regularly perform dust bathing when provided with a dusty substrate. In wire cages they show 'sham dust bathing', going through all of the motions of dust bathing in the absence of any dusty material. The appearance of these 'vacuum activities' led to the concept of behavioural deprivation and the question of whether animals suffer when they do not have the opportunity to perform some patterns of behaviour that may be important to them. Scientists have used studies of motivation to address these questions.

What Controls Behaviour?

Motivation is the concept used to describe the internal process or state that determines an animal's behavioural responses to environmental stimuli at different times. In other words, concepts of motivation explain why an animal does what it does when it does it. For example, a sleeping pig gets up, goes to the feed and begins to eat. What is it about the pig that has caused it to switch from sleeping to eating? Certainly the mechanism for this difference involves some complex interaction of neuronal, hormonal and physiologic processes (see Chapters 1 and 8 for research findings on the core emotional systems that drive behaviour). In this situation we usually say that the animal has simply 'become hungry' or that its 'feeding motivation' has changed. Hunger, thirst and libido are the common terms that we use to describe the motivational states for feeding, drinking and sexual behaviour, respectively.

Early theorists on motivation classified different systems along a continuum with behaviour motivated primarily by internal physiological factors on the one end and those motivated primarily by external factors or environmental stimuli on the other. Although we know now that most motivational systems depend on much more complex interactions between internal and external factors, it is often helpful to think about the different mechanisms for the control of behaviour as lying somewhere along this continuum. For behavioural systems that clearly serve in regulating homeostasis – feeding, drinking, thermoregulatory behaviour – internal factors such as blood glucose, osmotic pressure of the blood, or core body temperature, respectively, can predominate. However, there are other non-regulatory behaviours that also appear to depend greatly on physiological factors. Nest building in domestic sows is an example.

Sow nest building is motivated by internal stimuli

When given the opportunity, sows build nests before giving birth to their piglets. The day before farrowing the sow seeks out a nest site, excavates a shallow depression, and then gathers various nesting materials, such as branches, leaves and grasses that she uses to line the nest. Within a few hours after nest completion, the sow delivers her young in the nest. In intensive swine production systems many pre-parturient sows kept in farrowing crates still exhibit the motor patterns associated with nest building. During the 16 h or so prior to farrowing she becomes restless and roots and paws at the floor or bars of the crate. These changes in behaviour consistently and predictably occur during a fairly narrow time period regardless of the environment that the sow is in (Widowski and Curtis, 1990). Nest building in sows presumably has a strong internal component, because it appears in what seems to be a complete absence of appropriate external stimuli in the farrowing crate – little space and no materials for building. Some of the hormone changes that are also associated with labour and delivery, namely the release of prostaglandins, stimulate the onset of this behaviour (Widowski *et al.*, 1990; Gilbert *et al.*, 2002). In a natural environment, the external stimuli – grass, substrate, location of the nest site – then serve to guide the sequence of behaviour patterns, which finally results in nest construction (Jensen, 1993). Considering the function of the maternal nest – a warm, protected place that is necessary for offspring survival – and the critical importance of timing it right, it makes sense that a strong internal control system has evolved. The nest must be completed *before* the young are born, and its being triggered by hormonal changes otherwise associated with reproductive events of parturition ensures that a nest will be completed at the appropriate time.

Behaviours stimulated by external stimuli

In contrast, some behaviour patterns are stimulated primarily by external stimuli. Anti-predator behaviour – alarm calls, protective postures, avoidance behaviour – is the classic example because it generally occurs only in the presence of external stimuli that signal the approach of predators or something perceived as a predator. Animals may perceive rapid movement or a large looming object as threatening (see Chapter 5). Aggression is also usually due to primarily environmental factors since the presence of a competitor is necessary, although physiological factors such as testosterone in males or hunger may increase the tendency of an animal to engage in fighting. At any point in time, animals may be stimulated to do different things simultaneously so that different motivational systems are in essence competing for control. The behaviour that is exhibited is the one that is most strongly motivated at the time or the one that is not inhibited by other factors.

Do Animals Have Behavioural 'Needs'?

Emotional states are thought to play a role in motivation, because feelings such as fear or frustration or even pleasure make it more likely that the animal does the right thing at the right time (Dawkins, 1990). For well over 40 years, scientists have grappled with the problem of behavioural deprivation and whether animals have behavioural needs (Duncan, 1998). The term 'behavioural (or ethological) need' first arose in response to a report by a committee struck by the British government to address public concerns about the welfare of farm animals. The Brambell Committee report proposed that animals have 'natural, instinctive urges and behaviour patterns' and that animals should not be kept in conditions that suppress these behaviour patterns (Brambell, 1965). From its inception, the term was highly debated and often criticized for its lack of both clear definition and scientific foundation (Dawkins, 1983).

Over time, there was some consensus that the term 'behavioural need' should refer to specific behaviour patterns that may be important for animals to perform and that, when prevented, would result in frustration or some negative psychological state that would cause suffering and impair welfare (Dawkins, 1983; Hughes and Duncan, 1988; Jensen and Toates, 1993). Behavioural deprivation is generally considered to be more likely to reduce welfare when the factors motivating the behaviour are primarily internal and when the performance of the behaviour itself seems to be important to the animal (Duncan, 1998). Dawkins (1990) and Fraser and Duncan (1998) suggested that behavioural 'need situations', that is, behaviour associated with intense negative emotions, such as fear or frustration, probably evolved for those behaviours where immediate action is necessary to cope with a threat to survival of the individual (e.g. escape from a predator) or survival of offspring (e.g. nesting). They also suggested that other types of behaviour that are less immediately critical to survival and that can be performed when opportunity arises (e.g. play, grooming) are more likely to be associated with positive emotional states, such as pleasure and contentment.

Scientific Approaches to Measuring Animal's Feelings

Animal welfare scientists have developed a number of different approaches for assessing how animals feel about specific aspects of housing and handling. There are two general approaches that are used, and these have recently been reviewed (Kirkden and Pajor, 2006). The first general approach is to offer animals some control over specific resources or specific experiences, either by offering them choices or by providing opportunities to get to or get away from some option, and then observing what decisions the animals make. Basically these are methods of asking animals what they want and how much they want (or do not want) it. Examples of the sorts of options that can be tested are types of flooring or stall design, different substrates such as straw, wood shavings or sand, items such as nest boxes, perches or social companions and different types of handling and restraint. Three types of standardized tests are used: preference tests, tests for strength of motivation and aversion tests, which will be explained in more detail. The second general approach to measuring an animal's feelings is to keep it in a particular environment (e.g. a wire cage) or expose it to a specific type of experience (e.g. freeze branding) and then carefully observe and measure the animal's responses in order to identify indications of negative feelings such as frustration, fear or distress. Behavioural responses might include things such as trying to escape (Schwartzkopf-Genswein *et al.*, 1998), behaviour that appears out of context to the situation or stereotypic pacing (Yue and Duncan, 2003).

Preference tests

In standardized preference tests scientists use either a Y-maze or a T-maze with different options offered at the ends of the maze. Animals are trained to use the maze, they learn what options are available at the ends of the maze, and then their choices of going to the different options are counted over a series of tests. Alternatively, the animal may be given continuous access to different options over a longer period of time and the frequency of accessing the options and amount of time spent with the different options is measured by live observation or by video recording. Preference tests have been used to study a whole variety of different design features of housing in all species including environmental temperature for sows (Phillips *et al.*, 2000), different types and intensities of lighting for poultry (Widowski *et al.*, 1992; Davis *et al.*, 1999), different types of bedding and flooring for dairy cows (Tucker *et al.*, 2003) and even different levels of ammonia in the air (Wathes *et al.*, 2002). Preference tests can provide good information about housing alternatives, but the tests have to be carefully designed to ensure that the animals are making choices for the options of interest; for example, not just always choosing the left side. They also have to be carefully interpreted because a number of factors can affect the results, such as the previous experience of the animals (they might be hesitant to enter an area with a novel type of flooring, even though it is more comfortable to lie on) or the motivational state that the animal is in during the test (a pig might choose straw to lie on when it is cold but choose bare concrete when it is hot). Another concern with preference tests is that animals do not always make choices that are best for their long-term health and welfare and this has to be considered when using the results to make housing recommendations. One example may be choosing a floor type that may be comfortable to lie on but may be bad for long-term foot health.

Lighting preferences for poultry

One of the advantages of using preference tests is that they allow us to ask the animals directly about whether they perceive differences in the environment and which ones they find more comfortable rather than relying on human perceptions. In this regard, preference tests have sometimes led to some surprising results. One example comes from a question regarding the types of lighting to use in poultry houses. The visual system of birds is very different from that of humans. Birds have very rich colour vision and they can see different wavelengths of light than those seen by humans. Birds also have different abilities for motion detection and are able to perceive flickering light at a much higher frequency than humans – a feature called the critical fusion frequency (CFF). The CFF is the frequency at which motion can no longer be detected or the frequency at which a discontinuous light source (flicker) appears to be continuous. In humans the CFF is 60 Hz. This means that at frequencies higher than 60 Hz images blend together and appear to be continuous; this is the phenomenon on which the motion picture depends. Birds generally have much better motion detection than humans. The domestic chickens's CFF has been estimated to be at around 105 Hz (Nuboer *et al.*, 1992).

In the early 1990s when compact fluorescent bulbs were first developed, there were questions regarding the welfare implications of using them in poultry houses. Common fluorescent light sources powered by magnetic ballasts flicker at twice the frequency of the electricity supply. In North America the supply frequency is 60 Hz, in Europe it is only 50 Hz. This means that fluorescent lights flicker at 120 and 100 Hz, respectively, in those places. Most humans cannot detect this flicker because it is well above our CFF. Because birds' CFF is near the flicker frequency of fluorescent light sources, it was speculated that birds may be able to see the flicker of fluorescent lighting and may find it aversive if they had to live in an environment with flickering light (Nuboer *et al.*, 1992). In order to determine whether laying hens find fluorescent lighting aversive, they were allowed to choose between two rooms illuminated respectively by either standard incandescent lighting or a compact fluorescent source during 6 h long preference tests (Widowski *et al.*, 1992). The rooms were identical except for type of illumination. Light intensities were similar to those used in laying-hen houses.

Contrary to what we expected from what we knew of the bird's CFF, the hens spent around 73% of the time in fluorescent light and only 27% in incandescent light. This meant that either the birds did not perceive the flicker of fluorescent lights (at least in North America) or, if they did, they did not avoid it. In fact, they found some aspect of the fluorescent light more attractive than incandescent light and spent most of the time in it. Similar results have been obtained from preference tests conducted in the UK on turkeys (Sherwin, 1999). It is not clear why birds prefer fluorescent lighting over incandescent, but it probably is due to differences in the wavelengths of light emitted. The compact fluorescent source used emits more energy in the shorter wavelengths (UV and blue region of the spectrum) and domestic fowl are more sensitive to wavelengths in this region than are humans. Results of these studies on lighting preferences of birds emphasize how the sensory capacities and perceptions of animals differ from those of humans, and that it can be difficult for people to predict how animals may respond. One of the best ways to obtain answers is to use an experimental technique that allows us to ask the animal directly.

Aversion tests

Aversion tests are based on the idea that unpleasant feelings or negative emotions evolved to help animals learn to avoid situations that may cause them harm. When an animal experiences something that causes fear, pain or discomfort, it usually behaves in ways that serve to remove the source of those feelings (avoid and escape). If an animal repeatedly experiences something unpleasant, painful or distressing, it will learn to avoid the place or the situation associated with those unpleasant feelings. A number of studies have shown that cattle will learn to avoid rough handling by either hesitating or refusing to go down a runway where hitting and prodding have been experienced (Pajor *et al.*, 2000) or by choosing the arm of a maze associated with gentle handling rather than rough (Pajor *et al.*, 2003). Similar tests have shown that sheep quickly learn to avoid an arm of a maze in which they were subjected to electro-immobilization as a means of restraint (Grandin *et al.*, 1986). These tests indicate that cattle and sheep do, in fact, distinguish between different types of handling and restraint and that they find some types more aversive than others.

Although the majority of aversion tests have been used in the context of handling and restraint, the approach has also been used to determine the aversiveness of some gasses that are used for stunning at slaughter or euthanasia. Raj and Gregory (1995) trained pigs to put their heads in a chamber in order to eat apples from a box in the chamber. Over the course of a series of test days the chamber was filled with either air (control), 90% argon gas in air, 30% carbon dioxide (CO_2) in air or 90% CO_2 in air for a 3 min exposure period. On test days when the chamber was filled with air or argon, all of the pigs remained in the chamber for the majority of the 3 min test period eating the apples, and none of the pigs hesitated to enter the chamber on the subsequent test day. On days when the box was filled with 90% CO_2, on the other hand, nearly all of the pigs withdrew their heads immediately and spent very little time feeding in the box, even on test days when the pigs were fasted for 24 h. The response to 30% CO_2 was intermediate. A number of the pigs hesitated and one even refused to enter the box on the day following the experience of the 90% CO_2, when the chamber was subsequently filled with air, suggesting that the high concentration of CO_2 was highly aversive (see Chapters 9 and 10 for further research on gas stunning and euthanasia).

Strength of motivation tests

Although preference tests do provide us with good information about the relative merits of different choices to the animal, they do not give us any indication how important those choices are. When given the option, a hen may prefer to dust bathe in peat moss over wood shavings, so if we were going to offer them a substrate we know the one that she values more. However, the results of the preference test tell us nothing about *how much she wants* a dust-bathing substrate. This information is particularly important when we consider the more difficult questions regarding whether animals suffer from behavioural deprivation. Dawkins (1983) pioneered the concept of using another type of test, the demand test, to gauge the strength of motivation an animal has for a particular resource. Demand tests are based on techniques used by economists to determine what types of things people view as needs or luxury items based on their purchasing behaviour. Demand tests used for animals are designed to ask animals to pay for a resource through their behaviour. They may have to incur a cost to obtain a resource, for example, forgoing an opportunity to feed in order to spend time in a larger cage or with a social companion. More often the animal is asked to perform an operant technique – peck at a key, press a lever with their hoof or snout – in order to obtain the reward. Once the animals learn the task, the price they have to pay for getting to the item is increased – they have to pay more pecks at the key in order to obtain the reward. Animals' willingness to pay for a resource such as litter or straw is often compared with their willingness to pay for food; food is considered to be the gold standard of what an animal wants. Other types of tasks referred to as obstruction tests have also been used to determine how much an animal wants to get a resource – for example pushing through weighted doors that can be made heavier and more difficult to open (Duncan and Kite, 1987) or squeezing through gaps that are made narrower and more difficult to squeeze through (Cooper and Appleby, 1996). Sometimes animals are tested before and after being deprived of a resource to determine whether the value of the resource increases after the animal has been without it for some time or, alternatively, whether 'out of sight is out of mind' (Duncan and Petherick, 1991).

Using studies of motivation to determine nest-box and dust-bathing needs in laying hens

The battery cage system for laying hens has been highly criticized, in part because of the lack of opportunities for hens to perform nesting and dust bathing. Literally hundreds of studies have been conducted to help us understand the motivation for these behaviours and there are several recent comprehensive reviews of that work (Cooper and Albentosa, 2003; Olsson and Keeling, 2005). Comparisons of what we know about what controls behaviour and how hard hens are willing to get to a nest box or a dust bath indicate that there are some important differences between these behaviours with implications for hen welfare.

The pre-laying behaviour of hens has been studied extensively for well over 40 years, beginning with the studies of Wood-Gush and Gilbert, who showed that nesting is stimulated by the release of hormones during ovulation resulting in an organized sequence of behaviour prior to laying the egg on the following day (see Wood-Gush and Gilbert, 1964). Hens begin to show signs of searching for a nest site several hours prior to egg laying when they increase locomotion and examine potential nesting sites. The searching phase is followed by a period of sitting at the final site where the egg is laid; there the hen constructs a nest by rotating her body to form a hollow and arranges nesting materials (Duncan and Kite, 1989).

Most hens prefer to lay their eggs in a discrete enclosed nest box and the strength of hens' motivation to access a nest box has been demonstrated in a variety of ways (Fig. 15.1). Hens have been

Fig. 15.1. The hormones from ovulation cause the hen to begin searching for a nest site to lay her egg. Most hens prefer to lay their eggs in an enclosed nest box lined with shavings, straw or artificial turf.

shown to be willing to squeeze through narrow gaps (Cooper and Appleby, 1996), push open heavily weighted doors (Follensbee *et al.*, 1992) and pass through cages occupied by unfamiliar or dominant hens in order to gain access to a nest box (Freire *et al.*, 1997). Cooper and Appleby (2003) showed that hens' work rate (by pushing through a locked door) for a small pen furnished with a wooden nest box at 40 min before egg laying was equal to their work rate to return to their home pen after 4 h of confinement without food, and the work rate to access the nest was double that amount at 20 min prior to egg laying. Hens can be highly persistent in their efforts to get to a nest box. When hens were trained to push through a door to get to a nest box and then were prevented from doing so, light hybrid hens averaged 150 attempts to push through the door in an hour (Follensbee, 1992). Figure 15.2 shows a door apparatus that is for measuring the strength of motivation.

When a nest box is not available, hens are more active, engage in walking for a longer duration before laying their eggs, and often perform what has been described as stereotyped pacing; behavioural differences that have been interpreted as signs of frustration (Wood-Gush and Gilbert, 1969; Yue and Duncan, 2003). Yue and Duncan (2003) found that hens in cages without access to a nest box spent over 20% of the hour before egg laying walking and pacing, compared to around 7% by hens with a nest box. Nesting in a secluded place is a high-priority behaviour that can be accommodated by providing the hen with a nest box.

How important is dust bathing?

Domestic chicks begin dust bathing during the first weeks after hatching. The behaviour involves a sequence of distinct lying, fluffing and rubbing motor patterns that distribute dust under the feathers. Dust bathing is controlled by an interaction of internal and environmental factors, and the motivational system is quite different from that of nesting. Dust bathing occurs on average every 2 days and follows a diurnal rhythm, with most dust bathing occurring in the late morning and early afternoon (Lindberg and Nicol, 1997). A number of environmental factors also affect hens' motivation to dust bathe, including the sight of a dusty substrate (Petherick *et al.*, 1995), ambient temperature, light and a radiant heat source (Duncan *et al.*, 1998). External factors (like sunlight and peat moss) can be very potent stimuli for dust bathing and the behaviour can essentially be 'switched on' under the right conditions.

Fig. 15.2. A hen is trained to push through this door to measure how much she is willing to work to get to a resource, such as a nest box. After the hen has learned to open the door, increasingly heavy weights are attached to the ends of the horizontal crossbars. The amount of weight that the hen is willing to lift to open the door provides a measure of the strength of her motivation.

The willingness of hens to work for a dust-bathing substrate after a period of deprivation has been tested using a variety of operant and obstruction tests, but the results have highly variable (see reviews by Cooper and Albentosa, 2003; Olsson and Keeling, 2005). Widowski and Duncan (2000) found considerable individual variation in hens' willingness to push through a weighted door to access a dust bath in peat moss. Although the majority of hens tested pushed more weight after they had been deprived of litter, some hens pushed more weight after they had just completed a bout of dust bathing in their home pen. Others pushed through the door and then did not perform any dust bathing. The hens' behaviour in the test runway was clearly different in this study than in previous studies with nest boxes. We concluded that these results do not support a 'needs' model of motivation for dust bathing but rather that hens dust bathe when the opportunity presents itself. However, the behaviour is likely to be rewarding to the hens.

When birds have been kept without a substrate for dust bathing, they begin dust bathing more quickly and they perform longer and more intense dust bathing bouts when they are finally presented with a substrate. This is often referred to as a rebound effect and is considered to be evidence of a 'build up' of dust-bathing motivation after a period of deprivation (Cooper and Albentosa, 2003). However, any behaviour thought to be associated with frustration such as pacing, head flicking or displacement preening has rarely been reported on studies of dust-bathing deprivation. Although the appearance of vacuum dust bathing raises concern, hens in enriched cages supplied with a dust bath are often seen to dust bathe on the wire floors. Olsson and Keeling (2005) argue that the 'need' versus 'opportunity' models for dust bathing are more likely to depend on the internal state of the hen. If a hen is presented with a highly attractive substrate she may perform the behaviour even if she has recently dust bathed. However, for a hen deprived of substrate for some time, the need to dust bathe will supersede any external factors and she will perform sham dust bathing in the absence of substrate.

The various studies on nesting and dust bathing have led a number of scientists to conclude that lack of a suitable secluded nest box is one of the biggest welfare concerns in conventional cage systems and that hens may be frustrated when they do not have access to a nest (Duncan, 2001; also see Cooper and Albentosa, 2003; Weeks and Nicol, 2006). Nesting in a secluded place is a hard-wired behaviour pattern that helped survival of chicks in the wild. Dust bathing, on the other hand, seems to be less of a priority for laying hens (see Chapters 1 and 8). Nesting motivation can be satisfied by providing an enclosed nest box lined with wood shavings, straw or artificial turf (Struelens *et al.*, 2005, 2008). A variety of models of cages furnished with nest boxes, perches and dust baths have been developed that provide opportunities for hens to engage in a variety of behaviour and at the same time provide the health and hygiene benefits of cages (Tauson, 2005).

Consequences of Behavioural Deprivation

Although studies of motivation can tell us a great deal about what types of behaviour are highly motivated and may be important for animals, much less is known about the physiological and health consequences of depriving them. Some of the best examples of the link between behavioural deprivation and any effects on physiology or development of abnormal behaviour have to do with ingestive behaviour. Feeding and drinking are 'regulatory' behaviour patterns in that they directly impact on body function. Performance of the motor patterns associated with feeding and drinking contribute to feelings of satiety and help signal to the brain that the animal has eaten or had a drink well before digestion has occurred or osmotic balance of the blood is re-established. In production settings there is often a mismatch between how the animals are fed and their natural feeding behaviour.

Cross-sucking problems in calves

Because their survival depends on it, young mammals are highly motivated to perform sucking behaviour. Veal calves and dairy replacement heifers are usually weaned from their dams within a day or so after birth and are fed milk (or milk replacer) from buckets. When calves are kept in groups they often develop cross-sucking which involves sucking on the ears, muzzle, tail, prepuce, udder or scrotum of other calves. Cross-sucking can be detrimental to the health of calves. In some cases cross-sucking can lead to urine drinking and reduced body weight. In the absence of other calves, sucking is often directed at pen-fixtures.

Numerous studies have been done to determine the factors that control sucking behaviour in calves as well as the physiological consequences (de Passillé, 2001; also see Rushen *et al.*, 2008). In studies investigating the control of sucking in calves, de Passillé and colleagues (2001) found that cross-sucking within groups of calves occurs most often *after* calves have ingested a meal and that cross-sucking was reduced when calves were provided with an artificial teat. Their studies also indicated that sucking behaviour is motivated by the taste of milk. Even giving a small amount of milk or milk replacer (5 ml) by syringe in the mouth triggers sucking, and the more concentrated the milk replacer, the more sucking is performed. Increasing the volume of milk at a meal does not reduce motivation for sucking although the behaviour is not completely independent of hunger. Calves that missed a meal or were fed low levels of milk did increase non-nutritive sucking after the subsequent meal. Sucking motivation diminishes spontaneously within 10 min after feeding, whether or not calves have been able to suck. Also, the *performance* of sucking behaviour appears to be linked to satiety signals and digestive function of the calves. Providing an artificial teat after feeding milk from a bucket significantly increased plasma cholecystokinin (CCK) and serum insulin concentrations in the hepatic portal vein compared to calves that could not suck. Calves given a dry teat to suck after a meal also have lower heart rates and rest more. Providing either a dry teat or feeding calves from a bottle with a nipple reduces cross-sucking. Many progressive producers have found that feeding calves with a nipple (teat) bottle has many advantages compared to using buckets. The studies on sucking behaviour in calves are a good illustration of the close links between behaviour patterns and the physiological systems that they support.

Weaning problems in piglets

Weaning age and feeding systems in commercial production can also cause problems for young piglets. Early-weaned piglets often have trouble getting on to solid feed, spend a great deal of time at the drinker and develop belly nosing, a rhythmic massage directed at the bellies or navels of other piglets (see Widowski *et al.*, 2008). The behaviour appears very similar to massaging the sow's udder that piglets use to stimulate milk let down. Belly nosing and belly sucking can cause lesions on the recipient pigs (Straw and Bartlett, 2001) (Fig. 15.3). The incidence of belly-nosing behaviour is

Fig. 15.3. Belly nosing is abnormal behaviour that can cause damage to other pigs. Piglets that are weaned too early will often engage in more belly nosing.

significantly affected by weaning age, with younger-weaned piglets performing significantly more belly nosing. It appears to be a result of confusion among the control systems for sucking, feeding and drinking. Providing an artificial teat or artificial udder to massage can significantly reduce belly nosing as well as time spent at the drinker. Drinker design can also affect belly nosing: piglets with push-bowl drinkers perform significantly less belly nosing than piglets with 'nipple' drinkers (Torrey and Widowski, 2004). Even though the name 'nipple drinker' implies that the drinker allows for sucking, water actually pours into the piglet's mouth and it does not appear to satisfy sucking motivation. Although we know less about the physiological consequences of sucking (or sucking deprivation) in piglets than in calves, we do know that piglets who develop high levels of belly nosing grow more poorly and, in extreme cases, appear to be wasting. In a field study on commercial farms, we found that even short 2h video samples of weaned piglets are enough to detect when belly nosing is occurring (Widowski *et al.*, 2003). Many US and Canadian producers have stopped extreme early weaning and have switched back to weaning at 21–28 days. Weaning at a later age will help reduce belly nosing and other abnormal behaviour in pigs.

Oral stereotypies in adult ungulates

In adult ungulates, restricted feeding and the lack of opportunities to forage appear to contribute to the development of oral stereotypies (Bergeron *et al.*, 2006). The behaviour often peaks around feeding time or shortly thereafter, and factors that increase satiety appear to reduce them. Several hypotheses have been developed to explain why oral stereotypies occur. One hypothesis is that the diets provided simply are not satisfying – the diets are deficient in calories or some other nutrient. In restricted-fed sows, increasing feed allotment or fibre and bulkiness of the diet reduces behaviour patterns such as sham chewing and bar biting. Another hypothesis for development of stereotypies is that in horses and ruminants, the behaviour fulfils some function on the gut. For example, tongue rolling in cattle and cribbing in horses has been suggested to result in increased saliva production, which buffers the gut, and some studies do support this. Finally, there is the notion that lack of opportunities to engage in foraging and processing of feeds (reduced rumination time) can contribute to oral stereotypies. While our understanding of the mechanisms of oral stereotypies is still not complete, it does suggest that more attention needs to be given to matching the way livestock are fed with their natural feeding systems. Providing hay straw or other fibrous roughages can help prevent these abnormal oral behaviours in sows, cows and horses.

Assessing Aggression in Group-housed Sows

Solutions to some welfare problems can be accomplished by assessing behaviour on farms. Although behaviour observations are often too time consuming for practical welfare assessment, aggression and some other behaviour problems that cause injury can be assessed indirectly. Scoring systems for injuries due to fighting in pigs (Séguin *et al.*, 2006; Turner *et al.*, 2006; Baumgartner, 2007) or feather loss due to aggression and feather pecking in hens (Bilcík and Keeling, 1999) have been shown to be correlated to observations of actual fighting and pecking behaviour. Attention to the nature of the injury (where on the body it occurs) or the timing of injury in relation to management can help determine what type of behaviour and when the behavior is occurring.

In pigs, for example, injuries and scratches to the shoulders and head were correlated with reciprocal fighting and were more prevalent in aggressive animals while lesions to the flank or back end of pigs were more prevalent in pigs who were observed being attacked by other pigs (Fig. 15.4) (Turner *et al.*, 2006). Baumgartner (2007) also found that lesions on the head and shoulders were more prevalent in mixed pigs (pigs from different pens who had been fighting) than in non-mixed pigs, while lesions on the tails or ears were not. In group-housing systems for sows, aggression can be a problem. Some fighting is almost inevitable when new groups are mixed, because pigs engage in vigorous fighting for several hours in order to establish their dominance hierarchies. Chronic aggression can also occur if sows are competing for food or space.

Scratch scoring determines if fighting continues after initial mixing

Tracking the timing of lesions relative to mixing may help to determine when aggression is occurring

Fig. 15.4. An aggressive sow is biting the hindquarters of another sow. Some sows are subordinate and they are more likely to be bullied and attacked by other animals. Sows and pigs with scratches on the rear flank are being attacked by other animals.

and whether chronic aggression over a long period of time is a problem. In order to determine the welfare implications of converting part of the gestation barn from stalls to group housing at our research station, Séguin *et al.* (2006) scored superficial skin scratches on sows each week for 5 weeks after mixing in a group-housing system. The scoring system was fairly simple. The number of scratches was counted on each shoulder and assigned scores: 0 for no scratches; 1 for less than five scratches; 2 for five to ten scratches; and 3 for more than ten scratches (Fig. 15.5). The scores were summed for the two shoulders. Figures 15.6 and 15.7 show the percentages of sows in each group with moderate or severe scratches and the average scores for sows over time. Tracking the percentages of sows with severe scratch scores was most useful at identifying lesions due to fighting at mixing. Over 30% of sows had severe scores the day after mixing, but these decreased over time and by 3 weeks after mixing only around 5% had many scratches. The pens were stocked at fairly generous space allotments ranging from 2.3 to 3.5 m^2/sow and group sizes

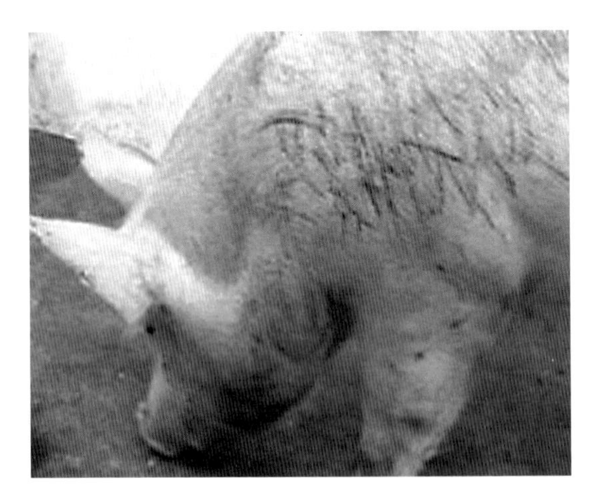

Fig. 15.5. Sow with severe scratches on the shoulder. Shoulder scratches due to fighting can be scored with a simple four-point scoring system: 0 for no scratches; 1 for less than five scratches; 2 for five to ten scratches; and 3 for more than ten scratches. Scratches on the shoulder are correlated with reciprocal fighting. Aggressive animals will usually have more shoulder scratches.

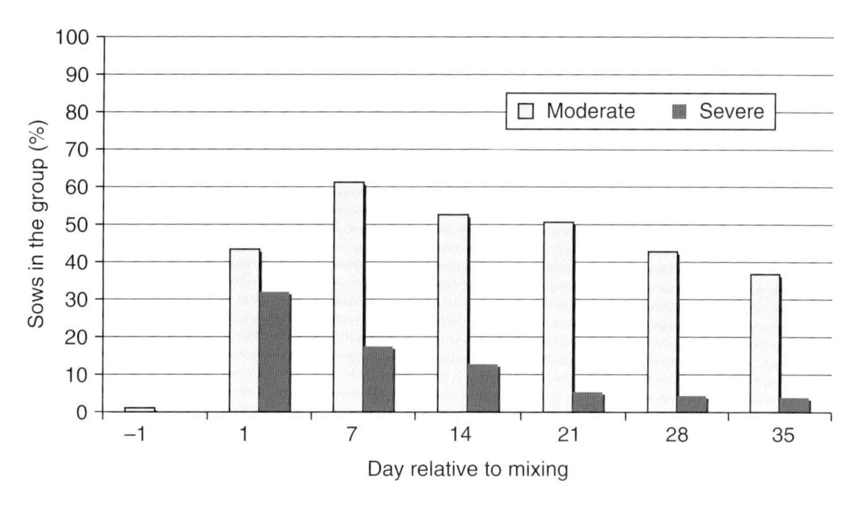

Fig. 15.6. Percentage of sows with scratch scores of 2 (moderate) or 3 (severe) on both of their shoulders after mixing in groups of 11–31 sows. During the first 7 days, 18–30% of the sows had severe scratches. By the 21st day, the percentage of sows with severe scratches dropped to less than 5%.

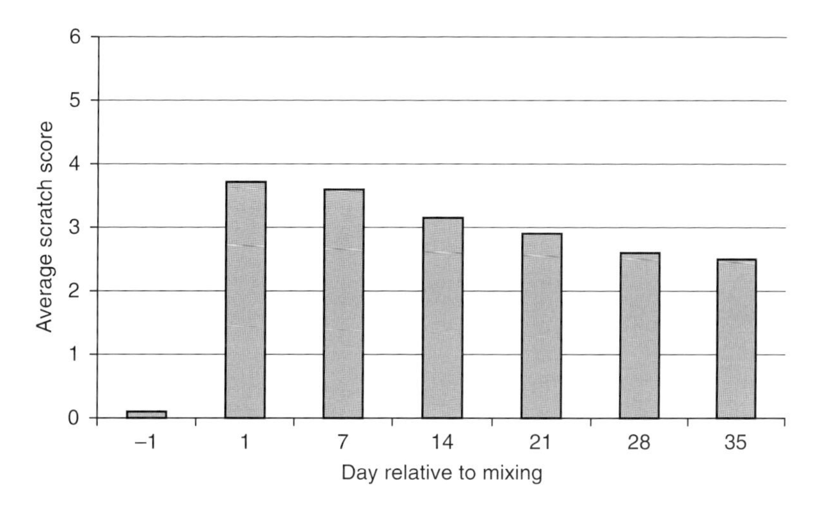

Fig. 15.7. Average shoulder scratch scores after mixing in group-housed sows. In this graph, severe, moderate and mild scratch scores are combined. It is much easier to see changes over time when tracking the percentages of sows in a group with severe and moderate scratches (mild scratches are difficult to score).

ranged from 11 to 31. Practical experience in the industry has shown that larger groups of over 60 sows may fight less (Chapter 14). Data were collected from a total of 15 groups. Sows were fed on the floor (Fig. 15.8). Litter sizes and piglet weights were higher for the sows in groups compared to sows still held in stalls on the same farm. Body condition scores were good and did not vary over time in the groups and there were no problems with illness or vulva biting. We concluded that welfare in the system was good and that scratches due to fighting were a short-term consequence of mixing sows in the groups. When using lesion scores for assessing aggression and welfare in group-housing systems it is important to consider the scores in relation to the severity and when the sows are mixed.

Fig. 15.8. Group-housed sows with a feeding system that drops feed on the floor. This type of feeding system may have increased levels of fighting due to competition for feed. The pens have partial partitions and large floor-space allowances, which possibly had a beneficial effect on providing places to escape. This farm had good management and the sows had larger litters in these pens compared to sow gestation stalls. When new facilities are being built, other types of feeding equipment that reduces competition over food is strongly recommended.

Conclusions

The motivation for different types of behaviour can be measured in a very scientific and objective manner. Research clearly shows that some natural behaviours are more highly motivated than others and that when animals are unable to perform them, there can be behavioural, physiological and welfare consequences. Most researchers agree that these types of highly motivated behaviours should be accommodated when livestock or poultry are housed in intensive systems.

References

Baumgartner, J. (2007) How to deal with complex data of skin lesions in weaner pigs. *Animal Welfare* 16, 165–168.

Bergeron, R., Badnell-Waters, A.J., Lambton, S. and Mason, G. (2006) Stereotypic oral behaviour in captive ungulates: foraging, diet and gastrointestinal function. In: Mason, G. and Rushen, J. (eds) *Stereotypic Animal Behaviour Fundamentals and Application to Welfare*, 2nd edn. CAB International, Wallingford, UK, pp. 19–57.

Bilcík, B. and Keeling, L.J. (1999) Changes in feather condition in relation to feather pecking and aggressive behaviour in laying hens. *British Poultry Science* 40, 444–451.

Brambell, R.W.R. (chairman) (1965) *Report of the Technical Committee to Enquire into the Welfare of Animals Kept Under Intensive Livestock Husbandry Systems*. Command paper 2836. Her Majesty's Stationery Office, London.

Buckner, L.J., Edwards, S.A. and Bruce, J.M. (1998) Behaviour and shelter use by outdoor sows. *Applied Animal Behaviour Science* 57, 69–80.

Cooper, J.J. and Albentosa, M.J. (2003) Behavioural priorities of laying hens. *Avian and Poultry Biology Reviews* 14, 127–149.

Cooper, J.J. and Appleby, M.C. (1996) Demand for nest boxes in laying hens. *Behavioural Processes* 36, 171–182.

Cooper, J.J. and Appleby, M.C. (2003) The value of environmental resources to domestic hens: a comparison of the work-rate for food and for nests as a function of time. *Animal Welfare* 12, 39–52.

Davis, N.J., Prescott, N.B., Savory, C.J. and Wathes, C.M. (1999) Preferences of growing fowls for different light intensities in relation to age, strain and behaviour. *Animal Welfare* 8, 193–204.

Dawkins, M.S. (1983) Battery hens name their price: consumer demand theory and the measurement of ethological 'needs'. *Animal Behaviour* 31, 1195–1205.

Dawkins, M.S. (1990) From an animal's point of view: motivation, fitness and animal welfare. *Behavioural and Brain Sciences* 13, 1–14.

Dawkins, M.S. (2004) Using behaviour to assess animal welfare. *Animal Welfare* 13, 3–7.

de Passillé, A.M. (2001) Sucking motivation and related problems in calves. *Applied Animal Behaviour Science* 72, 175–187.

Duncan, I.J.H. (1996) Animal welfare defined in terms of feelings. *Acta Agriculturae Scandinavica (Section A – Animal Science)* 27 (Supplement), 29–35.

Duncan, I.J.H. (1998) Behavior and behavioral needs. *Poultry Science* 77, 1766–1772.

Duncan, I.J.H. (2001) The pros and cons of cages. *World's Poultry Science Journal* 57, 381–390.

Duncan, I.J.H. and Kite, V.G. (1987) Some investigations into motivation in the domestic fowl. *Applied Animal Behaviour Science* 18, 387–388.

Duncan, I.J.H. and Kite, V.G. (1989) Nest site selection and nest-building behaviour in domestic fowl. *Animal Behaviour* 37, 215–231.

Duncan, I.J.H. and Petherick, J.C. (1991) The implications of cognitive processes for animal welfare. *Journal of Animal Science* 69, 5017–5022.

Duncan, I.J.H., Widowski, T.M., Malleau, A.M., Lindberg, A.C. and Petherick, J.C. (1998) External factors and causation of dustbathing in domestic hens. *Behavioural Processes* 43, 219–228.

Follensbee, M. (1992) Quantifying the nesting motivation of domestic hens. MSc thesis, University of Guelph, Guelph, Ontario, Canada.

Follensbee, M.E., Duncan, I.J.H. and Widowski, T.M. (1992) Quantifying nesting motivation of domestic hens. *Journal of Animal Science* 70(1), 164.

Fraser, D. (2003) Assessing animal welfare at the farm and group level: the interplay of science and values. *Animal Welfare* 12, 433–443.

Fraser, D. and Duncan, I.J.H. (1998) 'Pleasures', 'pains' and animal welfare: toward a natural history of affect. *Animal Welfare* 7, 383–396.

Freire, R., Appleby, M.C. and Hughes, B.O. (1997) Assessment of pre-laying motivation in the domestic hen using social interaction. *Animal Behaviour* 54, 313–319.

Gilbert, C.L., Burne, T.H., Goode, J.A., Murfitt, P.J. and Walton, S.L. (2002) Indomethacin blocks pre-partum nest building behaviour in the pig (*Sus scrofa*): effects on plasma prostaglandin F metabolite, oxytocin, cortisol and progesterone. *Journal of Endocrinology* 172, 507–517.

Grandin, T., Curtis, S.E., Widowski, T.M. and Thurmon, J.C. (1986) Electro-immobilization versus mechanical restraint in an avoid-avoid choice test for ewes. *Journal of Animal Science* 66, 1469–1480.

Hughes, B.O. and Duncan, I.J.H. (1988) The notion of ethological 'need'. Models of motivation and animal welfare. *Animal Behaviour* 36, 1696–1707.

Jensen, P. (1993) Nest building in domestic sows – the role of external stimuli. *Animal Behaviour* 45, 351–358.

Jensen, P. (2006) Domestication – from behaviour to genes and back again. *Applied Animal Behaviour Science* 97, 3–15.

Jensen, P. and Recén, B. (1989) When to wean – observations from free-raging domestic pigs. *Applied Animal Behaviour Science* 23, 49–60.

Jensen, P. and Toates, F.M. (1993) Who needs 'behavioural needs'? Motivational aspects of the needs of animals. *Applied Animal Behaviour Science* 37, 161–181.

Kirkden, R.D. and Pajor, E.A. (2006) Using preference, motivation and aversion tests to ask scientific questions about animals' feelings. *Applied Animal Behaviour Science* 100, 29–47.

LayWel (2009) Available at: www.laywel.eu (accessed 19 June 2009).

Lindberg, A.C. and Nicol, C.J. (1997) Dustbathing in modified battery cages. Is sham dustbathing an adequate substitute? *Applied Animal Behaviour Science* 55, 113–128.

Mason, G.J. and Latham, N.R. (2004) Can't stop, won't stop: is stereotypy a reliable indicator of welfare? *Animal Welfare* 13, S57–S69.

Mason, G. and Rushen, J. (2006) *Sterotypic Animal Behaviour Fundamentals and Application to Welfare*, 2nd edn. CAB International, Wallingford, UK, 367 pp.

McCarthy, S., Horan, B., Rath, M., Linnane, M., O'Connor, P. and Dillon, P. (2007) The influence of strain of Holstein-Friesian dairy cow and pasture-based feeding system on grazing behaviour, intake and milk production. *Grass and Forage Science* 62, 13–26.

Nuboer, J.F.W., Coemans, M.A.J.M. and Vos, J.J. (1992) Artificial lighting in poultry houses: do hens perceive the modulation of fluorescent lamps as flicker? *British Poultry Science* 33, 123–133.

Olsson, I.A.S. and Keeling, L.J. (2005) Why in earth? Dustbathing behaviour in jungle and domestic fowl reviewed from a Tinbergian and animal welfare perspective. *Applied Animal Behaviour Science* 93, 259–282.

Pajor, E.A., Rushen, J. and de Pasillé, A.M. (2000) Aversion learning techniques to evaluate dairy cattle handling practices. *Applied Animal Behaviour Science* 69, 89–102.

Pajor, E.A., Rushen, J. and de Pasillé, A.M. (2003) Dairy cattle's choice of handling treatments in a Y-maze. *Applied Animal Behaviour Science* 80, 93–107.

Petherick, J.C., Seawright, E., Waddington, D., Duncan, I.J.H. and Murphy, L.B. (1995) The role of perception in the causation of dustbathing behaviour in domestic fowl. *Animal Behaviour* 49, 1521–1530.

Phillips, P.A., Fraser, D. and Pawluczuk, B. (2000) Floor temperature preference of sows at farrowing. *Applied Animal Behaviour Science* 67, 59–65.

Price, E.O. (2003) *Animal Domestication and Behavior*. CAB International, Wallingford, UK, 297 pp.

Raj, A.B.M. and Gregory, N.G. (1995) Welfare implication of the gas stunning of pigs 1. Determination of aversion to the initial inhalation of carbon dioxide or argon. *Animal Welfare* 4, 273–280.

Rushen, J. (1996) Using aversion learning techniques to assess the mental state, suffering, and welfare of farm animals. *Journal of Animal Science* 74, 1990–1995.

Rushen, J., de Passillé, A.M., von Keyserlingk, M.A.G. and Weary, D.M. (2008) *The Welfare of Cattle*. Springer, Dordrecht, The Netherlands, 310 pp.

Schwartzkopf-Genswein, K.S., Stookey, J.M., Crowe, T.G. and Genswein, B.M. (1998) Comparison of image analysis, exertion force, and behavior measurements for use in the assessment of beef cattle responses to hot-iron and freeze-branding. *Journal of Animal Science* 76, 972–979.

Séguin, M.J., Barney, D. and Widowski, T.M. (2006) Assessment of a group-housing system for gestating sows: effects of space allowance and pen size on the incidence of superficial skin lesions, changes in body condition and farrowing performance. *Swine Health and Production* 14, 89–96.

Sherwin, C.M. (1999) Domestic turkeys are not averse to compact fluorescent lighting. *Applied Animal Behaviour Science* 64, 47–55.

Shütz, K.E. and Jensen, P. (2001) Effects of resource allocation on behavioural strategies: a comparison of red junglefowl (*Gallus gallus*) and two domesticated breeds of poultry. *Ethology* 107, 753–765.

Shütz, K.E., Forkman, B. and Jensen, P. (2001) Domestication effects on foraging strategy, social behaviour and different fear responses: a comparison between the red jungle fowl (*Gallus gallus*) and a modern layer strain. *Applied Animal Behaviour Science* 74, 1–14.

Spinka, M. (2006) How important is natural behaviour in animal farming systems? *Applied Animal Behaviour Science* 100, 117–128.

Straw, B.E. and Bartlett, P. (2001) Flank or belly nosing in weaned pigs. *Journal of Swine Health and Production* 9, 19–23.

Struelens, E., Tuyttens, F.A.M., Janssen, A., Leroy, T., Audoorn, L., Vranken, E., de Baere, K., Ödberg, F., Berckmans, D., Zoons, J. and Sonck, B. (2005) Design of laying nests in furnished cages: influence of nesting material, nest box position and seclusion. *British Poultry Science* 46, 9–15.

Struelens, E., Van Nuffel, A., Tuyttens, F.A.M., Audoorn, L., Vranken, E., Zoons, J., Berckmans, D., Ödberg, F., van Dongen, S. and Sonck, B. (2008) Influence of nest seclusion and nesting material on pre-laying behaviour of laying hens. *Applied Animal Behaviour Science* 112, 106–119.

Tauson, R. (2005) Management and housing systems for layers – effects on welfare and production. *World's Poultry Science Journal* 61 477–490.

Torrey, S. and Widowski, T.M. (2004) Effect of drinker type and sound stimuli on early-weaned pig performance and behaviour. *Journal of Animal Science* 82, 2105–2114.

Tucker, C.B., Weary, D.M. and Fraser, D. (2003) Effects of three types of free stall surfaces on preferences and stall usage by dairy cows. *Journal of Dairy Science* 86, 521–529.

Turner, S.P., Farnworth, M.J., White, I.M.S., Brotherstone, S., Mendl, M., Knap, P., Penny, P. and Lawrence, A.B. (2006) The accumulation of skin lesions and their use as a predictor of individual aggressiveness in pigs. *Applied Animal Behaviour Science* 96, 245–259.

Wathes, C.M., Jones, J.B., Kristensen, H.H., Jones, E.K.M. and Webster, A.J.F. (2002) Aversion of pigs and domestic fowl to atmospheric ammonia. *Transactions of the American Society of Agricultural Engineers* 45, 1605–1610.

Weeks, C.A. and Nicol, C.J. (2006) Behavioural needs, priorities and preferences of laying hens. *World's Poultry Science Journal* 62, 296–307.

Widowski, T.M. and Curtis, S.E. (1990) The influence of straw, cloth tassel, or both on the prepartum behavior of sows. *Applied Animal Behaviour Science* 27, 53–71.

Widowski, T.M. and Duncan, I.J.H. (2000) Working for a dust-bath: are hens increasing pleasure rather then reducing suffering? *Applied Animal Behaviour Science* 68, 39–53.

Widowski, T.M., Curtis, S.E., Dziuk, P.J., Wagner, W.C. and Sherwood, O.D. (1990) Behavioral and endocrine responses of sows to prostaglandin F2alpha and cloprostenol. *Biology of Reproduction* 43, 290–297.

Widowski, T.M., Keeling, L.J. and Duncan, I.J.H. (1992) The preferences of hens for compact fluorescent over incandescent lighting. *Canadian Journal of Animal Science* 72, 203–211.

Widowski, T.M., Cottrell, T., Dewey, C.E. and Friendship, R.M. (2003) Observations of piglet-directed behaviour patterns and skin lesions in eleven commercial swine herds. *Swine Health and Production* 11, 181–185.

Widowski, T.M., Torrey, S., Bench, C.J. and Gonyou, H.W. (2008) Development of ingestive behaviour and the relationship to belly nosing in early-weaned piglets. *Applied Animal Behaviour Science* 110, 109–127.

Wood-Gush, D.G.M. and Duncan, I.J.H. (1976) Some behavioural observations on domestic fowl in the wild. *Applied Animal Ethology* 2, 255–260.

Wood-Gush, D.G.M. and Gilbert, A.B. (1964) The control of the nesting behavior of the domestic hen, II. The role of the ovary. *Animal Behaviour* 12, 451–453.

Wood-Gush, D.G.M. and Gilbert, A.B. (1969) Observations on the laying behaviour of hens in battery cages. *British Poultry Science* 10, 29–36.

Yue, S. and Duncan, I.J.H. (2003) Frustrated nesting behaviour: relation to extra-cuticular shell calcium and bone strength in White Leghorn hens. *British Poultry Science* 44, 175–181.

Useful Web Sites

Animal Behaviour, Professional Societies and Journals

Animal Behavior Society
Excellent links and educational materials; publishes the journal *Animal Behaviour*
www.animalbehaviorsociety.org

Applied Animal Behaviour Science
Access behaviour abstracts in this journal
www.scirus.com

Association for the Study of Animal Behaviour
Links to many behaviour societies
http://asab.nottingham.ac.uk

Federation of Animal Science Societies
Contains the *Guide for Care and Use of Agricultural Animals in Research and Teaching* (Ag Guide)
www.fass.org

International Society of Applied Ethology
Many researchers in farm and animal behaviour belong to this society
www.applied-ethology.org

Journal of Applied Animal Welfare Science (JAAWS)
Archive of articles
www.psyeta.org/jaaws/index.html

Journal of Ethology
Access to abstracts, type article title into Google

Universities Federation Animal Welfare
Publishes the *Journal of Animal Welfare*
www.ufaw.org.uk

Industry and Veterinary Organizations

American Meat Institute
Contains guidelines for humane slaughter and welfare audit forms
www.animalhandling.org

American Veterinary Medical Association
Publisher of *Journal of the American Veterinary Medical Association*
www.avma.org

Index to veterinary associations all over the world
www.vetmedicine.about.com

Animal Transportation Association
Contains regulations and transport information
www.aata.animaltransport.org

Australian Meat Industry Council
Contains welfare guidelines
www.amic.org.au

British Veterinary Medical Association
Publisher of *Veterinary Record*
www.bva.co.uk; www.bvapublications.com

International Air Transport Association (IATA)
Contains international regulations on transport
www.iata.org

Government Web Sites

Agriculture, Fisheries and Forestry of Australia
Contains welfare and transport codes
www.daff.gov.au

Animal and Plant Health Inspection Service, US Department of Agriculture (USDA)
www.aphis.usda.gov

Canadian Food Inspection Agency
Contains humane slaughter and food safety regulations
www.inspection.gc.ca

Commonwealth Scientific and Industrial Research Organization (CSIRO)
Searchable database of Australian research on agriculture and livestock
www.csiro.au

Department for Environment, Food and Rural Affairs (Defra) in the UK
Contains welfare and transport codes
www.defra.gov.uk

European Union
Contains welfare and transport codes, type Europa animal welfare into Google
www.europa.eu

Provides information on EU Welfare Quality® Project
www.welfarequality.net

Food and Agriculture Organization of the United Nations
Type FAO animal welfare into Google
www.fao.org

Food Safety and Inspection Service, US Department of Agriculture (USDA)
Contains humane slaughter and food safety regulations
www.fsis.usda.gov

Instituto Nacional de Carnes, Montevideo, Uruguay
Animal welfare guidelines
www.inac.gub.uy

LayWel
Contains scoring system for evaluating laying-hen welfare
www.laywel.eu

National Agriculture Library, Animal Welfare Information Center (AWIC)
Database of animal welfare information
www.nal.usda.gov/awic

OIE Organisation Mondiale de la Sante Animale (World Organization for Animal Health)
Contains transport and humane slaughter codes
www.oie.int

Zinpro
Contains lameness scoring videos
www.zinpro.com

Information Sites on Behaviour, Transport and Welfare

Elsevier Journals Search Engine
Search the scientific literature
www.scirus.com

Michigan State University Animal Legal and Historical Center
Contains animal welfare laws from many countries
www.animallaw.info

NetVet – Mosby's veterinary guide to the Internet
Lots of links to livestock web pages
Type NetVet into Google or Yahoo
www.netvet.wustl.edu

Prairie Swine Centre in Canada
Research studies on pig behaviour
www.prairieswine.com

PubMed National Library of Medicine
Search the scientific literature
Type Pubmed into Google

Purdue Center for Food Animal Well-Being
Links to journals and animal behaviour research
Type the centre's name into Google

Temple Grandin's web site
Information on handling, humane slaughter transport and facility design
www.grandin.com

VetMed Resource (CABI)
Really good for author searching
www.cabi.org/vetmedresource

Virtual Livestock Library
Lots of good links run by the Animal Sciences Department, Oklahoma State University
Type Virtual Livestock Library into Google
www.ansi.okstate.edu/library

Western College of Veterinary Medicine Saskatchewan (Joe Stookey's site)
Lots of links to behaviour research and information
www.usask.ca/wcvm/herdmed/applied-ethology

Non-governmental Animal Welfare Organizations

American Livestock Breeds Conservancy
Type the name into Google for information on preserving rare breeds
www.albc-usa.org

Compassion in World Farming (CIWF)
www.ciwf.org.uk

Farm Animal Welfare Council (FAWC)
Publishers of animal welfare report
www.fawc.org.uk

Humane Slaughter Association
Publishers of training materials
www.hsa.org.uk

Humane Society of the USA
www.humanesociety.org

National Sustainable Agriculture Information Service
Useful information on sustainable agriculture
www.attra.org

People for the Ethical Treatment of Animals (PETA)
www.peta.org

World Society for the Protection of Animals
www.wspa-international.org

Index

Note: page numbers in *italics* refer to figures and tables, those in **bold** refer to boxes.